Communications
in Computer and Inform                    91

T0092441

Andrzej Kwiecień   Piotr Gaj
Piotr Stera (Eds.)

# Computer Networks

19th International Conference, CN 2012
Szczyrk, Poland, June 19-23, 2012
Proceedings

 Springer

Volume Editors

Andrzej Kwiecień
Piotr Gaj
Piotr Stera

Silesian University of Technology
Institute of Informatics
ul. Akademicka 16
44-100 Gliwice, Poland

E-mail: {andrzej.kwiecien, piotr.gaj, piotr.stera}@polsl.pl

ISSN 1865-0929      e-ISSN 1865-0937
ISBN 978-3-642-31216-8      e-ISBN 978-3-642-31217-5
DOI 10.1007/978-3-642-31217-5
Springer Heidelberg Dordrecht London New York

Library of Congress Control Number: 2012939645

CR Subject Classification (1998): C.2, H.4, D.2, H.3.4-5, C.4, K.4.4, G.3, I.6

*Typesetting:* Camera-ready by author, data conversion by Scientific Publishing Services, Chennai, India

Printed on acid-free paper

Springer is part of Springer Science+Business Media (www.springer.com)

# Preface

The contemporary technical world is based on informatics solutions. Most of them are based on functionality dispersion and use distributed processing. This would be impossible without appropriate methods and ways of data transmission. Fundamentals, in this case, are computer network technologies. Hence, there is a big role for engineers who are able to act as specialists in the domain of communication in computer and information science. It is a common view and opinion that almost everyone knows something about this area. However, deep knowledge and skills related to networking are very rare. Especially when one takes into consideration a wide and systemic point of view, not a particular skill.

Nowadays, information technologies are objects of regular changes and improvements. Communications questions, as a key in this matter, have been evolving constantly. On the one hand well-known solutions are not sufficient; on the other hand some of them are no longer developed and become obsolete in the context of current requirements. The main reason for such a state is the constant growth of expectations in almost every branch of human activity as well as the constant development of the human environment.

It necessitates the modernization of existing solutions as well as creating brand new ones. New methods and tools for designing, modelling, testing, and for other actions supporting researchers enable networking technologies to be continuously enriched and changed. First of all, a general development is possible thanks to the results of research and thanks to the proposals of modern applications delivered by a group of engineers and scientists whose eminent representatives are authors of this book. The contents include 48 chapters covering a broad spectrum of issues, problems and topics that are strongly connected to the following subjects, currently considered as valid and up to date:

- New and emerging technologies related to networking fields
- Fundamentals of computer networks
- Internet and internetworking
- Security and safety issues
- Industrial computer networks
- Wireless systems and sensor networks
- Theory of queues and queuing networks
- Applications and computer networks usage

Additionally, there are some topics referring to QoS issues, multiservice, cellular, high-speed, and mobile networks as well as quantum technologies.

Generally, the book is focused on the above-mentioned subjects in the presented order. However, we decided not to create separate parts because the contents of the chapters are not separated from one another but overlap partially.

The chapters related to the fundamentals and to the subjects of new approaches are presented at the beginning of the book and comprise among others: Web services, content-aware networks, data mining methods, and quantum technologies.

The next group concerns network and resource management, performance forecasting, flow analysis, efficiency consideration while streaming, and other important issues referring to internetworking.

The fourth group is related to security issues, particularly on various risks and methods of detection and prevention. Among others, topics about retrieving a program code based on voltage supply changes, analyses of malware activity from the honeypots viewpoint, and micrographics usage in information security domains are presented, as well as valid issues related to traffic anonymization and the analysis of IP storage security.

A very important field of computer communications is the industrial informatics area. It is considered in the next few chapters related to the new concept of data transmission with real-time constraints, performance estimation of data transfer based on the OPC UA model, as well as to authentication, management, and failure-detection topics.

A great effort in the research of the new-generation networks is focused on wireless solutions. Many common applications such as home and entertainment networks, and specialized ones such as sensor networks, are based on wireless technologies. The next part of the book presents topics on networking without a cable, e.g., energy consumption, modeling, simulations, and routing algorithms. Additionally, there are interesting chapters which refer to cellular technology. The first one refers to the evaluation of data transmission performance in cellular networks used in industrial computer systems and the second one refers to the influence of weather conditions on mobile phone usage.

Next, in view of the importance of the theory of queues, a few chapters related to this area are included in the book. At the end of the volume there are chapters with an evident application character. They are connected with the e-Bussiness area, vehicular sensor networks, earth science calculations in cluster architecture, efficiency of DCOM and CORBA standard techniques within distributed wireless environments, and last but not least, the weather to warning systems.

We would like to take this opportunity to express our thanks to all the authors for sharing the research results and for their assistance in creating this monograph. This book, in our belief, is a valuable reference on computer networks. We would also like to thank the members of international Program Committee for their participation in reviewing each paper twice.

April 2012                                                             Andrzej Kwiecień
                                                                              Piotr Gaj

# Organization

CN 2012 was organized by the Institute of Informatics, Faculty of Automatic Control, Electronics and Computer Science, Silesian University of Technology (SUT) and supported by the Committee of Informatics of the Polish Academy of Sciences, Section of Computer Network and Distributed Systems in technical cooperation with the IEEE and iNEER organizations.

Institute of Informatics
Silesian University of Technology
ul. Akademicka 16, 44-100 Gliwice, Poland
e-mail: cn@polsl.pl
web: http://cn.polsl.pl

## Executive Committee

All members of the Executing Committee are from the Silesian University of Technology, Poland.

| | |
|---|---|
| Honorary Member | Halina Węgrzyn |
| Organizing Chair | Piotr Gaj |
| Technical Volume Editor | Piotr Stera |
| Technical Support | Aleksander Cisek |
| Technical Support | Arkadiusz Jestratjew |
| Technical Support | Jacek Stój |
| Office | Małgorzata Gładysz |
| WEB Support | Piotr Kuźniacki |
| IEEE PS Coordinator | Jacek Izydorczyk |
| iNEER Coordinator | Win Aung |

## Program Committee

### Program Chair

| | |
|---|---|
| Andrzej Kwiecień | Silesian University of Technology, Poland |

### Honorary Members

| | |
|---|---|
| Klaus Bender | TU München, Germany |
| Zdzisław Duda | Silesian University of Technology, Poland |
| Andrzej Karbownik | Silesian University of Technology, Poland |
| Jerzy Rutkowski | Silesian University of Technology, Poland |
| Bogdan M. Wilamowski | Auburn University, USA |

## Program Committee Members

| | |
|---|---|
| Anoosh Abdy | Realm Information Technologies, USA |
| Iosif Androulidakis | University of Ioannina, Greece |
| Tülin Atmaca | Institut National de Télécommunication, France |
| Win Aung | iNEER, USA |
| Leszek Borzemski | Wrocław University of Technology, Poland |
| Markus Bregulla | University of Applied Sciences Ingolstadt, Germany |
| Tadeusz Czachórski | Silesian University of Technology, Poland |
| Andrzej Duda | INP Grenoble, France |
| Alexander N. Dudin | Belarusian State University, Belarus |
| Max Felser | Bern University of Applied Sciences, Switzerland |
| Jean-Michel Fourneau | Versailles University, France |
| Natalia Gaviria | Universidad de Antioquia, Colombia |
| Roman Gielerak | University of Zielona Góra, Poland |
| Adam Grzech | Wrocław University of Technology, Poland |
| Zbigniew Huzar | Wrocław University of Technology, Poland |
| Jürgen Jasperneite | Ostwestfalen-Lippe University of Applied Sciences, Germany |
| Jerzy Klamka | IITiS Polish Academy of Sciences, Gliwice, Poland |
| Demetres D. Kouvatsos | University of Bradford, UK |
| Stanisław Kozielski | Silesian University of Technology, Poland |
| Henryk Krawczyk | Gdańsk University of Technology, Poland |
| Wolfgang Mahnke | ABB, Germany |
| Kevin M. McNeil | BAE Systems, USA |
| Michael Pagano | University of Pisa, Italy |
| Nihal Pekergin | Versailles University, France |
| Piotr Pikiewicz | College of Business in Dąbrowa Górnicza, Poland |
| Bolesław Pochopień | Silesian University of Technology, Poland |
| Silvana Rodrigues | Integrated Device Technology, Canada |
| Akash Singh | IBM Corp, USA |
| Mirosław Skrzewski | Silesian University of Technology, Poland |
| Kerry-Lynn Thomson | Nelson Mandela Metropolitan University, South Africa |
| Oleg Tikhonenko | IITiS Polish Academy of Sciences, Gliwice, Poland |
| Bane Vasic | University of Arizona, USA |
| Sylwester Warecki | Freescale Semiconductor Inc., USA |
| Tadeusz Wieczorek | Silesian University of Technology, Poland |
| Józef Woźniak | Gdańsk University of Technology, Poland |
| Hao Yu | Auburn University, USA |
| Grzegorz Zaręba | University of Arizona, USA |

# Referees

Iosif Androulidakis
Tülin Atmaca
Leszek Borzemski
Tadeusz Czachórski
Andrzej Duda
Alexander N. Dudin
Max Felser
Jean-Michel Fourneau
Roman Gielerak
Adam Grzech
Zbigniew Huzar

Jürgen Jasperneite
Jerzy Klamka
Demetres D. Kouvatsos
Stanisław Kozielski
Henryk Krawczyk
Andrzej Kwiecień
Wolfgang Mahnke
Kevin M. McNeil
Michael Pagano
Piotr Pikiewicz
Bolesław Pochopień

Akash Singh
Mirosław Skrzewski
Kerry-Lynn Thomson
Oleg Tikhonenko
Bane Vasic
Sylwester Warecki
Tadeusz Wieczorek
Józef Woźniak
Hao Yu
Grzegorz Zaręba

# Sponsoring Institutions

Technical cosponsors: IEEE Poland Section, iNEER.

# Table of Contents

# The Interactions of SOAP-Based Web Services for Recording and Replaying Video Files

Miroslav Voznak, Lukas Kapicak, Martin Mikulec, Pavel Nevlud, and Jaroslav Zdralek

VSB – Technical University of Ostrava
17. listopadu 15, 70800 Ostrava, Czech Republic
{miroslav.voznak,lukas.kapicak,martin.mikulec,
pavel.nevlud,jaroslav.zdralek}@vsb.cz

**Abstract.** The paper deals with the extended capability of a multimedia server to control a video transaction remotely. The presented paper is a result of applied research and is a contribution into the INDECT portal which integrates the interactions of SOAP-based web services. The project aims at developing the tools for enhancing the security of citizens and protecting the confidentiality of recorded and stored information and is a part of the $7^{th}$ FP EU. We describe a multimedia handling controlled by the client application via web services. We have developed JAVA based software that allows remote control of SW PBX Asterisk through web services and AMI interface. Video files are recorded by smartphones and saved in a repository on the server. The list of stored files is generated automatically and users can replay a particular video file from the server. The names in this list are read to users by means of a speech synthesis. Subsequently, DTMF key input is used to enable making the right choice.

**Keywords:** IVAS, web services, multimedia functions, IVR/IVVR.

## 1 Introduction

The members of our team participate in an international project referred to as the INDECT project solved within $7^{th}$ Framework Programme EU [1]. The entire project is conducted by AGH Cracow and consists of ten work packages. Each package focuses on a specific problem. We participate in Work Package 6 which is conducted by Poznan University of Technology. The main focus of this package is developing the INDECT portal and the IVAS (Interactive Video Audio System), our team works on the IVAS system [2].

Users of the IVAS system are divided into two basic groups. Users in the office – they are the managers who evaluate the resulting events and users in the field – they accept commands from managers and use their mobile devices as a means of obtaining report information.

If there is an event, office user chooses the executive user group that become familiar with this incident. He chooses users depending on their state – online or

A. Kwiecień, P. Gaj, and P. Stera (Eds.): CN 2012, CCIS 291, pp. 1–13, 2012.
© Springer-Verlag Berlin Heidelberg 2012

offline. Chosen users are then notified by incoming call. If in-field-user accepts the call, user in office is informed.

Subsequently audio and video messages are played to in-field-user's device. They can respond either from IVR (Interactive Voice Response) or IVVR (Interactive Voice and Video Response) menu with DTMF choice. After end of the call and specific action performed by in-field-user, they can inform the user in office by calling back the same number they were called from. The in-field-user can respond by using DTMF in pre-prepared IVR/IVVR menu and they are able to stream video and audio content. Subsequently they can save the content to remote server storage. The user in the office is also informed via SOAP (Simple Object Access Protocol) message about the new recorded content.

The IVAS system provides a gateway between office users and field users. Office users are equipped with a personal computer and control Asterisk through web pages. Field users, on the other hand, use mobile devices such as smartphones and tablets to access multimedia video files.

Office users can control the whole session. They have capabilities to obtain information about all users, about online users. Web page access provides the opportunity to control any call session between two or more users or between a saved video file and field users. Office users have a supervisor status and they have to be informed about all changes in the field user's status. They should obtain information about call acceptance and about user's choices through DTMF. On the other hand, they have to have access to multimedia files recorded by field users. Field users use their smartphones to record video files directly to storage defined in the Asterisk dialplan [3].

Web pages cooperate with Asterisk through the web service requester. The web service requester sends SOAP (Simple Object Access Protocol) messages [4,5] with a specific task to the web service provider. This web service provider translates this message into Asterisk language.

## 2   IVAS System Architecture

The IVAS system provides gateway between office users and field users. It consists of two main parts. These are Asterisk SW PBX and a web service provider.

Asterisk SW PBX [6] is a VOIP (Voice Over Internet Protocol) gateway which in its default configuration offers a controlling function for audio and video calls between multiple users [7]. Control and management functions are implemented too [8]. We used these functions to manage Asterisk remotely and to gather call status information. We use AMI (Asterisk Management Interface) and CLI (Command Line Interface) [4] interfaces to perform these functions. Figure 1 shows architecture of the IVAS system with a field user. The SIP protocol [7] provides control functions.

The web service provider is JAVA based software and it comes out from the Asterisk JAVA project [9]. The web service provider listens for SOAP messages originated by the web service requester and these messages are translated into AMI or CLI commands. On the other side, the web service provider provides

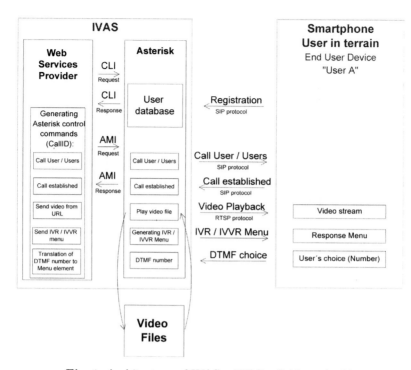

**Fig. 1.** Architecture of IVAS – IVAS – field user's side

system and user status information to the web service requester. This happens on web service requester's demand. Figure 2 shows the web service requester within the IVAS system.

## 2.1 The Asterisk Side

Figure 1 shows the connection between the IVAS system represented by Asterisk with end user devices. The primary role of the IVAS system is to perform multimedia functions for users in the terrain. For this reason, we have to presume that end user devices consist only of smartphones.

You can notice that Asterisk handles functions such as registration, establishing audio and video calls, creating IVR and IVVR menu and processes DTMF choices by field users. All of these choices could be performed by the web service provider through AMI and CLI interface.

Asterisk supports registration of users, establishing video calls, DTMF by default but it does not support video recording and replay. It was necessary to implement new modules into Asterisk to provide video recording and replay support. We used modules mp4play and mp4save comes from library app_mp4.c. This library was developed for Asterisk in version 1.4 but in our project we use Asterisk in 1.6 version. It was necessary to modify the module to support 1.6 version of Asterisk.

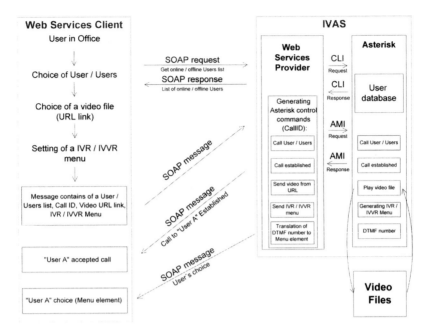

**Fig. 2.** Architecture of IVAS – web service requester – the IVAS side

The first tests were provided by hard coded records in the Asterisk dial plan which can be found in default in /etc/asterisk/extensions.conf. For replay videos, we used the following commands:

```
exten => XXXX,1,Answer
exten => XXXX,n,mp4play(/movies/movie.mp4)
exten => XXXX,n,Hangup
```

For record videos, we used the command with mp4save:

```
exten => XXXX,1,Answer
exten => XXXX,n,mp4save(/movies/movie.mp4)
exten => XXXX,n,Hangup
```

The hard coded record does not allow us to make changes to records due to the fact that requests come from the web service requester, so we used Asterisk CLI commands and developed an application to control Asterisk dialplan records through the web service provider. In the paragraph below, you can see control commands used to modify dynamic records in the dial plan.

```
dialplan add extension XXXX,1,Answer, into default
dialplan add extension XXXX,2,mp4play,/movies/movie.mp4 into default
dialplan add extension XXXX,3,Hangup, into default
```

These commands have the same function but they can be modified dynamically by our web service provider which can change XXXX number and it can also change the link to a specific video file; see Sect. 3.3.

At the moment we use the AMI interface to obtain information about the user and the call status. We developed a technique to catch DTMF codes via the AMI interface and to transfer these choices to the web service requester.

```
Event: Newexten
Privilege: dialplan,all
Channel: Local/Campbell@default-4f6d;1
Context: menu1
Extension: s
Priority: 1
Application: Answer
AppData:
Uniqueid: 1320702940.40
```

The web service provider has to translate DTMF choices into specific choices which had been made by the web service requester (the client) at the beginning of the connection. In Figures 1 and 2 you can see that software which provides the translation between DTMF codes into specific choices is a part of the IVAS system, or more precisely it forms a part of the web service provider.

## 2.2 The Web Service Provider

This part of our solution could be considered as a bridge or rather middleware between Asterisk and a client's application. The web service provider consists of a set of "public" methods (web services – WS) [10] which can be called from the client's application to achieve the desired result. Several results were mentioned in previous sections and we will describe another one in detail later in Sect. 3.

Web services need to have some kind of a web container where they can be hosted and accessed from the Internet. For our purposes, the best suitable container is the GlasshFish Server (ver. 3.1.1) [11] with full Java EE6 platform support that includes necessary functionality for our business logic as well as security.

You may ask why we decided to control Asterisk via web services instead of some kind of a custom communication protocol over TCP/IP. The answer is clear. Web services are platform independent so we can afford to re-use the same methods for different clients i.e. for a smart phone with OS Android or from an ASP.NET website. This multi-platform approach gives us many possibilities how to use Asterisk.

The web service layer does not communicate with Asterisk directly. It is responsible solely for catching requests, parsing the parameters passed-through and calling the appropriate lower-level library. These libraries contain the business logic. If, for example, you want to establish a connection, one of the libraries knows what steps have to be taken (what methods have to be called in the right order with the right parameters) to ensure the connection.

Since we have been cooperating with other teams, it is necessary to keep all web services available and running at all times. Architecture of WS allows us to swap the logic of certain methods without affecting the WS itself.

Another advantage of web services is that they can be consumed by the Intalio server (see Sect. 2.3).

As you know, web services are based on client-server architecture, i.e. you send a request to the server, the server "computes" a result and returns it back to the requestor. As you can imagine, in our case it would be impossible to handle services in a synchronous way. Let's mention some issues that could occur if we use the synchronous way. Imagine that we call a certain service from the website to establish a connection between two users. You choose users and click the "Connect" button. The website sends a request to our WS Provider. The business logic checks whether both users are on-line, available, etc. and forwards the request to Asterisk. Asterisk will try to establish a connection. The phone rings and Asterisk waits for the user to answer the call. Both of them answer and the call is in progress and can take un-estimated time. Before all these processes have been completed, our website is waiting for the web response. It cannot be sent earlier. Another issue can arise where an unexpected event occurs at any stage of the scenario.

To avoid these issues and to save server's capacity we decided to handle web services asynchronously. Accordingly, we had to divide communication into two parts. The first uses the pull approach, i.e. the WS Provider waits for a request from the client. The second uses the push approach to send requests back asynchronously. In this case, the client has to provide its own web services waiting for our response. It brings several benefits, in particular the possibility to cooperate with any BMPS Server (Business Process Management System) as an Intalio Server. Another advantage is that we are able to send more than one response for one request. It is suitable in case where one response is determined to be displayed in the client's user interface (i.e. "The call has been established.") and another response will kick off some event at the client's (i.e. after the call is finished the caller's DTMF choice may switch some business process to another state).

## 2.3   The Intalio Server

The Intalio server is responsible for running business process workflows which could be specified as input. These processes are modelled by means of special diagrams. Each diagram has several nodes and each of the nodes represent an event. These events can call zero to N web services or display certain views to the user. Once the diagram has been created and tested, it can run automatically. Using Intalio helps to unify the real life workflow since the server provides the user the very same workflow every time it is requested.

## 2.4   The Web Service Requester

In order to test web services, we created a web page that consumes all web services and provides the user with an easy interface. We used ASP.NET to program the website. Using ASP.NET also demonstrates the platform's independence of

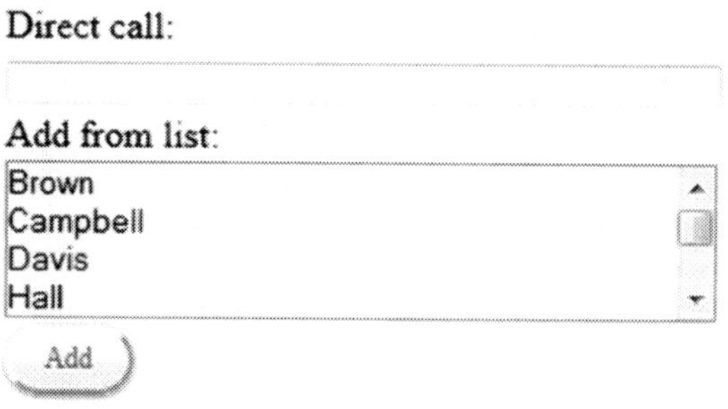

Direct call:

Add from list:

Brown
Campbell
Davis
Hall

Add

**Fig. 3.** The Web Services Client interface

our solution as the core and WSs have been programmed using Java, the client interface is depicted in Fig. 3.

We added a small part of the code responsible for creating a client for the web services and obtaining data from the server.

```
myUserClient= new UserWSServiceReference.UsersWSClient();
var list = myUserClient.GetOnLineUsers(true);
```

Once the data have been gathered from the server, they could be displayed in any way.

### 2.5  Field Users – Using Smartphones

The IVAS system is designed for end users in the field. They should have the opportunity to record video files and to replay them. We use mobile data networks to establish audio and video calls. Mobile end user devices have to be equipped with appropriate VOIP software. VOIP software has to support video codecs H.263, H.263p and H.264. Another criterion enabling the cooperation with the IVAS system is DTMF code support and video support.

We tested smartphones equipped with the Android operating system. VOIP software Vippie [12] ensures video support, see Fig. 4. It supports all necessary audio and video codecs and it also supports DTMF key input.

## 3  Controlling Asterisk via Web Services

As mentioned before, Asterisk provides PBX functionality for audio and video calls. There are many web-based controlling services such as FreePBX or Asterisk GUI. These web-based controlling services provide web management for Asterisk PBX. Web-based management tools offer web-based controlling but do not provide a gateway that would enable to control through the web service client – these are not web service providers [13]. This is why we decided to develop our own web services provider.

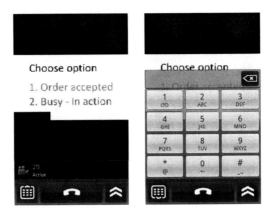

**Fig. 4.** Smartphone VOIP software with software keyboard

## 3.1 Dynamic Dial Plan in Asterisk

The dial plan in Asterisk should be generated automatically. We are using CLI scripts shown in Sect. 2.1. Scripts in Sect. 2.1 show the configuration to replay video. After a successful call, we have to remove these configuration commands with the following commands:

```
dialplan remove extension XXXX@default 1
dialplan remove extension XXXX@default 2
dialplan remove extension XXXX@default 3
```

All these commands are used to remove previous commands (see Sect. 2.1). Removing these commands is necessary to enable calling back for calls initiated by field users.

## 3.2 Calling Back the Same Number

If we use the command *originate* to establish a video call between the number with an assigned video file and with a specific user, the user can see the caller's name in Asterisk but he cannot see the caller's number assigned to the video file. It was necessary to implement a new command before commands shown in Sect. 2.1. We have to create a virtual number and link it with a particular user. The first command set YYYY as a virtual number linked with user Jones. XXXX is the extension linked with a video file.

```
dialplan add extension YYYY,1,Set(CALLERID(num)=XXXX) into default
dialplan add extension YYYY,2,Dial(SIP/Jones) into default
```

Through the CLI interface, these specific commands allow to add records to fix the problem with Asterisk caller name. We established a call via CLI with the *originate* command as shown below. The user with caller ID Jones sees the caller XXXX on his screen.

```
originate local/YYYY@default extension XXXX@default
```

After the call has been successfully established, we remove all dial plan records linked with numbers XXXX and YYYY. Removing these records is executed by means of commands shown in Sect. 3.1. All this configuration is necessary to allow for call back on a call established by a field user. He calls back number XXXX. We have to add a new record into the dial plan:

```
dialplan add extension XXXX,1,Goto(menu1,s,1) into default
```

This configuration command set XXXX number as a pre-defined scenario. If the field user chooses to call number XXXX back, he will see the pre-pared IVR/IVVR as defined below:

```
[menu1]
exten => s,1,Answer
exten => s,n,Wait(1)
exten => s,n,mp4play(/var/movie3x.mp4)
exten => s,n,Festival(Choose number one or two)
exten => s,n,WaitExten()
exten => 1,1,Goto(menu2,s,1)
exten => 2,1,Goto(menuX,s,1)
```

This is only an example of the dial plan menu. We use Festival for the speech synthesis for IVR and we prepared the IVVR menu into a video format supported by Asterisk and end user devices. We can navigate the user to record video files through the IVVR menu. A sample configuration command is shown in Sect. 2.1.

## 3.3   Java Example

This section contains several short examples of Java code that we have written in order to control Asterisk. We used libraries comes from Asterisk-Java project [9,14]. The first example shows the constructor of the *Notify Manager* class. The main purpose of this class is to work with the dial plan and extensions contained within it. The constructor creates an instance of the *ManagerConnectionFactory* class using location and login parameters. The factory is then responsible for establishing the connection to Asterisk.

```
public NotifyManager()
{
  ManagerConnectionFactory factory = new
      ManagerConnectionFactory(
      Globals.AsteriskIP,
      Globals.AsteriskUserName,
      Globals.AsteriskPassword);
  managerConnection = factory.createManagerConnection();
}
```

As we have already established the connection, we can now prepare and issue commands. In the following example, we will add a new extension to the dial plan. First, we have to replace key parameters of CLI command with values fed by the parent method.

```
private static final String cmdAddDialplanExtension = "
    dialplan_add_extension_\$exten,\ $priority,\$app,_into_\
    $context";
String action = cmdAddDialplanExtension.replace("$exten",
    extensionNum + "");
        action = action.replace("$priority", 1 + "");
        action = action.replace("$app", "Answer");
        action = action.replace("$context", "default");
```

The CLI command is prepared and well formatted, so we can send it to Asterisk through the *managerConnection* which we created earlier.

```
        cmdAction = new CommandAction(action);
        managerConnection.sendAction(cmdAction, TIMEOUT);
```

# 4    Video Quality Call Testing in Mobile Data Networks

The IVAS system is oriented on mobile video call transmission. Asterisk was equipped with app_mp4.c plugin, which provides recording and playback functions. We established testing environment with two Asterisk servers. One Asterisk established pre-recorded video call and second Asterisk received the call and saved it to a local storage. We prepared testing video sample with bandwidth 304 kbps.

We simulated mobile network parameters with Simena network emulator [15]. At first we decided to choose bandwidth 218 kbps (EDGE transmission rate). This technology does not offer transmission rate adequate to video sample. Later we simulated network (HSDPA R5) with bandwidth 1990 kbps and errors (simulated latency and packet loss), which could appear in real transmission.

The individual video files were compared with Video Quality Measurement Tool (MSU) [16]. This software offers technique to compare two or more video files. It is designed for comparison two or more video files in the codecs and lost frames field. MSU VQMT consists of many plugins and we applied plugin SSIM – Structural Similarity.

Structural Similarity [17] – is based on the measurement of 3 main criteria – similarity of brightness, contrast and similarity structure of a video file. Calculation of the total value is performed by the previously mentioned criteria. There are 2 implementations of this method – quick and precise. The difference between a quick and precise method is in used filter. While a quick method uses a filter box, the precise method uses Gauss blur filter [17].

## 4.1    SSIM Tests

SSIM value describes similarity between the original video frame and transmitted video frame. Tables 1 and 2 describes average SSIM value for specific transmission characteristics. Value 1.0000 corresponds with 100% similarity compare to original video frame.

**Table 1.** SSIM values for latency affecting video calls

| H263 | 218 kbps | | 1990 kbps | |
|---|---|---|---|---|
| Latency | Frames | Avg. SSIM | Frames | Avg. SSIM |
| 0 | 255 | 0.99954 | 391 | 0.99985 |
| 50 | 305 | 0.89954 | 391 | 0.99985 |
| 150 | 253 | 0.99947 | 391 | 0.99985 |
| 250 | 241 | 0.94062 | 391 | 0.99985 |
| 350 | 226 | 0.99911 | 391 | 0.99985 |
| 500 | 374 | 0.98782 | 391 | 0.99985 |

**Table 2.** SSIM values for packet loss affecting video calls

| H263 | 218 kbps | | 1990 kbps | |
|---|---|---|---|---|
| Packet Loss | Frames | Avg. SSIM | Frames | Avg. SSIM |
| 0% | 255 | 0.99954 | 391 | 0.99985 |
| 5% | 297 | 0.75648 | 366 | 0.35738 |
| 15% | 266 | 0.33385 | 319 | 0.32075 |
| 25% | 297 | 0.27950 | 298 | 0.28947 |
| 35% | 260 | 0.28148 | 263 | 0.25502 |
| 45% | 225 | 0.24798 | 230 | 0.24841 |
| 50% | 158 | 0.27858 | 122 | 0.28791 |

## 4.2 Test Results

Transmission errors could be divided into these categories:

- Video call file has the same size such as original video file. Transmission line covers transmission demands of the video calls. Transmission delay caused by latency is not found.
- Video call has smaller size compared with original video file and there are SSIM values equal 1. The transmission line does not cover demands of the video calls. Transmission delay caused by latency is not found.
- Video call has smaller size compared with original video file and there are low SSIM values. All broadcasted packet do not achieve destination and the packets are lost during transmission. Video contains many defects in video.

With these results we can describe link properties and also define the cause of transmission errors.

## 5 Conclusion

Our team participates in an international project referred to as the INDECT project. We are the principal developers of the IVAS system which is the main gateway between the INDECT portal and end user devices. End user devices include mostly smartphones and tablets. We have focused on end users which operate out of office, in particular those using the mobile internet connection.

The interaction with the INDECT portal is provided through the SOAP protocol. We developed JAVA-based web service provider which forms a part of the IVAS system together with Asterisk which ensures the PBX functionality.

The web service provider communicates with the INDECT portal which can be called as the web service requester (the client). The web service requester generates commands that enable to control Asterisk remotely. It also obtains information generated by the web service provider. The web service provider handles messages produced by Asterisk available on the AMI interface. We plan to implement new functions controlled by the web service provider e.g. sending text messages or URL link.

We modified Asterisk to support video recording and replay. These functions form the basis for multimedia handling.

We have not finished developing the IVAS system yet. We plan to integrate a fully automatic IVVR menu, we will also test performance of the IVAS system with full video support. At the moment, we are using data transmission in mobile networks. The next step is to use the UMTS video call service. It will be necessary to prepare the gateway between data and mobile networks and to convert the video format from H.263 and H.264 to H.324M video codec which is used for video calls in UMTS networks [18].

The main contribution of this paper are video services controlled by SOAP messages. The IVAS system has not any GUI interface. GUI interface is provided by other INDECT partners. The integration process between INDECT partners and IVAS system is ensured by SOAP protocol. IVAS system is ready to respond to commands coming from INDECT partner's GUI interface. IVAS ensures delivery of these orders or specific content to end users. The specific content means video and audio messages, IVR and IVVR menu. End users can then respond to commands by using DTMF choice or they may stream immediately video or audio content to a remote server.

**Acknowledgments.** The research leading to these results has received funding from the European Community's Seventh Framework Programme (FP7/2007-2013) under grant agreement no. 218086.

# References

1. INDECT Project, `http://www.indect-project.eu`
2. Tomala, K., Rozhon, J., Rezac, F., Vychodil, J., Voznak, M., Zdralek, J.: Interactive VoiceXML Module into SIP-Based Warning Distribution System. In: Dziech, A., Czyżewski, A. (eds.) MCSS 2011. CCIS, vol. 149, pp. 338–344. Springer, Heidelberg (2011)
3. Voznak, M., Kapicak, L., Zdralek, J., Nevlud, P., Rozhon, J.: Video files recording and playback with VoiceXML. In: Proceedings ISPRA 2011, Cambridge, pp. 350–354 (2011)
4. Kapicak, L., Nevlud, P., Zdralek, J., Dubec, P., Plucar, J.: Remote Control of Asterisk via Web Services. In: Proceedings TSP 2011, Budapest, pp. 27–30 (2011)

5. Weerawarana, S., Curbera, F., Leymann, F., Storey, T., Ferguson, D.F.: Web Services Platform Architecture: SOAP, WSDL, WS-Policy, WS-Addressing, WS-BPEL, WS-Reliable Messaging, and More, p. 47, 96, 112. Prentice Hall (2005)
6. Meggelen, J.P., Smith, J., Madsen, L.: Asterisk: The Future of Telephony. O'Reilly Media (2005)
7. Rosenberg, J., Schulzrinne, H., Camarillo, G., Johnston, A., Peterson, J., Sparks, R., Handley, M., Schooler, E.: RFC 3261 – SIP: Session Initiation Protocol. Technical report, IETF (2002)
8. Voznak, M., Rudinsky, J.: Alternative methods of intelligent network service implementation in IP telephony. In: International Conference on Communications – Proceedings, Corfu, pp. 204–207 (2010)
9. Asterisk JAVA project, `http://asterisk-java.org/`
10. Alonso, G., Casati, F., Kuno, H., Machiraju, V.: Web Services: Concepts, Architecture and Applications, 1st edn., pp. 185–197. Springer (2010); Softcover of orig. 2004 edition
11. Boudreau, T., Glick, J., Greene, S., Woehr, J., Spurlin, V.: Netbeans: The Definitive Guide, pp. 321–352. O'Reilly Media (2002)
12. Vippie: Video calls via internet, `http://voipswitch.com/en/`
13. Voznak, M., Kapicak, L., Zdralek, J., Nevlud, P., Plucar, J.: Multimedia services in Asterisk based on VoiceXML. International Journal of Mathematical Models and Methods in Applied Sciences 5(5), 857–865 (2011)
14. Reuter, S.: Asterisk-Java, The free Java library for Asterisk PBX integration, `http://asterisk-java.org/`
15. Simena, Simena network emulator, `http://www.simena.net/`
16. MSU, MSU Quality Measurement Tool: Metrics information, `http://www.compression.ru/video/quality_measure/info_en.html`
17. Wang, Z., Bovik, A.C., Sheikh, H.R., Simoncelli, E.P.: Image Quality Assessment: From Error Visibility to Structural Similarity. IEEE Transactions on Image Processing 13(4), 600–613 (2004)
18. Heine, G.: GSM Networks: Protocols, Terminology and Implementation, pp. 11–19. Artech House (1998)

# A Friendliness Study of TCP Linux Variants

Christian Callegari, Stefano Giordano, Michele Pagano, and Teresa Pepe

Dept. of Information Engineering, University of Pisa,
Via Caruso 16, I-56122, Italy
{c.callegari,s.giordano,m.pagano,t.pepe}@iet.unipi.it

**Abstract.** The Transmission Control Protocol (TCP) is used by the vast majority of Internet applications. Since its introduction in the 70s, a lot of variants have been proposed to cope with the different network conditions we can have (e.g., wired networks, wireless networks, satellite links) and nowadays Linux OS includes 12 different TCP variants.

The aim of this paper is to offer a detailed comparative analysis of the behavior offered by these variants, in terms of friendliness.

**Keywords:** TCP Linux, friendliness.

## 1 Introduction

The Transmission Control Protocol (TCP) [1] is used by the vast majority of Internet applications. The TCP is a connection-oriented transport protocol that provides a reliable byte-stream data transfer service between pairs of processes.

When two processes want to communicate, they must first establish a TCP connection (initialize the status information on each side). Since connections must be established between unreliable hosts and over the unreliable internet communication system, a handshake mechanism with clock-based sequence numbers is used to avoid erroneous initialization of connections.

The TCP provides **multiplexing** facilities by using source and destination port numbers. These port numbers allow the TCP to set up virtual connections over a physical connection and multiplex the data stream through that connection. Each connection is uniquely specified by a pair of sockets identifying its two sides and by specific parameters such as sequence numbers, window sizes, and status information (necessary for reliability and flow and congestion control mechanisms – see the following for more details). At the end of a communication, the connection is terminated or closed to free the resources for other uses.

The TCP is able to transfer a continuous stream of bytes, in each direction, by packaging some number of bytes into *segments* (TCP data unit). The size of these segments and the timing at which they are sent are generally left to the TCP module. However, an application can force delivery of segments to the output stream using a push operation provided by the TCP to the application layer. A push causes the TCP to promptly forward and deliver data to the receiver.

Apart from this *basic data transfer*, the TCP provides some more services. First of all it is able to recover from data that are damaged, lost, duplicated, or delivered out of order by the internet communication system (**reliability**).

A. Kwiecień, P. Gaj, and P. Stera (Eds.): CN 2012, CCIS 291, pp. 14–23, 2012.

Missing or corrupted segments are detected (and retransmitted) assigning a sequence number to each transmitted segment, and requiring a positive acknowledgment (ACK) from the receiver. If the ACK is not received within a timeout interval, the data are retransmitted.

At the receiver, the sequence numbers are used to correctly order segments that may be received out of order and to eliminate duplicates. Damage is handled by adding a checksum to each transmitted segment, checking it at the receiver, and discarding damaged segments.

The TCP also provides a means for the receiver to govern the amount of data sent by the sender (**flow control**). This is achieved by returning a "window" that indicates the allowed number of bytes that the sender may transmit before receiving further permission.

Moreover, TCP users may indicate the security and precedence of their communication. Provision is made for default values to be used when these features are not needed.

Finally, one of the most important features provided by the TCP is the **congestion control** that is used to make the TCP sensitive to the network conditions. The basic idea is that the rate of ACKs returned by the receiver determines the rate at which the sender can transmit data. The TCP uses several algorithms for congestion control, each of those controls the sending rate by manipulating a congestion window (*cwnd*) that limits the number of outstanding unacknowledged bytes that are allowed at any time. All the modern implementations of the TCP contain at least four intertwined algorithms: Slow Start, Congestion Avoidance, Fast Retransmit, and Fast Recovery [2]. Moreover, the more recent implementations of the TCP also include some other mechanisms and algorithms.

The aim of this paper is to offer a detailed comparative analysis of the performance, in terms of friendliness, offered by all the current TCP Linux variants. Indeed, at the best of our knowledge, such comparison is not reported anywhere in the literature.

For sake of brevity we skip the description of all the TCP Linux variants as well as of optional mechanisms, such as SACK [3], FACK [4], DSACK [5], and ECN [6].

The rest of the paper is organized as follows: in the next section we provide a brief description of the test scenario used in this work, while in Sect. 3 a friendliness analysis is offered. Finally Section 4 concludes the paper with some final remarks.

## 2   Test Scenario

In this section, we outline the experimental tests we have performed to evaluate the performance offered by the different TCP variants implemented in the Linux kernel 2.6.x, namely:

- TCP Reno [7],
- TCP Vegas [8],
- TCP Veno [9],

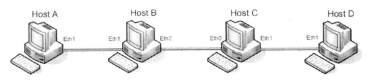

**Fig. 1.** Network topology

- TCP Westwood [10],
- BIC (Binary Increase Congestion Control) TCP [11],
- TCP CUBIC [12],
- TCP Low Priority [13],
- High-Speed TCP (HSTCP) [14],
- TCP Hybla [15],
- Scalable TCP [16],
- TCP Illinois [17],
- TCP YeAH (Yet Another Highspeed) [18].

Our test-bed was composed of four general purpose PCs, equipped with a 2.53 Ghz Intel CPU and 4 GB of RAM. The PCs have been interconnected with a 1 Gbps Ethernet network (as depicted in Fig. 1).

To perform our tests we have used several tools, all freely available for Linux OS. In more detail we have used:

- **Iperf:** it is a standard tool [19], commonly used for network performance measurements, and allows the creation of TCP data streams between two machines. It provides both client and server functionalities, and can measure the throughput between the two end-hosts, either uni-directionally or bi-directionally.
- **NetEm:** it is a tool [20] that permits to emulate a generic network environment on a point-to-point connection. We have used such tool to study the protocol performance in non-optimum conditions (RTT and loss probability). In this way, we have been able to emulate long distance networks in a lab environment.

  NetEm consists of two portions, a small kernel module for creating a queuing discipline and a command line utility to configure it. Communication between the command and the kernel is done via the Netlink socket interface. Requests are encoded into a standard message format that is decoded by the kernel.

  A queuing discipline exists between the TCP output and the network device. The default queuing discipline is a simple FIFO queue with no delay, loss, duplication, or reordering of packets. For emulating packet loss, delay, and duplication, the user specifies the parameters (average values) to the network emulator as arguments to the "tc" command.

The tests have been conducted considering a $RTT = 100$ ms and $P_l = 0.0001\%$.

Moreover, the value of $MSS$ is fixed and assumes the following value: $MSS = MTU - TCP\&IPheader$, where $MTU = 1500\,$B. Regarding the different TCP options, we have used the standard TCP Linux configuration, that is SACK, FACK, and DSACK are active for all the TCP variants, while ECN is deactivated.

## 3   Friendliness

**Friendliness** refers to the TCP ability of sharing the link capacity among the different sessions active on the link. It is assumed that the connections are using different versions of the TCP. The maximum level of friendliness is reached in the case when the $N$ TCP sessions sharing the same bottleneck link, get $1/N$ of the link capacity each.

In this analysis we have taken into account the friendliness of the different TCP variants with respect to two distinct TCP variants:

- TCP Reno: it is the standard TCP, and thus it is quite common that the TCP sessions will have to compete with TCP Reno for sharing the bandwidth. It is worth noticing that Windows OSs use TCP Reno as default TCP.
- TCP CUBIC: in this case the choice of the TCP variant is justified by the fact that TCP CUBIC is the default TCP variant used by Linux.

Figure 2 shows the throughput achieved by TCP Reno, when the background traffic is using TCP CUBIC: TCP Reno is much less aggressive than TCP CUBIC and thus achieves a throughput, which is much lower than that of TCP CUBIC.

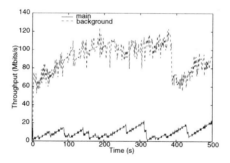

**Fig. 2.** Friendliness vs. CUBIC (TCP Reno)

In Figure 3(a) we can see that TCP Reno and TCP Vegas are able to friendly share the link bandwidth. This is reasonable since the two protocols almost behave equally. Indeed TCP Vegas should behaves differently in strong congestion situations that are not present in the considered scenarios. On the contrary TCP CUBIC (see Fig. 3(b)) appears to be much more aggressive than TCP Vegas.

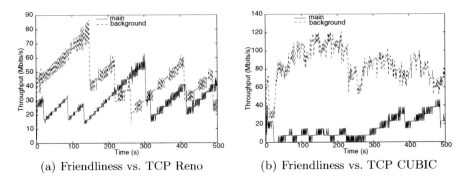

(a) Friendliness vs. TCP Reno          (b) Friendliness vs. TCP CUBIC

**Fig. 3.** TCP Vegas

The observations done for TCP Vegas are still valid for TCP Veno (see Fig. 4(a) and 4(b)). Also in this case the results are quite reasonable, since TCP Veno behaves similarly to TCP Reno and TCP Vegas.

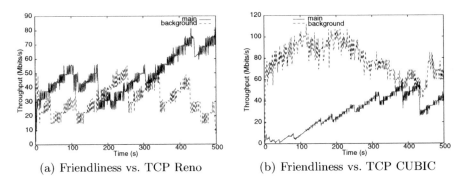

(a) Friendliness vs. TCP Reno          (b) Friendliness vs. TCP CUBIC

**Fig. 4.** TCP Veno

As shown in Fig. 5(a) and 5(b), TCP Westwood friendly shares the link bandwidth both with the connection using TCP Reno and with the one using TCP CUBIC. It is important to highlight that this is the only TCP variant, which appears to be friendly with respect to both the considered protocols.

Differently from all the previous TCP variants, TCP BIC results much more aggressive of both the background protocols. To be noted that in the first case (see Fig. 6(a)), it is quite obvious since, as already said (see Fig. 2), TCP Reno is much less aggressive than TCP CUBIC (which is almost the same of TCP BIC).

The situation shown in Fig. 7 is symmetric to the one shown in Fig. 2, hence the same consideration holds.

As it is obvious TCP-LP is much less aggressive than TCP Reno (see Fig. 8(a)) and TCP CUBIC (see Fig. 8(b)).

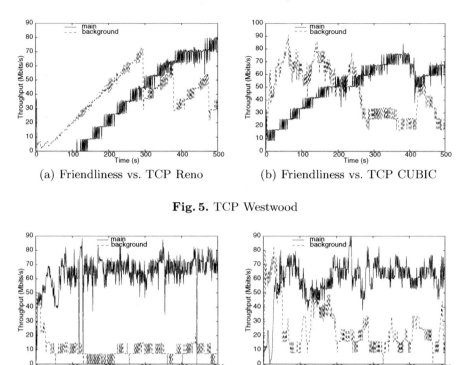

(a) Friendliness vs. TCP Reno          (b) Friendliness vs. TCP CUBIC

**Fig. 5.** TCP Westwood

(a) Friendliness vs. TCP Reno          (b) Friendliness vs. TCP CUBIC

**Fig. 6.** TCP BIC

As we can see from Fig. 9(a), HSTCP is able to achieve better performance when competing with TCP Reno, while it behaves friendly when sharing the link with TCP CUBIC (see Fig. 9(b)).

TCP Hybla presents a behavior that is very close to HSTCP. Thus, it appears to be more aggressive than TCP Reno (see Fig. 10(a)) and friendly when compared to TCP CUBIC (see Fig. 10(b)).

Scalable TCP appears to be the most aggressive TCP variant considered so far. Indeed, it is able to achieve a throughput much higher than that achieved by both TCP Reno (see Fig. 11(a)) and TCP CUBIC (see Fig. 11(b)).

The same considerations done for Scalable TCP still hold for the two other high-speed TCP variants (TCP Illinois and TCP YeAH) that achieve the same behavior of Scalable TCP. Indeed, as shown in Fig. 12(a), 12(b), 13(a), and 13(b), they both achieve better performance than TCP Reno and TCP CUBIC.

In general we can notice that, the total achieved throughput is almost the same (apart from the TCP-LP case) when using the different TCP variants; even though, it is a bit higher when TCP CUBIC is used for the background traffic and a bit lower when Reno is used. This can be justified by the fact that TCP CUBIC is more aggressive than TCP Reno, when probing for bandwidth.

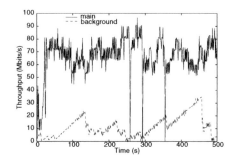

**Fig. 7.** Friendliness vs. TCP Reno (TCP CUBIC)

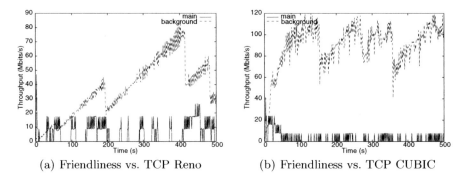

(a) Friendliness vs. TCP Reno          (b) Friendliness vs. TCP CUBIC

**Fig. 8.** TCP-LP

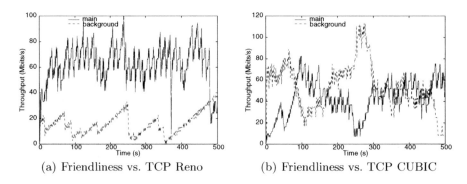

(a) Friendliness vs. TCP Reno          (b) Friendliness vs. TCP CUBIC

**Fig. 9.** HSTCP

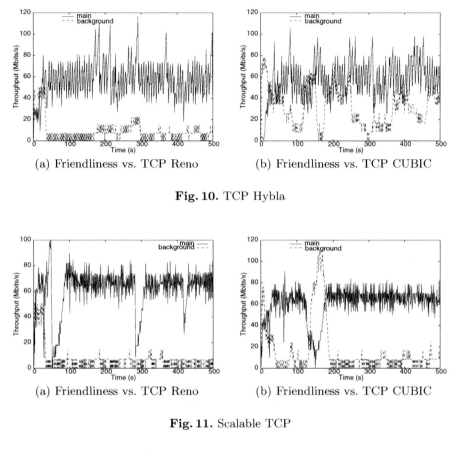

(a) Friendliness vs. TCP Reno          (b) Friendliness vs. TCP CUBIC

**Fig. 10.** TCP Hybla

(a) Friendliness vs. TCP Reno          (b) Friendliness vs. TCP CUBIC

**Fig. 11.** Scalable TCP

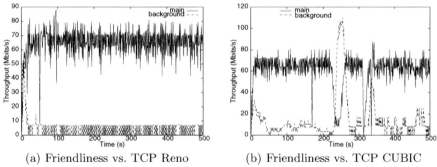

(a) Friendliness vs. TCP Reno          (b) Friendliness vs. TCP CUBIC

**Fig. 12.** TCP Illinois

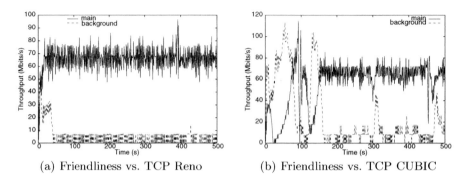

(a) Friendliness vs. TCP Reno         (b) Friendliness vs. TCP CUBIC

**Fig. 13.** TCP YeAH

## 4   Conclusions

The TCP is for sure the most used transport layer protocol in the Internet. In the years several variants have been proposed to overcome the limitations that the standard TCP has in particular scenarios, such as wireless networks or satellite networks.

As a results, the current kernel of the Linux OS (version 2.6.x) includes 12 different versions, going from the standard TCP (TCP Reno) and its improved version (TCP Vegas), to the variants for wireless networks (TCP Veno and TCP Westwood), high-speed networks (TCP BIC, TCP CUBIC), and satellite networks (HSTCP, TCP Hybla, TCP Illinois, Scalable TCP, and TCP YeAH), and also a low-priority version (TCP-LP).

In this paper we have offered an experimental comparison of the friendliness of all these variants with respect to two standard benchmarks, the "classical" TCP Reno and the Linux-standard TCP CUBIC.

## References

1. Postel, J.: RFC 793: Transmission Control Protocol (September 1981)
2. Allman, M., Paxson, V., Blanton, E.: TCP Congestion Control. RFC 5681 (Draft Standard) (September 2009)
3. Mathis, M., Mahdavi, J., Floyd, S., Romanow, A.: RFC 2018: TCP selective acknowledgment options. Status: PROPOSED STANDARD (October 1996)
4. Mathis, M., Mahdavi, J.: Forward acknowledgement: refining TCP congestion control. In: Proceedings of the SIGCOMM 1996. ACM, New York (1996)
5. Floyd, S., Mahdavi, J., Mathis, M., Podolsky, M.: An Extension to the Selective Acknowledgement (SACK) Option for TCP. RFC 2883 (Proposed Standard) (July 2000)
6. Kuzmanovic, A., Mondal, A., Floyd, S., Ramakrishnan, K.: Adding Explicit Congestion Notification (ECN) Capability to TCP's SYN/ACK Packets. RFC 5562 (Experimental) (June 2009)
7. Floyd, S., Henderson, T., Gurtov, A.: The NewReno Modification to TCP's Fast Recovery Algorithm. RFC 3782 (Proposed Standard) (April 2004)

8. Brakmo, L.S., O'Malley, S.W., Peterson, L.L.: TCP Vegas: New Techniques for Congestion Detection and Avoidance. In: SIGCOMM, pp. 24–35 (1994)
9. Fu, C.P., Liew, S.: TCP Veno: TCP Enhancement for Transmission Over Wireless Access Networks. IEEE Journal on Selected Areas in Communications, 216–228 (2003)
10. Mascolo, S., Casetti, C., Gerla, M., Sanadidi, M.Y., Wang, R.: TCP Westwood: Bandwidth estimation for enhanced transport over wireless links. In: MobiCom 2001: Proceedings of the 7th Annual International Conference on Mobile Computing and Networking, pp. 287–297. ACM, New York (2001)
11. Xu, L., Harfoush, K., Rhee, I.: Binary Increase Congestion Control (BIC) for Fast Long-Distance Networks. In: INFOCOM (2004)
12. Rhee, I., Xu, L.: CUBIC: a new TCP-friendly high-speed TCP variant. SIGOPS Oper. Syst. Rev. 42(5), 64–74 (2008)
13. Kuzmanovic, A., Knightly, E.W., Service, A.: TCP-LP: A Distributed Algorithm for Low Priority Data Transfer (2003)
14. Floyd, S.: HighSpeed TCP for Large Congestion Windows (2002)
15. Caini, C., Firrincieli, R.: TCP Hybla: a TCP enhancement for heterogeneous networks. International Journal of Satellite Communications and Networking 22 (2004)
16. Kelly, T.: Scalable TCP: improving performance in highspeed wide area networks. SIGCOMM Comput. Commun. Rev. 33(2), 83–91 (2003)
17. Liu, S., Başar, T., Srikant, R.: TCP-Illinois: a loss and delay-based congestion control algorithm for high-speed networks. In: Valuetools 2006: Proceedings of the 1st International Conference on Performance Evaluation Methodolgies and Tools, p. 55. ACM, New York (2006)
18. Baiocchi, A., Castellani, A.P., Vacirca, F.: YeAH-TCP: Yet Another Highspeed TCP. In: Proceedings of PFLDnet (2007)
19. NLANR/DAST: Iperf – The TCP/UDP Bandwidth Measurement Tool, http://sourceforge.net/projects/iperf/
20. Hemminger, S.: Network Emulation with NetEm. In: Linux Conf. Au. (April 2005)

# Admission Policy in Web Services Based on Auction Approach

Jolanta Wrzuszczak-Noga and Leszek Borzemski

Institute of Informatics, Wroclaw University of Technology, Wroclaw, Poland
{jolanta.wrzuszczak,leszek.borzemski}@pwr.wroc.pl

**Abstract.** In the paper a new web admission auction-based algorithm is presented and different pricing policies combined with inherent QoS characteristics are discussed. Performance indexes oriented on business profit – service income and client loyalty were discussed and measured during experiments. Investigations have been performed to validate the proposed mechanisms.

**Keywords:** auction mechanisms, admission control, scheduling algorithms in web service, quality of service.

## 1 Introduction

Nowadays in the era of globalization web services are commonly used for doing any kind of business, especially in specific cases such as e-shops, and auction services. Most of network users expect services delivering their functionalities on a proper service quality level, i.e. guaranteeing of maintaining specified transaction parameter or delivering resources in the proper time. Investigations show [1], that "a statistical user" can accept only a few second pause for reaction while working with the service browser. This is very important aspect which must be considered by designing modern e-commerce sites, which are oriented on assuring Quality of Service (QoS).

Nowadays the meaning of Internet has changed, it has evaluated from a network for forwarding information and servicing web clients on the same level, to the network for supporting business transactions and considering management groups of clients i. e. key-customers on the special level [2, 3]. Economical aspects of functioning of Internet involve managing of web services. In the literature many economical strategies have been presented to increase profit of web services by introducing various management plans [2, 4–9].

In this paper a new admission policy based on auction approach will be proposed and evaluated. Different pricing mechanisms and different performance indexes (service revenue and client loyaly) will be presented.

First we present the background, next the assumptions and problem formulation. Then the experiment conducted in the testing network laboratory environment is presented, and discussed. Finally, we conclude our research.

A. Kwiecień, P. Gaj, and P. Stera (Eds.): CN 2012, CCIS 291, pp. 24–31, 2012.

## 2   Background

Admission control of web services has been analyzed in various works [4, 5, 10, 11, 6–8, 3]. The most common policy developed for queuing in web services is based on the method employing the time stamp of arriving requests (known as First In First Out approach). The other proposed policy orders jobs to the queue according to the time of execution (known as Shortest Job First algorithm).

Modern web services are equipped with dynamic scheduling and pricing policies complying dynamic strategies like auctions [6, 7, 12]. Such designs are able to support business transactions as good as possible.

In the literature many works deal with the management of limited resources by means of an auction policy. In [13] a problem of delivering electrical energy based on the predictive demand was described. [7, 12] treat with delivering of bandwidth volume to providers. [10] presents managing of radio spectrum based on bidding procedure. In extended systems the agent-based solution was distinguished by supporting transactions of goods, which takes place between multiply buyers and multiply sellers [6].

Some researches concentrated on scheduling mechanisms considering pricing schemas [10, 7, 12, 3]. The most common policy by bidding process is an increasing auction, so that in every auction run the price increases. Another common mechanism doesn't exceed the submitted value and is determined by offers from another clients. The auction based on the second highest price determined by web clients is known as Vickrey auction [14, 8].

Our proposal considers providing web service resources for individual users for which charges are taxed. Regarding service level guarantees, web service assures the bandwidth parameter for transmission of resources.

The main contribution of this paper is a novel web service admission policy including new auction-based scheduling mechanism and different pricing policies. Our proposal concentrates also on QoS level and client satisfaction (client loyalty).

## 3   Assumptions and Problem Formulation

The aim of the web service is delivering resources for which charges are taxed. The architecture of web service is shown in Fig. 1. Web clients send offers to the web service and the scheduler makes decision about accepting or rejecting offers and assigns a bandwidth for scheduled requests to the queue. After downloading the resources the resulting charges are calculated according to the pricing policy.

Some assumptions were made to present the web service role. The scheduling process takes place in every defined time interval $T$ (for example, every 15 minutes).

The offer contains: resource name ($r$), resource price ($p$) and expected bandwidth pieces ($b$, optional), i.e, $n$-th offer is described by $o_n = \langle\ r_j,\ p_n,\ b_n\ \rangle$.

Many offers can be sent by user for given resource, so that for the increasing number of bandwidth pieces the increasing price is proposed (as in Fig. 2).

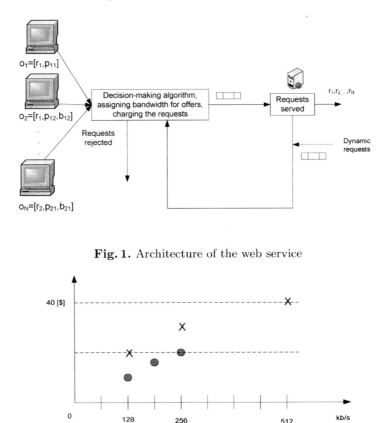

**Fig. 1.** Architecture of the web service

**Fig. 2.** Offers for two resources for different bandwidth pieces

In the Fig. 2 the sample offers for two resources are presented. The offers for the first resource are depicted by "x" and for the second one by circles. Offers without the expected bandwidth pieces contain the reverse (initial) price for every resource. The process of collecting offers is closed at the end of one time period and is valid for the next time period.

To measure and compare scheduling mechanisms three performance indexes were distinguished:

– the sum of submitted offers (the service revenue):

$$\varphi_p = \sum_n^N p_n^* \ , \tag{1}$$

– the number of offers, for which calculating charges were lower than declared $\varphi_l = \sum_n^N x_n$ by assumption $p_{nI}^* \neq p_n$,
– the mean value of RFM index (based on the transaction history for every web client) $\varphi_R = \sum_n^N \frac{1}{N} RFM(o_n)$,

where:

$N$ – the number of served offers,

$p_n^*$ – the price pay for serving $n$-th offer,

$p_{nI}^*$ – the exactly price (first price) offered by the $n$-th offer,

$x_n$ – $\{0, 1\}$ – decision variable, which determines if the price paid was lower than declared,

$RFM$ – (Recency Frequency Monetary) the loyalty customer index based on history of bought products.

The scheduling admission problem is formulated as follows, for given:

- $o_1, \ldots o_N$ – the set of offers,
- $R$ – the set of data files (resources), $r_j \in R$,
- $T$ – scheduling time interval,
- $B$ – total available bandwidth (i.e. 10 Gbps),
- the performance index defined as web service revenue (1)

a scheduling admission strategy should be found to select the set of offers $S_d$ for processing, which increases the performance index and assures the client satisfaction simultaneously in every time instant $(T)$.

## 4    Algorithms

It is worth to say that presented algorithms are recommend in such systems, where the overloading is noticed. Web service resources are limited and it is not possible to serve all requests at the same time. Some requests must be selected to serve them whereas the others have to be rejected because of lack of web service capacity.

The considered problem is NP complete [12, 14, 8]. To solve the problem a greedy knapsack algorithm is proposed. Existing index (defined as a relation of value to weight of the products) for calculating income of service offer is modified because of relations between resource and price aren't unambiguously (equivalent to the typical knapsack problem). The additional parameter the bandwidth pieces determine the real profit of choosing offers.

New "profit index" is proposed, which contains information about transmission time and feasibility of $n$-th offer:

$$\varphi_n = \frac{p_n}{t_n} \tag{2}$$

by assumption:

$$t_n = \frac{\text{sizeof}(r_j)}{b_n} \tag{3}$$

where:

$t_n$ – time of transmission of resource $r_j$ and $b_n$ bandwidth pieces for $n$-th offer,

$p_n$ – proposed price for the $n$-th offer,

$b_n$ – number of requested bandwidth pieces.

The problem is more advanced than typical knapsack problem, because resources don't reflect assigned offers and the available of free bandwidth. It is important that free bandwidth pieces must be available continuously.

### Scheduling Algorithm based on a knapsack algorithm
The pseudocode of the algorithm is the following:

```
calculate the profit index for every offer
sort offers according to profit index desc
for (n=0;n<N;n++){ //for every offer in the queue
 choose offer according to highest value of profit index

    while(available free time span || end of queue of offers){
    take requested bandwidth pieces for n-th offer
    check, if number of time spans (number_ts) available continuously
    to serve n-th request

        if(number of available free time spans continuously = number_ts){
            schedule the n-th request
            mark number_ts time spans as busy
    }
    } //while
  n=n+1//take next offer
} //for
```

### Scheduling Algoritm based on FIFO algorithm
The pseudocode of the algorithm is the following:

```
for (n=0;n<N;n++){ //for every offer in the queue
    while(available free time span || end of queue of offers){
    choose offer according to earliest time stamp in system
    take requested bandwidth pieces for n-th offer
    check, if number of time spans (number_ts) available continuously
    to serve n-th request

        if(number of available free time spans continuously = number_ts){
            schedule the n-th request
            mark number_ts time spans as busy
    }
    } //while
  n=n+1//take next offer
} //for
```

Three different pricing algorithms are distinguish:

- The first price policy $p_I^*(n) = p_n$.
- The second price policy $p_{II}^*(n) = \frac{p_n \varphi_{n+1}}{\varphi_n}$.
- The price policy build according to the regular customer ($RC$) status and the value of RFM index in the way as they are presented in Table 1, where $p_{rb}$ describes the mean price of the resource $r$ and bandwidth pieces $b$ known from history.

**Table 1.** Conditions for pricing algorithm based on RFM index for $n$-th offer

| Conditions | Pricing policy |
|---|---|
| $RC = 1$ and $RFM \geq 70\%$ | $\max(p_{rb}(n), p^*_{II}(n))$ when calculated price lower than declared. In other case $p^*_{II}(n)$ |
| $RC = 1$ and $RFM < 70\%$ | $\min(p_{rb}(n), p^*_{II}(n))$ |
| $RC = 0$ and $RFM \geq 70\%$ | $p^*_I(n)$ |
| $RC = 0$ and $RFM < 70\%$ | $\max(p_{rb}(n), p^*_{II}(n), 0.9p^*_I(n))$ when calculated price lower than declared. In other case $\max(p^*_{II}(n), 0.9p^*_I(n))$ |

## 5   Experiments and Results

The aim of the experiments is to compare our auction-based approach to the classic FIFO algorithm in scheduling requests for web services. Three performance indexes were measured to compare different pricing mechanisms.

The computations have been performed in the Distributed Computer Systems Laboratory at the Institute of Informatics, Wroclaw University of Technology. The experimental setup includes Apache Web Server and MySQL database.

A database of clients was generated for 10 000 instants of clients. A relational table was defined to store information of incoming transactions (i.e. sum of transactions, frequency and number of transactions) to calculate the RFM index. A generator of offers was implemented in the Matlab environment, and the generated data was imported into the MySQL database. Available bandwidth ($B$) was 10 Gbps and every bandwidth piece was a multiply of 64 kbps.

Every experiment ran for different number of offers, while another parameters were kept constant. For auction based scheduling algorithm and FIFO policy three various pricing mechanisms were measured.

The results of experiments are presented in Figs. 3–5. In Figure 3 the first price policy for auction and FIFO scheduling versus the number of offers are

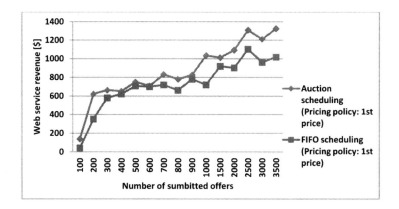

**Fig. 3.** Service income versus different number of submitted offers for first price policy

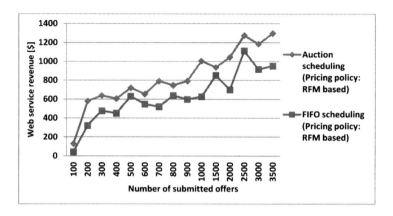

**Fig. 4.** Service income versus different number of submitted offers for RFM pricing policy

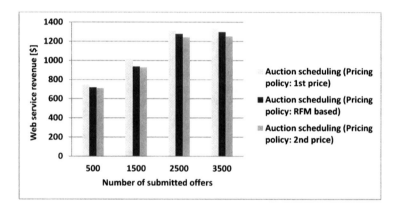

**Fig. 5.** Service income versus different number of submitted offers for auction based scheduling policy

presented. In Figure 4 the RFM-based policy for auction scheduling and FIFO scheduling are presented. In Figure 5 three pricing mechanism are compared in auction based scheduling versus the number of offers.

## 6  Conclusion

Experiments showed that the best value of considered performance index – the service revenue – was observed for the auction based scheduling algorithm with first price policy. However, in our opinion this pricing policy do not promise client's satisfaction and loyalty.

The income of the web service for the RFM-based pricing mechanism was a little bit worse comparing to first price policy, and a slightly better comparing to second price pricing policy. The results for RFM-based policy look promisingly. RFM-based approach takes into account the client's loyalty what can give back in lower prices than declared.

# References

1. Nah, F.: A study on tolerable waiting time: how long are Web users willing to wait? Behaviour and Information Technology 23(3), 153–163 (2004)
2. Borzemski, L., Suchacka, G.: Business-Oriented Admission Control and Request Scheduling for E-Commerce Websites. Cybernetics and Systems 41(8), 592–609 (2010)
3. Wrzuszczak-Noga, J., Borzemski, L.: An Approach to Auction-Based Web Server Admission Control. In: Mehrotra, K.G., Mohan, C., Oh, J.C., Varshney, P.K., Ali, M., et al. (eds.) Developing Concepts in Applied Intelligence. SCI, vol. 363, pp. 101–106. Springer, Heidelberg (2011)
4. Borzemski, L., Wrzuszczak, J., Kotowski, G.: Management of Web service delivering multimedia files based on the bid strategy. In: Information Systems Architecture and Technology, ISAT 2008, pp. 13–23 (2008)
5. Cherkasova, L., Phaal, P.: Peak Load Management for commercial Web servers using adaptive session-based admission control. In: Proceedings of the 34th Hawaii International Conference on System Sciences (2001)
6. Lubacz, J.: Auction and Stocks Mechanisms in Communication Resources E-commerce. WKL, Warszawa (2011) (in Polish)
7. Maille, P., Tuffin, B.: Pricing the Internet with multibid auctions. IEEE/ACM Trans. on Networking 14(5), 992–1004 (2006)
8. Wrzuszczak, J.: Auction mechanism in management of processing nodes in a computer cluster. Contemporary Aspects of Computer Networks 2, 259–265 (2008)
9. Wrzuszczak, J., Borzemski, L.: Management of Web Services Based on the Bid Strategy Using the User Valuation Function. In: Kwiecień, A., Gaj, P., Stera, P. (eds.) CN 2009. CCIS, vol. 39, pp. 19–25. Springer, Heidelberg (2009)
10. Cramton: The FCC Spectrum Auctions: An Early Assessment. Journal of Economics and Management Strategy 6(3), 431–495 (1997)
11. Lee, M., Lui, J., Yau, D.: Admission Control and Dynamic Adaptation for a Proportional-Delay DiffServ-Enabled Web Server. In: Proceedings of the 2002 ACM SIGMETRICS International Conference on Measurement and Modeling of Computer Systems, pp. 172–182 (2002)
12. Pérez-Bellido, Á.M., Salcedo-Sanz, S., Portilla-Figueras, J.A., Ortíz-García, E.G., García-Díaz, P.: An Agent System for Bandwidth Allocation in Reservation-Based Networks Using Evolutionary Computing and Vickrey Auctions. In: Nguyen, N.T., Grzech, A., Howlett, R.J., Jain, L.C. (eds.) KES-AMSTA 2007. LNCS (LNAI), vol. 4496, pp. 476–485. Springer, Heidelberg (2007)
13. Brazier, F., et al.: A multi-agent system performing one-to-many negotiation for load balancing of electricity use. Electronic Commerce Research and Applications 1, 208–224 (2002)
14. Vickrey, W.: Counterspeculation, Auctions, and Competitive Sealed Tenders. The Journal of Finance 16(1), 8–37 (1961)

# Decentralized Algorithm for Joint Data Placement and Rate Allocation in Content-Aware Networks

Dariusz Gąsior* and Maciej Drwal**

Institute of Informatics
Wroclaw University of Technology
Wyb. Wyspianskiego 27, 50-370 Wroclaw, Poland
{dariusz.gasior,maciej.drwal}@pwr.wroc.pl

**Abstract.** In this paper a mathematical model of data delivery in Content-Aware Network (CAN) is studied. Such network architecture assumes that users access data objects (content) replicated on a number of cache servers. A CAN operator decides on the placement of data objects on behalf of the content publishers, and decides on the transmission rate allocation of users (content consumers downloading content, and publishers uploading and updating their content). A problem of minimizing the mean transmission delay is considered and formulated as a mixed-integer programming problem. Decomposition-based heuristic is proposed, which utilizes projected gradient descent and greedy strategy. The algorithm is presented as a decentralized protocol, embedded in the network architecture.

**Keywords:** Content-Aware Network, flow control, facility location.

## 1 Introduction

The problem of optimal planning of data delivery networks in the Internet has been studied systematically in the last decade, in response to rapidly increasing popularity of the Web [1]. The various formulations of placement problems of servers, proxies and caches in computer networks have been studied in e.g.: [2,3]. The first Content Delivery Network was deployed by Akamai Technologies, as a result of the pioneering research in this area [4]. Since then, most of the major content providers entrusted their offered services to such systems. Recently, several new system architectures for future Internet have been proposed, such as Content Centric Network [5] and Content-Aware Networks, optimized for delivery of large volumes of content [6].

---

* Support provided by the European Union from the European Regional Development Fund within the Innovative Economy Operational Programme project number POIG.01.01.02-00-045/09-00 "Future Internet Engineering".

** Support provided by the scholarship co-financed by European Union within European Social Fund.

A. Kwiecień, P. Gaj, and P. Stera (Eds.): CN 2012, CCIS 291, pp. 32–44, 2012.

In this paper a mathematical model of Content-Aware Network (CAN) is presented. In such network the transit routers are supplied with special cache servers, which store the content replicas generated on behalf of the content publishers. The performance of the considered system is expressed as a mean transmission delay between users and servers. A decomposition-based heuristic algorithm is presented, which allows for sequential solving the rate allocation [7,8] and content placement [9,10] problems, utilizing projected gradient descent and greedy strategy, respectively.

## 2    Problem Statement

In the considered problem given is a set of users (called *clients*) who want to read data objects and a set of users (called *publishers*) who want to have their data content delivered to the clients. Clients never read data directly from the publishers' servers. Instead, publishers place the data objects inside public cache servers, connected to the network transit routers. These cache servers are under control of network *operator*, who manages the placement of content on behalf of publishers, as well as assigns transmission rates. Further in this paper the term *user* refers to the local area network (LAN) which generates a packet *flow*, and can be either client user LAN or publisher user LAN.

The description above illustrates the concept of data delivery in *Content-Aware Networks*. The following decisions are to be made by the network operator:

1. how to allocate transmission rates to the users (clients and publishers),
2. in which cache servers to place individual data objects,
3. how to redirect clients to the cache servers.

All those decisions influence the system's performance in terms of request delays perceived by users (clients and publishers). It is assumed that the objective of network operator's decisions is to minimize the sum of mean delays. The formulation is presented in the next subsections and the summary of notation is contained in Table 1.

### 2.1    Objective

The total delay, associated with $m$th user's requests concerning $n$th object on server $s$, is denoted by $\tau_{mns}(\mathbf{x}, \mathbf{y})$ and it can be given with the formula:

$$\tau_{mns}(\mathbf{x}, \mathbf{y}) = \tau_{mns}^{(1)}(y_{mns}) + \sum_{l=1}^{L} a_{mnsl} \left( \tau_l^{(2)}(\mathbf{x}, \mathbf{y}) + \tau_l^{(3)} \right)$$

where $\tau_{mns}^{(1)}(y_{mns})$ is a time it takes to push all $n$th object's bits onto link by the source node, $\tau_l^{(2)}(\mathbf{x}, \mathbf{y})$ is a queue sojourn delay (incorporating packet's waiting and serving time on the router's interface), $\tau_l^{(3)}$ are the rest of constant delays

**Table 1.** Summary of notation

| Symbol | Description |
|---|---|
| $m \in \{0, 1, \ldots, M\}$ | user's index, where $m = 0$ denotes publisher and $m \in \{1, \ldots M\}$ denotes client, |
| $n \in \{1, \ldots, N\}$ | data object's index, |
| $s \in \{1, \ldots, S\}$ | cache server's index, |
| $l \in \{1, \ldots, L\}$ | transmission link's index, |
| $x_{mns} \in \{0, 1\}$ | assignment of $m$-th client to $s$-th server for reading object $n$, |
| $x_{0ns} \in \{0, 1\}$ | assignment of $n$-th object the cache server $s$ |
| | $x_{ns} = [x_{0ns}], n = 1, 2, \ldots, N, s = 1, 2, \ldots, S$, |
| | $\mathbf{x} = [x_{mns}], m = 0, 1, \ldots, M, n = 1, 2, \ldots, N, s = 1, 2, \ldots, S$, |
| $y_{mns} \geq 0$ | rate allocated to the $m$-th client for downloading $n$-th object from server $s$, |
| $y_{0ns} \geq 0$ | rate allocated to the publisher of $n$-th object for uploading this object to server $s$, |
| | $\mathbf{y} = [y_{mns}], m = 0, 1, \ldots, M, n = 1, 2, \ldots, N, s = 1, 2, \ldots, S$, |
| $y_{mns,\max} \geq 0$ | maximum achievable rate for transmitting $n$-th object from/to $m$-th user to/from $s$-th server; here it is assumed to equal: $y_{mns,\max} = \min\limits_{l \in \{l: a_{mnsl} = 1\}} C_l$, |
| $a_{mnsl} \in \{0, 1\}$ | routing variable, which defines if user $m$ uses link $l$ while downloading (uploading) object $n$ from (to) server $s$, |
| $C_l$ | capacity of physical link $l$, |
| $b_n$ | size of object $n$ |

related to the link $l$ (e.g. processing delay, propagation delay). So, the objective function is given as:

$$Q(\mathbf{x}, \mathbf{y}) = \sum_{m=0}^{M} \sum_{n=1}^{N} \sum_{s=1}^{S} x_{mns} \tau_{mns}(\mathbf{x}, \mathbf{y}).$$

### 2.2   Constraints

The following requirements imposed by the problem may be formulated:

1. Each user must be able to download every object from one location:

$$\forall_{m \in \{1, \ldots, M\}} \forall_{n=1, \ldots, N} \quad \sum_{s=1}^{S} x_{mns} = 1 \ . \tag{1}$$

2. There must be a copy of an object located on the server for which this user is assigned to:

$$\forall_{m \in \{1, \ldots, M\}} \forall_n \forall_s \quad x_{mns} \leq x_{0ns} \ . \tag{2}$$

3. The links' capacities must not be exceeded:

$$\forall_{l=1, \ldots, L} \quad \sum_{m=0}^{M} \sum_{n=1}^{N} \sum_{s=1}^{S} a_{mnsl} x_{mns} y_{mns} \leq C_l \ . \tag{3}$$

Observe that constraints (1) and (2) imply that at least one copy of each object must be located in the network, i.e. $\forall_{n=1,...,N}$  $\sum_{s=1}^{S} x_{0ns} \geq 1$. In this paper, it is assumed, that storage capacity on every server is big enough to keep all the objects (i.e. $\forall_{s \in \{1,...,S\}} \sum_{n=1}^{N} b_n \leq B_s$, where $B_s$ is the storage capacity of server $s$). Without such assumption the problem becomes much more difficult and hard to approximate [11], however it appears that mass storage is the least deficient resource in practice.

## 2.3   Problem Formulation

The joint object placement and capacity allocation problem in Content-Aware Network is formulated as follows. For given values of $M, N, S, L, \mathbf{A} = [a_{mnsl}]$, $\mathbf{C} = [C_l]$ one should find:

$$(\mathbf{x}^*, \mathbf{y}^*) = \arg \min_{\mathbf{x}, \mathbf{y}} Q(\mathbf{x}, \mathbf{y})$$

such that constraints (1)–(3) hold.

This is nonlinear mixed integer programming problem (discrete-continuous). Even assuming that $\mathbf{y}^*$ is known in advance the problem is NP-hard with respect of $\mathbf{x}$. To obtain exact solution, algorithms such as branch-and-bound or outer approximation [12] may be applied. However, such approach leads to solving procedures with prohibitive running time (requiring exponential number of steps in terms of $M$, $N$, $S$ and $L$). Moreover, these procedures are hard to implement effectively in decentralized manner. In this paper, a particular heuristic approach based on problem decomposition is proposed and its decentralized version is presented. The general formulation of objective functions allows for plugging functions representing the time spent by packets inside routers' queues arbitrarily. However, in this paper it is assumed that all routers adhere to M/M/1 queueing characteristic, i.e. mean sojourn time of packet in link's $l$ router is $\tau_l^{(2)}(\mathbf{x}, \mathbf{y}) \approx (C_l - \sum_{m,n,s} x_{mns} y_{mns})^{-1}$. This is justified for FTP-type transactions, and it is assumed that such traffic dominates the transmissions in Content-Aware Networks. Thus solving the problem formulated in this paper can be seen as optimizing the network with respect to the static file transfer applications. For other types of traffic different queueing models need to be taken into consideration [13].

# 3   Solution Algorithms

An additional specificity of the problem is the decentralized architecture of the system. It is required that the solution algorithms can be distributed across the network devices, since both input data, as well as decision variables, are bound to spatially distributed network nodes. The implementation should assume the limited communication between entities cooperating in the computation of the global solution. In this section the idea of decomposition is presented, followed by the description of two-level solution algorithm, and its implementation in the decentralized model. Such implementation consists of routines which must be implemented on specified network nodes, communicating via fixed protocol.

## 3.1   Two-Level Heuristic Algorithm

Given the complexity of the problem, efficient approximate solution algorithms are desirable. An observation that the discrete subproblem (i.e. finding $\mathbf{x}$) and continuous subproblem (i.e. finding $\mathbf{y}$) alone are tractable motivates the following heuristic approach. First, the problem is decomposed into two subproblems:

1. (upper level problem) The rate allocation problem which is solved for a fixed content location, i.e. for given $M, N, S, L, \mathbf{A} = [a_{mnsl}], \mathbf{C} = [C_l], \mathbf{x}'$, find:

$$\mathbf{y}' = \arg\min \sum_{m=0}^{M} \sum_{n=1}^{N} \sum_{s=1}^{S} x'_{mns} \tau_{mns}(\mathbf{x}', \mathbf{y})$$

   subject to constraint (3) (assuming $\mathbf{x} = \mathbf{x}'$).
2. (lower level problem) The data placement problem which is solved for a particular capacity allocation in virtual networks, i.e. for given $M, N, S, \bar{\tau} = [\bar{\tau}_{mns}]$, find:

$$\mathbf{x}' = \arg\min \sum_{m=0}^{M} \sum_{n=1}^{N} \sum_{s=1}^{S} x_{mns} \bar{\tau}_{mns}$$

   subject to constraints (1)–(2), where $\bar{\tau}_{mns}$ is equal to $\tau_{mns}(\mathbf{x}, \mathbf{y})$, where $\mathbf{y}$ is calculated as:

$$y_{mns} = \begin{cases} y'_{mns} & \text{if } x_{mns} = 1, \\ y_{mns,\max} & \text{if } x_{mns} = 0, \end{cases}$$

   and $\mathbf{x}$ being the previous value of $\mathbf{x}$, i.e. those for which $\mathbf{y}'$ was obtained.

The data placement in cache servers influences the need for capacity in different links. At the same time, any particular allocation of transmission rates causes different load characteristics in the network and on cache servers. Since the solution of one problem gives the input data for another one, procedures for solving these both subproblems may be repeated alternately, as long as the value of objective function is decreasing. The upper and lower level problems may be solved with any known solution algorithm.

However, even if both problems are solved to optimality, it is not guaranteed in general that such approach will result in optimal solution to the global problem. Moreover, one should notice that even if the solution of the lower level was given with dependence on $\mathbf{x}$, the lower level would not provide the optimal solution of the problem. That is because the objective function for the lower level problem is linearized by assuming constant value of $\tau_{mns}$ (while it depends on $\mathbf{x}$). Nevertheless, such approach allows for elegant decomposition and decentralization of the solution, supporting practical implementation.

In this paper it is proposed to use greedy approximation algorithm on the lower level and asymptotically exact algorithm on the upper level. First, the centralized approach is presented, which serves as a basis of the decentralized version of the algorithms.

## 3.2  Centralized Algorithm Based on Two-Level Heuristic

**Rate Allocation Algorithm.** The presented algorithm for the rate allocation subproblem requires that the delay functions $\tau_{mns}(\mathbf{x}', \mathbf{y})$ are convex with respect to $\mathbf{y}$. Assume that $\mathbf{x}$ is fixed and equal to $\mathbf{x}'$. Let us define Lagrange function:

$$L_1(\mathbf{y}, \lambda) = \sum_{m=0}^{M} \sum_{n=1}^{N} \sum_{s=1}^{S} x'_{mns} \tau_{mns}(\mathbf{x}', \mathbf{y}) +$$

$$+ \sum_{l=1}^{L} \lambda_l \left( \sum_{m=0}^{M} \sum_{n=1}^{N} \sum_{s=1}^{S} a_{mnsl} y_{mns} x_{mns} - C_l \right)$$

where $\lambda = [\lambda_1, \ldots \lambda_L]^T$ is the vector of Lagrange multipliers.

The Langrange function's derivatives are as follows:

$$\frac{\partial L_1(\mathbf{y}, \lambda)}{\partial y_{mns}} = \sum_{i=0}^{M} \sum_{j=1}^{N} \sum_{k=1}^{S} \left( x'_{ijk} \frac{\partial \tau_{ijk}(\mathbf{x}', \mathbf{y})}{\partial y_{mns}} \right) + x'_{mns} \sum_{l=1}^{L} a_{mnsl} \lambda_l \qquad (4)$$

$$\frac{\partial L_1(\mathbf{y}, \lambda)}{\partial \lambda_l} = \sum_{m=0}^{M} \sum_{n=1}^{N} \sum_{s=1}^{S} a_{mnsl} x'_{mns} y_{mns} - C_l \ . \qquad (5)$$

Equating above derivatives to 0 and solving the set of equations with respect to $y_{mns}$ and $\lambda_l$ one can obtain the optimal solution of lower level problem (from KKT conditions [14], the only points where derivatives of convex function are zero coincide with the global optimum). But generally, due to nonlinearity, it is hardly possible to obtain analytical solution of such system. Consequently, the projected gradient method is proposed to be applied [15].

The following iterative solution algorithm is proposed. For each variable $y_{mns}$ and $\lambda_l$ a sequence of stepwise updates are applied, starting from any feasible point, according to the following formulas:

$$y_{mns}[t+1] = y_{mns}[t] - K_{y_{mns}} \left( \sum_{i=0}^{M} \sum_{j=1}^{N} \sum_{k=1}^{S} \left( x'_{ijk} \frac{\partial \tau_{ijk}(\mathbf{x}', \mathbf{y})}{\partial y_{mns}} \bigg|_{(\mathbf{y}, \lambda) = (\mathbf{y}[t], \lambda[t])} \right) + \right.$$

$$\left. + x'_{mns} \sum_{l=1}^{L} a_{mnsl} \lambda_l[t] \right) \qquad (6)$$

$$\lambda_l[t+1] = \lambda_l[t] - K_{\lambda l} \left( \sum_{m=0}^{M} \sum_{n=1}^{N} \sum_{s=1}^{S} a_{mnsl} x'_{mns} y_{mns}[t] - C_l \right) \qquad (7)$$

where $K_{y_{mns}}$ and $K_{\lambda l}$ are parameters controlling the rate of convergence (for sufficiently small values it is guaranteed that algorithms will converge, however may require a large number of steps).

It is clear that computation of (7) can be executed locally on each link (or rather on an appropriate router). Thus the decentralization of the whole algorithm relies on the possibilities of distributing iterative routine (6). This inherently depends on the particular form of the delay functions $\tau_{mns}(\mathbf{x}, \mathbf{y})$.

**Data Placement Algorithm.** Recall that in our considerations all cache server storages are large enough to hold any number of objects. Assuming fixed **y**, equal to **y'**, the problem becomes well-studied *uncapacitated facility location problem* [16]. In this subsection it is presented how to efficiently obtain a good solution using centralized algorithm. The approach is based on 2-approximation algorithm given in [17].

Relaxing the integrality constraints on **x**, the Lagrange function with respect to the remaining constraints (1)–(2) takes form:

$$L_2(\mathbf{x}, \phi, \psi) = \sum_{m=0}^{M} \sum_{n=1}^{N} \sum_{s=1}^{S} x_{mns} \bar{\tau}_{mns} + \sum_{m=0}^{M} \sum_{n=1}^{N} \phi_{mn} \left( 1 - \sum_{s=1}^{S} x_{mns} \right) +$$

$$+ \sum_{m=1}^{M} \sum_{n=1}^{N} \sum_{s=1}^{S} \psi_{mns} (x_{0ns} - x_{mns})$$

where $\bar{\tau}_{mns}$ is equal to $\tau_{mns}(\mathbf{x}, \mathbf{y})$ for given **y** (a solution of the lower level problem) and previous value of **x**. Using the Lagrange multipliers as decision variables the following dual problem can be obtained:

$$\text{maximize} \quad \sum_{m=1}^{M} \sum_{n=1}^{N} \phi_{mn} \tag{8}$$

subject to:

$$\forall_{m=1,\ldots,M} \forall_{n=1,\ldots,N} \forall_{s=1,\ldots,S} \quad \phi_{mn} - \psi_{mns} \leq \bar{\tau}_{mns} \ ,$$

$$\forall_{n=1,\ldots,N} \forall_{s=1,\ldots,S} \quad \sum_{m=1}^{M} \psi_{mns} \leq \bar{\tau}_{0ns} \ ,$$

$$\forall_{m=1,\ldots,M} \forall_{n=1,\ldots,N} \forall_{s=1,\ldots,S} \quad \phi_{mn} \geq 0, \psi_{mns} \geq 0 \ .$$

Applying greedy strategy to determine the solution of the dual problem allows for finding **x** satisfying complementary slackness conditions of the relaxed problem. Such solution can be shown to be close to the optimal of the original (discrete) problem (assuming metric weights in the input data). The algorithm is described in detail in Sect. 3.3.

**Initialization.** The application of two presented stages of heuristic requires an initial feasible solution to start with. Before the transmission rates can be computed, it is required to give a feasible data placement and user assignment. This can be accomplished by running the lower level algorithm first, assuming no traffic in the network. This corresponds to solving the data placement problem for the input data consisting of "best-case" delays, that is ones obtained by assuming all maximal rates and zero queuing delays (only constant link propagation delays). Thus initialization input data becomes:

$$\tau_{mns} = \frac{1}{y_{mns,\max}} + \sum_{l} a_{mnsl} \tau_l^{(3)} \ .$$

For the initial **x**, the heuristic continues by solving upper level subproblem.

### 3.3  Decentralized Algorithm for Special Case

**Decentralized Rate Allocation Algorithm.** Let $\alpha_{mns}$ defines fraction of object sent in one packet (i.e. $\alpha_{mns}b_n$ is the total number of packets to be sent). Then delay of file transfer, corresponding to the transmission with rate $y_{mns}$, is given by:

$$\tau_{mns}^{(1)}(y_{mns}) = \frac{\alpha_{mns}b_n}{y_{mns}} . \tag{9}$$

Assuming that all queues in routers are M/M/1, the queuing delay may be expressed with the following formula [18]:

$$\tau_l^{(2)}(\mathbf{x}, \mathbf{y}) = \frac{1}{C_l - \sum_{m=0}^{M} \sum_{n=1}^{N} \sum_{s=1}^{S} a_{mnsl} x_{mns} y_{mns}} . \tag{10}$$

Consequently, the total delay, associated with $m$th user's requests concerning $n$th object on server $s$ is given by the formula:

$$\tau_{mns}(\mathbf{x}, \mathbf{y}) = \frac{\alpha_{mns}b_n}{y_{mns}} + \sum_{l=1}^{L} a_{mnsl} \left( \frac{1}{C_l - \sum_{i=0}^{M} \sum_{j=1}^{N} \sum_{k=1}^{S} a_{ijkl} x_{ijk} y_{ijk}} + \tau_l^{(3)} \right) . \tag{11}$$

Since the function $f(z) = \frac{1}{z}$ is convex for $z > 0$, thus (9) is also convex. Moreover, taking into account the convexity preservation of affine mappings, (10) is also convex. Finally, the objective function (11) is also convex since it is linear combination (with positive coefficients) of convex functions [14].

Now the derivative (5) may be calculated, and the algorithm (6) may be rewritten as follows:

$$y_{mns}[t+1] = y_{mns}[t] - K_{ymns} \left( x'_{mns} \left( \frac{\alpha_{mns}b_n}{(y_{mns}[t])^2} + \right. \right. \tag{12}$$

$$\left. + \sum_{i=0}^{M} \sum_{j=1}^{N} \sum_{k=1}^{S} \sum_{l=1}^{L} \frac{a_{ijkl} a_{mnsl} x'_{mns}}{\left( C_l - \sum_{e=0}^{M} \sum_{f=1}^{N} \sum_{g=1}^{S} a_{efgl} x'_{efgs} y_{efg}[t] \right)^2} \right) + $$

$$\left. + x'_{mns} \sum_{l=1}^{L} a_{mnsl} \lambda_l[t] \right) .$$

The formula (12) may be rewritten using (9) and (10):

$$y_{mns}[t+1] = y_{mns}[t] - K_{ymns} \left( x'_{mns} \left( \frac{(\tau_{mns}^{(1)}(y_{mns}))^2}{\alpha_{mns}b_n} + \right. \right. \tag{13}$$

$$\left. + \sum_{i=0}^{M} \sum_{j=1}^{N} \sum_{k=1}^{S} \sum_{l=1}^{L} a_{ijkl} a_{mnsl} x'_{mns} (\tau_l^{(2)}(\mathbf{x}, \mathbf{y}))^2 \right) + x'_{mns} \sum_{l=1}^{L} a_{mnsl} \lambda_l[t] \right) .$$

Each source may easily determine the time $\tau_{mns}^{(1)}(y_{mns})$, but to calculate (13), the values of queuing delays (10) and congestion signals (7) along the route must be known. The congestion signals may be estimated e.g. basing on the loss probability (when some Active Queue Management procedures are implemented in the network [19]). The delays in the network may be monitored (e.g. using Internet Control Message Protocol Echo Request or other utilities).

---

**Algorithm 1.** User algorithm for upper level subproblem

---

It is assumed that user $m$ knows routing matrix entries $a_{mnsl}$ corresponding to its path to all servers, as well as its own current assignments $x'_{mns}$. Moreover, values $\alpha_{mns}$, $b_n$, $K_{ymns}$ are known.

The procedure is executed by source nodes of every active flow (i.e. $m$th user request for object $n$ from server $s$ if $x_{mns} = 1$):

1. Let $t = 0$, $\epsilon > 0$.
2. Compute $\tau_{mns}^{(1)}(y_{mns}[t])$ according to (9).
3. Receive $\tau_l^{(2)}$ from all links (routers) on the path to server $s$.
4. Receive the value $x'_{mns} \sum_l a_{mnsl} \lambda_l[t]$ from the network (routers).
5. Update transmission rate $y_{mns}[t + 1]$ according to (13).
6. If $|y_{mns}[t + 1] - y_{mns}[t]| < \epsilon$ then STOP.
7. Increase $t$. Go to 2.

---

---

**Algorithm 2.** Link algorithm for upper level subproblem

---

The procedure below is executed in link $l$ (routers' software):

1. Let $t = 0$.
2. Measure the total traffic on the link, i.e. $\sum_{m,n,s} a_{mnsl} x'_{mns} y_{mns}[t]$.
3. Update congestion signals $\lambda_l[t + 1]$ according to (7).
4. Send congestion signals to all users transmitting data through the link $l$.
5. Increase $t$. Go to 2.

---

**Decentralized Data Placement Algorithm.** Observe that the linear program (8) is equivalent to $N$ linear programs for each $n = 1, \ldots, N$, thus without the loss of generality one can assume that $n$ is fixed and solve for each $n$ independently. The complementary slackness conditions imply that the solution $\mathbf{x} = [x_{mns}]$ is optimal if and only if:

$$\forall_{m=1,\ldots,M} \forall_{n=1,\ldots,N} \forall_{s=1,\ldots,S} \quad x_{mns} = 0 \vee \phi_{mn} = \psi_{mns} + \bar{\tau}_{mns}$$

$$\forall_{n=1,\ldots,N} \forall_{s=1,\ldots,S} \quad x_{0ns} = 0 \vee \sum_{m=1}^{M} \psi_{mns} = \bar{\tau}_{0ns}$$

which suggests the following procedure. Starting from $\phi = \mathbf{0}$, $\psi = \mathbf{0}$, for each user $m$ increase the dual variable $\phi_{mn}$ until the "connection cost" to any server $s$ is paid (this cost equals to $\bar{\tau}_{mns}$). After this, increase the dual variable $\psi_{mns}$ for such server $s$, as long as the sum of contributions $\sum_m \psi_{mns}$ is equal to the "object placement cost", equal to $\bar{\tau}_{0ns}$. This can be translated into following greedy

---

**Algorithm 3.** User algorithm for lower level subproblem

---

The procedure below is executed by every user's node $m$ and for every data object $n$ requested by that user, until all clients cease to be unconnected:

1. (initialization) Let $\delta > 0$, $t = 0$, $\phi_{mn}[0] = 0$, $\psi_{mns}[0] = 0$, $x_{mns} = 0$ for all $s = 1, \ldots, S$.
2. Increase $t$.
3. Probe the neighbor servers to estimate the current value of $\bar{\tau}_{mns}$.
4. Set $\phi_{mn}[t] \leftarrow \phi_{mn}[t-1] + \delta$.
5. For all servers $s$:
   if $\phi_{mn}[t] \geq \bar{\tau}_{mns}$ then set $\psi_{mns}[t] \leftarrow \psi_{mns}[t-1] + \delta$ else $\psi_{mns}[t] = 0$.
6. Send the value $\phi_{mn}[t]$ and $\psi_{mns}[t]$ to the server $s$.
7. Receive messages $x_{0ns}$ from servers. If $m$ is unconnected, and if for some $s$, $x_{0ns} = 1$, then connect to the server, i.e. put $x_{mns} = 1$, and go to the step 8. Otherwise repeat from the step 2.
8. Increase $t$. Set $\phi_{mn}[t] \leftarrow \phi_{mn}[t-1]$ and $\psi_{mns}[t] \leftarrow \psi_{mns}[t-1]$.
9. Send the values $\phi_{mn}[t]$ and $\psi_{mns}[t]$ to the server $s$.
10. Receive messages $x_{0ns}$ from servers. If for some server $s'$, $x_{0ns'} = 1$ and the estimated value $\bar{\tau}_{mns} - \bar{\tau}_{mns'} > 0$, then reconnect to the server $s'$, i.e. put $x_{mns} = 0$ and $x_{mns'} = 1$.
11. Go to step 8.

---

scheme: for each object $n$ on server $s$ sort all unconnected clients $m$ by their connection costs $\bar{\tau}_{mns}$. Now consider all clusters consisting of a server and a subset of "cheapest" clients, denoting the set of such client indices by $D_{ns}$. Select the most cost-effective cluster, i.e. one minimizing the ratio $(\sum_{m \in D_{ns} \cup \{0\}} \bar{\tau}_{mns})/|D_{ns}|$. Such server is selected to cache the object and all clients from $D_{ns}$ are assigned to it. The process is repeated for the remaining unconnected clients, with the exception, that all already selected servers have cost $\bar{\tau}_{0ns}$ substituted by 0.

Such algorithm allows to find an object placement $x_0$ and users' assignment $\mathbf{x}$ to the object replicas, which is usually very close to the optimal solution. Under the assumption that the input data matrix $\bar{\tau}_{mns}$ satisfies the triangle inequality, it can be shown that such algorithm gives the solution no worse than 2 times the optimal value [17]. Decentralized implementation is presented as Algorithm 3 (users' routine) and Algorithm 4 (cache servers' routine).

## 4    Computational Example

Simulation study was carried out for an example network, consisting of two cache servers, two user LANs and one publisher. The network consists of two routers with all link capacities equal to 10 MB/s. Results of this experiment are presented in the Fig. 1. Although global minimum is not achieved (the final value is very close to optimal), the heuristic approach requires to search in only a fraction of feasible solution space to end up with satisfactory outcome. It can be seen that subsequent iterations gradually improve the solution. This example supports the legitimacy of presented idea.

**Algorithm 4.** Cache server algorithm for lower level subproblem

The procedure below is executed by every cache server's node $s$ and for every data object $n$:

1. (initialization) Let $t = 0$, $x_{0ns} = 0$, $\psi_{mns}[0] = 0$.
2. Receive messages $\phi_{mn}[t]$ and $\psi_{mns}[t]$ from all clients.
3. If $\sum_m \phi_{mn}[t] = \sum_m \phi_{mn}[t-1]$ then STOP.
4. If $\sum_m \psi_{mns}[t] \geq \bar{\tau}_{0ns}$ then cache object, i.e. put $x_{0ns} = 1$, and send this information to all such clients $m$ that $\psi_{mns}[t] > 0$.
5. Increase $t$. Go to 2.

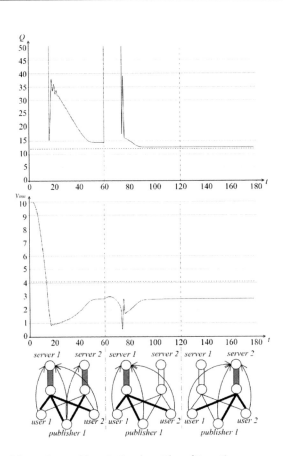

**Fig. 1.** Example: 3 iterations of heuristic algorithm (iterations are separated by dotted vertical lines). The value of objective function $Q$ (topmost plot) can be seen to converge very closely to optimal value (horizontal dashed line). The process of control of transmission rate $y_{mn} = \sum_s x_{mns} y_{mns}$ is presented in the middle plot. Object assignment configurations in all iterations are shown in the bottom plot.

# 5   Conclusions

Content delivery networks provide a number of challenges in the area of computer network optimization. In order to minimize the mean delay of transmitting data between user participants (client and publisher LANs) and caching servers, CAN operators need to efficiently approximate solutions to hard mixed-integer programming problems. Such task can be accomplished in the real-time by applying decentralized protocols, which incorporate adequate optimization algorithms. One such solution, based on two-level decomposition is given in this paper.

The presented algorithm optimizes the rate allocation policy with respect to transfer of static content. In case of small sized files the queuing delays on routers contribute a significant fraction of total transfer time, thus the optimal solutions tend to avoid assigning many flows to the same links. In case of large files the queuing delays are often negligible, and such flows require allocating links with large capacity in order to reduce the total transfer time.

Future works will include the incorporation of market-based and utility-based analysis to the model of CAN, in order to establish the pricing strategies for operators. Additionally, server storage capacities need to be taken into consideration, as the model presented in this paper simplifies the operational properties of cache servers.

# References

1. Leighton, T.: Improving performance on the internet. Communications of the ACM 52(2), 44–51 (2009)
2. Sivasubramanian, S., Szymaniak, M., Pierre, G., Steen, M.: Replication for web hosting systems. ACM Computing Surveys 36(3), 291–334 (2004)
3. Tang, X., Xu, J.: On replica placement for qos-aware content distribution. In: Twenty-third Annual Joint Conference of the IEEE Computer and Communications Societies, INFOCOM 2004, vol. 2, pp. 806–815. IEEE (2004)
4. Karger, D., Lehman, E., Leighton, T., Panigrahy, R., Levine, M., Lewin, D.: Consistent hashing and random trees: Distributed caching protocols for relieving hot spots on the world wide web. In: Proceedings of the Twenty-ninth Annual ACM Symposium on Theory of Computing, pp. 654–663. ACM (1997)
5. Jacobson, V., Smetters, D.K., Thornton, J.D., Plass, M.F., Briggs, N.H., Braynard, R.L.: Networking named content. In: Proceedings of the 5th International Conference on Emerging Networking Experiments and Technologies, pp. 1–12. ACM (2009)
6. Koumaras, H., Negru, D., Borcoci, E., Koumaras, V., Troulos, C., Lapid, Y., Pallis, E., Sidibé, M., Pinto, A., Gardikis, G., Xilouris, G., Timmerer, C.: Media Ecosystems: A Novel Approach for Content- Awareness in Future Networks. In: Domingue, J., Galis, A., Gavras, A., Zahariadis, T., Lambert, D., Cleary, F., Daras, P., Krco, S., Müller, H., Li, M.-S., Schaffers, H., Lotz, V., Alvarez, F., Stiller, B., Karnouskos, S., Avessta, S., Nilsson, M. (eds.) Future Internet Assembly. LNCS, vol. 6656, pp. 369–380. Springer, Heidelberg (2011)

7. Kelly, F.P., Maulloo, A.K., Tan, D.K.H.: Rate control for communication networks: shadow prices, proportional fairness and stability. Journal of the Operational Research Society 49(3), 237–252 (1998)

8. Gasior, D.: QoS rate allocation in computer networks under uncertainty. Kybernetes 37(5), 693–712 (2008)

9. Baev, I., Rajaraman, R., Swamy, C.: Approximation algorithms for data placement problems. SIAM Journal on Computing 38(4), 1411–1429 (2008)

10. Drwal, M., Józefczyk, J.: Decentralized Approximation Algorithm for Data Placement Problem in Content Delivery Networks. In: Camarinha-Matos, L.M., Shahamatnia, E., Nunes, G. (eds.) DoCEIS 2012. IFIP AICT, vol. 372, pp. 85–92. Springer, Heidelberg (2012)

11. Bateni, M.H., Hajiaghayi, M.T.: Assignment problem in content distribution networks: unsplittable hard-capacitated facility location. In: Proceedings of the Twentieth Annual ACM-SIAM Symposium on Discrete Algorithms, pp. 805–814. Society for Industrial and Applied Mathematics (2009)

12. Duran, M.A., Grossmann, I.E.: An outer-approximation algorithm for a class of mixed-integer nonlinear programs. Mathematical Programming 36(3), 307–339 (1986)

13. Grzech, A., Świątek, P.: Parallel processing of connection streams in nodes of packet-switched computer communication systems. Cybernetics and Systems: An International Journal 39(2), 155–170 (2008)

14. Boyd, S.P., Vandenberghe, L.: Convex optimization. Cambridge Univ. Pr. (2004)

15. Drwal, M., Gasior, D.: Utility-based rate control and capacity allocation in virtual networks. In: Proceedings of the 1st European Teletraffic Seminar, pp. 176–181 (2011)

16. Williamson, D.P., Shmoys, D.B.: The Design of Approximation Algorithms. Cambridge University Press (2011)

17. Jain, K., Mahdian, M., Markakis, E., Saberi, A., Vazirani, V.V.: Greedy facility location algorithms analyzed using dual fitting with factor-revealing LP. Journal of the ACM 50(6), 795–824 (2003)

18. Kleinrock, L.: Queueing Systems: Theory. Wiley-Interscience (1975)

19. Athuraliya, S., Low, S.H., Li, V.H., Yin, Q.: REM: Active queue management. IEEE Network 15(3), 48–53 (2001)

# Development of Service Composition by Applying ICT Service Mapping

Jakub M. Tomczak, Katarzyna Cieślińska, and Michał Pleszkun

Institute of Computer Science, Wrocław University of Technology,
Wybrzeże Wyspiańskiego 27, 50-370, Wrocław, Poland
{jakub.tomczak,katarzyna.cieslinska,michal.pleszkun}@pwr.wroc.pl

**Abstract.** In this paper, the problem of ICT service mapping in service composition process is highlighted. In many cases, especially for telecommunication operators, it is important to allow service providers to compose services which support data transfer and processing. In this work a general service composition process is outlined and the ICT service mapping task is described. Next, the solution is proposed. At the end the case study is presented and conclusions are drawn. The main contribution of this paper is the proposition of decision tables as a tool for ICT service mapping.

**Keywords:** service mapping, service composition, communication service, computational service.

## 1 Introduction

Recently, enterprises tend to collaborate and integrate their business cores in Web markets in order to maximize both clients' satisfaction and own profits [1]. Therefore, there is an increasing need to develop methods for combining existing services[1] together to enrich functionalities and decrease execution costs [2]. Hence old-fashioned way of developing composite services, i.e., human-based service designing, becomes very insufficient due to enormous number of available services and lack of computational resources. That is why the automatic or semi-automatic *service composition* method is a crucial element of any system that implements service-oriented architecture (SOA) [1,2].

In general, service composition process consists of two steps [4]. First, the required functionalities and their interactions, i.e., control and data flow, are identified. Second, the execution plan is established, i.e., for sets of functionalities an appropriate version of service is selected. The service selection is accomplished basing on non-functional properties which in general can be referred to as *Quality-of-Service* (QoS) attributes.

---

[1] *Services* are self-describing, stateless, modular applications that are distributed across the Web [1,2]. The services provide functionalities and are described by quality attributes called *Quality-of-Service* (QoS), e.g., price, reputation, reliability, response time [1,3]. A service that provide single functionality is called *atomic service*, and a composition of atomic services is called a *composite services*.

A. Kwiecień, P. Gaj, and P. Stera (Eds.): CN 2012, CCIS 291, pp. 45–54, 2012.

In most of proposed approaches for service composition the domain services are considered apart from *information and communication services* (ICT services) which are understood as physical means supporting execution of domain services. However, in order to allow service-oriented computing (SOC) [1] to be applied properly most of requirements need to be *mapped* to ICT services. Otherwise it is ambiguous how the domain services should be executed physically. For example, a requirement for a building monitoring service consists of the following operations: signal acquisition, coding, sending and decoding. All these services are in fact the ICT services. Hence, the ICT service mapping is a process of mapping requirements to a combination of atomic ICT services which facilitates the delivery of demanded functionalities.

There are two standard approaches to service composition, namely, workflow composition and Artificial Intelligence planning (AI planning) [2]. In both approaches the emphasis is put on identifying control and data flow. Generally speaking, functionalities of a composite service are represented as a *directed acyclic graph* (DAG) [5,6,3] in which arcs denote interactions among functionalities and nodes determines functionalities. Thus, the service composition task can be seen as a determination of a DAG structure, e.g., using AI methods [7,2], and then a selection of atomic or composite services that provide required functionalities. The service selection is formulated as an optimization task in which an objective function reflects QoS attributes aggregation, e.g., response time, price, and constraints concern structural dependencies in DAG [4,8]. There are several known approaches to QoS optimization such as graph-based methods [5,9,10,11,6] and mathematical programming (integer programming [3], stochastic programming [12]). Recent trends include also semantics modelling [13], and optimization via meta-heuristics, e.g., genetic algorithms [14]. However, the problem of ICT service mapping is usually omitted and lately only few papers pointed out the necessity of considering ICT services in the process of service composition [15,16]. Hitherto, due to authors' knowledge, the problem of ICT service mapping has not been fully stated and hence no solution has been proposed.

In this paper, we indicate the problem of ICT service mapping in the process of service composition and propose the solution. The presented solution applies the idea of *decision tables* [17,18] as an ICT service mapping tool. We utilize the decision table to associate ICT services with corresponding requirements. The decision table for ICT service mapping consists of two columns. In the first column a requirement is given and in the second one – a DAG of ICT services. In the step of service mapping the service selection through QoS optimization is completed. Our approach allows to perform the execution plans on physical machines seamlessly because ICT services describe computational and communication aspects of SOC. Furthermore, it is worth mentioning that the presented approach is domain independent and could be applied in any business area that includes teleinformatics aspects.

The rest of this paper is organized as follows. We first present the description of the service composition process in Section 2. We outline our approach to ICT service mapping in Section 3 and indicate the way of development of the service composition process. We present a case study of a building monitoring in Section 4, followed by a conclusion and discussion in Section 5.

## 2    General Service Composition Framework

Before going into details about ICT service mapping, we outline a general framework for service composition [2]. The purpose of this section is to introduce necessary ideas and concepts, without considering implementations details, i.e., programming languages, platforms or algorithms. Moreover, we would like to point out the issue of ICT service mapping and locate it in the service composition process.

Generally speaking, a solution of the service composition problem should satisfy customer's demands. The solution can be obtained using available software services that are accessible through a computer network. In order to compose a service the following steps should be considered, namely, requirements translation, service matching, service selection and an execution plan realization.

The composition system has two types of actors, namely, the service requester (the client) and the service provider. The service provider shares services and additionally describe their functional and non-functional properties in one of the following languages WSDL or OWL-S [13]. The service requester formulates requirements which are translated to the *service level agreement* (SLA) [5]. The important phase of the service composition process is the integration of external languages used by actors and the internal languages used by the execution system. The requirements are represented as DAG. Next, service matching is performed, i.e., each functionality is associated with possible set of services. Eventually services are selected, i.e., optimization of QoS is performed. After that step the execution plan is launched and the result is presented to the service requester. The described process is presented in Fig. 1.

In some cases the presented framework for service composition is not complete. The deficiencies of existing service composition methods are especially seen in applications in which clients are allowed to choose services that can be executed by different composite ICT services. Particularly, telecommunication operators support ICT services for transferring and processing data. For example, a functionality on safe data sending can be matched with ICT composite service that has a serial structure and consists of data encryption, transferring and decryption. In the case if such service is not available, another ICT service could be used, even if it does not fulfil all requirements. According to mentioned deficiencies the process of service composition needs to be developed and augmented by a ICT service mapping tool.

## 3    ICT Service Mapping

### 3.1    Problem Background

In general, the task of service mapping can be stated as follows. Find versions of ICT services which are to be matched with functionalities and which are optimal

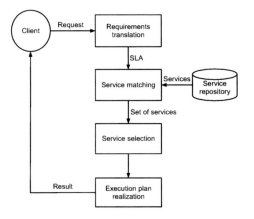

**Fig. 1.** The service composition process. The client's request is translated to SLA. Next, functionalities are matched with available and corresponding atomic services. Then, services are selected and the final execution plan is performed. At the end, the result is sent back to the client.

in the sense of the combination of QoS attributes. Thus, the mapping problem can be divided into three sub-problems. First, the functionalities for mapping have to be found. Second, found functionalities have to be matched with available ICT services. Third, one composite ICT service is selected.

In other words, the ICT service mapping can be seen as a service composition but performed only for functionalities that are identified to be mapped. However, the crucial step is how to determine which functionalities are supposed to be mapped.

In order to solve these problems following two assumptions are made. First, descriptions of functionalities and services are described in the same language in order to be identified by the same keys. For example, functionalities and services are described by XML-based language and by the same tags that are used for their content comparison. This simple and efficient approach is in general a special case of widely used ontology-based or semantic matching methods [13]. Second, the mapping between a functionality and a composite ICT service is given by an expert.

### 3.2 Proposed Solution

It is worth noting that the problem of ICT service mapping is strongly connected with the service composition process. Namely, functionalities are checked whether to be mapped to ICT services, or to be matched directly with services in a service repository. Therefore, the augmented service composition process can be represented by a sequence diagram presented in Fig. 2. The ICT service mapping consists of the following steps. In the first step all functionalities which are supposed to be mapped are found (the light grey rectangle in Fig. 2). In the next step for each functionality identified to be mapped a list of mappings is retrieved

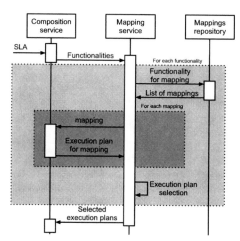

**Fig. 2.** The sequence diagram of interactions between modules of ICT service mapping and service composition. First, for all functionalities only those are considered which are supposed to be mapped. Second, for each mapping an execution plan is prepared. At the end, for each functionality, an execution plan is chosen.

from a repository of mappings. For each mapping in the list an execution plan is proposed by a service composition method due to given quality attributes (the dark rectangle in Fig. 2). At the end of the whole mapping process one execution plan is selected with respect to the non-functional (QoS) requirement stated in the SLA.

Analyzing the process of ICT service mapping leads to the following algorithm for mapping:

**Algorithm: ICT service mapping**

```
Input: functionalities - SLA represented as a DAG
       repository - a repository of functionalities
                    and corresponding mappings
Output: DAG with mapped ICT services

0. begin
1.  for each functionality in functionalities do
2.    if functionality in repository then
3.      mappingList = GetMappingList(functionality)
4.        for each mapping in mappingList do
5.          FormulateExecutionPlan(mapping)
6.        end
7.        ExecutionPlanSelection(QoSAttributes)
8.    end
9. end
```

In the above algorithms following functions are used:

- *GetMappingList( )* – the function retrieves a list of mappings which correspond to the considered functionality;
- *FormulateExecutionPlan( )* – an external function of the service composition that for the given mapping returns the execution plan;
- *ExecutionPlanSelection( )* – according to the QoS attributes the function chooses the best execution plan for the considered functionality.

There are two critical steps in the algorithm for ICT service mapping. First, how the repository is constructed. Second, how the final execution plan is selected. Here we focus only on the first problem because the second one can be solved using one of the methods known in the literature [4].

We take advantage of mentioned earlier assumptions. First, the mappings are given a priori. Second, functionalities and mappings can be matched using universal keys. Because of these advantages we have decided to apply decision tables [17,18] to represent the repository of mappings.

The decision table consists of two columns, the first column defines *condition*, and the second one – *decision*. In the considered application the condition is the functionality description, e.g., XML-based description, and the decision is the mapping. It is worth stressing that there are several possible mappings for one functionality. An exemplary decision table is presented in Fig. 3.

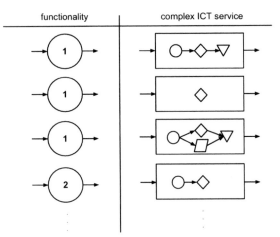

**Fig. 3.** An exemplary repository as a decision table. Dark and light gray triangles in circles represent functionalities. Circles, diamonds and triangles denote atomic ICT services. Rectangles represent composite ICT services. Notice that for the functionality number 1 there are several composite ICT services.

### 3.3   Technical Details

In order to test the proposed solution main functionalities are implemented in different programming languages. The *Composition Service* has been implemented in *Ruby* in order to easily manipulate XML files. The *Mapping Repository Service* has been written in *Java 6.0* and uses *prostgreSQL* database system. The

*Mapping Service* is implemented in *Python* and *PySimpleSoap* framework. This variety of programming languages corresponds to our secondary goal which was to check the integration abilities of SOAP protocol implementation in different programming technologies. Communications among each element of application are handled by SOAP protocol, and the *Mapping Service* is and integrator and main element of the proposed approach.

## 4    Case Study

In this work the ICT service mapping is presented on the building monitoring example. We consider only one domain service, namely, the building video camera surveillance service which is a crucial element in the monitoring process. The video camera surveillance is a domain service that needs to be mapped to ICT services that indicate a way of physical service execution. According to user's requirements the monitoring should regard security aspects and signal coding (compression). Decisions of a decision table for the domain service of the building video camera surveillance is presented in Fig. 4.

There are following mappings (see Fig. 4):

1. Data is transferred from place of monitoring to a monitoring company.
2. First data is coded, then coded data is transferred from a monitored building to a monitoring company. Data decoding is performed on a monitoring company's server.

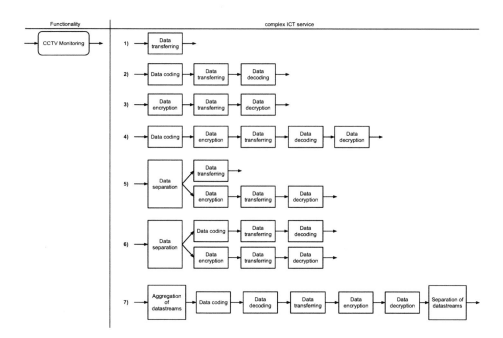

**Fig. 4.** Decisions of the decision table for the domain service of the building video camera surveillance

3. First data is encrypted, then encrypted data is transferred from a monitored building to a monitoring company. Data decryption is performed on
   a monitoring company's server.
4. First data is coded. Coded data is in a next step encrypted. Coded and
   encrypted data is transferred from a monitored building to a monitoring
   company. Data decryption and decoding are performed on a monitoring company's server.
5. Separation of the signal, due to type of data, into two paths. Paths are
   composite ICT services 1 and 3.
6. Separation of the signal, due to type of data, into two paths. Paths are
   composite ICT services 2 and 3.
7. After aggregation data is coded, then encrypted. Coded and encrypted data
   is transferred from a monitored building to a monitoring company. Data
   decryption and decoded are performed on a customer's server. Last step is
   separation of data streams.

Assume that the domain service is called "CCTV Monitoring". The "CCTV Monitoring" should be mapped to the ICT services. There are different possibilities
how the service can be executed (see Fig. 4). Assume that the user requests
*safe video transfer*. Moreover, assume that only one quality attribute is considered, e.g., the complexity of the service (number of nodes). The following ICT
services fulfil the requirement (see Fig. 4): service No. 3, service No. 4, service
No. 5, service No. 6, service No. 7. Next, for each ICT service optimal (due to
some criterion) atomic services are selected. Eventually, according to the service
complexity, the service No. 3 is chosen because it has the least number of nodes.
The result of the mapping is presented in Fig. 5.

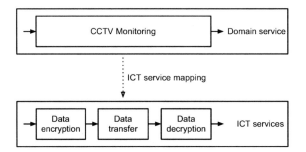

**Fig. 5.** An exemplary result of the mapping for requested functionalities

It is worth noting that we present simplistic example here only to highlight
the problem of the ICT service mapping in the service composition process. We
have considered a simple criterion to choose the execution plan. However, in
order to select an execution plan the QoS optimization problem needs to be
solved. Generally it is a non-trivial task [4] and hence going into its details is
beyond the scope of this paper.

## 5   Conclusion and Discussion

In this paper, the problem of ICT service mapping in the service composition process was considered. The ICT service mapping extends the service composition and provides an automatic mechanism for domain service translation into executable services.

In the ICT service task the following two problems arise. First, there is a danger that the QoS optimization technique can choose "simpler" ICT services. To show it let us consider requirements like in the case study (see Fig. 4) but the service number 3 is unavailable. Then the optimization algorithm can return the service number 1 instead of the service number 4 which fulfils demands but is more complex. In the proposed approach this issue was not considered but is strictly dependent on the QoS optimization algorithm. Second problem is connected with the decision table as a mapping tool. Decision making basing on the decision table requests considering all decisions. However, some or even most of them do not fulfil user's requirements. Hence time needed for the decision making can be reduced if the decisions are structured in a proper way. One possible solution is the tree structure in which each node represents a ICT service and any path from the root formulates a composite ICT service. Nevertheless, this issue needs further research.

**Acknowledgments.** The research presented in this paper has been partially supported by the European Union within the European Regional Development Fund program No. POIG.01.03.01-00-008/08.

## References

1. Papazoglou, M.P., Georgakopoulos, D.: Service-Oriented Computing. Communications of the ACM 46(10), 25–28 (2003)
2. Rao, J., Su, X.: A Survey of Automated Web Service Composition Methods. In: Cardoso, J., Sheth, A.P. (eds.) SWSWPC 2004. LNCS, vol. 3387, pp. 43–54. Springer, Heidelberg (2005)
3. Zeng, L., Benatallah, B., Ngu, A.H.H., Dumas, M., Kalagnanam, J., Chang, H.: QoS-Aware Middleware for Web Services Composition. IEEE Trans. on Soft. Eng. 30(5), 311–327 (2004)
4. Strunk, A.: QoS-Aware Service Composition: A Survey. In: Eighth IEEE European Conference on Web Services, pp. 67–74 (2010)
5. Grzech, A., Rygielski, P., Świątek, P.: Translations of Service Level Agreement in Systems Based on Service-Oriented Architectures. Cyb. and Syst.: An Int. J. 41, 610–627 (2010)
6. Yu, T., Lin, K.-J.: Service selection algorithms for Web services with end-to-end QoS constraints. Inf. Syst. E-Business Management 3, 103–126 (2005)
7. Prusiewicz, A., Zięba, M.: The Proposal of Service Oriented Data Mining System for Solving Real-Life Classification and Regression Problems. In: Camarinha-Matos, L.M. (ed.) DoCEIS 2011. IFIP AICT, vol. 349, pp. 83–90. Springer, Heidelberg (2011)

8. Stelmach, P., Grzech, A., Juszczyszyn, K.: A Model for Automated Service Composition System in SOA Environment. In: Camarinha-Matos, L.M. (ed.) DoCEIS 2011. IFIP AICT, vol. 349, pp. 75–82. Springer, Heidelberg (2011)
9. Grzech, A., Świątek, P.: Modeling and Optimization of Complex Services in Service-Based Systems. Cyb. and Syst.: An Int. J. 40, 706–723 (2009)
10. Grzech, A., Świątek, P.: Complex services availability in service oriented systems. In: ICSEng 2011, pp. 227–232. IEEE CS (2011)
11. Rygielski, P., Świątek, P.: Graph-fold: An Efficient Method for Complex Service Execution Plan Optimization. Systems Science 36(3), 25–32 (2010)
12. Wiesemann, W., Hochreiter, R., Kuhn, D.: A Stochastic Programming Approach for QoS-Aware Service Composition. In: Proc. of CCGRID 2008, pp. 226–233 (2008)
13. Liang, O.A., Chung, J.-Y., Miller, S.: Modeling semantics in composite Web service requests by utility elicitation. Knowl. Inf. Syst. 13, 367–394 (2007)
14. Canfora, G., Di Penta, M., Esposito, R., Villani, M.L.: An approach for QoS-aware service composition based on genetic algorithms. In: Proc. of GECCO 2005, pp. 1069–1075 (2005)
15. Grzech, A., Świątek, P., Rygielski, P.: Dynamic Resources Allocation for Delivery of Personalized Services. In: Cellary, W., Estevez, E. (eds.) I3E 2010. IFIP AICT, vol. 341, pp. 17–28. Springer, Heidelberg (2010)
16. Xu, D., Wang, Y., Li, X., Qiu, X.S.: ICT Service Composition Method Based on Service Catalogue Model. In: AIAI 2010, pp. 324–328 (2010)
17. Kohavi, R.: The Power of Decision Tables. In: Lavrač, N., Wrobel, S. (eds.) ECML 1995. LNCS, vol. 912, pp. 174–189. Springer, Heidelberg (1995)
18. Pawlak, Z.: Rough set theory and its applications. J. of Telecom. Inf. Tech. 3, 7–10 (2002)

# The Concept of Using Data Mining Methods for Creating Efficiency and Reliability Model of Middleware Applications

Kamil Folkert, Michał Bochenek, and Lukasz Huczala

Silesian University of Technology, Institute of Informatics
{kamil.folkert,michal.bochenek,lukasz.huczala}@polsl.pl

**Abstract.** Complexity of contemporary computer systems induces complicatedness of application models which implement those systems. In the same time, many systems consist of similar parts that may be defined globally and seamlessly configured to be used in specific systems. In many solutions this kind of application parts are defined as a separate middle layer of the application. With increasing demands on systems scalability and reliability, more and more applications were using so called middleware model. There are many applications that may be enhanced to the middleware model, but there is no methodology of determining the way of choosing proper environment, technology and implementation. Moreover, there is no research on how to increase application's reliability and performance using opportunities given by middleware. This article is a description of concept how data mining tools may be used in defining these factors.

**Keywords:** enterprise architectures, data mining, performance, high availability.

## 1 Introduction

Cloud computing is a very popular and promising direction in Information Technology alike for users and providers of miscellaneous services. However there are some areas that always need a lot of attention when speaking of cloud creation or development, such as availability, data lock-in, confidentiality, data transfer bottlenecks, performance, quick scalability, storage, licensing and many more. Problems are clearly visible, especially in terms of service fees, for complicated concurrent applications based on large data sets and complex communication patterns. The aim of this research is to create and develop models, methods and algorithms for optimizing parameters that would sustain required level of efficiency and reliability of each and every node of the cloud, in this particular case – the application server [1].

Enterprise systems, nowadays, in many branches of business stand at a very high level of complexity and their complexity still grows. Every part of such enterprise system is communicating with one part, and depends on the other, which creates a heterogeneous application environment, heart of which is the middleware

A. Kwiecień, P. Gaj, and P. Stera (Eds.): CN 2012, CCIS 291, pp. 55–62, 2012.
© Springer-Verlag Berlin Heidelberg 2012

software. Application servers make a quick developing and dynamically growing market, and the biggest software vendors want to participate in it, which can be seen in the names of vendors of the most popular Java Enterprise application servers. As enterprise applications become more and more sophisticated, there is a need of research on the most appropriate values of parameters responsible for efficient and reliable runtime environment. Every application that the business relies on, should present satisfactory performance, but also needs to be constantly available and there should be hardly any failure in its runtime environment.

Middleware may be considered as a mediation layer of the system. In many cases this is true. However in many systems middleware is the main part of the system, implementing not only business logic of the system, but also providing security mechanism, data layer's APIs and many other various features. For that reason, middleware must provide specialized mechanisms to increase scalability, efficiency and reliability of the application. Middleware is expected to be fault-tolerant, auto-scalable, robust and re-usable in many systems.

## 2   Middleware Architectures

The necessity to ensure the high availability of an application, independently of virtualization of the runtime environment resources is an important limitation to the usage of cloud architecture. One of the classic examples of extending the reliability of a particular system is the redundancy in communication layer. Such solutions applied in systems of known architecture frequently implies worsening of efficiency parameters and restricts application scalability, which is unacceptable in many cases. Recent researches show that for some classes of stand-alone applications based on services even redundancy does not give a lot of improvements in the area of reliability and because of that fact, searching for some new possibilities of reliability betterment is deeply reasonable. There is a possibility of efficiency and reliability parameters amelioration in customizing application structure so it would be possible to deploy it in an application server, working as a part of cloud. It would make it possible to transfer part of responsibility for high availability from application itself to the application server and a set of scalable components implementing business logic along with amenities offered by an application server including tuning of pools of resources, versioning in the runtime mode or distributing application to a cluster, load balancing and many more.

Research on influence of hardware layer architecture on efficiency of application servers systems ought to be performed in appropriate environments, that should reflect the need for resources which is present in enterprise systems. First necessary element is the application server architecture, effectively using given resources. Middleware software enforces to apply such environment architecture that would fulfill requirements of system itself, but also the load, generated by end-users, at the same time giving the warranty of high availability and efficiency of particular application. The master problem is the prediction of resource usage, based on different architectures and operating systems, working in various economical branches for the purpose of Java Enterprise Edition (JEE) business applications [2,3].

During work realization there will be an efficiency and reliability model created for the application server, containing relations to a particular runtime platform. Mentioned model will be verified through resources load statistics analysis, i.e. usage of RAM (Random Access Memory), CPU (Central Processing Unit) utilization, hard disk access time [4]. From the application server side, analysis will touch resources related with data source, messaging queues and web container. Each resource possesses a set of configurable parameters, which has a considerable influence on effective performance of any application. Some of them are: connection pool size, thread pool size, pool purge policy, heap size, size of heap areas: nursery (for newly created and short-living objects) and tenured (for long-living objects), etc. [5,6] Developing a model linking level of service required by an application with demand for resources provided by the cloud will make it possible to establish an optimal way of load balancing on each system node or tryouts for accessing application in Active-Active (an application is running simultaneously on several nodes) or Active-Passive (certain amount of nodes is a stand-by for active nodes) mode, as shown in Fig. 1. Developed model will give a full picture of application server behaviour as well as the software running in its environment in context of applied hardware architecture, which will eventually give a possibility to describe a relation between a set of parameters ensuring efficiency or availability and applied platform.

Designing new applications for future deployment in advanced environments does not seem to be a serious problem in contrary to performing analyses of existing systems in the area of their migration to cloud platform. The aim of the project will be to elaborate methodology allowing to describe such application

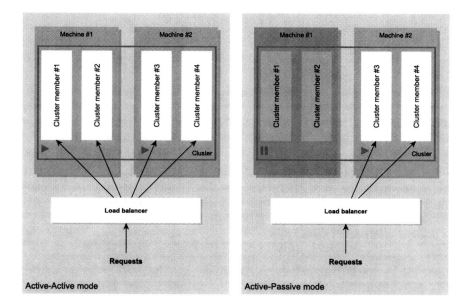

**Fig. 1.** High availability modes comparison

structure module, that will make it possible to distribute and deploy it to the modern runtime environments based on the cloud computing architecture. Correctness of a model will be verified in context of system efficiency and reliability parameters by comparison with outcome of research conducted in years 2010 and 2011. The object of the analysis will be the relation between required level of reliability of services, measured in reference to SIL 1 requirements, compliant with IEC61608 norm definitions. Practical verification subjects will be a number of applications, for which the unavailability time must be reduced to complete minimum, that will not endanger proper service of assisted processes. Operative research will be conducted on example of service oriented OPC Unified Architecture standard (IEC62541), which is currently used to unify communication in industrial systems [7]. Project work will make it possible to establish a new model that would replace current division into several corporation system layers, for which many individual tools were used in order to model characteristics as well as verify system parameters. Changes made to contemporary industrial IT systems indicate that this kind of a system model will encourage more effective usage of corporation resources. This particular model will be also helpful to estimate benefits of applying cloud computing architecture in relation to data processing tasks (where data is taken from distributed sources) along with required efficiency and reliability parameters of a system.

Proposed model will enable to obtain a certain estimation of a real environment used in corporate application servers. First verification that will allow to refine the model will consist of simulation research, which would approximate the model to real environment's requirements and limitations, in which the analysis of application server services will take place. Keeping in mind the concept of scalability, mobility and availability of the software, it is intended to perform experiments in an environment compliant with JEE standard, particularly using IBM WebSphere Application Server Network Deployment (WAS ND), which is considered as a representative of stable and efficient middleware software (architecture overview presented in Fig. 2). In hardware layer, it is planned to use the Power6 and Power7 architecture along with AIX and Suse Linux operating systems and Xeon architecture with Suse Linux and Windows 2008 Server R2. Lastly, research will be performed on IBM iSeries platform. One of the most useful tools that might be used is the Performance Monitoring Infrastructure (PMI) mechanism, embedded in the WebSphere Application Server. Additional element, supporting research on high availability, is the usage of WAS in the ND version, making creation of server clusters and distributed application deployment available along with configuration of a dedicated plug-in for a web server, which would then act as a load balancer, dispatching tasks to cluster members in a desired manner [8,9].

## 3    Middleware Optimization with Data Mining Methods

To support application server parameters optimization we use the algorithms based on data mining techniques related with middleware level data flow and the application data model.

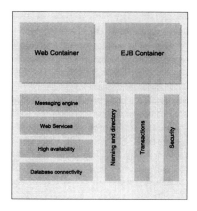

**Fig. 2.** WebSphere Application Server ND – architecture overview

The defined set of application parameters $M_i$ and the set of system resources parameters $P_i$ need to be complemented by a set of nondeterministic external factors affecting the cloud computing environment, such as disturbances of a communication layer or system load variation.

The aim is to find the relationship $S$ in the domain of application of models $M = M_1, \ldots, M_n$ and system resources, $P = P_1, \ldots, P_n$ associated with the observed external factors $Z$. It is obvious that the complexity of the described problem is too large to use analytical models. Our preliminary works suggest that this goal can be achieved with use of modern data mining methods. The presence of the nondeterministic factor and an anticipated discontinuity of the function $S$ in the domain of $M$ suggests using a neural network model with enhanced weights optimization [10].

Artificial neural networks can use knowledge obtained from several sources fed to network inputs by learning from examples the strength of the influence of given input data source on the quantity under consideration. Fundamental element of artificial neural network is neuron. Neuron is a cell with many inputs and one output. Each input value $x_k$ is multiplied by coefficient $w_{jk}$ signed to this input. The schema of neuron is presented in Fig. 3.

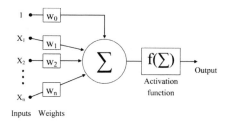

**Fig. 3.** Neuron schema

There is additional input value $x_0$, so called bias, with fixed value (usually 1) and coefficient $w_{j0}$ usually set at the beginning to $-1$. The artificial neuron output $y_j$ is given by an activation function $f$, which can have for example a form of sigmoidal like this:

$$y_j = \frac{1}{1 + e^{-t}} \qquad t = \sum_{k=0}^{i} w_{jk} x_k \ .$$

Neurons in networks are organized in layers. Single layer may contain one or more neurons. In consequence the outputs produce complex polynomial expressions, which can resolve complex multidimensional problems.

Size of a network (amount of layers and neurons) is significant, if network is too small it might be unable to achieve desired global error value during learning. Networks with too many neurons or layers might remember how to recognize all learning samples and lost generalization capabilities.

As presented in Fig. 4, inputs of first layer are fed by samples coefficients, values on the input should be normalized. Network can have many neurons in last layer, each such neuron is one element of output vector. Depending on activation function of the last layer we can get discrete or continuous value on the network output.

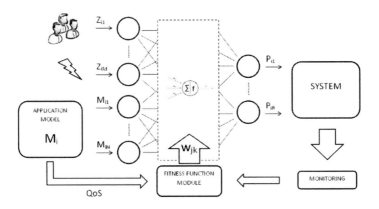

**Fig. 4.** ANN based self lerning system schema

To train networks we can use supervised or unsupervised learning methods. In our model supervised learning method should be used to control the parameters values. Learning samples on the network input and modified weights $w_{jk}$ will be applied using the descend method (back propagation) until we reach desired error value or number of iterations. It is important to create correct learning set. Samples in learning set should cover uniformly all cases. This requires in our case a well-designed learning process based on the monitoring data in the cloud environment. Also a proper fitness function is needed here to optimize

ANN parameters [11,12]. If some rule has significant more samples in learning set than other rules, it might have negative influence on the network, because during training network can learn how to recognize this predominant rule faster than other rules, therefore achieve low global error, but then the network wouldn't be able to recognize all other rules correctly.

Such combination of ANN and dedicated fitness function module is meant to behave as a regulator. Regulatory applications of ANN are widely recognized and used in medicine [13], industry [14] and communications [15,16].

The specific feature of neural networks is relation discovery between the input nodes $x_i$ and output nodes $y_i$, through "learning" process. This process is based on observation and evaluation of partial steps and that is why this model is appropriate here. In our case the input consists of the set $M_i$, while the outputs produce the parameter values $P_i$. The model let us discover the impact of external factors and also restrict the parameters values range by a proper fitness function in order to preserve the hardware and application platform from lowering its performance or even damage. Therefore the model is supported by a restricted fitness function optimization module. The feedback information for the fitness function are selected indicators of system performance, while the reference level is based on QoS parameters. Selection of an appropriate fitness and activation functions is a subject of the research.

In order to reduce the number of the system parameters the ranking algorithms of filtering rules, coverage, and similarity of the impact parameter to the parameters of services need to be executed after each neural model evolution cycle [17]. Assessment of the impact parameters on cloud computing efficiency and reliability of provided services will be available via an interactive interface with filtering rules and parameters, taking into account the subjective preferences.

## 4  Summary

This article presents a concept of using data mining methods for defining applications parameters in middleware environment. This is a proposition of realization of a test system and research on methods of increasing reliability and scalability of the applications operating in the middle layer. Further research will concern application servers, load balancing issues, virtualization problems, parametrization and distribution of the applications. The goal is to define and implement a data mining method for determining the best possible environment for a given class of applications, but also for determining how the applications could be enhanced using middleware model to increase the efficiency and reliability.

Moreover, it is expected to achieve an unique methodology that would enable estimations of possible benefits in systems performance, scalability and reliability for many various middleware environments. Since it is not trivial to predict possible profits from enhancing the application's architecture, the results may find implementation in many tools used for software designing. It may automate, ease or simply enable process of defining an environment that is optimal for best reliability or efficiency. For that reason the results of the research might be also important for maintenance analysis of currently existing systems.

**Acknowledgement.** This work was supported by the European Union from the European Social Fund (grant agreement number: UDA-POKL.04.01.01-00-106/09).

# References

1. Iyer, R., Srinivasan, S., Tickoo, O., Zhen, F., Illikkal, R., Zhang, S., Chadha, V., Stillwell, P.M., Lee, S.: CogniServe: Heterogeneous Server Architecture for Large-Scale Recognition. IEEE Micro 31 (2011)
2. Linthicum, D.S.: Enterprise application integration. Addison-Wesley Longman Ltd., Essex (2000)
3. Cecchet, E., Marguerite, J., Zwaenepoel, W.: Performance and scalability of EJB applications. In: OOPSLA 2002 Proc. of the 17th ACM SIGPLAN Conf. on Object-Oriented Programming, Systems, Languages, and Applications, New York (2002)
4. Khanna, G., Beaty, K., Kar, G., Kochut, A.: Application Performance Management in Virtualized Server Environments. In: 10th IEEE/IFIP Network Operations and Management Symposium, NOMS 2006 (2006)
5. The WebSphere Application Server architecture and programming model. IBM Systems Journal 37, 336–348 (1998)
6. Barcia, R., Hines, B., Alcott, T., Botzum, K.: IBM Websphere: deployement & advanced configuration. IBM (2008)
7. OPC Foundation: OPC UA Specification: Parts 1–13 (2009)
8. Cardellini, V., Colajanni, M., Yu, P.S.: Dynamic load balancing on Web-server systems. IEEE Internet Coumputing, 28–39(1999)
9. Dinker, D.: System and method for enabling failover for an application server cluster. US Patent 6.944.788 B2 (2005)
10. Nirkhi, S.: Potential use of Artificial Neural Network in Data Mining. In: Computer and Automation Engineering, ICCAE (2010)
11. Bishop, C.M.: Neural Networks for Pattern Recognition. Department of Computer Science and Applied Mathematics. Aston University Birmingham, UK (1995)
12. Grzechca, D.: Simulated annealing with artificial neural network fitness function for ECG amplifier testing. In: Circuit Theory and Design, ECCTD (2011)
13. Silvano Zanutto, B., Cernuschi Frías, B., Valentinuzzi, M.: Blood pressure long term regulation: a neural network model of the set point development. BioMedical Engineering OnLine (2011)
14. Tai, H.M., Wang, J., Ashenayi, K.: Motor speed regulation using neural networks. In: 16th Annual Conf. of IEEE IECON 1990, Industrial Electronics Society (1990)
15. Liu, Y.C., Douligeris, C.: Rate Regulation with Feedback Controller in ATM Networks – A Neural Network Approach. IEEE J. on Sel. Areas in Comm. 15(2) (1997)
16. In: Palmer-Brown, D., Draganova, C., Pimenidis, E., Mouratidis, H. (eds.) Proc. 11th International Conference on Engineering Applications of Neural Networks, EANN 2009, London, UK (2009)
17. Hindawi, M., Allab, K., Benabdeslem, K.: Constraint Selection-Based Semi-supervised Feature Selection. In: Data Mining, ICDM (2011)

# Transfer of Quantum Continuous Variable and Qudit States in Quantum Networks

Marek Sawerwain and Roman Gielerak

Institute of Control & Computation Engineering
University of Zielona Góra, ul. Podgórna 50, Zielona Góra 65-246, Poland
{M.Sawerwain,R.Gielerak}@issi.uz.zgora.pl

**Abstract.** The Bose lattice quantum gas based protocols for perfect coherent states transfer are discussed. Additionally, the perfect transfer of qudit states protocol based on Heisenberg-like interactions and in 1D-chains is discussed together with its computer implementation within Zielona Góra Quantum Computer Simulator.

**Keywords:** quantum information transfer, quantum continuous variables, qudits chains, numerical simulations.

## 1 Introduction

In many quantum information processing protocols the necessity of transferring quantum states (in the classical context information transfers also play important role as well e.g. [1]) with sufficiently high fidelity from one location to another spatially separated region is of major importance. Attempts to build solid state based quantum computers or quantum repeater technologies are good examples for this. Using optical channels the transfer can be achieved by the use of flying single photons or an appropriate prepared bunch of them. However, it is quite non-trivial problem to find physical system that provide sufficiently robust quantum state transmission channels communicating in between spatially separated parts of them.

Only a few papers [2,3,4,5,6,7] were addressed to the question of transmission of the continuous variable quantum states. The best known examples of such states are coherent and squeezed states [8]. It is worthwhile to stress that the so-called continuous variable quantum states can be used to perform any computational task of quantum information processing [8].

One of the main topics considered in this paper is the problem of supplying a protocol for the perfect transferring coherent states of the group $SU(1)$ on a selected geometry of the underlying quantum network. The so-called quasi-free, lattice based Bose-gases will be served for demonstrating that such perfect coherent state transfer on a special graphs and with appropriate choice of quasi-free dynamics is possible. These topics will be discussed in Section 3. In Section 5 some numerical experiments with the perfect transfer qutrit states transmission together with investigation of the influence of a noise on the fidelity of the transmission are presented. These experiments were performed within Zielona Góra

A. Kwiecień, P. Gaj, and P. Stera (Eds.): CN 2012, CCIS 291, pp. 63–72, 2012.

Quantum Computer Simulator environment [9]. Section 2 introduces the reader into some general framework for quantum state transfer protocols. In particular, an abstract theorem on the existence of an unitary circuits for transmitting quantum states is presented.

## 2   Transferring Quantum States Networks

Let $G = (\mathbb{V}, \mathbb{E})$ be a simple, connected graph where $\mathbb{V}$ stands for the set of vertices and $\mathbb{E}$ denotes the set of edges of $G$. The number $|\mathbb{V}|$ of vertices is assumed to be equal $N$. To each vertex $i \in \mathbb{V}$ a quantum system the Hilbert space of states of which is given $\mathcal{H}_i$ is associated. For a given Hilbert space $\mathcal{H}$ the corresponding space of states is denoted as $E(\mathcal{H})$ and the subset of pure states as $P(\mathcal{H})$. For any $X \subseteq \mathbb{V}$ the corresponding space of states $\mathcal{H}_X$ is given by the tensor product $\mathcal{H}_X = \bigotimes_{i \in X} \mathcal{H}_i$. A given unitary quantum dynamics on $G$ is given by an unitary group $U_t$ acting on $\mathcal{H}_{\mathbb{V}}$. From the Stone theorem it follows that for any such $U_t$ there exists an uniquely defined self-adjoint generator $H$, called Hamiltonian and such that $U_t = \exp itH$. A given dynamics $U_t$ is called quasi-local iff $H = \sum_{x \subset \mathbb{V}} H_x$, where $H_X$ are self-adjoint operators acting in the local spaces $\mathcal{H}_X$. In particular, if $H_X = 0$ for any $|X| > 2$ then the dynamics $U_t$ is called two-particle dynamics.

A given dynamics $U_t$ is called perfectly one-particle state transferring iff for any locally, in the site $i \in \mathbb{V}$ prepared state $\rho_i$, any site $j \neq i$, there exists a global state $\rho_G \in E(\mathcal{H}_G)$ such that $\mathrm{Tr}^c_{\{i\}}(\rho_G) = \rho_i$ and such that after some time $t^*$, $U_{t^*}(\rho_G) = \exp -it^* H(\rho_G) \exp it^* H = \rho_G^{t^*}$ where $\mathrm{Tr}_{\{j\}^c}(\rho^{t^*}_G) = \rho_i$.

A one-particle state transferring properties were discussed in the case of $\frac{1}{2}$-spin chains [10,11,12,13] and $\frac{1}{2}$-spin systems located on a special graphs, the so-called distance-regular graphs [14,15] the cube-like graphs are best known examples of such graphs. Complete characterisation of quantum $\frac{1}{2}$-spin networks with nearest-neighbouring couplings and located on circulant graphs is given in [16].

Let us enumerate vertices of $G$ by integers $i = 1 : N$. Then the problem of transferring of a given (pure) state $\psi_i$ in $i \in \mathbb{V}$ to vertex $j \in \mathbb{V}$ can be seen as the problem of finding appropriate form of unitary operation $U_{ij}$ defined by

$$U_{ij}(\psi_1 \otimes \ldots \otimes \psi_i \otimes \ldots \otimes \psi_j \otimes \ldots \otimes \psi_N) = \psi_1 \otimes \ldots \otimes \psi_j \otimes \ldots \otimes \psi_i \otimes \ldots \otimes \psi_N \ . \ (1)$$

More generally let $\mathrm{Sym}(N)$ be a symmetric group of $N$-th element set $\mathbb{V}$ and let $U_N(0, 1)$ be the unitary faithfully representation of $\mathrm{Sym}(N)$ by 0-1 valued unitary matrices of size $N \times N$. The group $\mathrm{Sym}(N)$ consists of all permutations $\pi$ of the $N$-th elements set of indices and contains $N!$ elements. In accordance to Bernstein theorem, $\mathrm{Sym}(N)$ as a finite group possess a set(s) of generators consisting of at most $O(\ln(N!))$ generators. The very convenient system of generators is that consisting of $(N-1)$ adjacent transpositions $T_i$, $i = 1, \ldots, N-1$. In fact it is well known that the group $\mathrm{Sym}(N)$ is generated by $T_1$ and identity $\mathbb{I}$.

Let $U_n(d)$ be any universal library of quantum gates of the system under consideration. For example, if in each site of $G$ is attached a qudit of dimension

$d$ then according to Brylinski [17] theorem any set of the form $\{U(d), \mathcal{E}_d\}$, where $\mathcal{E}$ is any entangling two-qudit unitary gate forms such a library. In particular any transposition $T_i = U_{i,i+1}(N)$ can be synthesised from gates belonging to a chosen library $U_n(d)$.

**Proposition 1.** *Let a quantum system of d-dimensional qudits will be located on a graph $G$ having $N$-vertices. Then for $\pi \in \mathrm{Sym}(N)$, there exits an unitary circuits implementation the following generalised swapping operations*

*Remark 1.* The most important aspect of this proposition is that the complexity of the underlying circuit grows polynomially in the size $N$ of the system.

*Remark 2.* There are many papers devoted to the equivalence in between adiabatic quantum calculations and the unitary circuits calculations [18]. It is not difficult to conclude in spite of the analysis the existence of an unitary dynamics which did some work as the circuit from Proposition 1.

## 3    Perfect Transfer of Coherent States

By a quantum Bose gas on a given sample (undirected) connected graph $G = (\mathbb{V}, \mathbb{E})$ we mean a quantum system for which to any site $i \in \mathbb{V}$ there is associated quantum oscillator the space of states of which is given by the space $L^2(\mathbb{R})$ which can be seen as a bosonic Fock space build over the complex plane $\mathbb{C}$. In the space $L^2(\mathbb{R})$ there live canonical representation of CCR-algebra given by creation $a^\dagger$ (resp. annihilation $a$) operators. The canonical Hermite orthogonal basis of n-excited states $|n\rangle$ is given by the formula $|n\rangle = (n!)^{-1/2}|0\rangle$ where $|0\rangle$ is vacuum state.

For any $z \in \mathbb{C}$ the coherent state $|z\rangle$ is defined by the equation

$$|z\rangle = e^{(-|z|^2)/2} \sum_{n=0}^{\infty} \frac{z^n}{\sqrt{n!}}|n\rangle = e^{(-|z|^2)/2}e^{za^\dagger}|0\rangle \ . \tag{2}$$

The global Hilbert space is then $\mathcal{H} = \bigotimes_{i\in\mathbb{V}}(L^2(\mathbb{R})) = L^2(\mathbb{R}^{\mathbb{N}})$ the canonical basis of which is given by vectors

$$|n_1, \ldots, n_N\rangle = (n_1!, \ldots, n_N!)^{-\frac{1}{2}} \left(a_N^\dagger\right)^{n_N} \ldots \left(a_1^\dagger\right)^{n_1} |0\ldots0\rangle \ . \tag{3}$$

For $z = (z_1, \ldots, z_N) \in \mathbb{C}^N$ the product coherent state $|z_1, \ldots, z_N\rangle$ is given as $|z_1, \ldots, z_N\rangle = |z_1\rangle \otimes \ldots \otimes |z_N\rangle$.

The gauge-invariant, quasi-free dynamics is given by the following two-particle Hamiltonian

$$H_G = \sum_{e \in \mathbb{E}} \left[ J_e a^{\dagger}_{i(e)} a_{f(e)} \right] + \sum_{i \in \mathbb{V}} n_i \tag{4}$$

where $n_i = a^{\dagger}_i a_i$ is the local number operator. As the hamiltonian $H_G$ is commuting with the global number operator $\hat{N} = \sum_{i \in G} a^{\dagger}_i a_i$ it follows that the following orthogonal decomposition holds

$$\mathcal{H}_G = \bigoplus_{p=0}^{\infty} \mathcal{H}^p_G , \tag{5}$$

where

$$\mathcal{H}^p_G = \oplus_{n_1 + \ldots + n_N = p} |n_1 \ldots n_N\rangle , \tag{6}$$

is preserved under the dynamics defined by (4). In particular the subspace of $\mathcal{H}^1_G$ corresponding to one-quanta excitations of the analysed harmonic system is an invariant subspace of the considered dynamics. Restricting the attention to the subspace $\mathcal{H}^1_G$ only the model arising is in a natural way isomorphic to $\frac{1}{2}$-spin model on a graph $G$ and therefore all results obtained for the perfect quantum states transfer in $\frac{1}{2}$-spin system are applicable to this case.

Our question in the following: do there exist graphs $G$ based Bose gases and a quasi-free dynamics of the form (4) such that preparing the initial state of the form $|c_1\rangle = |z0\ldots0\rangle$, i.e. in the first vertex $a$ coherent state $|z\rangle$ is being created, it is possible after some time $t$ to transfer this coherent state to the final vertex $N$.

**Theorem 1.** *Let $\mathcal{G}$ corresponds to the association scheme of some finite, with non-trivial centre group $G$ obeying the conditions of the paper [15]. Then there exists quasi-free dynamics of the form (4) which perfectly transfers one-mode coherent states.*

*Proof.* (The main points of the proof)
The $M$-mode trancutation of the initial state $|z0\ldots0\rangle$ is given

$$|z0\ldots0\rangle^{(M)} = \sum_{k=0}^{M} \frac{z^k}{\sqrt{k!}} \left( a^{\dagger}_1 \right)^k |0\ldots0\rangle . \tag{7}$$

Then, obviously $|z0\ldots0\rangle^M \to_{M \to \infty} |z0\ldots0\rangle$ is the norm and as the total number of excitations is preserved it follows that for any $t > 0$, $U_t(|z0\ldots0\rangle^{(M)}) \in \mathcal{H}^M_G$ and therefore

$$U_t \left( |z0\ldots0\rangle^{(M)} \right) = \sum_{k_1 + \ldots + k_N = M} C(t,z)_{k_1,\ldots,k_N} \left( a^{\dagger}_1 \right)^{k_1} \ldots \left( a^{\dagger}_1 \right)^{k_N} |0\ldots0\rangle . \tag{8}$$

It follows from [15] that under the conditions on $G$ and $\mathcal{H}_G$ as stated above there exists a time $t_M$ such that

$$U_{t_M} \left( |z0\ldots0\rangle^{(M)} \right) = e^{i\phi}|0\ldots z\rangle^{(M)} , \tag{9}$$

for some real phase $\phi \in [0, 2\pi]$.

It follows from the analysis of [15] that $\sup_M\{t_M\} = t^\star < \infty$ and therefore there exists an accumulation point of the set $\{t_M\}$. Let $(M') \subset (M)$ be any subsequence such that $\lim_{M'} t_{M'} = t^\star$. Then there exists a sequence of quasi-free dynamics

$$H^{M'} = \sum_{e \in \mathbb{V}(G)} \left[ J_e^{(M')} a^\dagger_{i(e)} a_{f(e)} \right] \tag{10}$$

such that

$$U^{M'} |z0\ldots0\rangle^{M'} = |0\ldots z\rangle^{M'} . \tag{11}$$

The proof is finished by showing that $\lim_{M'} U_{M'}$ do exists in the strong topology and in the space of unitaries acting on the space $L^2(\mathbb{R}^N)$.

## 4 The XY-like Hamiltonian for Quantum Information Transfers in Qudits Chain

A generators of Lie's algebra of a group $SU(d)$ for $d \geq 2$ are necessary to define a XY-like Hamiltonian. The generators of $SU(d)$ are constructed as follows: in the first step, the following set of projectors is defined:

$$\left( P^{k,j} \right)_{v,\mu} = |k\rangle\langle j| = \delta_{v,j}\delta_{\mu,k}, \quad 1 \leq v, \mu \leq d . \tag{12}$$

The first suite of $d(d-1)$ operators from the group $SU(d)$ is determined as

$$\begin{aligned} \Theta^{k,j} &= P^{k,j} + P^{j,k}, \\ \beta^{k,j} &= -i(P^{k,j} - P^{j,k}), \quad 1 \leq k < j \leq d . \end{aligned} \tag{13}$$

The remaining $(d-1)$ generators are defined in the following way

$$\eta^{r,r} = \sqrt{\frac{2}{r(r+1)}} \left[ \left( \sum_{j=1}^{r} P^{j,j} \right) - r P^{r+1,r+1} \right], \quad 1 \leq r \leq (d-1) , \tag{14}$$

in total, the $d^2 - 1$ operators belonging to the $SU(d)$ group are obtained.

It is known that the $SU(d)$ generators found many applications, we only mention e.g. the theory of quantum state tomography [19] and [20].

In our first example an unknown pure state of qudit will be transferred,

$$|\psi\rangle = \alpha_0|0\rangle + \alpha_1|1\rangle + \ldots + \alpha_{d-1}|d-1\rangle \quad \text{and} \quad \sum_{i=0}^{d-1} |\alpha_i|^2 = 1, \text{ where } \alpha_i, \in \mathbb{C} . \tag{15}$$

The transfer process (or transfer protocol) in an one-dimensional chain of $n$ qudits is regarded as a transformation of the state $|\Psi_{\text{in}}\rangle$ into the state $|\Psi_{\text{out}}\rangle$

$$|\Psi_{\text{in}}\rangle = |\psi\rangle|\underbrace{000\ldots0}_{n-1}\rangle \implies |\Psi_{\text{out}}\rangle = |\underbrace{000\ldots0}_{n-1}\rangle|\psi\rangle . \tag{16}$$

The transfer operation can be performed directly by the use of permutation operators as described in Sect. 2, which is direct consequence of Equation (16). For a two qudit system with level $d = 3$, the operation of transfer is performed by permutation of the following three probability amplitudes:

$$
\begin{array}{cc}
|\Psi_{\text{in}}\rangle & |\Psi_{\text{out}}\rangle \\
\hline
\alpha \quad 00 & \alpha \quad 00 \\
01 & \beta \quad 01 \\
02 & \gamma \quad 02 \\
\beta \quad 10 & 10 \\
11 \quad \rightarrow & 11 \\
12 & 12 \\
\gamma \quad 20 & 20 \\
21 & 21 \\
22 & 22
\end{array}
\tag{17}
$$

The transfer operation can be easily expressed with a suitable permutation matrix, what is shown in Fig. 1.

Dynamics of the transfer operation can be also described in terms of a XY-like Hamiltonian and its definition arose from the work [10] and was discussed in [13,11] and [12]. The XY-like Hamiltonian for perfect transfer with additional couplings for qubit chains is defined as follows:

$$
H_{XY} = \sum_{(i,i+1)\in\mathcal{L}(G)} \frac{J_i}{2}\left(\sigma^x_{(i)}\sigma^x_{(i+1)} + \sigma^y_{(i)}\sigma^y_{(i+1)}\right), \text{ where } J_i = \frac{\sqrt{i(N-i)}}{2}
\tag{18}
$$

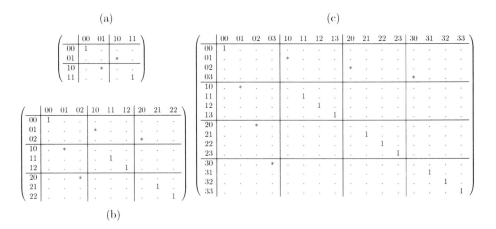

**Fig. 1.** Matrices of transfer operations for two qudits with levels (a) $d = 2$, (b) $d = 3$ and (c) $d = 4$. The dot symbol represents zero values, whereas the star symbol refers to the elements which are responsible for change of a local phase. The local phase can be fully restored with a Pauli generalised operators for qudits.

$|\psi_0\rangle \qquad |\psi_1\rangle \qquad |\psi_2\rangle \qquad |\psi_{n-2}\rangle \qquad |\psi_{n-1}\rangle$

$||$ $\qquad\qquad\qquad\qquad\qquad\qquad\qquad\qquad\qquad\qquad\qquad$ $||$

$|\Psi_{\text{in}}\rangle \qquad\qquad\qquad\qquad\qquad\qquad\qquad\qquad\qquad |\Psi_{\text{out}}\rangle$

**Fig. 2.** The realisation of information transfer in qudits chains, interactions between qudits are performed only between neighbourhood qudits

where $\sigma_{(i)}^x$ or $\sigma_{(i+1)}^y$ denotes Pauli's gate applied to $i$-th and $(i + 1)$-th qubits. An appropriate unitary operator for the transfer operation is determined by the equation given below

$$U(t) = e^{-itH_{XY}} \ . \tag{19}$$

In this work we propose the following Hamiltonian $H_{XY}^d$ to realise the transfer of quantum information in qudits chains where each qudit has the same level:

$$H_{XY}^d = \sum_{(i,i+1)\in\mathcal{L}(G)} \frac{J_i}{2} \left( \Theta_{(i)}^{k,j} \Theta_{(i+1)}^{k,j} + \beta_{(i)}^{k,j} \beta_{(i+1)}^{k,j} \right) \ , \tag{20}$$

and $J_i$ is defined as follows $J_i = \frac{\sqrt{i(N-i)}}{2}$ and $1 \le k < j < d$ and $\Theta_{(i)}^{k,j}$, $\beta_{(i)}^{k,j}$ are $SU(d)$ group operators defined by (13) applied to the $i$-th and $(i + 1)$-th qudit. Figure 2 gives graphical representation of a chain where the transfer protocol is performed.

## 5   Numerical Experiments

The numerical experiments were prepared and performed with Zielona Góra Quantum Computing Simulator (QCS). A suitable script in Python language with the use of the QCS package was prepared. The main elements of mentioned script which are directly responsible for numerical experiments are as follows

```
import qcs
q1 = qcs.QuantumReg(3, 3) ; q1.Reset()
q2 = qcs.QuantumReg(1, 3) ; q2.Reset()
q2.RandomRealAmpl()
q=qcs.CombineRegisters(q2,q1)
op = qcs.CreateTransferOperator(4, 3, 6) ; p=0;
while p < 1:
    for i in range(1, 17, 1):
        mm = q.GenDenMat() ;    mm.DepolariseChannel( p )
        v = mm.CCNRTest() ;              # collect data v, p for CCNR
        v = mm.Nagativity([1,2]) ;  # collect data v, p
        q.UserGlobalOperator( op ) ;  #  for Negativity
    p = p + 0.1
```

In the recalled script a noise is represented by a quantum depolarisation channel. The value of CCNR criterion and value of Negativity are calculated after realisation of the partial transposition operation.

In the first experiment, the test of entanglement presence was performed with the use of CCNR and Negativity. The realisation of protocol of information transfer in four qutrit chain was divided into sixteen discrete steps. According to expectations in the first and the last step the amount of entanglement is equal to zero. Similar to the transfer performed in $\frac{1}{2}$-spin chains, the highest amount of entanglement is detected in the middle step of the information transmission, which is confirmed in Fig. 2.

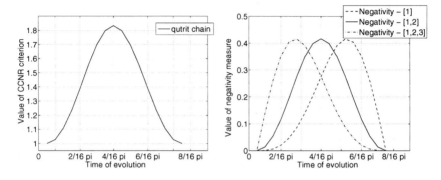

**Fig. 3.** Level of entanglement (calculated by the use of CCNR criteria and Negativity measure) during the simulation of $\mathcal{U}(t)$ for $t \in \langle 0, \pi/2 \rangle$ for chain built from four qutrits. The process of simulation of transfer has been divided into sixteen discrete steps.

The use of Negativity measure where the partial transposition is performed at the first qudit or at the first and the second qudit and finally at the first, the second and the thrid qudit clearly shows how the amount of entanglement raises and falls during the realisation of information transfer protocol. Additionally, the shape of the measure Negativity values in the case of partial transposition of the first and second qutrits is the same as for the CCNR criterion.

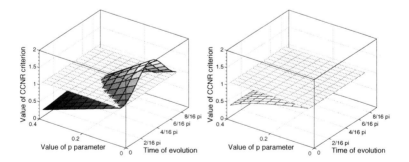

**Fig. 4.** Level of entanglement for state with the noise modelled as depolarisation of quantum channel and the value of the Fidelity during the simulation evolution of a transfer protocol

When the noise, represented by the quantum depolarisation channel,

$$\mathcal{E}(\rho) = \frac{p \mathbb{I}_d}{d} + (1 - p)\rho \quad \text{and} \quad 0 \leq p \leq 1 \; , \tag{21}$$

is added into the system then the transmission can not be finished. This can be observed by using the CCNR criterion, where the value of CCNR falls below the unity suggests the absence of entanglement (or the very weak entanglement can not be properly detected because in the case of CCNR only a value above an unity guarantees the presence of entanglement). This fact is also confirmed by the Fidelity value, although for larger values of $p$ the Fidelity indicates that the similarity of given states is still high, however the presence of noise means a loss of entanglement, which in turn out make impossible the proper transmission of information.

## 6 Conclusions

A general abstract results on the existence of unitary circuits constructed from any universal library of one and two qudit gates which implements the perfect qudit states transfers are being presented. Additionally, the constructive scenarios for the perfect transfer of continuous variable states with the use of quasi-free lattice Bose gases models were presented.

In the particular case of qutrits and geometry of chain the perfect state transferring dynamics of Heisenberg type was simulated using the tools of Zielona Góra Quantum Computer Simulator. This virtual implementation of corresponding physical systems enables us to begun to study the robustness of this transfer against sufficiently small noise otherwise problem which seems to be hardly analysed by any analytical tools.

**Acknowledgments.** We acknowledge useful discussions with the *Q-INFO* group at the Institute of Control and Computation Engineering (ISSI) of the University of Zielona Góra, Poland. We would like also to thank to anonymous referees for useful comments on the preliminary version of this paper. The numerical results were done using the hardware and software available at the "GPU Mini-Lab" located at the Institute of Control and Computation Engineering of the University of Zielona Góra, Poland.

## References

1. Walkowiak, K.: Anycasting in connection-oriented computer networks: models, algorithms and results. Int. J. Appl. Math. Comput. Sci. 20(1), 207–220 (2010)
2. Plenio, M.B., Hartley, J., Eisert, J.: Dynamics and manipulation of entanglement in coupled harmonic systems with many degrees of freedom. New J. Phys. 6, 36 (2004)
3. Plenio, M.B., Semiao, F.L.: High efficiency transfer of quantum information and multi-particle entanglement generation in translation invariant quantum chains. New J. Phys. 7, 73 (2005)

4. Yung, M.H., Bose, S.: Perfect state transfer, effective gates, and entanglement generation in engineered bosonic and fermionic networks. Phys. Rev. A 71, 032310 (2005)
5. Feder, D.L.: Cooling ultra-cold bosons in optical lattices by spectral transform. Phys. Rev. A 79, 013604 (2009)
6. Paz-Silva, G.A., Rebić, S., Twamley, J., Duty, T.: Perfect mirror transport protocol with higher dimensional quantum chains. Phys. Rev. Lett. 102, 020503 (2009)
7. Wu, L.A., Miranowicz, A., Wang, X., Liu, Y., Nori, F.: Perfect function transfer and interference effects in interacting boson lattices. Phys. Rev. A 80, 012332 (2009)
8. Braunstein, S.L., Pati, A.K.: Quantum Information with Continuous Variables. Springer (2003)
9. Sawerwain, M.: Parallel Algorithm for Simulation of Circuit and One-Way Quantum Computation Models. In: Wyrzykowski, R., Dongarra, J., Karczewski, K., Wasniewski, J. (eds.) PPAM 2007. LNCS, vol. 4967, pp. 530–539. Springer, Heidelberg (2008)
10. Bose, S.: Quantum Communication through an Unmodulated Spin Chain. Phys. Rev. Lett. 91, 207901 (2003)
11. Burgarth, D., Giovannetti, V., Bose, S.: Efficient and perfect state transfer in quantum chains. Phys. A: Math. Gen. 38(30), 6793 (2005)
12. Burgarth, D., Bose, S.: Perfect quantum state transfer with randomly coupled quantum chains. New J. Phys. 7, 135 (2005)
13. Christandl, M., Datta, N., Ekert, A., Landahl, A.J.: Perfect state transfer in quantum spin networks. Phys. Rev. Lett. 92(18), 187902 (2004)
14. Jafarizadeh, M.A., Sufiani, R.: Perfect state transfer over distance-regular spin networks. Phys. Rev. A 77, 022315 (2008)
15. Jafarizadeh, M.A., Sufiani, R., Taghavi, S.F., Barati, E.: Optimal transfer of a d-level quantum state over pseudo-distance-regular networks. J. Phys. A: Math. Theor. 41, 475302 (2008)
16. Basić, M.: Characterisation of circulant graphs having perfect state transfer. Available at arXiv:1104.1825v1 (2011)
17. Brylinski, J., Brylinski, R.: Universal quantum gates. In: Chen, G., et al. (eds.) Mathematics of Quantum Computation, pp. 110–125. A CRC Press Company (2002)
18. Aharonov, D., van Dam, W., Landau, Z., Lloayd, S., Reger, O.: Adiabatic quantum computing in equivalent to standard quantum computations. Available at arXiv:quant-ph/0405098v2 (2005)
19. Thew, R.T., Nemoto, K., White, A.G., Munro, W.J.: Qudit quantum-state tomography. Phys. Rev. A 66(1), 012303 (2002)
20. Kimura, G., Kossakowski, A.: The Bloch-vector space for N-level systems – the spherical-coordinate point of view. Open Sys. Information Dyn. 12(3), 207–229 (2005)

# Quantum Computer Network Model for a Decision Making Algorithm

Joanna Wiśniewska

Military University of Technology, Faculty of Cybernetics,
Institute of Information Systems, Kaliskiego 2, 00-908 Warsaw, Poland
jwisniewska@wat.edu.pl
http://www.wat.edu.pl

**Abstract.** The author's aim is to present how to construct circuits, built of known quantum gates, solving some decision making problems.

A decision making algorithm [1] shows how to solve a binary problem with use of quantum computer by constructing quantum operator's matrix form for analyzed problem. This matrix is always unitary, additionally its elements are binary. In this chapter some sets of basic operations are described, which allow to obtain binary and unitary matrices.

Finally, for operations from one of these sets, the implementation with known quantum gates is shown.

**Keywords:** quantum computing, unitary matrix, universal quantum gates, constructing quantum circuits.

## 1  Introduction

In [1] a decision making algorithm is described, designed for solving binary decision problems [2] with the use of a quantum computer [3]. In this algorithm we assume that the initial state of a quantum register is transformed to the final state (solution of the problem) by a quantum unitary operator. This operator can be presented by a unitary matrix $U$, i.e. matrix, for which equation (1) is true:

$$UU^* = U^*U = I \ , \tag{1}$$

where $U^*$ is a conjugate transpose (also called a Hermitian adjoint) of $U$ and $I$ is an identity matrix.

Additionally, in the considered decision making algorithm unitary matrices $U$ (which present the quantum operator) are binary matrices with precisely one value "1" in every row and column – other elements are equal to "0". It means that columns of U are orthogonal and it is consistent with the definition of the unitary matrix (1) – proof in [4]. Every quantum operator, for a $n$-bit quantum circuit, may be represented as a matrix $U$. According to the mentioned decision making

A. Kwiecień, P. Gaj, and P. Stera (Eds.): CN 2012, CCIS 291, pp. 73–81, 2012.

algorithm, $U$ is a permutation of $2^n$ columns:

$$
\begin{bmatrix} 1 \\ 0 \\ \vdots \\ 0 \end{bmatrix}, \begin{bmatrix} 0 \\ 1 \\ \vdots \\ 0 \end{bmatrix}, \ldots, \begin{bmatrix} 0 \\ 0 \\ \vdots \\ 1 \end{bmatrix} . \tag{2}
$$

We may assume that solving decision problems, with use of the decision making algorithm, needs all $(2^n)!$ matrices $U$, which are permutations of columns from (2). This tells us that for a 1-bit quantum circuit we need two matrices (identity matrix and negation matrix), for a 2-bit circuit we would like to have 24 matrices and for a 3-bit circuit there are 40,320 matrices needed $((2^3)!)$. To implement this decision making algorithm on a quantum computer, it is obvious that it is impossible to directly provide every matrix form of a quantum operator. It is necessary to specify a set of universal matrices (in the number described by the complexity function of polynomial magnitude [5]), from which we can construct all $(2^n)!$ binary matrices $U$ for a $n$-bit quantum circuit.

## 2 Basic Operations Sets

The quantum computer's mathematical model is a network model – transforming any initial state of quantum register with the use of quantum gates. Before we will present a set of basic operations and algorithms, for the known problem of synthesizing circuits for the quantum operator represented as a binary unitary matrix $U$, let us remind some useful lemmas and propositions.

**Proposition 1.** *If the $n$-qubit series circuit is built out of $m$ $n$-bit unitary quantum gates and gates' matrix representations $U_1$, $U_2$, ..., $U_m$ are known, then the matrix $U$, which represents the whole quantum circuit, may be calculated as a product of $U_i$ $(1 \leq i \leq m)$ : starting from the matrix, which is the nearest to circuit's output, following with the consecutive (adjacent) matrices, ending with the matrix, which is the nearest to the circuit's input.*

**Lemma 1.** *If any $2^n \times 2^n$ matrix $A$ is right multiplied by a binary unitary matrix $U$, of the same dimension:*

$$
AU = B , \tag{3}
$$

*then the matrix $B$ is a following permutation of the matrix's $A$ columns: if in the $j^{th}$ column of the matrix $U$ the value "1" occurs in an $i^{th}$ row, then in the $j^{th}$ column of matrix $B$, the $i^{th}$ column of matrix $A$ will be placed $(0 \leq i, j \leq 2^n - 1)$.*

**Lemma 2.** *If any $2^n \times 2^n$ matrix $A$ is left multiplied by a binary unitary matrix $U$, of the same dimension:*

$$
UA = B , \tag{4}
$$

*then the matrix $B$ is a following permutation of the matrix's $A$ columns: if in the $i^{th}$ row of the matrix $U$ the value "1" occurs in a $j^{th}$ column, then in the $i^{th}$ row of matrix $B$, the $j^{th}$ row of matrix $A$ will be placed $(0 \leq i, j \leq 2^n - 1)$.*

**Definition 1.** *A quantum circuit's part, built out of parallel placed gates (without any serially connected gates), will be called a quantum circuit's layer.*

Lemma 1 and Lemma 2 help illustrate, how the matrix form of circuit's quantum operator changes, when two or more network's gates/layers are connected serially. Using these lemmas it is easy to explain why the matrices' sets, shown below, may be treated as the sets of basic operations.

Matrices representing quantum operators, which solve a decision problem according to the algorithm mentioned in the Introduction, are binary and unitary like in (2). We may suggest some sets of basic matrices, which are also binary and unitary, to construct every quantum operator for the considered algorithm using: Proposition 1, Lemma 1 and Lemma 2.

Every $n$-bit quantum circuit for the decision making algorithm may be built out of one of the following sets containing $(2^n - 1)$ binary unitary matrices, which are constructed of $2^n \times 2^n$ identity matrix's $I$ columns, by:

1. changing every adjacent pair of columns – this action gives three matrices for every 2-qubit circuit, like in (5)

$$
\begin{bmatrix} 0&1&0&0 \\ 1&0&0&0 \\ 0&0&1&0 \\ 0&0&0&1 \end{bmatrix}, \quad
\begin{bmatrix} 1&0&0&0 \\ 0&0&1&0 \\ 0&1&0&0 \\ 0&0&0&1 \end{bmatrix}, \quad
\begin{bmatrix} 1&0&0&0 \\ 0&1&0&0 \\ 0&0&0&1 \\ 0&0&1&0 \end{bmatrix}
\tag{5}
$$

(we may construct every binary unitary matrix $U$ by changing the adjacent columns of matrix $I$ until we will obtain $U$ – see Example 1);
2. moving matrix's $I$ columns one column to left, inserting the first column as the last one; in other matrices from this set we leave without changing every time one more column in the front and the rest of columns will be moved like in the first matrix – this action gives matrices like in (6) for every 2-qubit circuit

$$
\begin{bmatrix} 0&0&0&1 \\ 1&0&0&0 \\ 0&1&0&0 \\ 0&0&1&0 \end{bmatrix}, \quad
\begin{bmatrix} 1&0&0&0 \\ 0&0&0&1 \\ 0&1&0&0 \\ 0&0&1&0 \end{bmatrix}, \quad
\begin{bmatrix} 1&0&0&0 \\ 0&1&0&0 \\ 0&0&0&1 \\ 0&0&1&0 \end{bmatrix}
\tag{6}
$$

(we will obtain every binary unitary matrix $U$ by: changing the first column of matrix $I$ with use of the first matrix from (6), then we can fix every column's value by using consecutive matrices from this set);
3. changing the one chosen column (e.g. first column) with each another – this action gives matrices like in (7) for every 2-qubit circuit

$$
\begin{bmatrix} 0&1&0&0 \\ 1&0&0&0 \\ 0&0&1&0 \\ 0&0&0&1 \end{bmatrix}, \quad
\begin{bmatrix} 0&0&1&0 \\ 0&1&0&0 \\ 1&0&0&0 \\ 0&0&0&1 \end{bmatrix}, \quad
\begin{bmatrix} 0&0&0&1 \\ 0&1&0&0 \\ 0&0&1&0 \\ 1&0&0&0 \end{bmatrix}
\tag{7}
$$

(we will construct every binary unitary matrix $U$ by fixing all columns using the first column as an element, which can directly obtain every "column value" and relay it to any other "column place").

These three sets shown above are exemplary and it is possible to propose other sets of $(2^n - 1)$ universal operations.

*Example 1.* Using the first set of basic operations (5), we are able to obtain matrix $V$, shown in (8).

$$V = \begin{bmatrix} 0\,1\,0\,0 \\ 1\,0\,0\,0 \\ 0\,0\,0\,1 \\ 0\,0\,1\,0 \end{bmatrix} \tag{8}$$

Step no. 1:

$$\begin{bmatrix} 1\,0\,0\,0 \\ 0\,1\,0\,0 \\ 0\,0\,1\,0 \\ 0\,0\,0\,1 \end{bmatrix} \cdot \begin{bmatrix} 0\,1\,0\,0 \\ 1\,0\,0\,0 \\ 0\,0\,1\,0 \\ 0\,0\,0\,1 \end{bmatrix} = \begin{bmatrix} 0\,1\,0\,0 \\ 1\,0\,0\,0 \\ 0\,0\,1\,0 \\ 0\,0\,0\,1 \end{bmatrix} , \tag{9}$$

step no. 2:

$$\begin{bmatrix} 0\,1\,0\,0 \\ 1\,0\,0\,0 \\ 0\,0\,1\,0 \\ 0\,0\,0\,1 \end{bmatrix} \cdot \begin{bmatrix} 1\,0\,0\,0 \\ 0\,1\,0\,0 \\ 0\,0\,0\,1 \\ 0\,0\,1\,0 \end{bmatrix} = \begin{bmatrix} 0\,1\,0\,0 \\ 1\,0\,0\,0 \\ 0\,0\,0\,1 \\ 0\,0\,1\,0 \end{bmatrix} = V . \tag{10}$$

Note that: used "universal" matrices may be components (layers/gates) of a quantum circuit, but they have to be placed in the right order (see: Proposition 1); matrices multiplication in step no. 1 is redundant, because one of these matrices is $I$, but it helps to illustrate the process of a quantum circuit's construction (we treat matrix $I$ as an initial state).

Let's notice that, using first (5) or second (6) basic operations set, in the worst case we have to perform the number of operations given in (11).

$$W(n) = 1 + 2 + \ldots + (2^n - 1) = 2^{2n-1} - 2^{n-1} . \tag{11}$$

This situation will occur when we will be constructing a circuit performing $n$-qubit NOT operation. If we will use the matrices from the third set (7), the most complex operation will be:

$$U_{2^n \times 2^n} = \begin{bmatrix} 0\,1 \\ 1\,0 \end{bmatrix} \otimes I_{2^n \times 2^n} \tag{12}$$

where $I$ is an identity matrix. In this case we have to perform the number of operations given in (13).

$$W(n) = 2(2^n - 1) + 2^{n-1} = 2^n + 2^{n-1} - 2 \tag{13}$$

We should observe that synthesizing binary and unitary matrices $U$ in the fastest way, we are able to obtain using operations from the basic set number 3.

The amount of performed operations is exponential, but it is caused by the fact that the number of columns in analyzed matrices is $2^n$ (even if we will propose a $(2^{2n-1} - 2^{n-1})$-element set, containing operations which allow to change places of every two columns in only one step (hypercube structure), there will be always a matrix which construction will take $(2^n - 1)$ steps).

# 3   Minimized Basic Operations Set

We can show that we need only two operations to construct a circuit performing any operation which can be described as binary and unitary matrix $U$.

The Toffoli gate [4,6] is a known universal gate, because it can perform the NAND operation and, according to electronic circuits synthesis theory [7], every logical operation may be presented as a network built of NAND gates. In this chapter we are not trying to construct a quantum network for a decision function performed by the mentioned decision making algorithm, but we are considering the possibilities of the quantum operator's matrix form modeling – anyway the Toffoli gate, its matrix form is shown in (14), will be useful for us.

$$T = \begin{bmatrix} 1 & 0 & \cdots & 0 & 0 & 0 \\ 0 & 1 & \cdots & 0 & 0 & 0 \\ \vdots & \vdots & \ddots & \vdots & \vdots & \vdots \\ 0 & 0 & \cdots & 1 & 0 & 0 \\ 0 & 0 & \cdots & 0 & 0 & 1 \\ 0 & 0 & \cdots & 0 & 1 & 0 \end{bmatrix} \tag{14}$$

The Toffoli gate changes two last columns/rows in matrix representations of other gates connected serially in the same circuit (according to Proposition 1 and Lemma 1/Lemma 2).

The unitary "rolling" operation $R$, presented in this chapter, moves all columns around the matrix:

$$R = \begin{bmatrix} 0 & 0 & \cdots & 0 & 0 & 1 \\ 1 & 0 & \cdots & 0 & 0 & 0 \\ 0 & 1 & \cdots & 0 & 0 & 0 \\ \vdots & \vdots & \ddots & \vdots & \vdots & \vdots \\ 0 & 0 & \cdots & 1 & 0 & 0 \\ 0 & 0 & \cdots & 0 & 1 & 0 \end{bmatrix}. \tag{15}$$

The matrix $R$ places the first column as the last one and the other columns are moved one column to the left (or the last row becomes the first one and the other rows go down). Having $R$ we are also able to perform this column/row operation in the other direction. The reverse "rolling" operation will be performed by $R^*$, which for every $n$-qubit gate $R$ we may obtain by joining serially $(2^n - 1)$ gates $R$ together.

$$R^* = R^{2^n - 1} \tag{16}$$

**Proposition 2.** *To construct every $n$-qubit quantum circuit $(n > 1)$, which performs an operation represented by a binary and unitary matrix $U$, we only need two basic operations: the $n$-qubit Toffoli gate and the $n$-qubit unitary "rolling" gate $R$.*

*Remark 1.* For 1-qubit circuits only 1-qubit negation gate NOT is necessary.

To prove that the $n$-qubit Toffoli and "rolling" $R$ gates may be universal operations for every $n$-qubit quantum circuit performing $U$, let us present a very simple algorithm of obtaining matrices changing adjacent columns, from which we are able to obtain every binary and unitary $2^n \times 2^n$ matrix $U$. Having the Toffoli gate (in the algorithm presented below, its matrix form is an initial value of variable $A$), which changes the last two columns, we want to obtain the other $(2^n - 2)$ gates changing adjacent columns – it means, that we need to perform $(2^n - 2)$ times two matrix multiplications:

1. $R^*A = B$;
2. $BR = A$.

Every matrix $A$ is a matrix changing the other adjacent columns (the basic operations set like in (5)).

Anyway synthesizing a circuit for binary and unitary matrix $U$ with use of $n$-qubit Toffoli gate and the $n$-qubit "rolling" gate $R$ is complex, because using only these two or three operations (it is possible to use as a third operation: $R^*$ – explanation will be shown in Sect. 4) forces more steps to obtain any $U$, than are needed in the case of using $(2^n - 1)$ elements from sets presented in (5)–(7) – a small variety of basic elements causes more operations while constructing $U$.

## 4    Constructing Circuits of Known Quantum Gates

The efficient implementation of Toffoli gate with use of qudits [8] is known, so we treat it as a single circuit's layer (Definition 1) and we focus on $R$ operation's quantum implementation.

The $R$ operation can be presented as a circuit at the Fig. 1, with use of an operation $\mathrm{NOT_{COND}}$ – changing columns: first and $(2^{n-1} - 1)^{\mathrm{th}}$. A binary and unitary matrix for the 2-qubit $\mathrm{NOT_{COND}}$ operation in (17).

$$\mathrm{NOT_{COND}} = \begin{bmatrix} 0 & 0 & 1 & 0 \\ 0 & 1 & 0 & 0 \\ 1 & 0 & 0 & 0 \\ 0 & 0 & 0 & 1 \end{bmatrix} \tag{17}$$

The $n$-qubit $(n > 1)$ circuit of known quantum gates, performing $\mathrm{NOT_{COND}}$ operation is a five-layer circuit as at the Fig. 2 (the number of layers doesn't rise, while $n$ increases).

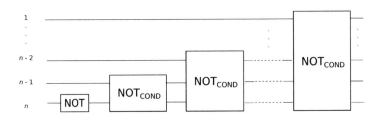

**Fig. 1.** A circuit performing $n$-qubit $R$ operation

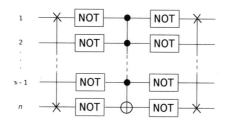

**Fig. 2.** A circuit performing $n$-qubit $\mathrm{NOT_{COND}}$ operation $(n > 1)$

We have previously noticed, that constructing circuits for $U$ operation with use of Toffoli gate, $R$ and $R^*$ operation, is more complex, than constructing $U$ with sets' elements presented in (5)–(7). Moreover we assume that Toffoli gate is single operation – this lets us calculate, that the $n$-qubit $R$ operation needs a circuit built of $(5n-4)$ layers. Fortunately the $R^*$ operation doesn't need to be implemented as a sequence of $(2^n - 1)$ $R$ operations (according to (16)). Upon the Proposition 1, we can present $R^*$ as circuit reversible to $R$ (with the same number of layers: $(5n - 4)$).

To minimize the number of layers in a quantum circuit, performing $U$ operation, we will present a basic operation set with use of which we can calculate any $U$ with the best (i.e. smallest), described in this chapter, complexity. Additionally this set's elements can be presented as a circuits of known quantum gates with constant number of layers.

As we previously noticed, sets of basic operations shown in the Sect. 2 are exemplary and we can define more cases when a set of $(2^n - 1)$ operations is a basic operations set for unitary and binary matrices $U$. Analyzing the matrices from the third set, we can see that there are no known quantum gates with matrix representation like any of (7), but we may slightly modify this set's content. Let us choose as the "central" column the penultimate column, so for the 2-qubit circuit we will obtain the following set of basic operations:

$$
\begin{bmatrix} 0\,0\,1\,0 \\ 0\,1\,0\,0 \\ 1\,0\,0\,0 \\ 0\,0\,0\,1 \end{bmatrix}, \;
\begin{bmatrix} 1\,0\,0\,0 \\ 0\,0\,1\,0 \\ 0\,1\,0\,0 \\ 0\,0\,0\,1 \end{bmatrix}, \;
\begin{bmatrix} 1\,0\,0\,0 \\ 0\,1\,0\,0 \\ 0\,0\,0\,1 \\ 0\,0\,1\,0 \end{bmatrix} . \tag{18}
$$

Now we can see, that the last one matrix, changing last two columns, is a known XOR gate and for the $n$-qubit circuits $(n > 2)$ it will be Toffoli gate (for 2-qubit circuit exceptionally the second matrix is a SWAP gate). It means that we have to show, how to construct $(2^n - 2)$ basic operations for $n$-qubit circuit. Let us assign the symbols to every matrix for this basic set (with the penultimate "central" column): $M_{2^n-1}$ changes first and next to last column, $M_{2^n-2}$ changes second and next to last column,..., $M_1$ changes next to last and last column ($n$-qubit Toffoli gate).

Experimentally we have specified a rule telling that the number of layers for one actual basic operation is constant even if the number of qubits raises.

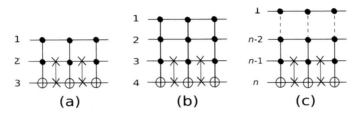

**Fig. 3.** A circuit performing operation $M_2$, built of SWAP and Toffoli gates: (a) 3-qubit circuit; (b) 4-qubit circuit; (c) $n$-qubit circuit $(n > 4)$

We have synthesized the 3-qubit and 4-qubit circuits of quantum gates for the operations $M_i$ $(2 \leq i \leq 2^n - 1)$ and we have noticed that once we calculate $n$-qubit and $(n + 1)$-qubit circuit $(n > 2)$ for any matrix $M_i$, than we can observe some properties of $M_i$ and use them to define $M_i$ of grater size.

*Example 2.* For matrix $M_2$ (changing columns $(2^n - 2)$ and $(2^n - 1)$) we have specified 3-qubit and 4-qubit circuits and we are able to define a pattern of circuit for $M_2$ for circuits with more qubits than four. As we can see at the Fig. 3, the circuit for $M_2$ has always five layers (and also is built of five gates – there is only one gate in every layer).

There is no point to present circuits for next $M_i$, but it is important that we have synthesized all 3-qubit and 4-qubit circuits for $M_i$ operations with use of: SWAP gate, XOR gate, 1-qubit NOT gate, 3-qubit and 4-qubit Toffoli gate. The number of layers for $M_i$ operations raises (irregularly) with increasing $i$ value, e.g. circuits for $M_3$ and $M_5$ have got four layers, for $M_2$ and $M_4$ – five layers, for $M_6$ – eight layers and for $M_7$ – seven layers. Ending the chapter let us present an example for constructing a matrix $U$ with use of operations $M_i$ and then with known quantum gates.

*Example 3.* Let's suppose that mentioned decision making algorithm produces matrix $U$ for some decision problem:

$$U = \begin{bmatrix} 1 & 0 & 0 & 0 & 0 & 0 & 0 & 0 \\ 0 & 1 & 0 & 0 & 0 & 0 & 0 & 0 \\ 0 & 0 & 0 & 0 & 0 & 0 & 1 & 0 \\ 0 & 0 & 0 & 0 & 0 & 0 & 0 & 1 \\ 0 & 0 & 0 & 0 & 1 & 0 & 0 & 0 \\ 0 & 0 & 0 & 0 & 0 & 1 & 0 & 0 \\ 0 & 0 & 0 & 1 & 0 & 0 & 0 & 0 \\ 0 & 0 & 1 & 0 & 0 & 0 & 0 & 0 \end{bmatrix} . \tag{19}$$

We can present $U$ as multiplication of $8 \times 8$ $M_i$ matrices:

$$U = M_4 \cdot M_1 \cdot M_5 . \tag{20}$$

According to Proposition 1 a quantum circuit performing $U$ is shown at the Fig. 4.

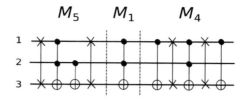

**Fig. 4.** A circuit performing exemplary operation $U$

## 5  Summary

The mentioned in the Introduction decision making algorithm, to solve $n$-variable task, needs a quantum circuit, which can be represented by unitary and binary $2^n \times 2^n$ matrix. We assume that solving all $n$-variable tasks, which may be posed by the user, needs $(2^n)!$ different circuits. In the Section 3 we have shown that only two universal operations are necessary to solve every decision task of $n$ variables, but we use a $(2^n - 1)$-element basic operation set to reduce the number of needed operations. In the last section we have presented the results of an experiment, which shows that the basic operations $M_i$ can be implemented with use of known quantum gates, moreover the number of layers within specified circuit performing $M_i$ is constant.

## References

1. Wiśniewska, J.: Fast method of calculating unitary matrix for quantum decision algorithm. Polish Journal of Environmental Studies 18(3B), 381–385 (2009)
2. Ripley, B.D.: Pattern Recognition and Neural Networks. Cambridge University Press, Cambridge (1996)
3. DiVincenzo, D., Loss, D.: Quantum information is physical. [@:] arXiv.org: cond-mat/9710259 (1998)
4. Chudy, M.: Introduction to Quantum Informatics. Akademicka Oficyna Wydawnicza EXIT, Warszawa, Poland (2011) (in Polish)
5. Aho, A.V., Ullman, J.D.: Foundations of Computer Science: C Edition. W.H. Freeman, New York (1994)
6. Kaye, P., Laflamme, R., Mosca, M.: An Introduction to Quantum Computing. Oxford University Press, Oxford (2007)
7. Cogdell, J.R.: Foundations of Electronics. Prentice Hall, Upper Saddle River (1999)
8. Gilchrist, A., Ralph, T.C., Resch, K.J.: Efficient Toffoli Gates Using Qudits. [@:] arXiv.org: 0806.0654 v1 [quant-ph] (2008)
9. Gawron, P., Klamka, J., Miszczak, J., Winiarczyk, R.: Structural programming in quantum octave. Bulletin of the Polish Academy of Sciences. Technical Sciences 58(1), 77–88 (2010)
10. Hardy, Y., Steeb, W.H.: A Sequence of Quantum Gates. [@:] arXiv.org: 1202.2259v1 [quant-ph] (2012)

# Comparison of AQM Control Systems with the Use of Fluid Flow Approximation

Adam Domański[1], Joanna Domańska[2], and Tadeusz Czachórski[2]

[1] Institute of Informatics, Silesian Technical University
Akademicka 16, 44-100 Gliwice, Poland
adamd@polsl.pl
[2] Institute of Theoretical and Applied Informatics, Polish Academy of Sciences
Baltycka 5, 44-100 Gliwice, Poland
{joanna,tadek}@iitis.gliwice.pl

**Abstract.** In the article we study a model of TCP/UDP connection with Active Queue Management in an intermediate IP router. We use the fluid flow approximation technique to model the interactions between the set of TCP/UDP flows and AQM. We model a system where RED or NLRED are the AQM policy. The obtained results confirm the advantage of our NLRED algorithm.

**Keywords:** fluid flow, AQM, RED, NLRED.

## 1 Introduction

Design technology for TCP/IP networks is one of the most important topics in the field of telecommunications networks. The main problem is still the modeling of congestion control mechanisms. The development of new active queue management (AQM) routers allow to increase the performance of Internet applications.

A number of analytical models of AQM in IP routers in open-loop scenario – because of the difficulty in analyzing AQM mathematically – was already presented [1,2]. In this article we try to use the nonlinear dynamic model of TCP [3,4] to analyze the AQM systems. This model enables application of control rules to address the basic feedback nature of AQM.

We use the fluid flow modeling methodology based on mean value analysis. This analytical method of modeling has a great potential in analyzing and understanding various network congestion control algorithms [5]. The models based on fluid flow approximation are able to capture the dynamics of TCP flows [6] and allow to analyze networks with a large number of flows. Here, we use this method to compare routers having different active queue management principles and transmitting TCP/UDP flows. The model allows to study not only the steady-state behavior of the network, but also the transient behavior when a set of TCP flows start or finish transmission. We concentrate on transient average router queue length for different AQM strategies.

A. Kwiecień, P. Gaj, and P. Stera (Eds.): CN 2012, CCIS 291, pp. 82–90, 2012.

The rest of this article is organized as follows. The Section 2 describes the fluid flow model of AQM router supporting TCP/UDP flows. The Section 3 shows obtained results. The conclusions are presented in Section 4.

## 2    Fluid Flow Model of Mixed TCP and UDP Behavior

This section presents a fluid flow model the AQM router supporting TCP/UDP flows.

The model presented in [5] demonstrates TCP protocol dynamics. This model ignores the TCP timeout mechanisms and allows to obtain the average value of key network variables. This model is described by the following nonlinear differential equations [7]:

$$W'(t) = \frac{1}{R(t)} - \frac{W(t)W(t - R(t))}{2R(t - R(t))}p(t - R(t)) \tag{1}$$

$$q'(t) = \frac{W(t)}{R(t)}N(t) - C \tag{2}$$

where:

- $W$ = expected TCP sending window size [packets],
- $q$ = expected queue length [packets],
- $R$ = round-trip time = $q/C + T_p$ [s],
- $C$ = link capacity [packets/s],
- $T_p$ = propagation delay [s],
- $N$ = number of TCP sessions,
- $p$ = packet drop probability.

The maximum values of $q$ and $W$ (queue length and congestion window size) depend on the buffer capacity and maximum window size. The dropping probability $p$ depends on the queue algorithm.

The traffic composed of TCP and UDP streams has been considered in [8]. For this model a single router supports N sessions and each session is assumed to be either a TCP or UDP session. Each TCP stream is a TCP-Reno connection and each UDP sender is a CBR source. The rate of UDP sessions is denoted by $\lambda$. Fluid flow equations of TCP and UDP mixed traffic become:

$$W'(t) = \frac{1}{R'(t)} - \frac{W(t)W(t - R(t))}{2R'(t - R(t))}p(t - R'(t)) \tag{3}$$

$$q'(t) = \frac{W(t)}{R(t)}N_\gamma(t) - (C - \lambda) \tag{4}$$

where $R'$ = round-trip time = $q/(C - \lambda) + T_p$ [s].

In *passive* queue management (e.g. FIFO scheduling), packets coming to a buffer are rejected only if there is no space in the buffer to store them, the drop probability $p$ (see Equation (3)) takes values 0 (there is space in buffer) or 1 (buffer is full). In RED mechanism at arrival of each packet the average queue size $x$ is calculated as an exponentially weighted moving average using the following formula: $x_i = (1 - \alpha)x_{i-1} + \alpha q_{inst}$ where $q_{inst}$ is the current queue length. Then the RED drop function is applied: there are two thresholds $Min_{th}$ and $Max_{th}$; if $x < Min_{th}$ all packets are admitted, if $Min_{th} < x < Max_{th}$ then dropping probability $p$ is growing linearly from 0 to $p_{max}$:

$$p = p_{max}\frac{x - Min_{th}}{Max_{th} - Min_{th}} \tag{5}$$

and if $x > Max_{th}$ then all packets are dropped. The value of $p_{max}$ has also a strong influence on the RED performance: if it is too large, the overall throughput is unnecessarily choked and if it is too small, the danger of synchronization arises; [9] recommends $p_{max} = 0.1$. The problem of the choice of parameters is still discussed, see e.g. [10,11]. In NLRED (Non Linear RED) algorithm (proposed by authors in [12,13]) the probability of packet dropping function is defined as follows:

$$\tag{6}$$

$$p(x, a_1, a_2, p_{max}) = \begin{cases} 0 & \text{for } x < Min_{th} \\ \varphi_0(x) + a_1\varphi_1(x) + a_2\varphi_2(x) & \text{for } Min_{th} \le x \le Max_{th} \\ 1 & \text{for } x > Max_{th} \end{cases}$$

where basis functions are defined:

$$\varphi_0(x) = p_{max}\frac{x - Min_{th}}{Max_{th} - Min_{th}}, \tag{7}$$
$$\varphi_1(x) = (x - Min_{th})(Max_{th} - x), \tag{8}$$
$$\varphi_2(x) = (x - Min_{th})^2(Max_{th} - x) \tag{9}$$

and $x$ is the weighted moving average queue size.

A set of functions $p$ is presented in Fig. 1.

## 3  Results

The computations were made with the use of PyLab (Python numeric computation environment) [14], a combination of Python, NumPy, SciPy, Matplotlib, and IPython. The graphs shown below present transient system behavior, the time axis is drawn in seconds.

The parameters of AQM buffer:

- $Min_{th} = 10$,
- $Max_{th} = 15$,
- buffer size (measured in packets) = 20,
- weight parameter $\alpha = 0.007$.

The parameters of TCP connection:

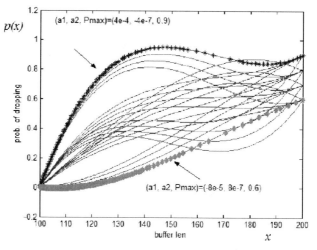

**Fig. 1.** A set of functions $p$

- transmission capacity of AQM router: $C = 0.075$,
- propagation delay for $i$-th flow: $T_{p_i} = 2$,
- initial congestion window size for $i$-th flow (measured in packets): $W_i = 1$.

The obtained mean queue lengths for TCP connections are:

- for RED $(p_{\max} = 0.1) = 9.9931$,
- for RED $(p_{\max} = 0.6) = 8.3359$,
- for NLRED $(p_{\max} = 0.6, a_1 = 0.00008, a_2 = 0.0000008) = 8.3327$,
- for NLRED $(p_{\max} = 0.9, a_1 = 0.0004, a_2 = 0.000004) = 7.9111$.

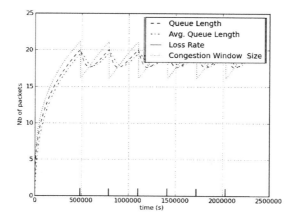

**Fig. 2.** FIFO queue, one TCP flow

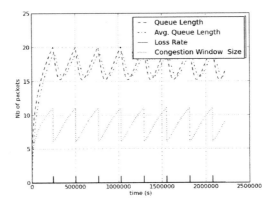

**Fig. 3.** FIFO queue, two TCP flows

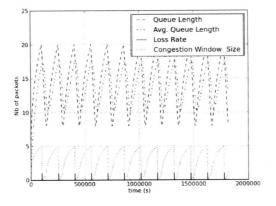

**Fig. 4.** FIFO queue, five TCP flows

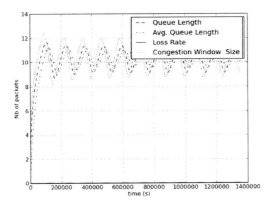

**Fig. 5.** RED queue, one TCP flow

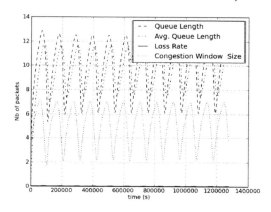

**Fig. 6.** RED queue, two TCP flows

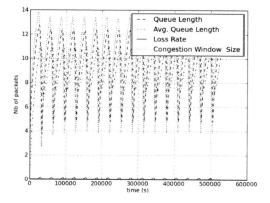

**Fig. 7.** NLRED queue, one TCP flow

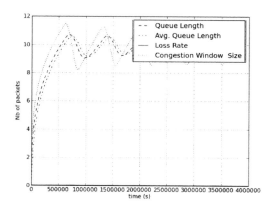

**Fig. 8.** RED queue, one mixed TCP/UDP flow

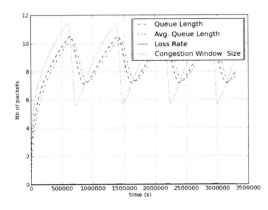

**Fig. 9.** NLRED queue, 1 mixed TCP/UDP flow

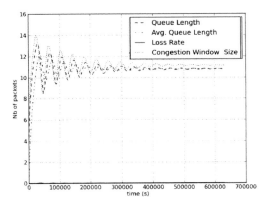

**Fig. 10.** RED queue, 1 mixed TCP, UDP (big) flow

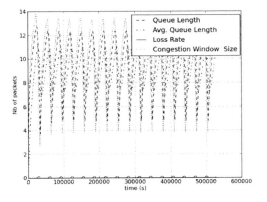

**Fig. 11.** NLRED queue, 1 mixed TCP, UDP (big) flow

For mixed TCP/UDP connection:

- for RED $(p_{max} = 0.1) = 10.6779$,
- NLRED $(p_{max} = 0.9,\ a_1 = 0.0004,\ a_2 = 0.000004) = 8.4566$.

Figure 2 shows the queue behavior in case of one TCP flow and FIFO queue. The size of congestion window increases until the buffer becomes full. Packets are dropped, the size of congestion window decreases causing a slow decrease of the queue length – this pattern is repeated periodically. Figure 3 presents the analogous situation in case of two TCP flows. The curves are similar to the previous, the only difference is that the congestion window in the case of two TCP flows never achieves such large values as in the case of one TCP flow. Figure 4 presents the same pattern for five TCP flows. Figures 5 and 6 present the queue evolution when RED mechanism is implemented. The loss rate is in this cases smaller. Queue length and weighted moving average queue length oscillate between their mean values. Figure 7 shows the behavior of the queue with NLRED algorithm. The characteristic of the plots is the same to the RED queue. But the mean values of queue length presented earlier confirm the advantage of NLRED algorithm (also for mixed TCP/UDP traffic). Figures 8 and 9 show the behavior of the RED/NLRED queues for mixed TCP/UDP traffic. For such kind of traffic decreasing of congestion window causes not as slowly decreasing of queue length as for TCP traffic. This decreasing of queue length is faster for NLRED queue. Figures 10 and 11 show the case of large amount of UDP traffic. The NLRED queue reacts faster than RED queue (the function of packet rejection probability increases faster). The congestion window is subject of large fluctuation.

## 4    Conclusions

A number of authors studied already fluid flow nonlinear dynamic model of TCP connections [15], but only a few of them have taken into account the mixed TCP/UDP traffic [8,16,17]. This article confirms the advantage of proposed NLRED algorithm over standard RED, prevously studied in open loop scenario [12,13]. It confirms also the advantage of the algorithm in presence of mixed TCP/UDP traffic.

Our future works will focus on the integration our the model with packet level simulation, this approach was proved to be very efficient [6,18]. In traditional packet-level discrete event simulators the time consumption increases rapidly with the number of TCP flows being simulated and the use of mixed models can increase the scalability of modeling.

**Acknowledgements.** This research was partially financed by Polish Ministry of Science and Higher Education project no. N N516479640.

## References

1. Liu, C., Jain, R.: Improving explicit congestion notification with the mark-front strategy. Computer Networks 35(2-3) (2000)

2. Domańska, J., Domański, A., Czachórski, T.: The Drop-From-Front Strategy in AQM. In: Koucheryavy, Y., Harju, J., Sayenko, A. (eds.) NEW2AN 2007. LNCS, vol. 4712, pp. 61–72. Springer, Heidelberg (2007)
3. Hollot, C.V., Misra, V., Towsley, D., Gong, W.-B.: On Designing Improved Controllers for AQM Routers Supporting TCP Flows. In: IEEE INFOCOM (2002)
4. Rahme, S., Labit, Y., Gouaisbaut, F.: An unknown input sliding observer for anomaly detection in TCP/IP networks. In: Ultra Modern Telecommunications & Workshops (2009)
5. Misra, V., Gong, W.-B., Towsley, D.: Fluid-based Analysis of a Network of AQM Routers Supporting TCP Flows with an Application to RED. In: ACM SIGCOMM (2000)
6. Yung, T.K., Martin, J., Takai, M., Bagrodia, R.: Integration of fluid-based analytical model with Packet-Level Simulation for Analysis of Computer Networks. In: SPIE (2001)
7. Hollot, C.V., Misra, V., Towsley, D.: A control theoretic analysis of RED. In: IEEE/INFOCOM (2001)
8. Wang, L., Li, Z., Chen, Y.-P., Xue, K.: Fluid-based stability analysis of mixed TCP and UDP traffic under RED. In: 10th IEEE International Conference on Engineering of Complex Computer Systems (2005)
9. Domańska, J., Domański, A., Czachórski, T.: Implementation of modified AQM mechanisms in IP routers. Journal of Communications Software and Systems 4(1) (March 2008)
10. Feng, W.-C., Kandlur, D.D., Saha, D., Shin, K.G.: Adaptive packet marking for maintaining end to end throughput in a differentiated service internet. IEEE/ACM Transactions on Networking 7(5), 685–697 (1999)
11. May, M., Diot, C., Lyles, B., Bolot, J.: Influence of active queue management parameters on aggregate traffic performance. Research Report, Institut de Recherche en Informatique et en Automatique (2000)
12. Augustyn, D.R., Domański, A., Domańska, J.: Active Queue Management with non linear packet dropping function. In: 6th International Conference on Performance Modelling and Evaluation of Heterogeneous Networks HET-NETs (2010)
13. Augustyn, D.R., Domański, A., Domańska, J.: A Choice of Optimal Packet Dropping Function for Active Queue Management. In: Kwiecień, A., Gaj, P., Stera, P. (eds.) CN 2010. CCIS, vol. 79, pp. 199–206. Springer, Heidelberg (2010)
14. http://www.scipy.org
15. Chen, J., Paganini, F., Wang, R., Sanadidi, M.Y., Gerla, M.: Fluid-flow Analysis of TCP Westwood with RED. In: GLOBECOM (2004)
16. Czachórski, T., Grochla, K., Pekergin, F.: Stability and Dynamics of TCP-NCR(DCR) Protocol in Presence of UDP Flows. In: García-Vidal, J., Cerdà-Alabern, L. (eds.) Euro-NGI 2007. LNCS, vol. 4396, pp. 241–254. Springer, Heidelberg (2007)
17. Mao, P., Xiao, Y., Hu, S., Kim, K.: Stable parameter settings for PI router mixing TCP and UDP traffic. In: IEEE 10th International Conference on Signal Processing, ICSP (2010)
18. Kiddle, C., Simmonds, R., Williamson, C., Unger, B.: Hybrid packet/fluid flow network simulation. In: Seventeenth Workshop on Parallel and Distributed Simulation, pp. 143–152 (2003)

# Testing and Scalability Analysis of Network Management Systems Using Device Emulation

Krzysztof Grochla and Leszek Naruszewicz

Proximetry Poland Sp. z o.o.,
Al. Rozdzienskiego 91, 40-203 Katowice, Poland
{kgrochla,lnaruszewicz}@proximetry.pl
http://www.proximetry.com

**Abstract.** The network management systems support the network operators by providing automation of common management tasks, fault management and automation of network reconfiguration. The testing and performance analysis of such system is a challenging engineering issue. In this paper the methodology for testing and performance analysis of the network management systems is proposed, based on a software device emulator. The emulator fully represents the managed devices and implements the management protocols to provide the same load on the server as the real devices. The sample results achieved by the proposed framework are presented.

**Keywords:** computer networks, perfomance evaluation, emulation, network management systems, Erlang.

## 1 Introduction

The management of very large computer network is a big engineering challenge. The current telecommunications networks consist of thousands of networking devices: routers, switches, access points, base stations etc. The network operator must monitor their status, manage its configuration, provide regular firmware updates etc. The problem becomes even more complicated with the introduction of Smart Grid network and machine-to-machine communication, as the size of the managed network grows by two or three orders of magnitude.

To support the network operators in every day management tasks the Network Management Systems (NMS) have been created. The NMS is a combination of hardware and software used to monitor and administer a computer network. Device management includes faults, configuration, accounting, performance and security (FCAPS) management. Management tasks include discovering and keeping the track of network inventory, monitoring device health and status, providing alerts to conditions that impact system performance and identification of problems, their source(s), and possible solutions. An NMS uses network management protocols to execute these tasks. For example, SNMP [1], CAPWAP [2], netconf [3] or TR.069 [4] protocols can be used to gather the information from devices in the network hierarchy. The NMS collects device statistics and may

A. Kwiecień, P. Gaj, and P. Stera (Eds.): CN 2012, CCIS 291, pp. 91–100, 2012.
© Springer-Verlag Berlin Heidelberg 2012

maintain an archive of previous network statistics including problems and solutions that were successful in the past. If faults recur, the NMS can search the archive for the possible solutions.

There are few most commonly used NMS architectures:

− centralized,
− distributed,
− hierarchical.

The centralized NMS architecture is based on a single NMS server, which communicates with the devices and executes the management tasks. It is typically equipped with a database which stores the device configuration, network state and statistics. The graphical user interface may be realized by a separate application communicating with the server or as a web application. The distributed architecture assumes that the management of the network nodes is distributed among multiple places in the network (e.g. to multiple servers managing different segments of the network), without a single coordination point. In hierarchical NMS systems the management is coordinated by a one single entity, but there are multiple nodes of the NMS that perform the management tasks. In all those three architectures the network management protocols are used to communicate the status and state of the managed devices to the entities of the NMS.

The performance and stability of the NMS is crucial for the network operation. Without the NMS the operator will not be informed about the failures of the network elements (or will get this information with large delay). Since the NMS software needs to be thoughtfully tested and verified. The size of the managed network together with the network operator response to failure limits imposes also strict performance requirements. These elements need to be verified, but it is almost impossible to perform such test in a real-life environment, as stopping the operation of the managed network is not an option. Thus the novel methods of testing and performance evaluation of NMS need to be developed. Apart from the reliability of the system they should also allow to estimate the response time and maximum number of devices the system is able to manage.

Most of the previous works on network management systems concentrate on the testing of the state of the network elements, not on the NMS itself. Lee et al in [5] present the passive method of locating faults during network operation by monitoring input/output behaviour. This can be also applied to the NMS by monitoring it as a black box, but does not provide much information about its performance. There are multiple papers on the evaluation of network management protocols – see e.g. [6,7] and [8] and the results provided there can be used to assess the amount of network traffic that will go to and from the NMS. The general methods of evaluation of the scalability of distributed systems, as those presented by Tanenbaum [9], Zhang et al [10] or Jogalekar and Woodside [11] can be also applied to NMS evaluation, but will not capture the specific of the interactions between the managed devices, the operator and the management system.

The emulators have been used previously to verify the operation of the networking protocols and algorithms in conditions hard to reproduce in the laboratory. The Seawind [12] emulator allows to represent the wireless devices using wired infrastructure. The Empower tool [13] has been created to allow to evaluate the scalability of networking protocols. The recent works use the network virtualization to execute the real unmodified protocol implementations within a configurable synthetic environment [14]. The approach of using the emulation to evaluate the network protocols shall be joined with the protocols performance evaluation techniques to create an emulation tool dedicated to validate the Network Management System.

The rest of the paper is organized as follows: the Sect. 2 describes the architecture and methods for evaluations of the NMS. The next section describes the architecture, functionality of the emulator and the gains of using it as a NMS systems evaluation tool.

## 2   Methods for NMS Testing

### 2.1   Laboratory Testing

The testing and evaluation of the network management system requires the reproduction of the interactions between the managed devices and the NMS. The most natural way of testing such system is to connect some devices in the laboratory and reproduce behaviour of the network using the traffic generation tools (such as DiTG [15] or iperf [16]) and verify that the NMS behaves accordingly to generated test case. The main problem with such approach is the large costs of reproduction of the single scenario, as it may require connecting multiple devices and require large amount of manpower to prepare. The size of the network tested this way is limited by the amount of devices available in the laboratory and the reproduction of some, even simple, scenarios is very hard.

### 2.2   Discrete Events Simulation

The Discrete Event Simulation (DES) allows to represent the operation of the system as the chronological sequence of events processed in the memory of a computer. It can be used to simulate the behaviour of the network management protocols and algorithms, but requires porting the code to a simulation environment. The simulation of the networking protocols typically operates on the packet level, when a single event represents a transmission of a data packet, but it is also possible to simulate the system at the flow level or at lower level of granularity. The discrete event simulation is very effective in testing and evaluating the performance of a single algorithm within the network management system, however adaptation of the whole system to be executed within the DES framework would be very costly. Some of the network simulators, such as NS3 or OMNeT++, allow to integrate the simulation with the real networking environment [17] and use the simulation as the source of the traffic to test the software. From the perspective of the testing and evaluating of the network management system, the DES may be used as the representation of the part of the managed networking devices and emulate the protocols and procedures executed by the devices. The challenge in integration of the simulation with real network is to synchronize the simulation clock with the wall clock and to properly implement all time relation. The main advantage of using the DES comparing to the laboratory network is the ability to easily simulate network of virtually any size and to script the behaviour of the devices.

### 2.3   Device Emulator

The emulation of networking devices uses the software processes which work and behave from the network perspective as the emulated device, but are executed on a different machine. Comparing to the DES the emulator operates always using the wall clock, while the simulators use different simulation clock and the execution of the code jumps instantaneously from one event to another. The emulation is typically simplified and a single PC machine can emulate multiple network devices. Comparing to the simulation it is easier to port the code from the actual device to the emulation and have the same timing of responses as on the emulated device, but the emulation does not allow to accelerate the execution time of the test.

## 3   Using the Device Emulator to Evaluate the NMS

### 3.1   Architecture of the Analyzed System

We propose to use the device emulator as the tool for the evaluation of the network management system. We consider a network consist of a set of networking devices which generate or consume data transmission. The devices can be managed, or unmanaged. The managed devices have an agent or other software module that implements the network management protocol and maintains the communication with the server. The unmanaged devices are only passively monitored by the NMS using e.g. ping probes. The network devices can be wired or wireless. The managed devices provide the statistics about their operation to the NMS and may contain e.g. the traffic rate, signal strength, modulation used, CPU usage etc. The NMS stores the statistics and presents to the user. The NMS also reports the faults to the user when detects some configured states or exceeding some thresholds – e.g. when a device is not responding, the signal level on wireless link is to low or the CPU is overloaded for a longer period of time. The NMS allows to manage the configuration of the devices by the user and upgrade of the devices' firmware. It also keeps accounting information about the device activity time and traffic transmitted, and logs important events in the network. The network management system is considered to be a single entity using the networking infrastructure to communicate with the devices, but internally may apply any of the architectures described in Sect. 1.

The emulator of the managed devices have been prepared as a software application running in Erlang virtual machine [18]. The Erlang environment has been

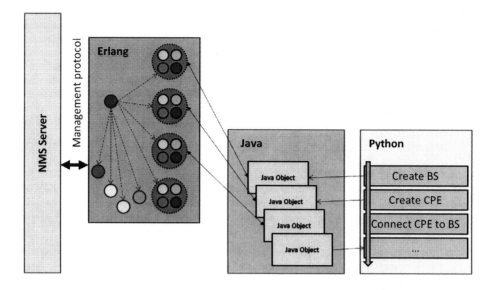

**Fig. 1.** The architecture of the device emulator

selected because it is oriented on development of telecommunication software, and it allows isolating and controlling large number of processes. Thanks to that emulator development may focus mainly on modeling of device state machines instead of spending time on building concurrent execution environment to run thousands or millions of emulated devices. The structure of the emulator is presented on Fig. 1 A single device is represented by a set of concurrent Erlang processes. The single Erlang process represents a single device. The Java classes were created to allow to control the behaviour of the processes and to provide the programming API to read the state of the emulator. The interface between Java and Python has been used to provide the easy scripting of the device state changes.

Typically the NMS use the software agent on the device that implements a management protocol, such as SNMP, CAPWAP or NETCONF and manages the device. The Erlang code emulates the NMS agent working on the device. It contains the same state machines as within the real agent, the same protocol implementation and simplified representation of the device state. It sends to the NMS all the information a regular device would send – statistics with some configured values, state changes, faults etc. The time and parameters of particular event reported by the device is coded using the Python or Java. By changing the attributes of the emulated devices the scripts allow to emulate different device types, models, network properties etc.

### 3.2   Functionality of the Emulator

The emulation framework provides the following functions during design and development of a complex NMS for heterogeneous network:

- testing of the whole NMS as a black box,
- unit testing,
- performance testing of the NMS,
- protocol validation and verification,
- proof of concept implementation,
- pre-deployment testing.

The main function of the device emulator is the evaluation of the Network Management System as a black box. The emulators are installed on one or multiple PCs and scripted to create the network of configured size. Next the scripts to represent the particular test scenario are executed and the system operator verifies if the NMS behaved according to the scenario. Using the emulator one can verify a single unit of the code (e.g. the execution of the proper fault management report), but also validates the whole process of interaction between the device and the NMS.

The performance testing targets the verification of the maximum size of the network that the NMS is capable to manage and the measurement of the processing time required to execute common operations in different networks. The profiling and time measurement tools are used to monitor the state of the NMS and the emulation scripts are executed with different number of emulated devices. By programming different number of events in the network the operator is also able to verify if the system is responsive even in some disaster scenarios.

The device emulator implements the same protocol as it is used to communicate with the real devices. The same data elements and protocol state machine are used. If any changes are introduced to the protocol, they shall be first implemented within the NMS server and the emulator, prior to deploying on devices. When errors or incompatibility is found on the server side the implementation may be corrected, without performing complicated tests in real world.

The emulation can be also used as an effective tool for quick verification and presentation of new NMS functionality. Before a new feature is developed on both the server and the devices, it is possible to use the emulator to present the user potential benefits. The emulator can be also programmed to represent the topology and structure of a network to which the management system is about to be deployed, and the network engineers can certify what are the requirements of the computing power for servers on which the NMS will be installed.

### 3.3   Testing and Validation of the Emulator and Scenario Correctness

The device emulator must itself be well tested to correctly verify the NMS. It is executed by the comparison of the emulator with results, logs, and NMS behaviour in a small scale laboratory network and by the cross-validation of the protocol with the server. The validation of the logs also verifies, if the emulator is performing accordingly to the scenario defined, by verification if the events programmed in the emulator are noted in NMS logs. The proper implementation of the networking protocol in the emulator is tested by the cross checking the server implementation with the real devices – the protocol updates are implemented independently on the devices and in the emulator by different persons, when the server works with both implementation correctly there are no errors. The same protocol state machines are used first in the implementation of the emulator and next in device code, what simplifies the implementation and allows to find errors in the protocol design faster. During the emulation tests it is important to validate if the performance of the machines that execute the emulation does not limits the execution of the device emulators – this is done by measurements of the CPU, memory and network load.

## 4   Sample Resutls

The proposed emulation framework has been used in the performance tests of the Grid AirSync system, the network management system designed to manage the wireless networks in Smart Grid environment. For the tests a network consist of 24 000 managed devices was emulated. It was assumed that there are 3 000 WiMAX base stations and 21 000 CPE devices, there are 9 service flows and that the traffic statistics are transmitted to the server every 15 s. The devices were using the CAPWAP [2] protocol with custom extensions to communicate with the server.

In the first test the emulator has been used to measure the amount of management traffic generated by the devices. The results are presented on Figs. 2 and 3. It can be observed that the traffic initially grows, as the devices are connecting to the NMS and after few tens of minutes, when all devices are registered and provisioned, the traffic reaches a stedy point, containing mainly statistics transmitted from the devices. The summary management traffic is relatively small, as it reaches only few tens of Mb/s for such a large network.

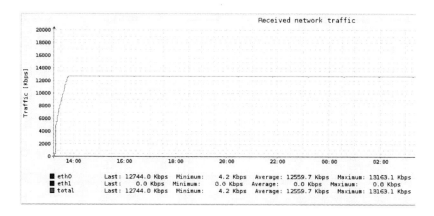

**Fig. 2.** Traffic received by the NMS for 24 000 devices with agent

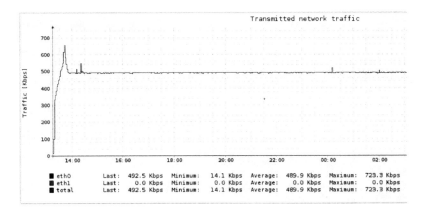

**Fig. 3.** Traffic transmitted by the NMS for 24 000 devices with agent

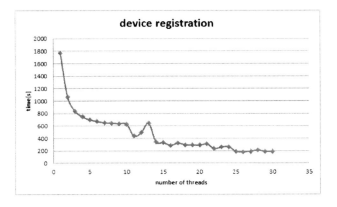

**Fig. 4.** Time to register 5 000 devices for different number of VM threads on the NMS server

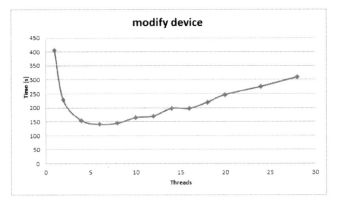

**Fig. 5.** Time to modify configuration of 5 000 devices for different number of VM threads on the NMS server

The multithreading performance analysis can be an example of how the emulator can be used to select optimal configuration of the network management system. The device registration time and status updates were measured for the emulated network, with different number of VM thread configured. The results are shown on Fig. 4 and Fig. 5.

# 5   Conclusions

The described emulation framework has been successfully used to test and evaluate the performance of the sample network management system. It provides a very effective and flexible tool to analyze the maximum number of devices the system is able to manage, profile it and to execute everyday tests of the system. Also some emerging similarity between emulator and NMS server architecture has been observed. Both parts must provide efficient execution of concurrent state machines. Hence emulator served as an incubator of ideas that had been later implemented in architecture of NMS server what resulted in improvments of whole system performance and reliability.

**Acknowledgement.** This work was partially supported by the "Grid AirSync – The Network Management System for Intelligent Grid" project subsidized by Polish Agency of Enterprise Development by the grant no POIG.01.04.00-24-022/11.

# References

1. McCloghrie, K., Schoenwaelder, J., Perkins, D. (eds.): Structure of Management Information Version 2 (SMIv2)., http://tools.ietf.org/html/rfc2578
2. Stanley, D., Montemurro, M.P., Calhoun, P.R.: Control and Provisioning of Wireless Access Points (CAPWAP) Protocol Binding for IEEE 802.11., http://tools.ietf.org/html/rfc5416
3. Bierman, A., Enns, R., Bjorklund, M., Schoenwaelder, J.: Network Configuration Protocol (NETCONF), http://tools.ietf.org/html/rfc6241
4. CPE WAN Management Protocol. TR-069 Amendment 4. Broadband Forum (July 2011) (retrieved February 16, 2012)
5. Lee, D., Netravali, A.N., Sabnani, K.K., Sugla, B., John, A.: Passive testing and applications to network management. In: Proceedings of International Conference on Network Protocols, pp. 113–122 (1997)
6. Pras, A., Drevers, T., van de Meent, R., Quartel, D.: Comparing the performance of SNMP and web services-based management. IEEE Transactions on Network and Service Management 1(2), 72–82 (2004)
7. Franco, T.F., Lima, W.Q., Silvestrin, G., Pereira, R.C., Almeida, M.J., Tarouco, L.M., Granville, L.G., Beller, A., Jamhour, E., Fonseca, M.: Substituting COPS-PR: an evaluation of NETCONF and SOAP for policy provisioning. In: Seventh IEEE International Workshop on Policies for Distributed Systems and Networks, Policy 2006, pp. 195–204 (2006)

8. Müller, M.: Performance evaluation of a failure detector using SNMP. Semester project report, École Polytechnique Fédérale de Lausanne, Lausanne, Switzerland (2004)
9. Tanenbaum, A.S., van Steen, M.: Distributed Systems. Addison Wesley (2004)
10. Zhang, X., Freschl, J.L., Schopf, J.M.: A performance study of monitoring and information services for distributed systems. In: Proceedings 12th IEEE International Symposium on High Performance Distributed Computing, pp. 270–281 (2003)
11. Jogalekar, P., Woodside, M.: Evaluating the scalability of distributed systems. IEEE Transactions on Parallel and Distributed Systems 11(6), 589–603 (2000)
12. Kojo, M., Gurtov, A., Manner, J., Sarolahti, P., Alanko, T., Raatikainen, K.: Seawind: a wireless network emulator. In: Proceedings of 11th GI/ITG Conference on Measuring, Modelling and Evaluation of Computer and Communication Systems, pp. 151–166 (2001)
13. Zheng, P., Ni, L.M.: Empower: A cluster architecture supporting network emulation. IEEE Transactions on Parallel and Distributed Systems 15(7), 617–629 (2004)
14. Maier, S.D.: Scalable computer network emulation using node virtualization and resource monitoring (2011)
15. Avallone, S., Guadagno, S., Emma, D., Pescapè, A., Ventre, G.: D-ITG distributed internet traffic generator. In: Proceedings of First International Conference on the Quantitative Evaluation of Systems, QEST 2004, pp. 316–317 (2004)
16. Tirumala, A., Qin, F., Dugan, J., Ferguson, J., Gibbs, K.: Iperf: The TCP/UDP bandwidth measurement tool (2005)
17. Zeigler, B.P., Praehofer, H., Kim, T.G.: Theory of modeling and simulation: Integrating discrete event and continuous complex dynamic systems. Academic Pr. (2000)
18. Armstrong, J., Virding, R., Wikstrom, C., Williams, M.: Concurrent Programming in ERLANG, 2nd edn. (1996)

# Resource Management in Grid Systems

Dariusz Czerwinski[1], Slawomir Przylucki[1], and Przemyslaw Matejczuk[2]

[1] Lublin University of Technology, 38A Nadbystrzycka Str, 20-618 Lublin, Poland
`d.czerwinski@pollub.pl`
[2] TP S.A. Telecommunication Company, Chodzki 10 a, 20-093 Lublin, Poland

**Abstract.** The management of resources in the grid networks becomes complex task. The reason for this is that grids are heterogeneous, distributed and owned by different organizations environments. Each solution has own policies of resource management and different cost models. This paper contains a discussion of the issues related to resource management in grid systems on the basis of XtreemOS grid. The purpose of this study is the analysis of cost model of network management in distributed environments. The results of performed tests demonstrate how the Resource Management System (RMS) behaves under different operating conditions.

**Keywords:** computer networks, grid networks, XtreemOS, resource management.

## 1 Introduction

The aim of a grid technology is to create a simple, but having great capabilities, virtual computer. Resource management systems are crucial to the operation of distributed processing environments, built on the grid model. Based on the distributed components, which are shared among different units, these systems require the use of the management mechanisms of included items.

Currently, the term grid network is determined as the underlying technologies to connect heterogeneous systems sharing resources in order to create a distributed supercomputer. In terms of user-visible the grid is a single machine with a huge computing power and memory, accessible from almost every place on earth thanks to the Internet [1]. Ian Foster has defined three basic features of grid environment [2]. This environment should:

- coordinate and consolidate resources and users, who are ruled by different administrative units, businesses, community policies and are not centralized,
- use open, standard protocols and interfaces that support authentication, authorization, finding resources, and access to them,
- provide advanced QoS, ensuring service levels for parameters such as response time, throughput, availability, security, allocation of resources for the group of complex requirements.

A. Kwiecień, P. Gaj, and P. Stera (Eds.): CN 2012, CCIS 291, pp. 101–110, 2012.

## 1.1   Structure of a Grid System

Architecture of a grid environment is organized in the form of layers. That allows associating components with similar characteristics and using the potential of lower layers [1,3,4]. Physical layer (called fabric layer) provides access to the resources used by the grid environment. These are computational resources, hard drives, network infrastructure, etc. Connectivity layer defines a way of communication and a set of protocols for network transactions. Resource layer allows users of grid environment to communicate with distributed services and resources located within grid network nodes. Collective layer contains protocols and services, incorporating interactions between sets of resources. The application layer protocols use the lower layers to provide tools that give users the ability to easily and efficiently carry out the tasks.

## 1.2   Resources in a Grid System

A resource can be called every element of a computer system, used to carry out the tasks for which the computing environment has been designed. There are three groups of resources used in computing environments, depending on their destination [3]:

- computing resources such as the computing power of the environment, allow for the implementation of mathematical tasks,
- data storage resources, the space available for the collection and storage of data, enabling the implementation of disk operations,
- network resources, such as communication bandwidth, allowing for efficient communication between processes running on different physical devices.

Applications implemented inside a grid environment utilize the available resources for the execution of tasks. Diversity of the application requirements and a multitude of users implies the existence within grid environment the Resource Management System (RMS). List of requirements to be met by the ideal RMS system, can be defined as a list of services that should have a resource management system. A set of such services was determined by a group of GSA-RG (Grid Scheduling Architecture Research Group) as part of an GGF organization (Global Grid Forum) and it can be described as: resources discovery, access to the information on available resources, monitoring the status of tasks and the environment, mediation and planning, support for the reservation and SLA (Service Level Agrement) contracts, managing jobs implementation, accounting and reporting [5]. Grid systems are gathering a large number of heterogeneous resources belonging to different entities or organizations and the choice of a suitable model for Resource Management plays an important role. There are three main models of Resource Management Systems in the Grid networks [6]:

- hierarchical model,
- abstract owner model,
- computational market or economy model.

**Table 1.** Models of Resource Management Systems for Grid environments

| Model name | Remarks | Implementation |
|---|---|---|
| Hierarchical | Contains the most modern RMS model architecture | Globus, Legion, Ninf [7], NetSolve/GridSolve [8] |
| Abstract owner | Includes model of commissioning and delivering the resources, which ignores the physical organization of existing resources, focusing on long-term goals. | Is expected |
| Computational Market/Economy Model | Includes model of discovering and planning resources. Model can coexist with current systems, captures the essence of both the hierarchical and abstract owner. | Nimrod/G [9], JaWS [10], Mariposa [11], JavaMarket [12] |

Grid systems architectural models affect the construction and planning of management systems, because they define the way in which user requests are mapped into a set of resources. Table 1 shows models with examples of systems whose architecture meets the specified models.

## 2    Tests of the Grid Network

The XtreemOS [13] grid network has been used as a platform for evaluating the performance of Resource Management System mechanisms for selected sets of jobs and their submission. In order to implement the experimental task, the testing methodology was developed and proposed. The evaluation criteria and set up test environment, in which the study was carried out, were worked out. For the purpose of carrying out the measurements as a set of tools was created. These tools allowed for automate the process of testing, data collection and pre-treatment for further analysis of the results.

### 2.1    Tests Carriage

Studies were divided into three independent experiments, carried out under the five test scenarios. Each of the experiments, performs the verification of system properties in the case of the commission by the user a particular operation. Operations, characteristic for the grid environment, were: resource reservation requests, requests for execution of job with specified requirements regarding use of grid resources, obtaining information on commissioned and launched by the user jobs. Each scenario was repeated several times in order to avoid random errors and count average time. Implementation of the XtreemOS system was realized on a dedicated hardware and software infrastructure, which consisted of a set of desktop PCs and network infrastructure linking them. Test scenarios studies have not assumed a geographically dispersed environment evaluation, focusing on aspects of the individual local environment, which may be involved in the construction of a large, geographically distributed Grid environment. For this reason, the test environment was implemented in a LAN network that provides fast communication and allows to study the characteristics of the management system, ignoring the influence of network parameters (delay, packets loss etc).

Hardware configuration of PC units used for tests was as follows:

– CPU: Intel® Pentium® E5400 (2M Cache, 2.70 GHz, 800 MHz FSB),
– RAM: 2 GB,
– HDD: WDC WD3200AAJS-60M0A1 ATA Device (298 GB, IDE),
– NIC: Realtek RTL8102E/RTL8103E Family PCI-E Fast Ethernet.

Testbed consists of core node and four resource nodes. Since the default configuration of the core node in XTreemOS forced the node to work as also as resource node, the grid environment was composed of five resource machines, with the stipulation that one of them also served as a core node. All nodes were defined as a virtual machines run under VMWarePlayer hypervisor and the host operating system was Ubuntu 10.04 LTS x86.

Virtual machines configuration was as follows:

– CPU: host CPU,
– RAM: 512 MB of host RAM,
– HDD: 20 GB of host disk space,
– NIC: host network interface (bridged mode).

## 2.2   Tests Scenarios

Resource and tasks management systems have been tested by determining the average time of execution of the job ordered by the user. Each of the experiments verified the behavior of the management system in one operation: experiment No. 1 used booking resource request without performing in any job under it; experiment No. 2 used a request to perform a specific job by the grid environment, leaving the issue of resource reservation to Resource Management System of XTreemOS; experiment No. 3 used to obtain information about the running jobs, which have been commissioned by the user (Fig. 1).

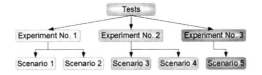

**Fig. 1.** Tests methodology

The study focused primarily on differences in the waiting times of obtaining information on running jobs in dependency on:

- the number of users (one user per specified number of nodes), requesting an operation at a time – scenarios No. 2 and 4,
- momentary system load by already made reservations or the executing jobs – scenarios No. 1 and 3,
- the physical distribution of the jobs on the nodes of grid system in the case of the jobs requested and running by the user – scenario No. 5.

To obtain deliberate objective five scenarios were used as follows:

- Scenario No. 1 was based on running a script that in a sequential execution commissioned a large number of reservations and for each measuring the time of their service, since the launch of a request to obtain the ID of the reservation, was made.
- Scenario No. 2 consisted of verification of the preservation of the environment, where the sequential reservation commissioning was opened from a larger number of nodes than in scenario 1, i.e. two, three, four and five at a time.
- Scenario No. 3 was based on sequential commissioning of the single-user specified type job execution. Obtained times allowed to define the characteristics of the service time freelance requests sequentially added by a single user, depending on the amount of other jobs carried out already in the system.
- Scenario No. 4 was based on sequential commissioning of jobs execution and measuring the time from the request to service until receiving the ID of job from a larger number of nodes. Obtained times helped determine the relationship between the time of a job requesting and the number of nodes.
- Scenario No. 5 determined the relationship between time to obtain information about the started by the user jobs, number of jobs running in the environment and the number of nodes on which the jobs had been started.

At the time of testing the performance of applications running within a distributed environment has not been studied, but the performance and scalability of the management system was under investigation. Therefore, applications that were used to create test scenarios in a very small extent used the computing performance of nodes.

Two kinds of jobs have been defined and used during all experiments: "easy" is a job which is not computationally intensive for CPU (about 6% of CPU utilization), and second called "difficult" which is computationally intensive for CPU (about 90% of CPU utilization).

## 3   Tests Results

Evaluation of the resource management system performance in XtreemOS system requires the determination of the characteristics of the basic operations in the grid environment. Resource reservation time measurements provide a benchmark for measuring the execution times of other operations. Figure 2 shows the results obtained in scenario 1. This scenario included the characteristics of the time needed to service the requests of a resource reservation. The average time of booking for 600 samples was approximately 2.39 s, while the minimum and maximum time to service the request amounted to: 1.72 s and 4.25 s.

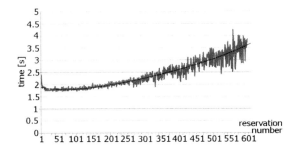

**Fig. 2.** Service request time versus reservation number

The results of comparison of scenario 2 to scenario 1 presents the Fig. 3. Characteristics of the response times obtained in scenario No. 2 deviates significantly from the characteristic of reservation time sequence shown in Fig. 2. Measuring times gain greater differences in the comparison of subsequent measurements. This is due to heavy load of reservation service manager running as a service on the main host. While adding more reservation requests, handling times lengthen along with increasing effort of the environment on the organization already set up reservations and finding free resources to use terms consistent with the description of the task.

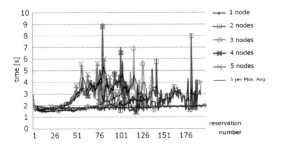

**Fig. 3.** Time versus parallel reservations

The aim of experiment 2 was evaluation of the job ordering process. Figure 4 shows the characteristics of the handle time for the "easy" job (job which is not computationally intensive for CPU). Time to handle the request to activate the operation stabilizes at the value of about 0.6 s. It should be noticed here, that the actual job start time may be longer, because the experiment measures the time needed for communication services XTreemOS API (XOSAGA) of the user and the jobs service manager. For this reason, the time needed for the resource reservation is not verified, as confirmed by comparison with the results contained in Fig. 2. The job itself takes a very short time, therefore the system load during the sequential addition of jobs does not increase. The average runtime of the job for 1100 iterations is approximately 0.59 s, while the minimum and maximum values are equal respectively: 0.35 s and 1.17 s.

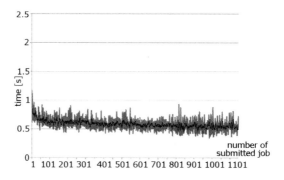

**Fig. 4.** Response time of sequential addition of "easy" job for one node

The time variation of adding the "easy" job depending on the number of nodes shows Fig. 5. As in the case of one node (see Fig. 4) request handling time is the established level, however this level is respectively higher for larger number of nodes requiring completion of the job.

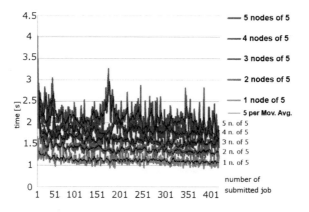

**Fig. 5.** Response time of sequential addition of "easy" job for many nodes

Figure 6 illustrates the execution times of the "difficult" job for one and two nodes. While Fig. 7 shows the times of execution of the "difficult" job for three, four and five nodes.

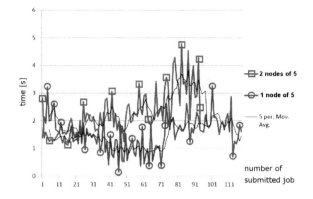

**Fig. 6.** Response time of sequential addition of "difficult" job for one and two nodes

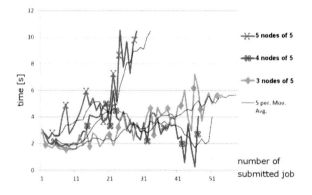

**Fig. 7.** Response time of sequential addition of "difficult" job for three, four and five nodes

In the case of "difficult" job runtimes lengthen with subsequent job orders. This is due to occupation of the computing node resources by running jobs and increase the management services running effort by the service manager and jobs manager. In the case of the "difficult" Response times vary much higher between successive samples, it can be observed especially for the final runtimes. This was due to occupation of the increasing number of memory by running jobs, which resulted in run of the daemon kswapd on nodes. Swap deamon reorganized the main memory, transferring the part of the process to the swap partition.

The purpose of the experiment No. 3 was to determine the response time characteristics of the environment on demand for supplying information of the job started by the user. All runtimes are obtained under scenario No. 5 and present the impact of the amount and placement of jobs on the speed of collecting of the information about the tasks requested by the user. Fig. 8 shows the results of measurements comparing the times of obtaining the information about the job running in the grid environment, depending on the number of nodes and the amount of active jobs.

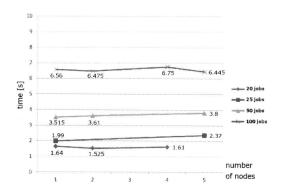

**Fig. 8.** Time of obtaining of the job status versus the number of nodes

Figure 9 shows the results of research on waiting times of jobs status, if the user orders the collection of jobs with the increasing number of nodes, and then queries the grid environment for the jobs status. It should be noted that in the case of twofold ordered jobs, the runtime of the request increases with the number of jobs, but is not exceeding twice the previous measurement.

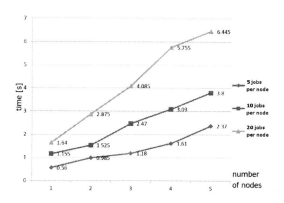

**Fig. 9.** Waiting times of jobs status versus the number of nodes

## 4   Conclusions

In this paper the authors presented the results of the test of RMS system performed in grid network build on the basis of XtreemOS. Carried out tests demonstrated how the XTreemOS RMS, which is based on the economy model, behaves under different operating conditions. Presented results, demonstrate how much more effective is commissioning of the tasks necessary to complete the job submitted by the user, if the job is carried on more nodes.

The results of the proposed scenarios shows that a grid environment presents great flexibility in terms of scalability. Growth of the tested grid environment does not decrease jobs runtimes rapidly. This suggests that, even for large grid environments, the influence of number of organization and management services on available computing power will not affect significantly for the viability of such solutions. Comparable results of waiting times of obtaining information on running jobs commissioned by the user (Fig. 9) show that the load balancing algorithms implemented in the XtreemOS system meet requirements. Although the individual response times lengthen with the growing number of nodes that request the resources at the same time, the total time of operations for the entire jobs set is shorter. This leads to the conclusion that in order to increase the efficiency of solving the job, the more efficient is to decompose a complex job and then transfer it to the higher number of nodes in the grid environment.

On the other hand, from the viewpoint of a distributed environment administrator, if the solution of the problem posed to grid network is less scattered on individual users, then the environmental impact of requests handling is lower.

# References

1. Foster, I., Kesselman, C., Tuecke, S.: The Anatomy of the Grid. Enabling Scalable Virtual Organizations. Int. Journal of Supercomputer Applications 15 (2001)
2. Foster, I.: What is the Grid? A Three Point Checklist. GRIDtoday (6) (2002)
3. Foster, I., Kesselman, C.: The Grid: Blueprint for a New Computing Infrastructure. Morgan-Kaufmann (1998)
4. Płaczek, W.: Grid – Nowa Rewolucja Informatyczna? (April 2011),
   http://th-www.if.uj.edu.pl/~placzek/referaty/grid.pdf
5. Project homepage: The Global Grid Forum. Grid Scheduling Architecture Research Group, https://forge.gridforum.org/sf/projects/gsa-rg
6. Buyya, R., Chapin, S., DiNucci, D.: Architectural Models for Resource Management in the Grid. In: Buyya, R., Baker, M. (eds.) GRID 2000. LNCS, vol. 1971, pp. 18–35. Springer, Heidelberg (2000)
7. Ninf project homepage: A Global Computing Infrastructure,
   http://ninf.apgrid.org/
8. Hurault, A., YarKhan, A.: Intelligent Service Trading and Brokering for Distributed Network Services in GridSolve. In: Palma, J.M.L.M., Daydé, M., Marques, O., Lopes, J.C. (eds.) VECPAR 2010. LNCS, vol. 6449, pp. 340–351. Springer, Heidelberg (2011)
9. Buyya, R., Abramson, D., Giddy, J.: Nimrod/G: An Architecture for a Resource Management and Scheduling System in a Global Computational Grid. In: 4th Int. Conf. on High Performance Computing in Asia-Pacific Region, HPC Asia (2000)
10. Lalis, S., Karipidis, A.: An Open Market-Based Framework for Distributed Computing over the Internet. Springer Verlag Press, Germany (2000)
11. Stonebraker, M., et al.: An Economic Paradigm for Query Processing and Data Migration in Mariposa. In: 3rd International Conference on Parallel and Distributed Information Systems, Austin, Texas. IEEE CS Press, USA (1994)
12. Al-Ali, R., et al.: QoS Support for High-Performance Scientific Grid Applications. CS518 Project Paper, Cluster Computing and tge Grid. In: IEEE International Symposium on Grid Computing (2004)
13. Project homepage: The XTreemOS, http://www.xtreemos.eu/

# Spatio-temporal Web Performance Forecasting with Sequential Gaussian Simulation Method

Leszek Borzemski and Anna Kamińska-Chuchmała

Institute of Informatics, Wroclaw Uniwersity of Technology,
Wroclaw, Poland
{leszek.borzemski,anna.kaminska-chuchmala}@pwr.wroc.pl

**Abstract.** In this article, we describe a novel proposal for the spatio-temporal Web performance forecasting using one of the geostatistical methods – Sequential Gaussian Simulation (SGS). Necessary data were obtained from Multiagent Internet Measurement System – MWING, which monitored web transactions issued by MWING's agent located in Las Vegas, US and targeting web servers in Europe. Data contains the measurements, which were taken every day at the same time: at 06:00 a.m., 12:00 a.m. and 6:00 p.m. during the period of May, 2009. First, the preliminary analysis of measurement data was conducted. Next, the structural analysis, which includes directional variogram approximated with the theoretical model, was performed. Subsequently, the spatial forecast (from one week in advance) of total time of downloading data from Web servers was calculated. The analysis of server activity on a particular weekday for a period of a few weeks in selected time intervals and considered forecasted errors was performed. Finally, the results of forecast were analyzed in detail, followed by the determination of subsequent research directions to improve Web performance forecasts.

**Keywords:** web performance forecast, spatio-temporal data mining, geostatistics, Sequential Gaussian Simulation.

## 1 Introduction

Web users expect good performance of services. The prediction of how particular servers ("seen" from a given client) could perform is crucial for effective Web performance, especially when services can be mirrored on many servers. Such information could be obtained with spatio-temporal predictions of performance, which allow to forecast required parameters on desired time basis. In this paper, the use of the Sequential Gaussian Simulation geostatistical method was proposed to predict user-perceived Web performance while downloading resources from Web servers by means of HTTP protocol. This method allows to study Web performance phenomenon that varies in space and time.

In our reaserch, we propose a novel approach to use the Sequential Gaussian Simulation geostatistical method to forecast spatio-temporal Web performance. This method considers the existence of a functional relationship between what

A. Kwiecień, P. Gaj, and P. Stera (Eds.): CN 2012, CCIS 291, pp. 111–119, 2012.

happens at one geographical location and what happens elsewhere. SGS method uses kriging, geostatistical technique, to interpolate a studied parameter; namely, available observations of nearby locations are used to calculate values of a studied parameter at unmonitored locations. Thus, this method gives as the result, not only forecasted performance for studied Web servers, but also for their neighborhoods. Of course, the forecasting results for locations where no server occurs, are just simulated, not factual. Nevertheless, it could be helpful, for example, to optimize network traffic or to suggest adding new servers to analyzed locations. This method was also chosen, because it requires a minimum number of input data to perform forecasting. SGS method have already been successfully used by the authors to forecast loads in power transmission and distribution networks [1].

## 2    Sequential Gaussian Simulation Method

The first step in geostatistical simulation is modeling of the process variables; next step is the simulation of these variables using an elementary grid. The bases of sequential conditional simulation are Bayes' theorem and Monte Carlo (stochastic) simulation [2]. The simulation is said to be conditional if the realizations take into account the data at their $n$ sample locations:

$$Z^\ell (u_\alpha) = z (u_\alpha), \forall \ell, \alpha = 1, \ldots, n \ . \tag{1}$$

A random function is a set of random variables over the study area. Let $\{Z (u'_j), j = 1, \ldots, N\}$ be a set of random variables at the $N$ grid cells with location $u'_j$. The objective of a sequential conditional simulation is to generate several joint realizations of the $N$ random variables: $\{z^\ell (u'_j), j = 1, \ldots, N\}$, $\ell = 1, \ldots, L$, conditional to the data set $\{z^\ell (u_\alpha), \alpha = 1, \ldots, n\}$. This gives an $N$-point conditional cumulative distribution function that models the joint uncertainty at $N$ locations $u'_j$:

$$F (u'_1, \ldots, u'_N; z_1, \ldots, z_N| (n)) = P \{Z (u'_1) \leq z_1, \ldots, Z (u'_N) \leq z_N| (n)\} \ . \tag{2}$$

By recursively applying Bayes' theorem, an $N$-point conditional cumulative distribution function is formulated as a product of $N$ one-point conditional cumulative distribution functions:

$$F (u'_1, \ldots, u'_N; z_1, \ldots, z_N| (n)) = \\ F (u'_N; z_N| (n + N - 1)) \times F (u'_{N-1}; z_{N-1}| (n + N - 2)) \times \ldots \times F (u'_1; z_1| (n)). \tag{3}$$

This decomposition allows to make a realization in $N$ successive or sequential steps. At the first location $u'_1$, the cumulative distribution function is modeled conditionally to the $n$ original sample data:

$$F (u'_1; z| (n)) = P \{Z (u'_1) \leq z| (n)\} \ . \tag{4}$$

Draw from the conditional cumulative distribution function an estimate, $z^\ell (u_1)$, which becomes part of the conditioning data for all subsequent drawings for the first realization. This process is repeated until all of the $N$ nodes have simulated values. This is one realization; the second would start with the original conditioning data and visit the $N$ nodes in a different sequence. The sequential simulation algorithm requires the determination of a conditional cumulative distribution function at each simulated node. In practice, the conditional cumulative distribution function is determined by the mean and variance from either simple kriging or ordinary kriging using either a multi-Gaussian or an indicator random function.

The algorithm for Sequential Gaussian Simulation [3]:

1. Define a random path through all nodes that are to be simulated, where each node is included only once.
2. Use simple kriging or ordinary kriging to determine the mean and variance of the Gaussian conditional cumulative distribution function at a node. Retain a specified number of neighboring data to be used as conditioning data. Both previously simulated node values and original data are included.
3. Draw randomly from the conditional cumulative distribution function and assign that value to a node being simulated.
4. Repeat steps 2 and 3 for all simulation locations.
5. Transform back the simulated normal values into the values for the original attribute.
6. Repeat steps 1–5 for multiple realizations.

Across the $L$ realizations, the means, variances, frequencies, etc., can be spatially assessed. The appropriateness of the Gaussian distribution must be tested before simulation, often calling for a prior transformation of original data into a new data set with a standard normal cumulative distribution function. After carrying out a simulation, it is necessary to calculate Gaussian anamorphosis one more time in order to return to the original (raw) values. SGS methods and their applications are described in detail in [4,5,6,7], and [8].

## 3    Preliminary Data Analysis

The database was created on the basis of active measurements made by MWING system – the Internet measurement infrastructure developed in our Institute [9,10], and [11]. The database used in this paper includes the measurements collected by the HTTP agent located in Las Vegas for target servers located in Europe. The measurements encompassed the period between 1st and 31th May 2009 and they were taken every day at the same time: at 06:00 a.m., 12:00 a.m., and 6:00 p.m. The input database, necessary to make a forecast, contains the information about a server's (node's) geographical location which the Las Vegas agent targeted, $Z$ – the web perfomance index which is the total downloading time of rfc1945.txt file, which is the main performance index studied here, and the timestamp of taking a measurement.

Basic statistics of web performance for considered servers are presented in Table 1. Minimum and maximum values show the span of data values particularly

114     L. Borzemski and A. Kamińska-Chuchmała

**Table 1.** Elementary statistics of download times from Web servers between 1–31.05.2009

| Statistical parameters | 6:00 a.m | 12:00 a.m. | 6:00 p.m. |
|---|---|---|---|
| Minimum value $Z_{\min}$ [s] | 0.28 | 0.29 | 0.29 |
| Maximum value $Z_{\max}$ [s] | 16.57 | 16.44 | 31.87 |
| Average value $Z$ [s] | 2.58 | 2.85 | 2.80 |
| Standard deviation S [s] | 2.96 | 3.21 | 3.31 |
| Variability coefficient V [%] | 114.73 | 96.14 | 118.21 |
| Skewness coefficient G | 3.33 | 2.74 | 3.65 |
| Kurtosis coefficient K | 14.15 | 10.23 | 21.29 |

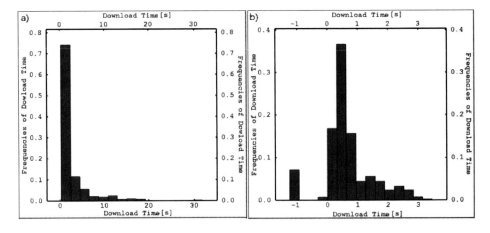

**Fig. 1.** The histogram of download time distribution from Web servers in May before (a) and after (b) the calculation of logarithms respectively

for 06:00 p.m. Additionally, there is a large value of standard deviation and variability coefficient which proves the changeability of the examined process. High values of both skewness (more than 3) and kurtosis coefficients indicates big right side asymmetry of performance distribution. Therefore, for data obtained for 06:00 a.m. and for 06:00 p.m. logarithmic values were calculated.

For analyzing changability of input data, we present histogram of download times before and after the calculation of logarithms on Figs. 1(a) and 1(b) respecitively. The histogram of web performance index distribution on the Web servers, presented on Fig. 1, is asymmetric, single-wing, and positively skewed. After calculating logarithms, the histogram has a shape slightly similar to symmetric distrubution.

## 4   Structural Analysis of Data

The first step after the preliminary data analysis is Gaussian anamorphosis calculation [12]. During the calculations of Gaussian transformation frequency,

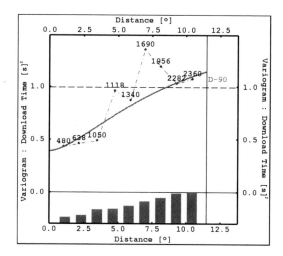

**Fig. 2.** Directional variogram along time axis for download times, approximated with the theoretical model of nuggets effect and $K$-Bessel

the inversion model was used. The number of adopted Hermite polynomials in anamorphosis was equal to 100.

The next step is modeling a theoretical variogram function. During the variogram model approximation, the nuggets effect function was used to consider Web performance at 06:00 a.m. A directional variogram was calculated along the time axis (for 90° direction). The distance class for this variogram was 1.16°. Figure 2 presents the directional variogram approximated by the theoretical model of the nuggets effect and K Bessel. The variogram function indicates a gentle rising trend.

The next two directional variograms of Web performance, measured at 12:00 a.m. and 6:00 p.m., were also approximated by the theoretical model of the nuggets effect and $K$-Bessel.

## 5  Spatio-temporal Forecast of Web Servers Performance Calculated with the Sequential Gaussian Simulation Method

The forecasting models – the above discussed variogram models and Gaussian anamorphosis – were used to predict the total download time of a given resource from the Internet. In the simulation, the sequential neighbourhood type was adopted for 6:00 a.m., and 6:00 p.m. Moving neighbourhood type was used for 12:00 a.m. The forecast of the download time was determined on the basis of 100 simulation realizations. In the simulation, the punctual type was used. Three-dimensional forecast was calculated with a one-week time advance, i.e. it encompassed the period between 1st and 7th June, 2009. The Table 2 presents

**Table 2.** Global statistics of forecasted Web download times with one-week time advance, calculated with SGS method

| Geostatistic parameter | Average value $Z$ [s] | Minimum value $Z_{min}$ [s] | Maximum value $Z_{max}$ [s] | Variance $S^2$ [s]$^2$ | Standard deviation [s] | Variability coefficient $V$ [%] |
|---|---|---|---|---|---|---|
| Mean forecasted value $Z$, for 6:00 a.m. | 1.86 | 0.28 | 16.57 | 0.27 | 0.52 | 27.96 |
| Mean forecasted value $Z$, for 12:00 a.m. | 2.97 | 0.29 | 11.84 | 2.45 | 1.57 | 52.86 |
| Mean forecasted value $Z$, for 6:00 p.m. | 1.76 | 0.29 | 26.13 | 0.60 | 0.77 | 43.75 |

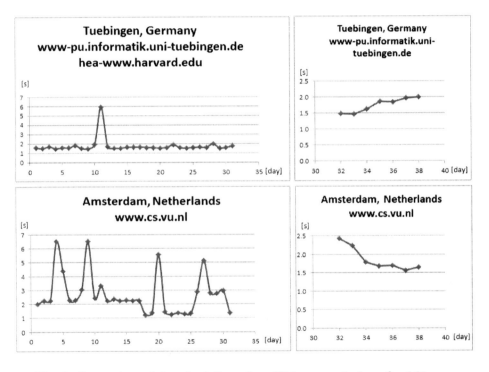

**Fig. 3.** Comparison of download times from Web servers in June for 6:00 a.m.

global statistics of the average forecasted download times for all Web servers in one week advance for 6:00 a.m., 12:00 a.m. and 6:00 p.m.

Because for the measurements taken for 12:00 a.m. no logarithms were calculated, the obtained results may be characterized not only by high variance coefficient, but also by variance, and standard deviation. However for measurements taken for 6:00 a.m. and 6:00 p.m., obtained results have low variance coefficient, so no large dispersion of values occurs. It also means that for such data we can forecast web performance with better accuracy.

**Table 3.** Sample results of performance forecasting for Web server located in Osnabrueck in Germany in June 2009

| Day of forecast | 01.06 | 02.06 | 03.06 | 04.06 | 05.06 | 06.06 | 07.06 |
|---|---|---|---|---|---|---|---|
| Real download time, for 06:00 a.m. [s] | 1.69 | 1.73 | 1.72 | 1.73 | 1.70 | 1.74 | 1.69 |
| Forecasted download time, for 06:00 a.m. [s] | 1.80 | 2.06 | 2.06 | 1.96 | 2.00 | 1.86 | 1.89 |
| Real download time, for 12:00 a.m. [s] | 2.12 | 1.70 | 1.73 | 1.69 | 1.75 | 5.93 | 1.94 |
| Forecasted download time, for 12:00 a.m. [s] | 2.32 | 2.73 | 2.65 | 3.10 | 3.14 | 2.65 | 2.71 |
| Real download time, for 06:00 p.m. [s] | 1.69 | 1.73 | 1.72 | 1.73 | 1.70 | 1.74 | 1.70 |
| Forecasted download time, for 06:00 p.m. [s] | 1.60 | 1.76 | 1.86 | 1.92 | 1.87 | 1.67 | 1.78 |

Figure 3 presents two web servers located in Germany and Netherlands; for both servers, not only historical (from 1 to 31 May, 2009), but also forecasted (from 1st to 7th June, 2009, which correspond to 32–38 day in figure) data are presented in the charts. As we see in the charts with historical data, server in Tuebingen had a very moderate values with only one peak that happened on 11th day. The server in Amsterdam, as opposed to that of Germany, had differential download times having unpredictable peaks. Taking into account such variation of measurement data, it is very hard to make an accurate forecast. For the sake of comparison, averaged mean forecasted error for one-week forecast for the server in Tuebingen equals 14% and for the server in Amsterdam equals 39%.

For analyzing web performance in different hours, we presented in Table 3 real and forecasted download times for 6:00 a.m., 12:00 a.m., 06:00 p.m. The measurements were taken between 1st and 7th June, 2009, for one Web server in Osnabrueck, Germany. The highest changeability of download time is characteristic for 12:00 a.m. All forecasted values are slightly overestimeted due to dispersion of data. Mean forecasted error *ex post* for one week for 6:00 a.m, 12:00 a.m., and 06:00 p.m. equals 13.41%, 54.20%, and 6.39% respectively. Despite the fact that skewness coefficient for the measurements taken for 12 a.m. is slightly less than 3, obtained data should be expressed as logarithms; this will make forecast error much smaller.

The final effect of the forecast calculations is presented in Fig. 4. This is a raster map for the 5th day of prognosis (05.06.2009), which presents the download time from the Web serevers. The size of the cross corresponds to the download time from a given Web server (the download time is given in seconds). Geostatistics methods could give information about performance not only for considered servers, but for a whole considered area. In this map, a server with the largest download time is located in Budapest, Hungary; what is more, this difference is readily visible when compared to other servers.

## 6 Summary

In this research, an approach for predicting Web performance by the innovative application of the Sequential Gaussian Simaltion method was proposed. On the basis of conducted research in this and earlier ([13,12], and [14]) papers, the

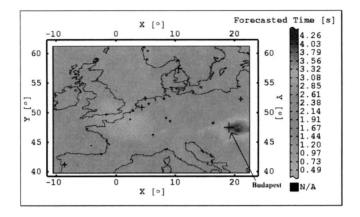

**Fig. 4.** Sample raster map of download time values from the Internet on 05.06.2009 at 06:00 p.m.

authors claim that there is a need to work on the improvement of forecast accuracy. This could be achieved by analyzing Web peroformance using various measurements such as data, timestamp of data, and prediction horizon lenghts.

# References

1. Kamińska-Chuchmała, A., Wilczyński, A.: Spatial electric load forecasting in transmission networks with Sequential Gaussian Simulation method. Rynek Energii 1(92), 35–40 (2011) (in Polish)
2. Tarantola, A.: Inverse Problem Theory and Methods for Model Parameter Estimation. In: SCIAM Society for Industrial and Applied Mathematics, Philadelphia (2005)
3. King, S.L.: Sequential Gaussian simulation vs. simulated annealing for locating pockets of high-value commercial trees in Pennsylvania. Annals of Operations Research 95, 177–203 (2000)
4. Bohling, G.: Stochastic simulation and reservoir modeling workflow. C&PE, 940 (2005)
5. Gertner, G.: Comparison of Computationally Intensive Spatial Statistical Methods for Generating Inputs for Spatially Explicit Error Budgets. In: IUFRO 4.11 Conference, Greenwich (2001)
6. Hansen, T.M., Journel, A.G., Tarantola, A., Mosegaard, K.: Linear Inverse Gaussian theory and geostatistics. Geophysics 71(6) (November-December 2006)
7. Hicks, P.J.: Unconditional sequential Gaussian simulation for 3-D flow in a heterogeneous core. Journal of Petroleum Science and Engineering 16, 209–219 (1996)
8. Wang, G., Gertner, G.Z., Parysow, P., Anderson, A.B.: Spatial prediction and uncertainty analysis of topographical factors for the Revised Universal Soil Loss Equation (RUSLE). Journal of Soil and Water Conservation (3rd Quarter 2000)
9. Borzemski, L.: The experimental design for data mining to discover web performance issues in a Wide Area Network. Cybernetics and Systems: An International Journal 41, 31–45 (2010)

10. Borzemski, L., Cichocki, Ł., Fraś, M., Kliber, M., Nowak, Z.: MWING: A Multiagent System for Web Site Measurements. In: Nguyen, N.T., Grzech, A., Howlett, R.J., Jain, L.C. (eds.) KES-AMSTA 2007. LNCS (LNAI), vol. 4496, pp. 278–287. Springer, Heidelberg (2007)
11. Borzemski, L., Nowak, Z.: Using autonomous system topological information in a Web server per-formance prediction. Cybernetics and Systems 39(7), 753–769 (2008)
12. Borzemski, L., Kamińska-Chuchmała, A.: 3D Web Performance Forecasting Using Turning Bands Method. In: Kwiecień, A., Gaj, P., Stera, P. (eds.) CN 2011. CCIS, vol. 160, pp. 102–113. Springer, Heidelberg (2011)
13. Borzemski, L., Kamińska-Chuchmała, A.: Spatial Web Performance Forecasting with Sequential Gaussian Simulation Method. In: Information Systems Architecture and Technology, pp. 37–46. Oficyna Wydawnicza PWr, Wrocław (2011)
14. Borzemski, L., Kamińska-Chuchmała, A.: Knowledge Discovery about Web Performance with Geostatistical Turning Bands Method. In: König, A., Dengel, A., Hinkelmann, K., Kise, K., Howlett, R.J., Jain, L.C. (eds.) KES 2011, Part II. LNCS, vol. 6882, pp. 581–590. Springer, Heidelberg (2011)

# Efficiency of IP Packets Pre-marking for H264 Video Quality Guarantees in Streaming Applications

Slawomir Przylucki

Technical University of Lublin, 38A Nadbystrzycka Str, 20-618 Lublin Poland
spg@politechnika.lublin.pl
http://www.pollub.pl

**Abstract.** For the last few years, many classifications and marking strategies have been proposed with the consideration of video streaming applications. According to IETF recommendation, two groups of solutions have been proposed. The firts one assumes that applications or IP end points pre-mark their packets. The second solution applies the router which is topologically closest to video source. It should perform Multifield Classification and mark all incomming packets. This paper investigates the most popular marking strategies belonging to both mentioned above groups of solutions. The pre-marking strategies based on H264 coder extensions are simulated based on NS-2 network simulator and Evalvid-RA framework. The results are compared with the IETF recommendations for video traffic shaping in the IP networks and marking algorithms proposed by other researchers.

**Keywords:** video streaming, IP packet marking, H264 video coding, DiffServ architecture.

## 1   Introduction

Video streaming over IP is a challenging task since IP was developed as a transmission environment providing no built-in quality of service. Consequently, IP based networks have no guarantee of resources in term of bandwidth, packet losses, delay variation (jitter), and transfer delay. To face these situation, the standardization bodies such as ITU (International Telecommunication Union) and IEFT (Internet Engineering Task Force) have developed a range of standards and recommendations [1,2]. One of the most fundamental one, ITU-T Y.1221 distinguishes three basic types of transmission: Dedicated Bandwidth (DBW), Statistical Bandwidth (SBW), and Best Effort (BE) [3,4]. Moreover, ITU-T Y.1541 [5] stands out for six QoS service classes. QoS classes 0 and 1 are determined on the basis of the delays, jitter and packet loss. Classes 2, 3, and 4 define the limits of the loss of packages, but not the jitter limits. Class 5 does not guarantee the QoS parameters.

The concept of classes requires a flexible mechanism for packets marking. In an IP network, one of the most popular solution of this kind is traffic differentiation

A. Kwiecień, P. Gaj, and P. Stera (Eds.): CN 2012, CCIS 291, pp. 120–129, 2012.

also called as differentiated services (DiffServ). The idea of differentiation is based on a simple model, which classifies and assigns the streams of IP packets to the appropriate aggregates. Basic DiffServ functional techniques have been described in the RFC2475 [6]. When classifying and marking packets according to DiffServ rules, the Differentiated Services (DS) field is used. It is located in the IP packet header. The first three bits of the DS field are used to determine the traffic class, while the next three define the packets rejection probability. Last two bits are left unused [7]. All these 6 bits create so called Differentiated Services Code Point (DSCP). The packet classification and marking on the edge of the DiffServ network leads to distinguish four basic classes [7,8], each represented by specific DSCP code value. These classes are respectively:

- Best Efford – BE,
- Assured Forwarding – AF,
- Expedited Forwarding – EF,
- Class Selector – CS.

Table 1 presents the relations among classes described in the ITU-T and IETF recommendations.

**Table 1.** Association of Y.1541 classes with Y.1221 and DiffServ PHB

| ITU-T Y.1541 QoS class | ITU-T Y.1221 transmission | IETF Diffserv PHBs |
|---|---|---|
| class 5, unspecified | Best Effort (BE) | BE |
| class 2, 3 or 4 | Statistical Bandwith (SBW) | AF |
| class 0 or 1 | Dedicated Bandwith (DBW) | EF |

The IETF RFC4594 [7] recommends AF3x subclass for applications that require near-real-time packet forwarding of variable rate elastic traffic sources that are not delay sensitive. These characteristics are consistent with the requirements of video streaming applications. IETF also recommends that the applications or IP end points should pre-mark their packets with DSCP values or the router topologically closest to video source should perform Multifield Classification and mark all packets as AF3x [7,8]. However above mentioned recommendations do not apply to the hierarchical structure of the video stream and to the relative importance of the different frames transmitted over the IP network. Therefore, they do not take into account the specific characteristics of codecs like H264 On the other hand, the recommendations of the IETF suggests the possibility of pre-marking of packets. Both solutions have advantages and disadvantages. Therefore, further part of article will be devoted to assessing their effectiveness in the process of video streaming.

## 2   Coloring Principles

Generally, packets coloring is a way of packets marking by setting of particular DSCP bits in the IP header. Packets can be colored in green, yellow and red.

As stated in the introduction, attention should be paid to DiffServ AF class [9]. The AF class offers different levels of forwarding assurances for IP packets, while accomplishing a target throughput for each network aggregate. This class defines IP packet forwarding for $N$ AF subclasses with $M$ dropping precedence. The standard values for $N$ and $M$ are 4 and 3, respectively. In the $M = 3$ case, packet drop priorities are usually identified by colors: green for the lowest drop precedence, yellow for the middle and red for the highest one.

The most popular three algorithms of packets coloring were defined by Internet Engineering Task Force (IEFT):

– Single Rate Three Color Marker (srTCM) [10],
– Two Rate Three Color Marker (trTCM) [11],
– Rate Adaptive Shapers (RAS) [12],
– The Time Sliding Window Three Color Marker (TSW3CM) [13].

From the perspective of a typical video streaming system configuration, the most analyzed solutions are the first two algorithms [14]. They are also the basis for IETF recommendations presented in the RFC4594. This recommended hybrid coloring algorithm is presented in Fig. 1.

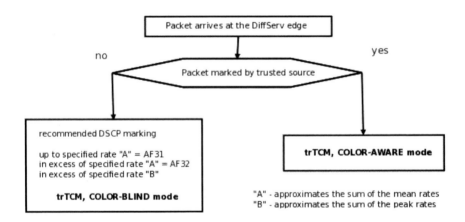

**Fig. 1.** IETF recommended marking and coloring algorithm

The performance analysis of above meter/markers can be found in [15]. Detailed analysis of markers models can be found in [16] and overview of methods for video packet marking in DiffServ networks are presented in [17,3]. The results from [14] are used for comparison purposes later in this article.

For the last few years, many other classifications and marking strategies have been proposed with the consideration of specific characteristics of video streams. Table 2 contains a summary list of the extensions of standard IETF solutions.

The last two proposed solution represent the state-of art in the packet marking for video streaming applications. That is why both of them will be compared with the results of our simulations.

**Table 2.** List of the most frequently cited extensions of IETF standards

| Proposed solution | Source | Remarks |
|---|---|---|
| Proportional Differential Model with four queues | [18] 2001 | coloring accounts the loss impact of frames or slices on video quality |
| score based marker | [3] 2001 | packet gets a score that is a funcion of internal structure of transmited frames |
| frame aware marking | [19] 2002 | coloring accounts information on the type of video frames in the packet classification |
| same as above | [20] 2003 | solution for streaming MPEG4/H264 video content |
| cost-distortion optimization | [21] 2003 | joint source-coding and classification model for H264 streaming was introduced |
| model-based distortion estimation | [22] 2004 | classification and marking packets based on the identification of perceptually important video regions |
| Two Markers System (TMS) | [23] 2005 | based on two operation modes of marker (blind and color-aware mode) |
| ASC-TS marker | [24] 2010 | Adaptive Significance Classification mechanism in Temporal and Spatial domains |
| Priority-Aware Two Rate Three Color Marker (PATRTCM) | [25] 2010 | introduces the concept of three marking probabilities, which are calculated in terms of the relative importance of the video packet and the traffic conditions in the network |
| Improved Two Rate Three Color Maker (ITRTCM) | [26] 2011 | maps pre-marked packet (based on importance of each of frame) to a fixed number of source marks at the real-time transport protocol layer |

# 3    H264 Video Coding

Nowadays, codecs perform compression by removing redundant information from the original signal within a single frame (spatial redundancy removal) and between consecutive frames (temporal redundancy removal). That is the case of ITU H.264/AVC codec which is the latest MPEG standard for video encoding [27,2]. In codecs like H264, there are three types of frames: I, P and B. These frames are obtained as a result of the redundancy reduction process mentioned above. I (Intra) frames are coded as still images and serve as reference for previous and future predicted frames. P (Predicted) frames are predicted from past I frames. Finally, B (Bidirectionally predicted) frames are predicted from both past and future I frames. Such a temporal dependence among frames creates a frame hierarchy, in the sense that some frames are more important than other ones to reconstruct the original video stream. In the context of IP networks (in which packet losses may occur), indiscriminated packet discard may seriously degrade a H264 video stream because of its hierarchical structure.

## 3.1  H264 Extensions

Two extensions play an important role in ensuring satisfactory quality of video for streaming applications. These extensions are Flexible Macroblock Ordering (FMO) and Data Partitioning (DP) [1]. Both are related to error concealment and resilience features of H264 codec.

FMO ordering allows to assign macroblocks to slices in an any order [1,28,29]. In general, a slice contains contiguous macroblocks (scan order). FMO permits to create slices with non-contiguous macroblocks and assign these slices to different groups. The H.264/AVC standard defines more complex FMO macroblock classification patterns, among which the most commonly used is the checker board. Apart from predefined patterns, fully flexible macroblock ordering (explicit mode) is also allowed. This flexibility allows dynamically change the macroblock classification schema based on the video content. Figure 2 shows graphically different ordering schemes.

Data partitioning (DP) is an error resilience feature that relies on the hierarchical separation of coded video data. According to H264 standard, all symbols of a macroblock are coded together in a single bit string. That string forms a slice. When the data partitioning is applied, it creates more than one bit strings per slice, These strings are called partitions. All symbols that have a close semantic relationship with each other are allocated to an individual partition. There are three types of the partitions created in DP process [28]:

- partition A – containing headers, macroblock types, quantization parameters and motion vectors,
- partition B – containing intra coefficients,
- partition C – containing inter coefficients.

Partition A is the most important one. B-type partitions require the availability of the corresponding A-type partitions. C-type partitions are the least important ones and they require A-type but not B-type partitions.

## 3.2  Applying the H264 Extensions

The simplest strategy of packets pre-marking at the video source, is to take the type of slices into account. In this case I slices are encapsulated in green packets,

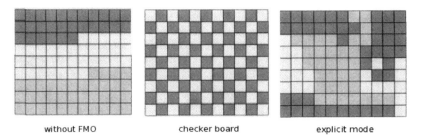

without FMO        checker board        explicit mode

**Fig. 2.** Different ways of macroblock ordering in the H264 coder

P slices in yellow and B in red ones, respectively. Another possibility is to use the H264 extensions in the process of packets pre-marking. Data partitioning is a natural candidate for AF DiffServ mapping. It offers a very granular semantic separation of the coded video, and this separation is done in three hierarchical levels, which can be directly mapped to the three AF precedences.

More sophisticated mapping methods are needed when FMO extension is to be used. The simplest approach is to assign the different groups of slices to the different packet discard priorities. This requires the inclusion of features analysis of the video content in the process of mapping the macroblocks into slides before pakiet pre-marking. This approach offers potentially greater flexibility of shaping the priorities assigned to each IP packet. For instance, the some part of an image may be more relevant to the viewer than other parts. Therefore, macroblocks at the important part of a frame might be assigned to a slice group which is forwarded with the highest priority [29]. From other hand, even checker board FMO mode should has influence on quality of transmited video stream. This FMO mode do not requires pre-marking but neither requires additional algorithms for frame analysis at the coder side.

## 4   Testing Scenarios

For the purposes of comparison, we selected three solutions that use H264 coder features and its extension. The first one was the pre-marking algorithm based on slice type, second used pre-coloring based on DP type and the last solution has genereted video stream with slices which have been created by FMO checker board mode. They are compared with two latest algorithm proposed in literature, pre-marking solution ITRTCM [26] and priority-aware algorith PA-TRTCM [25], respectively. All solutions were tested at the comparable network configuration and used the same video source. The aim of comparison is to asses if the pre-marking aproach improve the quality guarantees in streaming applications. Table 3 shows the description of the tests scenarios.

**Table 3.** Description of the tests scenarios

| Scenario | Codec feature | Pre-marking | Marker / mode |
|----------|---------------|-------------|---------------|
| BE | no | no | no |
| S3C | slices | I slices – green<br>P slices – yellow<br>B slices – red | trTCM / color-aware |
| FMO | FMO checker board | no | trTCM / blind |
| DP | DP partitions | A partitions – green<br>B partitions – yellow<br>C partitions – red | trTCM / color-aware |

## 4.1    Testbed Configuration

The coding part of the testbed system is built up based on a modified version of the test mode coder JM18 which is provided by JVT. Modification of H264 codec and implementation of DP and FMO extensions were based on the JVT documents [30]. The network part of the testbed consists of two basic elements: a NS2 simulator and a video quality evaluation framework Evalvid-RA [31]. The trTCM coloring scheme and WRED based AQM were implemented on the NS2 ingress node. The coding and quality assesment modules were developed based on Evalvid-Ra framework.

The test video source was a Highway CIF format sequence, which has 2000 frames. The encoded stream has a mean bit rate of 412 kbps and a peak rate of 1116 kbps. Each frame was fragmented into packets of 1000 bytes before transmission. This video flow competed with one ON-OFF traffic flow, which has an exponential distribution with mean packet size of 1000 bytes, burst time of 500 ms, idle time of 0 ms, and one background FTP traffic flow of 64 kbps. During the simulations, the bit rate of the ON-OFF source was changed in order to obtain link overflow. Total capacity of the transmission link was 512 kbps. The link overload ranged from 0 to 30%. Table 4 presents the parameters of trTCM meter and WRED queue management mechanism used during simulations.

**Table 4.** Adaptive queue management and marking parameters

| WRED parameters (unit:queue length) | | | |
|---|---|---|---|
| Packet | Threshold | | |
| | minimum | maximum | maximum probability |
| green | 0.6 | 0.8 | 0.025 |
| yellow | 0.4 | 0.6 | 0.05 |
| red | 0.2 | 0.4 | 0.1 |
| trTCM parameters | | | |
| CIR | CBS | PIR | PBS |
| 512 kbps | 1000 bytes | 1100 kbps | 1000 bytes |

## 4.2    Simulations Results

The final assessment of the video quality based on the PSNR metric [32]. Each test was repeated 20 times and average values of received PSNR are presented. Also, we assumed that a frame is only considered decodable if and only if all of the frames upon which it depended are also decodable. To justify our research, simulation results of our proposed method are compared with most resent methods, ITRTCM [26] and PATRTCM [25], as shown in Table 5.

In the case of the use of packets pre-marking (see the results for BE and S3C scenarios in Table 5) video quality improvement increases with packets losses. This is due to better protection of green packets in case of S3C. Using FMO extension, even in the simplest checker board scheme leads to vast improvement in received

**Table 5.** Comparison of PSNR V component in different pre-marking scenarios

| Scenario | Link overload | | |
|---|---|---|---|
| | 10% | 20% | 30% |
| BE | 31.8 dB | 26.6 dB | 24.8 dB |
| S3C | 34.9 dB | 31.7 dB | 28.0 dB |
| FMO | 37.5 dB | 33.6 dB | 26.9 dB |
| DP | 38.0 dB | 31.1 dB | 28.1 dB |
| ITRTCM L.Chen [25] | 38.5 dB | 33.5 dB | 28.0 dB |
| H.Wang [26] | 36.1 dB | 34.2 dB | not tested |
| PATRTCM L.Chen [25] | 38.5 dB | 37.0 dB | 33 dB |

video quality. It is particularly noticeable for relatively small values of congestion. Reason for that is since every lost macroblock has several spatial neighbors that belong to the other slice, an error-concealment mechanism has a lot of information it can employ for efficient concealment. Unfortunatelly, this feature is limited to relativelly low and average congestion. The next tested extension of the H264 was the DP mechanism. Data partitioning, like the FMO, can clearly improve the perceived video quality. The improvement, however, is smaller and is limited to the range of small overloads. At that range, due to the availability of the macroblock types and the motion vectors, a comparatively high reproduction quality can be achieved, as it is only the texture information that is missing.

## 5  Conclusion

In this paper the impact of packet pre-marking on a quality of the H264 video has been studied. the H.264 cedec, in addition to all its coding-efficiency oriented features, introduces a set of new tools that improve the error resilience. The application of the extensions like FMO and DP have been shown to significantly improve perceived video quality in the case of streaming applications in IP environments. At the same time, the tests carried out confirm that the application of traffic engineering mechanisms consistent with the model of DiffServ and the H264 extensions are comparable with other proposed solutions such as ITRTCM or PATRTCM. However, in contrast to these solutions, tested aproach is based entirely on industrial standards and does not require additional modules for priority assesment or deep analysis of video stream structure.

## References

1. Wenger, S.: H.264/AVC over IP. IEEE Trans. Circuits and Systems for Video Technology 13(7), 645–656 (2003)
2. Wiegand, T., Sullivan, G.J., Bjontegaard, G., Luthra, A.: Overview of the H.264/AVC Video Coding Standard. IEEE Trans. Circuits and Systems for Video Technology 13(7), 560–576 (2003)

3. Shin, J., Kim, J., Kuo, C.-C.: Quality of service mapping mechanism for packet video in differentiated services network. IEEE Trans. Multimedia 3(2), 219–231 (2001)
4. Recommendation ITU-T Y.1221: Traffic control and congestion control in IP based networks. ITU-T (2003)
5. Recommendation ITU-T Y.1541: Applicability of the Y.1221 transfer capabilities and IETF differentiated services to IP QoS classes. Appendix VI, ITU-T (2006)
6. Blake, S., Black, D., Carlson, M., Davies, E., Wang, Z., Weiss, W.: An architecture for differentiated services. IETF RFC 2475 (1998)
7. Babiarz, J., Chan, K., Baker, F.: Configuration Guidelines for DiffServ Service Classes. IETF RFC 4594 (2006)
8. Baker, F., Polk, J., Dolly, M.: A Differentiated Services Code Point (DSCP) Capacity-Admitted Traffic. IETF RFC 5865 (2010)
9. Heinanen, J., Baker, F., Weiss, W., Wroclawski, J.: Assured Forwarding PHB Group. IETF RFC 2597 (1999)
10. Heinanen, J., Guerin, R.A.: A single Rate Three Color Marker. IETF RFC 2697 (1999)
11. Heinanen, J., Guerin, R.A.: A Two Rate Three Color Marker. IETF RFC 2698 (1999)
12. Bonaventure, J., Cnodder, S.: Rate Adaptive Shaper for Differentiated Services. IETF RFC 2963 (2000)
13. Fang, W., Seddigh, N.: Time Sliding Window Three Color Marker. IETF RFC 2859 (2000)
14. Przylucki, S., Sawicki, D.: Coloring VBR Streams Inside the DiffServ Domain. In: Kwiecień, A., Gaj, P., Stera, P. (eds.) CN 2011. CCIS, vol. 160, pp. 154–163. Springer, Heidelberg (2011)
15. Lee, S.-H., Seok, S.-J., Lee, S.-J., Kang, C.-H.: Study of TCP and UDP Flows in a Differentiated Services Network Using Two Markers System. In: Al-Shaer, E.S., Pacifici, G. (eds.) MMNS 2001. LNCS, vol. 2216, pp. 198–203. Springer, Heidelberg (2001)
16. Stankiewicz, R., Jajszczyk, A.: Analytical models for DiffServ meter/markers. In: IEEE International Conference on Communications, vol. 1, pp. 158–162. IEEE Conference Publications (2005)
17. Zhang, F., Macnicol, J., Pickering, M.R., Frater, M.R., Arnold, J.F.: Efficient Streaming Packet Video over Differentiated Services Networks. IEEE Trans. Multimedia 8(5), 1005–1010 (2006)
18. Tan, W., Zakhor, A.: Packet classification schemes for streaming MPEG video over delay and loss differentiated networks. In: Packet Video Workshop, Kyongju, Korea (2001)
19. Ziviani, A., de Rezende, J.F., Duarte, O.C., Fdida, S.: Improving the delivery quality of MPEG video streams by using differentiated services. In: 2nd European Confeence on Universal Multiservice Networks (ECUMN 2002), Colmar, France (2002)
20. Zoi, S., Loukatos, D., Sarakis, L., Stathopoulos, P., Mitrou, N.: Extending an Open MPEG-4 Video Streaming Platform to Exploit a Differentiated Services Network. In: 6th IEEE International Conference on High Speed Networks and Multimedia Communications, HSNMC, Estoril (2003)
21. Zhai, F., Luna, C.E., Eisenberg, Y., Pappas, T.N., Berry, R., Katsaggelos, A.K.: A novel cost-distortion optimization framework for video streaming over differentiated services networks. In: IEEE International Conference on Image Processing, Barcelona, Spain (2003)

22. De Vito, F., Quaglia, D., De Martin, J.C.: Model-based distortion estimation for perceptual classification of video packets. In: IEEE Int. Workshop on Multimedia Signal Processing (MMSP), Siena, Italy (2004)

23. Ke, C.H., Shieh, C.K., Hwang, W.S., Ziviani, A.: A Two Markers System for Improved MPEG Video Delivery in a DiffServ Network. IEEE Communications Letters 9(4), 381–383 (2005)

24. Lee, C.C., Yu, Y.-J., Chang, P.C.: Adaptive Significance Classification Mechanism for Streaming Video over Differentiated Service Networks. Journal of Computers 21(2), 3–13 (2010)

25. Chen, L., Liu, G.: A Delivery System for Streaming Video Over DiffServ Networks. IEEE Trans. Circuits and Systems for Video Technology 20(9), 1255–1259 (2010)

26. Wang, H., Liu, G., Chen, L., Wang, Q.: A novel marking mechanism for packet video delivery over DiffServ Networks. In: IEEE International Conference on Multimedia and Expo. (ICME), pp. 1–5 (2011)

27. ITU-T and ISO/IEC JTC-1: Advanced Video Coding for generic audiovisual services. ITU-T Rec. H.264 and ISO/IEC 14496-10 (MPEG-4 part 10) AVC (2005)

28. Stockhammer, T., Wenger, S.: Standard-Compliant enhancement of JVT coded video for transmission over fixed and wireless IP. In: IWDC (2002), 
http://citeseerx.ist.psu.edu/viewdoc/
download?doi=10.1.1.70.4668&rep=rep1&type=pdf

29. Thomos, N., Argyropoulos, S., Boulgouris, N.V., Strintzis, M.G.: Error-resilient transmission of H.264/AVC streams using flexible macroblock ordering. In: 2nd European Workshop on the Integration of Knowledge, Semantics and Digital Media Technology (EWIMT 2005), London, pp. 183–189 (2005)

30. Ye, J.C., Chen, Y.: Flexible Data Partitioning Mode for Streaming Video. Joint Video Team (JVT) of ISO/IEC MPEG and ITU-T VCEG, 
http://wftp3.itu.int/av-arch/jvt-site/2002_07_Klagenfurt/JVT-D136.doc

31. Fitzek, F.H.P., Reisslein, M.: MPEG-4 and H.263 Video Traces for Network Performance Evaluation. IEEE Network 15(6), 40–54 (2001)

32. Winkler, S.: Digital Video Quality: Vision Models and Metrics. John Wiley & Sons Ltd. (2005)

# Universal Web Pages Content Parser

Piotr Pawlas, Adam Domański[1], and Joanna Domańska[2]

[1] Institute of Informatics
Silesian University of Technology
Akademicka 16, 44-100 Gliwice, Poland
`adamd@polsl.pl`
[2] Institute of Theoretical and Applied Informatics
Polish Academy of Sciences
Baltycka 5, 44-100 Gliwice, Poland
`joanna@iitis.gliwice.pl`

**Abstract.** This article describes the universal web pages content parser – cross-platform application enhancing the process of data extraction from the web pages. In this implementation user friendly interface, possibility of significant automation and reusability of already created patterns had been the key elements. Moreover, the original approach to the issue of parsing the not well-formed HTML, stating the application's core, is precisely presented. Universal web pages content parser shows that the simplified web scrapping utility may be available to masses and not well-formed HTML sources may feed useful tree-like data structures as well as the well-formed ones.

**Keywords:** content parser, web, HTML, XML.

## 1 Introduction

For the past two decades Web became ubiquitous, being the main source of information for the significant part of the society. However, with the constant tremendous increase of available data, individual may became overwhelmed and simply feel lost in this spate [1].

A definite necessity arose for providing some automation in data extraction from multiple sources and storing it for further analysis and presentation to end users [2]. As an answer to this need, the concept of web scrapping had been introduced [3]. Although Web pages parsing has been recently became an important task, relatively little work has been done in this field [4]. A few of scientists tried to get an answer how to get rid of human intervention and create extraction rules of subject information from a large number of web pages [1,5]. Generally, it refers to the set of techniques enhancing the process of more or less unattended acquisition of the data accessed usually in the raw HTML format.

The majority of currently available solutions in this field may be divided into two categories [6]:

- free of charge, but demanding specialist knowledge – text grepping, regular expressions matching, DOM (Document Object Model) parsing, HTML parsing, etc.,

A. Kwiecień, P. Gaj, and P. Stera (Eds.): CN 2012, CCIS 291, pp. 130–138, 2012.

– user friendly, but commercial software that does not demand any additional
specialist knowledge.

Universal web pages content parser had been developed to fill the gap, being
the free of charge complex tool that allows the advanced data extraction from
the previously parsed web pages with all the mechanics hidden behind the graphical user friendly interface [7,8].

## 2    HTML Parsing

Two different approaches to the data acquisition from the web pages sources were
considered. Since there are multiple open source regular expression libraries, like
python's re module, initially such solution had been put to the test. However, it
appeared that such concept has two significant drawbacks. First of all, end users
would have to be familiarized with the composition of regular expressions and
moreover with each modification of web page's source, new regular expression
should be created. Second promising idea was the utilisation of the Document
Object Model [9]. Idea of the accessing pieces of source data as leaves of tree-like
structure appeared to be an ideal base of the user friendly interface. However,
one extremely problematic issue emerged – since DOM parsers work perfectly
with the XML files, they could not process significant part of the HTML sources
due to the fact that they were not well-formed documents [10].

Well-formed element in web page design can be described as one that adheres
to the following rules:

– each start tag must posses matching end tag,
– there can be no end tag without start tag,
– elements cannot overlap.

Since there was no reasonable possibility to make easy and compact transformation from not well-formed into the well-formed documents, instead of DOM
method, it had been decided to parse all pages sequentially, providing functions
that were responsible for creation of artificial source structures. Such construc-
tions, while being significantly simplified in comparison to Document Object
Model, were on the other hand absolutely immune to the not well-formed elements problem. Source element and source structure terms became the essential
concept in the final application. General approach was to create artificial class
describing single tag, storing information about its attributes, text content and
other nested source elements. In such manner tree-like structure with HTML tag
as a root element could always be created.

*SourceElement* had been designed as a Python class with following data fields:

– *head* – name of the HTML tag described by current element,
– *attributes* – list of key, value pairs of tag's attributes,
– *content* – element's content understood as either plain text placed between
  starting and ending HTML tag or additionally one or more other *SourceElements* representing HTML tags nested inside the one currently analysed,

- *content_ index* – index defining the tag's placement in the *content* list of its parent,
- *trace* – list of consecutive *content_ index* data field values uniquely defining certain *SourceElement*'s position with respect to the first, root element inside which all remaining ones are nested.

Basing on the *HTMLParser* class, taken from the Python's standard library [11], *BasicHTMLParser* class had been derived by authors. It is used to create proper structures of *SourceElements* from the raw HTML by means of specially designed functions called during the sequential access to the analysed source. Those functions are presented in the form of pseudo-code in the Listing 1.1.

**Listing 1.1.** Crucial *BasicHTMLParser*'s functions in pseudo-code

```
handle_starttag
   make new_trace equals to trace of current
   append content_index of current to new_trace
   create element being the SourceElement that references
      encountered tag
   make element trace equals to new_trace
   increment content_index of current
   append element to content of current
   put current on stack
   set element as current

handle_endtag
   pop object from stack and make it current
   do nothing when stack is empty

handle_data
   if data is not whitespace
      increment content_index of current
      append data to content of current
```

There are three things, that should be additionally clarified:

- initially current element is set to be an empty *SourceElement*,
- while handling end tag, gentle behaviour in case of empty stack is provided to deal with not well-formed HTML structures,
- previously mentioned and frequently used in further considerations *Source-Structure* is simply the first *SourceElement* for currently analysed HTML source.

## 3    Patterns Concept

For the end user *SourceStructrue* is presented in the form of tree view with tag name in the first column, its attributes in the second and text content in the third. This view reflects exactly the tag nesting in the raw HTML source.

Such construction is the starting point for the further selection of data to be extracted. Universal web pages content parser introduces the concept of parse pattern – an easily modified and extensible collection referencing *SourceStructure* elements' attributes or text values. User works with previously mentioned tree view, selecting *SourceElements* that contain desired pieces of information and by means of simple interface appends them to the pattern.

There are two complementary strategies of the data extraction from Source elements that should be discussed.

### 3.1 Single Element Data Extraction

*SourceElement* is directly indicated in the tree view. Single *PatternObject* references single *SourceElement* by means of its *trace* data field (path from the source root).

### 3.2 Multiple Elements Data Extraction

The true power of the Universal web pages content parser comes with *Pattern-Blocks* usage. Following issue should be considered – many contemporary web pages, especially the ones displaying large sets of data are usually internally created from some repeatable blocks of HTML code, differing in the text content only. For example each movie repertoire on certain cinema web page is enclosed in single ⟨div⟩ tag stored in some larger structure; news' paragraphs on certain Internet portal are simply ⟨p⟩ tags also placed in some structure and so on.

General idea of block usage is about selecting certain element and defining which of its children (nested exactly in it) tags are in scope of user's interests.

In single element data extraction strategy, pattern objects are created, referencing some desired data which placement is resolved with respect to the root of *SourceStructure* (and since there is always only one root, single pattern object could reference exactly one piece of data). In case of *PatternBlocks* one can treat each of block members as new root of source sub structure. So when user adds *PatternObject* to the block it will create references to the data with respect to each of the members.

## 4   Parse Task and Its Execution

When *ParsePattern* is prepared it can be exported to the binary file and used during the actual data extraction. To stress the aspect of reusability, it should be mentioned that pattern created based on specific web page's source may be successfully used for the data extraction from other pages possessing similar internal structure.

Actual mechanics responsible for the data acquisition is covered by the utility named Task manager, which operates on specially designated *ParseTask* class object. Core of this class is the scope list. Its every element is an inner list of strings:

- source – URL of the web page to be downloaded and converted to *SourceElement*,
- pattern – path to the previously composed *PatternStructure* that should be applied to the obtained *SourceElement*,
- type – file / database – notification of the parse result final destination,
- destination – path to the directory or name of MongoDB database and collection, where parse result should be stored,
- name – name of the text file or MongoDB document containing parse result.

Those elements can be divided into three parts due to the task execution stages:

- source, used for source handling,
- pattern, used for second level parsing,
- remaining ones, used for parse result storage.

Those stages constitute logical execution chain repeated for each task element defined in the scope list. Figure 1 shows the general schema how Universal web pages content parser extracts data from multiple sources.

To supplement the general idea of task execution, three issues referenced already in the Fig. 1 should be discussed.

### 4.1   Source Handler

This class, initialized only once per application is responsible not only for downloading web pages sources but additional preparatory actions – to name a few:

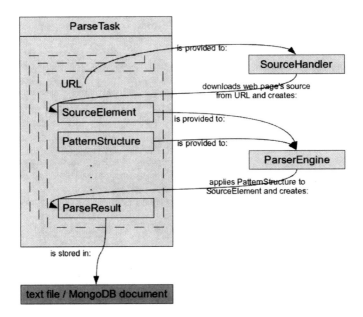

**Fig. 1.** General task execution schema

- proper source's encoding assignment,
- fault provoking CDATA sections deletion,
- fault provoking comments deletion,
- fault provoking script sections deletion.

Such processing is an additional protection against issues caused by not well-formed elements. In consequence, every web page source is prepared for the first level parsing from the HTML to *SourceElement*.

## 4.2 Parser Engine

This class, similarly to the *SourceHandler* is initialized only once per application and is the heart of the Universal web pages content parser. Generally it may be simply treated like a black-box which takes *SourceElement* and *PatternStructure* and returns *ParseResult* object, but it definitely deserves a deeper consideration.

*ParserEngine* class consists of a few methods responsible for the extraction of all desired pieces of data and storing them in formalized fashion. Three of them are strictly auxiliary ones and should be only mentioned for further comprehension of actual algorithm:

- *find_element* – method responsible for locating desired *SourceElement* with respect to given *SourceElement* root, using provided trace chain,
- *find_block_members* – method locating initially *SourceElement* being block's root and then traversing all its children, returning the ones fulfilling the conditions imposed by provided *PatternBlock*,
- *is_block_member* – method testing precisely provided *SourceElement*, verifying whether it fulfils conditions imposed by provided *PatternBlock*.

The actual second level parsing understood as extraction of desired data by means of application *ParsePattern* to *SourceStructure* is realised by calling the *parse_structure* method with mentioned objects as arguments. Further extraction process is done by means of iteration over top level pattern's elements and further recursive calls. Necessary methods are shown in pseudo-code in the Listing 1.2.

**Listing 1.2.** *ParserEngine*'s methods constituting the core of second level parser functionality written in pseudo-code

```
parse_element
  store data from provided SourceElement in element_data
  if desired content_type is 'text'
    return 'content' from element_data
  else if desired content_type is 'attribute'
    find in 'attributes' of element_data value of attribute
      with desired key
    return value of desired attribute

parse_block
```

```
for each member in provided block_members
  create member_parse_result list
  for each child in children of provided block_pattern
    if child is instance of PatternObject
      use find_element to locate member_element in member
        using trace from child
      use parse_element to obtain child_parse_result
        applying child's pattern to member_element
      append child_parse_result to member_parse_result
    else if child is instance of PatternBlock
      use find_block_members to locate child_block_members
        in member using trace and conditions from child
      call parse_block to obtain child_block_parse_result
      append child_block_parse_result to member_parse_result
  append member_parse_result to provided block_parse_result


parse_structure
  for each child in children of pattern_structure
    if child is instance of PatternObject
      use find_element to locate source_element in
        pattern_structure using trace from child
      use parse_element to obtain parse_result
        applying child's pattern to source_element
      append parse_result to final_result list
    else if child is instance of PatternBlock
      use find_block_members to locate child_block_members in
        pattern_structure using trace and conditions from
        child
      call parse_block to obtain block_parse_result
      append block_parse_result to final_result list
```

### 4.3   MongoDB Support

Final stage of each extraction task is of course successful storage of the desired data in the location chosen by end user. Besides the simple text files, Universal web pages content parser offers additionally support for the storage of results inside MongoDB collections [12]. It is an open source high-performance document-oriented database released for the first time in 2009. While providing many features of traditional relational databases such as indexes, dynamic queries and sorting, it allows application to store data in natural way matching their original structure. Since it was assumed that data extracted from web pages according to parse patterns will have no imposed structure and moreover, there is no requirement to have them connected by anything more than source from which they come, it is not necessary and even ill-advised to use a relational database.

# 5   Towards Automation

Each parse task may be saved as a binary file for further re-execution, for example to periodically update extracted data. Universal web pages content parser provides command line utility which, connected with system job scheduler, allows to perform unattended task execution.

Application can be invoked in command line with following parameters:

- -h, --help – shows command line utility help,
- -q, --quiet – loads application without GUI enabled,
- -t TASK, --task=TASK – loads automatically provided TASK,
- -e, --execute – executes all loaded tasks.

Individually, those options are not especially useful. However, combined together may provide complex desired functionality. The most practical and recommended application invocation for unattended parsing could look like (depending on operating system and tasks to be loaded):

```
./parser -q -t /home/pyjter/tasks/tutorial.tsk -e
```

Finally, all that user needs to do, is add such line to the system scheduler and check from time to time log file produced by the application. Everything else will be done automatically.

# 6   Tests and Further Development

Tests carried out revealed the fact that relatively small fraction of the tested web pages could not be properly processed due to some strange HTML tag glitches in their sources (for example $\langle / \rangle$ closing tag). Responses to the majority of such fault provoking elements had been correctly introduced and fraction of unhandled sources became significantly reduced.

Nevertheless, to provide HTML parsing mechanism totally immune to all kinds of not well-formed elements and different glitches, first level sequential access parser should probably be written from scratch (instead of inheritance from Python's standard library class).

Further development should also cover direct extraction of binary data (pictures, files, etc.) and storing data inside XML files.

# 7   Conclusions

This article presents the application that should fill the gap between commercial complex applications and free of charge but demanding specialist knowledge tools in the field of web scrapping. Moreover, original approach to the issue of parsing not well-formed HTML sources by means of sequential access but resulting in DOM-like structures is described.

Project is constantly developed and possibilities of further enhancements have been mentioned.

# References

1. Hu, J., Zhou, X., Shu, J., Xiong, C.: Research of Self-Adaptive Web Page Parser Based on Templates and Rules. In: International Conference on Management and Service Science (2009)
2. Zhang, W., van Engelen, R.: An Adaptive XML Parser for Developing High-Performance Web Services. In: IEEE Fourth International Conference on eScience (2008)
3. Malik, S.K., Rizvi, S.A.M.: Information Extraction Using Web Usage Mining, Web Scrapping and Semantic Annotation. In: International Conference on Digital Object Identifier (2011)
4. Li, Y., Yang, J.: A Novel Method to Extract Informative Blocks from Web Pages. In: International Joint Conference on Artificial Intelligence (2009)
5. Jianyi, N., Zhengqiu, Y.: An Automated Test Tool of Web Application Based on Struts. In: International Symposium on Computer Science and Computational Technology (2008)
6. Feng, D., Shanahan, J., Murray, N., Zajac, R.: Learning a Query Parser for Local Web Search. In: IEEE Fourth International Conference on Semantic Computing (2010)
7. Chen, K.-Y., Huang, C.-C., Hou, T.-W., Lee, T.-C., Yang, S.-F., Cheng, P.-W.: A quick XML parser for extracting signatures of secure Web services. In: The Fifth International Conference on Computer and Information Technology (2005)
8. Liu, F., Deng, S., Wang, N., Li, X.: A CS Grammar Based Query Form Information Extraction Method. In: International Symposium on Computer Science and Computational Technology (2008)
9. Takase, T., Tajima, K.: Lazy XML Parsing/Serialization Based on Literal and DOM Hybrid Representation. In: IEEE International Conference on Digital Object Identifier (2008)
10. Qin, J., Zhao, S., Yang, S., Dou, W.: Efficient storing well-formed XML documents using RDBMS. In: International Conference on Digital Object Identifier (2005)
11. Simple HTML and XHTML parser. Python v3.2.2 documentation, http://docs.python.org/py3k/library/html.parser.html
12. Introduction, MongoDB, http://www.mongodb.org/display/DOCS/Introduction

# Using Oracle 11.2g Database Server in Social Network Analysis Based on Recursive SQL

Lukasz Wycislik and Lukasz Warchal

Silesian University of Technology, Institute of Informatics,
16 Akademicka St., 44-100 Gliwice, Poland
{lukasz.wycislik,lukasz.warchal}@polsl.pl

**Abstract.** The article describes the possibility of using Oracle 11.2g database server in social networks analysis. Nowadays, when data about human's relationships, thanks to Internet technologies, are often gathered, it is becoming more and more interesting to process that data. Authors show how to compute some metrics (degree centrality, local clustering) describing networks based on data stored in a relational database. The use of recursive SQL resulted in compact formulas that can be process at the database server side in an efficient way.

**Keywords:** Oracle, social network analysis, relational model, recursive SQL.

## 1 Introduction

Due to continuous growth in the development of Internet technologies the social network analysis is becoming an increasingly popular subject. Relations between individuals may be modeled as graph-like structures thus to persist and to process this structures more efficiently new concepts are constantly inventing – e.g. NOSQL databases and especially graph databases. These platforms tend to store data in a more natural way which can result in easier and more efficient processing of that data in comparison to a *traditional* relational model. On the other hand many vendors of relational databases extend the standard SQL syntax of mechanisms to facilitate the processing of graph-like structures. These capabilities, combined with the undisputed possibilities of scalability and security makes relational databases still very good choice especially in situations where graph-like structures are only subset of a bigger relational schema. According to the most common approach, social network is modeled as a graph $G = (V, E)$, where $V$ is a set of nodes and $E$ is a set of edges. Edge $e_{ij}$ connects node $v_i$ with $v_j$, so $E$ can be defined as $E = e_{ij} : v_i, v_j \in V, i \neq j$.

Each node of defined graph represents an individual (role, actor, person). Relationships between each two individuals, which can be common activities or interactions, became the edges connecting (linking) two nodes. To model the relation strength (e.g. the more messages exchanged between two persons the bigger relation strength), each edge can have a weight. Set $W$ denotes then edges weights: $\lor w_{i,j} \in W, w_{ij} >= 0$ and whole graph is defined as $G = (V, E, W)$.

A. Kwiecień, P. Gaj, and P. Stera (Eds.): CN 2012, CCIS 291, pp. 139–143, 2012.

## 2   Domain Model

The UML class diagram on the Fig. 1 presents the basic model of a directed graph that is both easy and efficient to implement in a relational database. Each edge is defined by source (*from*) and destination (*to*) node. This forms basis of directed graphs but for social networks modeling it is often required to distinguish types of actors (e.g. teachers, students) and types of relations (e.g. taught by, liked). Each of actor/relation type can encapsulate several attributes (e.g. taught by relation can encapsulate discipline and language of communication). *From_ match* and *to_ match* classes define the possibility of assigning given node type to the source and destination sides of the edge type.

## 3   Implementation

### 3.1   Recursive SQL

There are several approaches to graphs processing in a relational database. Formerly, as SQL was based upon relational algebra and tuple relational calculus, the only possibility was to use server SQL statements to select atomic records representing one graph edge at one time and the rest of processing was being done at the client side. This often led to unacceptable loss of efficiency and the only way to prevent it was to implement all the processing in a stored procedure at the server side.

Possibility to build a recursive SQL query, which would allow for processing of graph structures was only introduced in the ANSI SQL-99 standard. By this time vendor specific extensions were sometimes introduced to database engines. Oracle for instance introduced SQL extension *connect by* in early version 8.1.7 but the possibility of cyclic graphs traversal (which is common issue in SNA) was introduced in 10g version (thanks to *nocycle* clause).

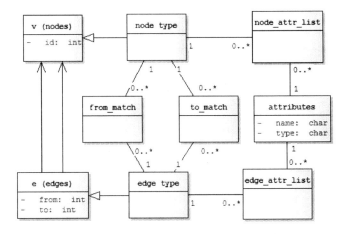

**Fig. 1.** Sample directed graph class model

In the latest 11.2g version also compatibility with ANSI SQL-99 recursive queries was implemented what is giving the possibility of defining recursive queries in two different ways.

## 3.2   Degree Centrality

Degree centrality is a simple and easy to calculate measure commonly used at the early stage of network studying [1]. It describes the involvement of node in network and can be defined as total number of nodes connected to examined node [1,2]. In weighted graphs this measure can be developed to comply not only node degree but also its strength (meant by a sum of weights on relationships to other nodes). Recently, Opsahl [1] proposed generalized degree centrality measure defined as:

$$C_D^{\omega\alpha} = k_i^{(1-\alpha)} \times s_i^{\alpha} \tag{1}$$

where $k_i$ is number of nodes connected to $i$-th node, $s_i$ is sum of weights from relationships to other nodes and $\alpha$ is a positive tuning parameter. If $\alpha$ is between 0 and 1, high node degree have priority, whereas if it is grater then 1 low degree is preferred. In relational database, measure form equation (1) can by calculated by a SQL query presented on Listing 1.1.

**Listing 1.1.** SQL query calculating degree centrality measure

```
select id ,
    (select  power(count(*),(1-L))*power(sum(w),L)
    from e where e."from" = v.id)  ks
from v order by ks desc;
```

As we can see it is quite simple to calculate the degree centrality for all nodes of the network. The outer select iterates on all nodes of network while the inner one takes into account edges starting from a given nodes.

## 3.3   Local Clustering

The degree to which nodes in a graph tends to cluster together is another popular and informative metric.

Many researchers find out, that especially in case of social networks, nodes tend to cluster into a smaller groups, which are heavy interconnected inside [3]. To observe this tendency, a global and local clustering coefficient measures were introduced [4,5]. First describes the overall level of clustering in particular network, second gives information about density of connections in node's neighborhood.

Local clustering coefficient for $i$-th node $v$ can be defined as:

$$C_{\text{local}}(i) = \frac{\lambda_G(v)}{\tau_G(v)} \tag{2}$$

where $\lambda_G(v)$ is number of triangles on $v \in V(G)$ on graph $G$, and $\tau_G(v)$ is number of paths of length 2 centered on $v$ node. This measure in relational database can by calculated by query shown on Listing 1.2.

**Listing 1.2.** SQL query calculating local clustering coefficient

```
select v.id ,
 (
    select count(*)
    from e
    where CONNECT_BY_ROOT "from" = "to" and level = 3
    start with "from" = v.id
    connect by nocycle prior "to" = "from" and level <= 3
 ) /
 (select count(*) from e e1 , e e2 where e1."to" = v.id
                                and e2."from" =    v.id
 ) as lcc
from v ;
```

As we can see it is quite simple to calculate the local clustering for all nodes
of network. The outer select iterates on all nodes of network. The number of
triangles is computed thanks to oracle *connect by* SQL extension. All paths in
graph for the given node starting from this node and maximum length of 3
is taking into account while counted are that of length exactly 3 and having
the same starting and ending edge. To compute the denominator the self-join
technique is involved. Every combination of two edges where the first one points
to given node and the second one goes out from the given node is counted.

The same computation but using the ANSI SQL-99 style can be done in a little
bit more complicated way presented on Listing 1.3.

**Listing 1.3.** Query calculating local clustering coefficient with ANSI SQL-99 syntax

```
with graph
( "from", "to", w, lvl , root )
as
(
select parent."from", parent."to",
       parent.w, 1, parent."from"
from e parent
union all
select child."from", child."to", child.w, lvl+1 , parent.root
from e child , graph parent
where parent."to" = child."from"  and lvl < 3
)
cycle "from" set is_cycle to '1' default '0'
select root , count(*) /
  (select count(*)
    from e e1 , e e2
    where e1."to" = root and e2."from" = root) as lcc
from graph w  where lvl = 3 and "to" = root
group by root ;
```

As we can see, an example given above is not as compact as that involving *connect by* clause, but its advantage is that it can be implemented in every database server that is compatible to ANSI SQL-99.

# 4    Summary

The enduring tendency of switching part of human's life to the Internet, both in private and professional areas, causes a large accumulation of social network data, which is interesting subject for scientists. These data are usually modeled as graphs so several measures like for example degree centrality or clustering coefficient can be applied. There are several experimental NOSQL database platforms to store and process graph-like data in a more intuitive and efficient way. However vendors of relational databases extend the standard SQL syntax of mechanisms to facilitate the processing of graph-like structures. These capabilities, combined with reliability and performance of industry-known relational databases make still very good choice especially in situations where graph-like structures are only subset of a bigger relational schema. The article shows on an example of degree centrality and local clustering measures the possibility to calculate them in a concise and efficient manner both in oracle-specific and ANSI SQL-99 style. As we can see, the oracle-specific style is much more intuitive and brief. The first study showed that oracle-specific SQL is also more efficient but on the other hand ANSI SQL-99 style can be applied to most modern database servers. Further studies will concentrate both on performance tuning (e.g. using of IOT Oracle mechanisms and moving indexes for different nodes/edges type to dedicated partitions) and efficiency comparison to graph databases (e.g. Neo4j).

**Acknowledgments.** This work was partially supported by the European Union from the European Social Fund.

# References

1. Opsahl, T., Agneessens, F., Skvoretz, J.: Node centrality in weighted networks: Generalizing degree and shortest paths. Social Networks 32, 245–251 (2010)
2. Freeman, L.C.: Centrality in social networks: Conceptual clarification. Social Networks 1, 215–239 (1978)
3. Holland, P.W., Leinhardt, S.: Transitivity in structural models of small groups. Comparative Group Studies 2, 107–124 (1971)
4. Watts, D.J., Strogatz, S.H.: Collective dynamics of small-world networks. Nature 393, 440–442 (1998)
5. Opsahl, T., Panzarasa, P.: Clustering in weighted networks. Social Networks 31, 155–163 (2009)

# Estimation of Web Page Download Time

Krzysztof Zatwarnicki and Anna Zatwarnicka

Department of Electrical, Control and Computer Engineering,
Opole University of Technology, Opole, Poland
{k.zatwarnicki,anna.zatwarnicka}@gmail.com

**Abstract.** This paper presents a study of a model of Web page download time. According to the proposed model the download time can be modeled on a base of knowledge of round trip time, bandwidth and a concurrency factor specifying the number of HTTP object downloaded concurrently. Through conducted experiments we designate a mean value of the concurrency factor for modern Web browsers: Mozilla Firefox, Internet Explorer and Google Chrome. Analysis of the results of the experiments confirms the serviceability and quality of the proposed model.

**Keywords:** Internet measurement, page download time estimation.

## 1 Introduction

Over the last several years the Internet has become the most innovative source of information and data. The omnipresence of the Web has become commonplaceness. The rapid development of systems using the WWW technology gives rise to the need for research on the effectiveness of the whole system delivering the required contents to the user.

The main semantic components forming the service of the Web system are: Hyper Text Markup Language (HTML), Hypertext Transfer Protocol (HTTP) and the Uniform Resource Identyfier (URI). The sources of HTML pages (Web pages) downloaded by the clients (Web browsers) are Web servers. Web pages are transferred from Web servers to clients with use of the HTTP protocol. In the interaction between the client and the Web service the client sends the HTTP request to the Web server and the server sends in the response HTTP object.

A Quality of Web Service (QoWS) can be evaluated by users in many different ways. On one hand, interesting content presented in the service causes growth in the number of users, however on the other hand it is very important to deliver the requested content within a certain time after requesting. Users will identify low quality of service if the time necessary to download even the most interesting content is too long. Therefore, the quality of web service is often evaluated primarily in the category of time required to download entire pages [1,2,3].

Delays in downloading Web pages are mainly related with delays in transmitting data over the Internet, and delays in processin HTTP requests by Web servers. In our previous work we have already dealt with the problem of modeling and estimating delays related with servicing HTTP requests by Web servers

A. Kwiecień, P. Gaj, and P. Stera (Eds.): CN 2012, CCIS 291, pp. 144–152, 2012.

[4,5]. In this paper we will propose a model enabling estimation of delays in transmitting data over the Internet. The proposed model can find its aplication in simulation experiments or in decision algorithms guaranteeing page response times.

The paper is divided into four sections. Section 2 presents the related work and a description of solutions enabling the estimation of delays in the Internet. Section 3 describes the Web page download time model proposed. Section 3 contains description of the testbed and results of experiments. The final Section 4 presents concluding remarks.

## 2   Related Work

The problem of measuring and modeling the World Wide Web resulted in many scientific works, experiments and measurements. Several measurement projects have been presented on the Internet [6,7,8,9]. Most of them are aimed to deal with the performance problems related to whole or the part of the Internet where measured data regards, for instance, the round trip delay among node pairs or the throughput [10]. The prediction of network performance is a challenging and topical problem [11,12,13,14,15]. Engineers and scientists need either a short-term or long-term forecast to be able to administer or deploy new solutions. There are many applications of methods and models enabling the prediction of bandwidth with the use of artificial intelligence and statistical approaches [16]. Huang and Subhlok have developed the framework for an intelligent real-time bandwidth predictor [11]. Mirza et al. utilize a machine learning approach [13]. There are also some applications of statistical models using an auto-regressive moving average, the Markov-modulated Poisson process [15] as well as based on an exponential weighted moving average (DualPats system) [12].

In the group of solutions enabling the estimation of the download time in the network there are only few approaches enabling the estimation of Web page download time [17,18]. In this paper an appropriate model of page download time will be presented.

## 3   Page Download Time Estimation

Web pages are composed of many different elements called HTTP objects. The main object is the HTML frame containing text, information on the way of displaying the text and information of embedded objects belonging to the page. Embedded objects are in most cases CSS files, java script programs, pictures and multimedia objects like video and audio files.

According to HTTP protocol each of the objects embedded in the Web page is downloaded from the server separately. The HTTP protocol allows, however, the downloading of many objects with the use of the same TCP connection. Web browsers nowadays open many concurrent TCP connections with the server to download single Web page. Internet Explorer 9.0, for example, opens up to

16 connections when it detects that the connections are done though a broadband network, otherwise it opens less conncetions. Opening many connections can significantly reduce the page download time and also the utiisation of the network.

As was mentioned before, the main aim of our work is to develop a model enabling the estimation of Web page download time. The page download time is measured from the moment the Web browser starts sending the HTTP request concerning the first object of the page to the moment the browser receives the last object of the page.

The delays connected with downloading the Web page have many sources related to delays in the browser, the server and the network. Figure 1 presents the process of fetching a Web object with the use of TCP and HTTP protocols. The request response time experienced by a user (person sitting in front of the

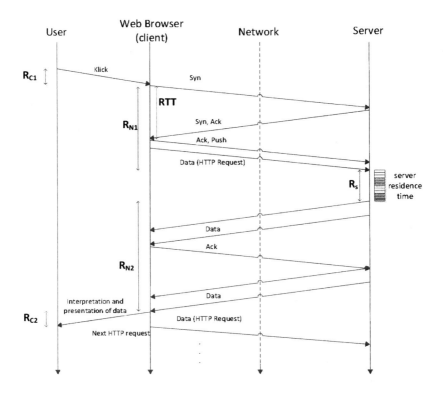

**Fig. 1.** Anatomy of HTTP connection delays

computer) can be calculated in the following way: $R_i^0 = R_{c1} + R_{c2} + R_{N1} + R_{N2} + R_s$, where $R_{c1} + R_{c2}$ is the time the browser takes to analyze and interpret the data, $R_s$ is the server residence time and $R_{N1} + R_{N2}$ are delays imposed by the network and associated with sending the request to the server and receiving the response. The delays connected with the process of analyzing the data by the browser depend on the kind of browser applied and the performance of the user's computer. The model of server residence time was already discussed in our previous works. In this work we will deal with delays connected with the network, and in the process of calculating the page response time we will only take network delays into account. The mean HTTP object download time for single object can be roughly calculated in following way:

$$R_i^O = RTT + \frac{O_i}{B} \; , \tag{1}$$

where $RTT$ is the Round Trip Time – the time of trip of a single IP datagram between the source and the destination and back, $O_i$ is the size in bytes of $i$-th object plus the size of the HTTP response header, $B$ is an effective bandwidth between the browser and the Web server observer for a single TCP connection. The RTT time in formula (1) is the time to send the HTTP request, and the $O_i/B$ is the time to download the response.

The page download time $R_\mathrm{p}$ for the Web browser downloading HTTP objects with the use of only one TCP connection can be calculated as follow:

$$R_\mathrm{p} = \sum_{i=0}^{N} R_i^O \; , \tag{2}$$

where $N$ is the number of objects embedded in the page and the object with the index 0 is the frame of the page. As was pointed out previously, modern Web browsers download many HTTP objects belonging to the same page concurrently. Figure 2 presents the process of downloading HTTP objects from the main Web page from the service www.elko.k12.nv.us with the use of three the most popular Web browsers: Internet Explorer, Mozilla Firefox, Google Chrome. A typical modern Web browser starts to download the HTML frame of the page, after getting the header of the frame it downloads the objects indicated in the header and after downloading the frame as a whole it downloads the rest of the objects. The page download time for the modern Web browser can be estimated according to formula:

$$R_\mathrm{p} = \alpha \sum_{i=0}^{N} R_i^O \; , \tag{3}$$

where $\alpha$ is the concurrency factor and depends on the number of objects being downloaded concurrently.

Concluding our discussion it can be noticed that having up to date information about RTT time, the effective bandwidth and knowing the concurrency factor it is possible to estimate the page download time. In the next section of the paper the testbed and the experiments results indicating if formula (3) and (1) can be applied to estimate the page download time will be presented.

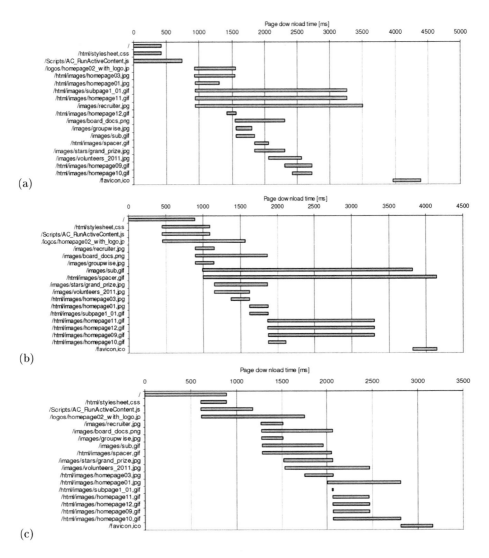

**Fig. 2.** Page download time in Web browser: (a) Chrome, (b) Firefox, (c) Internet Explorer

## 4 Testbed and Experiments

During the experiments the way the Web browsers operated was observed. In order to follow the flow of TCP segments sent between the Web browser and the Web server a special application was developed. The application let us record the RTT time measured in the phase of establishing the TCP connection, the first byte time (time between the moment of sending the HTTP request and receiving the first TCP segment with response) and the time to send the response. The effective bandwidth $B$ was calculated on the base of the response sending time and the size of response $O_i$. The application was developed with use of pcap library [19] enabling collecting ethernet frames before the frames are processed by the TCP/IP stock of the operating system.

Experiments were conducted for nine Web pages belonging to different Web services. In order to measure page download time on which server residence time have had very low influence, services belonging to universities and research centers from all over the world were chosen. The experiments showed that the first byte time was approximately equal to the RTT time which confirms that the server residence time had a very low impact on the page download time during experiments.

In Table 1 the information about the Web services, localizations and the downloaded pages are presented.

**Table 1.** Addresses, location and sizes of Web pages

| Name of server | IP Address | Country | Continent | Number of objects | Size [B] |
|---|---|---|---|---|---|
| www.italiamo.net | 77.232.81.14 | Switzerland | Europe | 29 | 103790 |
| www.fachschule-hauswirtschaft.de | 82.165.116.126 | Germany | Europe | 32 | 81288 |
| www.borhaug.skole.no | 79.125.17.4 | Norway | Europe | 26 | 390926 |
| www.ville-st-girons.fr | 195.114.115.194 | France | Europe | 40 | 399391 |
| www.greenhouse.cl | 190.54.31.131 | Chile | South America | 44 | 173876 |
| www.elko.k12.nv.us | 207.197.77.195 | USA | North America | 19 | 234980 |
| www.preston.k12.id.us | 96.5.146.8 | USA | North America | 58 | 237162 |
| www.hp.fujita-hu.ac.jp | 202.236.169.22 | Japan | Asia | 49 | 188448 |
| www.dghs.co.za | 196.22.132.14 | Republic of South Africa | Africa | 47 | 1265360 |

Experiments were conducted for three Web browsers: Internet Explorer 9.0, Mozilla Firefox 4.0 and Google Chrome 12.0. The operating system of all the computers hosting Web browsers was Windows 764 bits. During experiments the the Web browsers did not cache the content of the pages and there was no caching proxy servers on the way between browsers and Web server. Every 25 seconds each of the browsers downloaded one of the pages listed in Table 1. Experiments were conducted in Opole University of Technology, started at 19th of May 2011 and lasted for 30 h. For each Web browser and Web page about 1000 measurements were done.

After collecting and analyzing the data the mean value of the concurrency factor $\alpha$ for each of the pages was determined on the base of formula (3) where the HTTP object download time $R_i^O$ was a time measured during experiments. Table 2 presents the obtained values of the factor. The mean value of the concurrency factor for Google Chrome is 0.283, for Mozilla Firefox 0.246, for Internet

Explorer 0.272 and the mean value for all browsers is 0.267. As can be noticed the values of the concurrency factor for each of the browsers do not differ much. The mean number of concurrently downloaded HTTP objects is 3.74. The next step in the analysis was to estimate the page download time on the base of formulea (3) and (1), having knowledge of the size of downloaded objects, RTT time, effective bandwidth and the concurrency factor. Figure 3 presents

**Table 2.** Obtained values of concurrency factor

| Browser | Web service | www.borhaug.skole.no | www.dghs.co.za | www.elko.k12.nv.us | www.fachschule-hauswirtschaft.de | www.greenhouse.cl | www.hp.fujita-hu.ac.jp | www.italiamo.net |
|---|---|---|---|---|---|---|---|---|
| **Google Chrome** | Concurrency factor | 0.409 | 0.265 | 0.332 | 0.26 | 0.271 | 0.214 | 0.232 |
| | Standard deviation | 0.15 | 0.06 | 0.08 | 0.14 | 0.18 | 0.02 | 0.09 |
| **Mozilla Firefox** | Concurrency factor | 0.293 | 0.279 | 0.199 | 0.28 | 0.257 | 0.189 | 0.230 |
| | Standard deviation | 0.11 | 0.06 | 0.04 | 0.11 | 0.07 | 0.03 | 0.07 |
| **Internet Explorer** | Concurrency factor | 0.35 | 0.278 | 0.278 | 0.318 | 0.26 | 0.194 | 0.232 |
| | Standard deviation | 0.14 | 0.04 | 0.08 | 0.14 | 0.18 | 0.04 | 0.11 |

a diagram of estimated and measured page response times in a function of the number of downloaded pages for each browser and for example `www.dghs.co.za` Web service As one can notice the estimated page download time follows well the measured page response time. In each of the conducted experiments the mean relative error was not greater than 20% and in most cases it was lower then 10%. The results of the experiments indicate that the adopted model can estimate the page response time well and can be used in research.

## 5   Summary

In the article the model enabling the estimation of page download time was proposed. Obtained results indicate that the model estimates page download times well. The proposed model can be successfully used in simulation experiments or to estimate page download time in decision algorithms, for example in Web services guaranteeing page response times. According to the model, the page download time for modern Web browsers can be modeled having knowledge of the sizes of HTTP objects belonging to the page, Round Trip Time between the

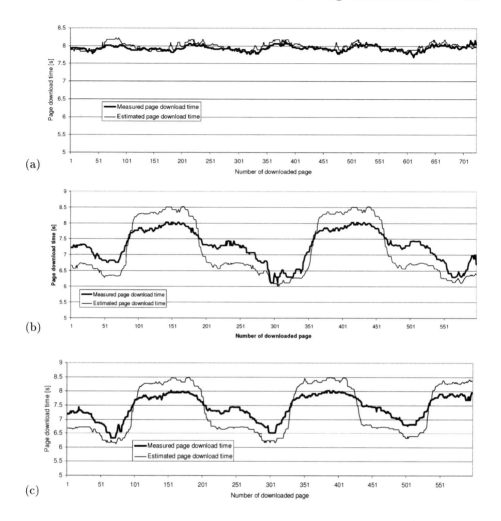

**Fig. 3.** Measured and estimated page download time for Web browsers: (a) Google Chrome, (b) Mozilla Firefox, (c) Internet Explorer

client and the Web server, the effective bandwidth of a single TCP connection and the concurrency factor. The obtained mean value of the concurrency factor is 0.267 which indicates that the mean number of concurrently downloaded HTTP objects is 3.74. The Round Trip Time and the effective bandwidth in simulation experiments can be modeled on a base of observed values in the Internet or with the use of appropriate probability distributions.

In order to confirm the quality of the proposed model further experiments have to be conducted. It also would be worthwhile to study the dependency of the concurrency factor on the bandwidth and the number of objects embedded in a Web page.

# References

1. Borzemski, L., Zatwarnicki, K.: Fuzzy-Neural Web Switch Supporting Differentiated Service. In: Gabrys, B., Howlett, R.J., Jain, L.C. (eds.) KES 2006. LNCS (LNAI), vol. 4252, pp. 195–203. Springer, Heidelberg (2006)
2. Mazzolla, M., Mirandola, R.: QoS Analysis For Web Service Applications: A Survey Of Performance-Oriented Approaches From An Architectural Viewpoint. Technical Report UBLCS-2010-05, University of Bologna (2010)
3. Zatwarnicki, K.: Providing Web Service of Established Quality with the Use of HTTP Requests Scheduling Methods. In: Jędrzejowicz, P., Nguyen, N.T., Howlet, R.J., Jain, L.C. (eds.) KES-AMSTA 2010. LNCS(LNAI), vol. 6070, pp. 142–151. Springer, Heidelberg (2010)
4. Zatwarnicki, K.: Proposal of a neuro-fuzzy model of a WWW server. In: 5th International Conference on Intelligent Systems Design and Applications, Poland, Wrocław, pp. 141–146. IEEE Press (2005)
5. Zatwarnicki, K.: Identification of the Web Server. In: Kwiecień, A., Gaj, P., Stera, P. (eds.) CN 2011. CCIS, vol. 160, pp. 45–54. Springer, Heidelberg (2011)
6. Brownlee, N., Claffy, K.C., Murray, M., Nemeth, E.: Methodology for passive analysis of a university Internet link. In: PAM 2001: Passive and Active Measurement Workshop, Amsterdam, pp. 22–24 (2001)
7. Brownlee, N., Claffy, K.C.: Internet measurement. IEEE Internet Computing 8(5), 30–33 (2004)
8. CAIDA. The Cooperative Association for Internet Data Analysis, http://www.caida.org (accessed 2010)
9. Keynote service, http://www.keynote.com (accessed 2010)
10. Borzemski, L.: Towards Web Performance Mining. In: Ting, I.-H., Wu, H.-J. (eds.) Web Mining Appli. in E-Commerce and E-Services. SCI, vol. 172, pp. 81–101. Springer, Heidelberg (2009)
11. Huang, T., Suhhlok, J.: Fast Pattern-Based Throughput Prediction for TCP Bulk Transfers. In: Proceedings of the Fifth IEEE International Symposium on Cluster Computing and the Grid, Washington, D.C, USA, vol. I, pp. 410–417 (2005)
12. Lu, D., Qiao, Z., Dinda, P.A., Bustamante, F.: Characterizing and Predicting TCP Throughput on the Wide Area Network Distributed Computing Systems. In: Proceedings of 25th IEEE International Conference, pp. 414–424 (2005)
13. Mirza, M., Sommers, J., Barford, P., Zhu, X.: A Machine Learning Approach to TCP Throughput Prediction. In: ACM SIGMETRICS 2007 Proc., vol. 35, pp. 97–108 (2007)
14. Qiao, Y., Skicewich, J., Dinda, P.: An Empirical Study of the Multiscale Predictability of Network Traffic. In: Proc. IEEE HPDC (2003)
15. Sang, A., Li, S.: A Predictability Analysis of Network Traffic. In: Proc. of the 2000 IEEE Computer and Comm. Societies, Conf. on Computer Communications, pp. 342–351 (2000)
16. Borzemski, L., Kliber, M., Nowak, Z.: Application of Data Mining Algorithms to TCP throughput Prediction in HTTP Transactions. In: Nguyen, N.T., Borzemski, L., Grzech, A., Ali, M. (eds.) IEA/AIE 2008. LNCS (LNAI), vol. 5027, pp. 159–168. Springer, Heidelberg (2008)
17. Kurose, J.F., Ross, K.W.: Computer Networking: A Top-Down Aproach Featuring the Internet. Addison Wesley, Boston (2001)
18. Menasce, D.A., Almeida, V.A.F.: Capacity planning for Web performance. Metrics, models, and methods. Prentice Hall PTR, New Jersey (2002)
19. Pcap library documentation, http://www.tcpdump.org/pcap.html (accessed 2010)

# Improving Packet Reception and Forwarding within Virtualized Xen Environments

Tomasz Fortuna and Blazej Adamczyk

Silesian University of Technology, Institute of Informatics
Akademicka 16, 44-100 Gliwice, Poland
{tomasz.fortuna,blazej.adamczyk}@polsl.pl
www.polsl.pl

**Abstract.** Popularization of virtualization techniques in networking makes, among the well-known advantages, the packet manipulation process to be more complicated and CPU-intensive. In this paper we study techniques for improving the performance of the packet reception and forwarding within virtualized Xen environments. Namely, in a set of experiments we demonstrate the possibility for decreasing the number of packet losses by proper IRQ balancing and discuss the possibility to exploit the coalesce functionality built-in modern network adapters.

**Keywords:** Linux, XEN, IRQ balancing, packet reception, performance.

## 1 Introduction

For several years now computer clusters are being widely deployed, not only to serve as fast computation centres, but also to improve availability and scalability of the computer systems. This clusters were most commonly built using computers, connected with each other via a high-speed Ethernet, by the use of one or more network switches.

Recently this pattern has been amended by the introduction of virtualization. Without sacrificing availability or scalability most of the machines in the cluster can be run from the same, multi core (SMP), physical hardware, and later, in case of hardware problems or cluster extension, migrated to the different machines. Generally, each of the machines running virtualized environment, can be though as a separate computer cluster in which physical machines are represented by guest operating systems and connections between those by a level 2 software network bridges working analogously to the aforementioned hardware switches.

One thing that may suffer in such setups is the performance. Even if all the virtual machines will be provided with an excessive number of CPU cores there are much more potential bottlenecks, in particular – a shared network card (NIC).

The network adapters communicate with the operating system with the help of a DMA transfers and the interrupts (IRQs). Some time after the card receives a packet, it raises the interrupt and thus informs the system about a packet reception. The kernel interrupt handler copies packet data into the memory and

A. Kwiecień, P. Gaj, and P. Stera (Eds.): CN 2012, CCIS 291, pp. 153–160, 2012.

performs additional preliminary operations before the packet is send into the userspace or is forwarded to other machines. This creates a much greater challenge for an operating system than the packet transmission, where the OS can buffer outgoing packets and dispatch them in larger chunks. Failure to respond fast enough for an incoming receive interrupt will effect in dropping the packets. The default IRQ handling methods utilized by the Linux kernel may limit the CPU usage, but do not maximize the packet reception performance.

The purpose of this article is to analyze drawbacks and side effects of the default behaviour in the plain and virtualized environments and propose possible solutions. We do not suggest any modifications to the Linux kernel or other system software. Rather we focus on how to improve the network performance using existing tools and configuration parameters.

For virtualization purposes, we use Xen hypervisor, [1], which is one of the most popular virtualization platforms. It has been released in an Open Source formula, which makes it perfect for scientific purposes. Xen hypervisor is a software running directly on the computer hardware, substituting the default operating system, which allows to run multiple independent instances of guest operating systems. The first guest system to boot is selected as a Privileged Domain (Dom0) – with direct access to the hardware and permissions to run further guests [1]. Our experiments were performed using stable Xen version 4.0.0.

As for the related work, the subject of Linux networking has been studied many times in the past. One can find several comprehensive books regarding Linux kernel as a whole ([2]) but also literature focused strictly on networking implementation, like [3] and [4]. There are also books related to the configuration and Linux network management, [5]. Some literature, regarding virtualized network resources is available as well (see [6]). Bottlenecks of networking in Xen was described well in [7]. The authors, also present some configuration pointers to achieve better throughput.

Nevertheless, all of the above do not clearly show how different optimization methods work in Linux with and without virtualization. We have also come to a conclusion that an average system throughput is not the most important factor of determining Linux networking efficiency. Our research shows that the first effect of performance problems are packet drops.

The remaining part of the paper is structured as follows. In Section 2, a detailed description of the testbed is given. In Section 3, the results of the experiments with IRQ balancing are presented and discussed. Then, in Section 4, other possible improvements based on the *coalesce* feature of the network adapter is presented. Finally, the paper is concluded in Sect. 5.

## 2    Description of the Testbed

Test environment consists of three devices:

- a HP server running a distribution of the GNU/Linux operating system,
- a Spirent TestCenter with four 1 Gb/s Ethernet ports, [8],
- and a reference hardware, a 1 Gb/s Cisco SG 100D-08 Ethernet switch.

The server has twelve 2.8 GHz Intel Xeon X5660 cores and is equipped with three network cards:

- 1x Quad port Intel 82571EB,
- 2x 1-port Broadcom NetXtreme II BCM5709.

Installed operating system uses two separate 2.6.38 Linux kernels – a vanilla (non-modified, generic) kernel, and one with the addition of Xen hypervisor patches.

Initial reference test was performed by connecting the Spirent Testcenter to the ports of the Ethernet switch using four, 1-meter long, twisted-pair 5E copper cables. After the reference measurements were taken, the switch was replaced with the server – two ports of the Intel NIC and two Broadcom ports were used in further tests.

During all performed tests, the Spirent was configured to transmit two independent data streams at a rate of 1 Gb per second. Each test was run until either two minutes elapsed, or $10^8$ packets were transmitted by each stream. Additional two low-rate, one packet per second (pps), streams were being transmitted in opposite directions, so that the tested devices could learn Spirent MAC addresses connected to each of their ports.

All tests were conducted with a constant data rate while altering the packet size, therefore the tested devices had to cope with a different number of packets per second in each test. The reported packet sizes don't include the Ethernet preamble (8 bytes total).

## 3   Test Results

Initial experiments with a the hardware Ethernet switch confirmed both the ability of the TestCenter to operate the required number of packets and the switch ability to forward 2 Gb/s data stream of 64 byte packets without dropping any of them.

Tests of the x86_64 Linux server started with the configuration mimicking the switch behaviour: The two pairs of the system interfaces (ethX) where bridged together by the use of the kernel bridge implementation (bridge-utils). This configuration, which features no virtualization techniques, is schematically shown in Fig. 1.

Two tests were performed in this setup; in the first one the CPU affinity of the network drivers receive interrupts (IRQs) was set to the same CPU core, and in the latter IRQ balancing techniques were used.

Most common approach to IRQ balancing is to let the system balance automatically. Although this tactic gives much better results than no balancing at all, it is not a perfect solution. Figure 2 presents the results of six, exactly the same, experiments. Each of them took two minutes and was separated from the others with at least half a minute of silence. Depending on how the network IRQs were balanced the results differ very much: in some cases the system drops at most 8% of the stream packets, in other cases – a half.

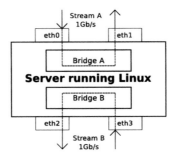

**Fig. 1.** Testbed configuration in the case without virtualization

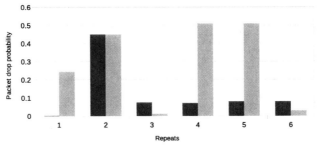

**Fig. 2.** Non-deterministic behaviour caused by automatic IRQ balancing

There are at least two separate methods of automatic balancing available: the simple built-in kernel balancing and more sophisticated userspace *irqbalance* (see [9]) software. Both were evaluated and produced similar non-deterministic behaviour which would severely affect the measurements. It was therefore necessary to manually balance the interrupts prior to gathering the test data.

Balancing was ultimately obtained by enabling the irqbalance under the network load, waiting until the results became satisfactory, and then disabling irqbalance – freezing the IRQ configuration. The same result could be obtained by manually setting CPU affinity masks in */proc/irq/\*/smp_ affinity* configuration files.

The results of the first test are presented in Fig. 3. In the cases where the drop probability values differed between the streams, the higher value was selected for plotting. The impact of the correct IRQ balancing becomes even more apparent in the virtualized environment where the CPU overhead grows significantly for each received packet.

The sketch of the next examined setup is shown in Fig. 4. It is a direct extension of the previous case, but this time the packet streams are forwarded into the virtual machines which are responsible for handling the traffic. In this setup the

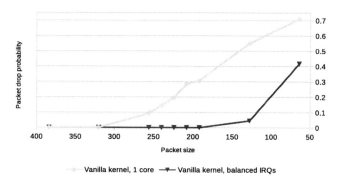

**Fig. 3.** Balanced IRQ / single core comparison (no virtualization)

packets are, as previously, forwarded back to the Spirent Testcenter. Xen itself uses Linux bridges to bind together physical network interfaces in dom0/driver domain and the virtual interfaces of guest operating systems (vifY.Z) so that there is a total of three bridges at the path of both streams.

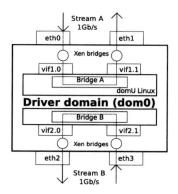

**Fig. 4.** Testbed configuration in the case with virtualization

The additional overhead is visible in Fig. 5 which compares the performance of these two setups during reception of 2 Gb/s stream of 256 byte packets with IRQs handled by a single core. The tested Xen kernel, in contrast to the vanilla kernel, did not include kernel-level IRQ balancing. The difference made by manually balancing the interrupts is shown in Fig. 6. It is important to notice a difference in scale of the packet size axis – the systems starts to drop packets much earlier compared to the first setup.

In order to achieve optimal network performance, it was necessary to disable automatic balancing and bind the interrupts to separate cores manually. In the virtualized environment the same problem occurs at one more level. The guest

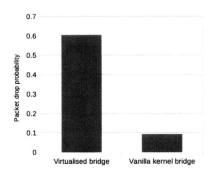

**Fig. 5.** Comparison of packet manipulation overhead in testbed configurations

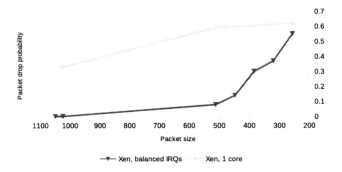

**Fig. 6.** Balanced IRQ / single core comparison (with virtualization)

operating systems run by the Xen operate the virtual processors (VCPUs). The Xen hypervisor uses a scheduler to balance the VCPUs across physical processors of an SMP host. Even if the IRQs are optimally bound to the virtual processors inside the driver domain, it is certainly possible that the hypervisor will share this virtual processor with another virtual machine. In the independent tests of an IIP system scheduler (defined in [10] and implemented according to [11]) this effect had a significant impact on the stability and determinism of the system behaviour. Xen allows to take the same precautions as with the IRQs and pin important VCPUs to certain physical cores (*xm vcpu-pin* command).

   In most network cards each interrupt corresponds to the single queue inside the adapter. To minimize the CPU cache misses the Intel proposes [12] for their Ethernet controllers to bind the receive and transmit interrupt pair to the same CPU. Within the Xen environment, it may be further effective to bind the net-back processes (part of a para-virtualized network adapter), operating a virtual interface (vif), to the same processor to which an interrupt of the physical interface bridged with the vif is bound to. The another drawback of an automatic balancing can be noticed when the network transmission is almost saturated. From time to time the kernel might rebalance the IRQs causing a sudden peak of dropped packets which otherwise appear sparsely. This case also benefits from the manual balancing.

After the packet is picked up by the kernel driver, the RPS (Receive Packet Steering) decides which CPU will be used to handle it [13]. In newer kernels this mechanisms is enabled by default and in some cases it might be beneficial to limit CPUs available for an interface to the one running the netback process.

## 4   Other Possible Improvements

Recent network adapters and their drivers have a built-in functionality for limiting the number of raised interrupts called *coalesce*. The utility able to control the feature parameters in the GNU/Linux system is the ethtool. There are three important settings:

- rx-usecs – maximal time to wait after a packet reception before the interrupt is raised (increase for bulk traffic, decrease for low-latency),
- rx-frames – number of frames to group before raising the interrupt,
- rx-frames-irq – maximal number of frames handled by a single interrupt.

Not all drivers support this feature and not all network cards implement it in the same way. Many different parameter configurations were tested, after manually balancing the interrupts, on both used NICs types, and none setting improved measurably the forwarding performance.

It is also possible to *coalesce* packets in the software after their reception, creating a much larger packet which is then forwarded into the virtual machine with much less overhead. The kernel achieves this for TCP streams using a GRO (Generic receive offload) mechanism written initially for 10 GbE adapters [14]. As it significantly alters the structure of a stream it is only possible to use it for incoming, non-forwarded data.

In limited cases where the driver domain doesn't need to filter, or otherwise alter the packets incoming into the guests, it is possible to use multi-queue network adapters to decrease the processing overhead. This can be done by assigning the MAC addresses directly to the NIC queues and therefore shift the packet demultiplexing, done by the software bridge, into the network card [15]. This approach also allows driver to store packet directly in the memory of a guest operating system, limiting the need to copy buffers between the running kernels.

## 5   Conclusions

In this paper we first demonstrated that virtualization may cause performance problems even with 1 GbE traffic, especially when small packet sizes are involved. Then we showed that proper IRQ balancing may improve the performance by far. However, by "proper balancing" we do not mean the standard, build-in automatic balancing algorithms, which, as presented, may behave in an indeterministic way. Instead of automatic balancing, what should be done is to enable the irqbalance under the network load, wait until the best results are obtained and then disable irqbalance – freeze the IRQ configuration. This procedure, which is now performed manually, indicates clearly that a new automatic IRQ balancing algorithm, prioritizing performance of the network, is needed.

**Acknowledgments.** This work is partially funded by the European Union, European Funds 2007–2013, under contract number POIG.01.01.02-00-045/09-00 "Future Internet Engineering".

# References

1. The Xen Hypervisor, http://xen.org/
2. Love, R.: Linux Kernel Development. Addison-Wesley Professional (2010)
3. Wehrle, K., Pählke, F., Ritter, H., Müller, D., Bechler, M.: The Linux Networking Architecture: Design and Implementation of Network Protocols in the Linux Kernel. Prentice Hall (2004)
4. Benvenuti, C.: Understanding Linux network internals. O'Reilly Media, Inc. (2006)
5. Kirch, O., Dawson, T.: Linux Network Administrator's Guide. O'Reilly Media, Inc. (2000)
6. Menon, A., Cox, A.L., Zwaenepoel, W.: Optimizing Network Virtualization in Xen. In: Proceedings USENIX Annual Technical Conference, pp. 15–28 (2006)
7. Zhang, H., Bai, Y., Li, Z., Du, N., Yang, W.: Achieving High Throughput by Transparent Network Interface Virtualization on Multi-core Systems. In: Proceedings of IEEE Second International Conference on Cloud Computing (2010)
8. Spirent Communications plc, http://www.spirent.com/
9. IRQ Balance, http://irqbalance.org/
10. Burakowski, W., Tarasiuk, H., Bęben, A., Góralski, W., Wiśniewski, P.: Idealne urządzenie umożliwiające wirtualizację infrastruktury sieciowej w Systemie IIP. In: Proceedings of the KSTiT Conference, Łódź, Poland (2011)
11. Zwierko, P., Tarasiuk, H., Rawski, M., Wiśniewski, P., Parniewicz, D., Juszczyk, A., Adamczyk, B., Kaliszan, A.: Platformy wirtualizacji dla Systemu IIP. In: Proceedings of the KSTiT Conference, Łódź, Poland (2011)
12. Intel Corporation: Assigning interrupts to processor cores, http://www.intel.com/content/www/us/en/ethernet-controllers/82575-82576-82598-82599-ethernet-controllers-interrupts-appl-note.html
13. Corbet, J.: Receive packet steering, http://lwn.net/Articles/362339/
14. Corbet, J.: Generic receive offload, http://lwn.net/Articles/358910/
15. Ram, K.K., Santos, J.R., Turner, Y., Cox, A.L., Rixner, S.: Achieving 10 Gb/s using Xen Para-virtualized Network Drivers. Xen Summit, Redwood City, USA (2009)

# Virtual Networks with the IPv6 Addressing in the Xen Virtualization Environment

Krzysztof Chudzik, Jan Kwiatkowski, and Kamil Nowak

Institute of Informatics, Wrocław University of Technology,
27 Wybrzeże Wyspiańskiego St, 50-370 Wrocław, Poland
{krzysztof.chudzik,jan.kwiatkowski,kamil.nowak}@pwr.wroc.pl

**Abstract.** There is a common agreement that virtualization allows more efficient use of the computational and communication resources. In most cases virtualization is used on the operating systems level. Occasionally virtualization is considered on the computer network level. The paper is devoted to the virtualization of the network connections at the Layer 2 and 3 within virtualizers and real devices. For this aim the Xen 4.0 virtualizer is used. VLANs and VLAN in VLAN tunnelling (802.1ad, QinQ) are considered at Layer 2. The static as well as the dynamic routing for the IPv6-in-IPv6 tunnelling are examined at Layer 3. The aim of the paper is to present the implementation of various technologies supporting the network virtualization in the real and virtual environments.

**Keywords:** virtualization, virtual networks, Xen virtualizer.

## 1 Introduction

Currently, virtualization of computational and communication resources is widely used. Services, operating systems and networks are virtualized using various methods of virtualization [1,2]. The virtualization, as a mechanism of abstraction and isolation of available resources, can be implemented at different levels of the computer network environment, communication as well as operating system levels. The network virtualization mostly is used at Layer 2 of the ISO/OSI model which is related to the physical addressing and switching or at Layer 3 associated with the logical addressing and routing. At Layer 2, for example, virtual local area networks (VLANs) can be created and each VLAN can posses a separate address space. At Layer 3, one can create virtual private networks and then the IP protocol can be used both as a transport protocol (tunnelling) and as a transported protocol (tunnelled). Moreover, the virtual devices which are equivalent to physical devices can be created, for example, multiple routers and/or switches, which are connected by virtual networks can be created within a single router, switch or virtualizer.

Cases of complex virtual networks implemented within virtualizers and connected to real networks do not often appear in the literature. The aim of the paper is to present the implementation of the various technologies supporting

A. Kwiecień, P. Gaj, and P. Stera (Eds.): CN 2012, CCIS 291, pp. 161–170, 2012.

the network virtualization in the real and virtual environments. The virtualization at Layer 2 and 3 of the ISO/OSI model is exploited. VLANs, the trunk connections and the 802.1ad (QinQ) standards are used at Layer 2 as well as the IPv6-in-IPv6 tunnelling, the static and the dynamic routing with tunnels at Layer 3. The Xen virtualizer has been selected for virtualization [3,4,5].

In the paper, the process of building more and more complex structures of the virtual networks is presented step by step. The practical aspects of each step are considered. The designed and implemented environment is described in Sect. 2. In the first step of experiments, the VLANs have been implemented. It is presented in Sect. 3. Based on VLAN structure, the static routing for the IPv6 protocol has been configured and tested. This process is described in Sect. 4. In the next step the IPv6-in-IPv6 tunnels have been configured and examined. They can be used for the construction of the complex virtual network structure, which abstracts from the physical network connections. Section 5 is devoted to the IPv6-in-IPv6 tunnels. The effective routing mechanisms have to be implemented for the packets forwarding in complex virtual networks. The static routing is not sufficient due to the manual configuration of the routing tables, therefore the dynamic routing have to be deployed. Section 6 is devoted to the OSPFv3 routing protocol implementation. From practical point of view, there is a need to carry VLANs inside another VLAN in real environment. Hence, Section 7 covers the 802.1ad (QinQ) standard for the VLAN in VLAN tunnelling. Finally, Section 8 concludes the paper and presents the future works.

## 2   Design Overview

The basic configuration of the designed and implemented environment is presented in Fig. 1, in Table 1 the network addressing is shown. The environment consist of the Cisco Catalyst 2900 switch, two servers running under the Debian 6 (Squezze) Linux distribution with Xen 4.0 as the virtualization environment (*debian1*, *debian2*). Two additional servers have been used for the VLAN isolation tests (*debian101*, *debian102*). The router has been configured for the access to the Internet.

Two VLANs, number 10 and 20 have been defined on the switch. The *debian1* and *debian2* servers are connected to the switch ports (23, 24) configured in the trunk mode. The ports working in the trunk mode can transmit data for multiple VLANs via a single connection (a network cable). The packets are distinguished by tags. The ports not configured for neither VLAN 10, 20 nor the trunk mode have been assigned to VLAN 1. VLAN 1 works as the native VLAN. The traffic from native VLAN is not tagged on the port working in the trunk mode.

The *debian1* and *debian2* servers have been configured with Xen 4.0 virtualization environment installed from the standard Debian 6 distribution repository. The *virt-manager* has been used for the Xen environment management. Additionally, the *vlan* package has been installed on *debian1* and *debian2* servers for

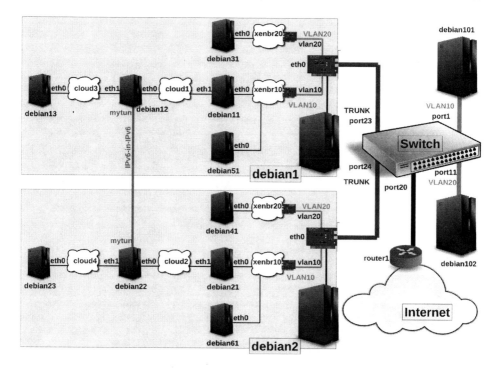

**Fig. 1.** The designed and implemented network environment

VLANs configuration. The interfaces for VLANs and the virtual bridges have been configured.

The virtual networks (noted as *clouds* in Fig. 1) have been defined to enable the creation of the complex network structures composed of virtual machines. The sample configuration of *cloud1* is as follows.

```
auto cloud1
iface cloud1 inet6 manual
  bridge_ports dummy1
  bridge_stp off
```

The virtual machines have been installed and configured in the *debian1* and *debian2* virtualization environment. The location and the connection of the virtual machines is presented in Fig. 1. Some of them are connected with two interfaces, for example, *debian11* with the *eth0* and *eth1* interfaces.

The servers *debian101* and *debian102* are connected to the switch ports configured for VLAN 10 (port 1) and VLAN 20 (port 11), respectively.

Although the IPv6 protocol is the main used protocol, the IPv4 protocol is also configured for the software installation from the Internet and for the device management purposes.

The address configuration for the physical devices and virtual machines is gathered in Table 1. If not stated otherwise, the address is the IPv6 address.

**Table 1.** The configuration of the physical devices and the virtual machines

| Device/Virtual Machine | Interface/VLAN | Address |
|---|---|---|
| *router1* | eth0 | IPv4: 192.168.1.254/24 |
| *switch* | vlan1 – management vlan | IPv4: 192.168.1.210/24 |
| *debian1* | eth0 | 2001:db8::1000:1/64 |
| | | IPv4: 192.168.1.201/24 |
| | vlan10 / VLAN 10 | 2001:db8::10:1/64 |
| | vlan20 / VLAN 20 | 2001:db8::20:1/64 |
| *debian2* | eth0 | 2001:db8::1000:2/64 |
| | | IPv4: 192.168.1.202/24 |
| | vlan10 / VLAN 10 | 2001:db8::10:2/64 |
| | vlan20 / VLAN 20 | 2001:db8::20:2/64 |
| *debian11* | eth0/VLAN10 | 2001:db8:0::11//64 |
| | route -6 add 2001:db8:2::/64 gw 2001:db8:0::21 | |
| | eth1 | 2001:db8:1::11/64 |
| *debian12* | eth0 | 2001:db8:1::12/64 |
| | Gateway: 2001:db8:1::11 | |
| | eth1 | 2001:db8:3::12/64 |
| *debian13* | eth0 | 2001:db8:3::13/64 |
| | Gateway: 2001:db8:3::12 | |
| *debian21* | eth0/VLAN 10 | 2001:db8:0::21/64 |
| | route -6 add 2001:db8:1::/64 gw 2001:db8:0::11 | |
| | eth1 | 2001:db8:2::21/64 |
| *debian22* | eth0 | 2001:db8:2::22/64 |
| | Gateway: 2001:db8:2::21 | |
| | eth1 | 2001:db8:2::22/64 |
| *debian23* | eth0 | 2001:db8:2::23/64 |
| | Gateway: 2001:db8:4::22 | |
| *debian31* | eth0/VLAN 20 | 2001:db8:0::31/64 |
| *debian41* | eth0/VLAN 20 | 2001:db8:0::41/64 |
| *debian51* | eth0/VLAN 10 | 2001:db8:0::51/64 |
| *debian61* | eth0/VLAN 10 | 2001:db8:0::61/64 |
| *debian101* | eth0/VLAN 10 | 2001:db8:0::101/64 |
| *debian102* | eth0/VLAN 20 | 2001:db8:0::102/64 |

# 3    Implementation of VLANs

The experiments related to the VLAN implementation have been conducted using the proposed in Sect. 2 environment. The following issues have been taken into account during the experiments:

- the support offered for links working in the trunk 802.1q standard by the virtualization environment based on Debian 6 with Xen 4.0,
- the communication between the virtualization environments via a trunk link using the hardware switch,
- the isolation between VLANs,
- the VLAN distribution within the virtualization environment.

Tests for VLAN have been carried out in two stages: the VLAN configuration and the tests of the VLAN operation. The virtual interfaces named *vlanXX*, where *XX* is the VLAN number, have been created. The packets with the relevant tags have been directed to the corresponding interfaces. The tests confirmed that it is possible to configure network interfaces in Debian 6 in such a way that the interfaces can receive and correctly interpret the VLAN tagged packets. The sample configuration for *debian1* and VLAN 10 is presented bellow.

```
auto vlan10
iface vlan10 inet6 static
  address 2001:db8::10:1
  netmask 64
  mtu 1500
  vlan_raw_device eth0
```

Then, the software bridges have been created to distribute VLANs to the virtual machines in the virtualization environment. One bridge has been created for each VLAN for every physical interface on the *debian1* and *debian2* servers. The sample configuration of the virtual bridge for VLAN 10 on the *debian1* is shown bellow.

```
auto xenbr10
iface xenbr10 inet6 manual
  bridge_ports vlan10
  bridge_stp off
```

The network interfaces of the virtual machines working in the specific VLAN have been virtually connected to the appropriate virtual bridge with the relevant VLAN number (Fig. 1). This operation has been performed using *virt-manager* during the virtual machine configuration. It is important that any additional configuration steps have not been taken inside the operating system of the configured virtual machines and that connection to the specific VLAN does not differ from the real (not virtual) environment.

All of the virtual machines connected to the virtual bridges on both physical servers (*debian1* and *debian2*) have been configured within the same logical address space. If the virtual machines are in the same VLAN then the direct communication between them would be possible. However, since the virtual machines are not in the same VLAN, the traffic from one VLAN should be isolated from the traffic from the second one. The *ping6* command has been used for the tests. The performed tests confirmed that the computers located in the same VLAN can communicate and the computers located in different VLANs cannot. The above is true regardless of whether tested on the same physical machine or between two different. The *wireshark* sniffer has been used for the monitoring of the network traffic and the correct use of tags have been confirmed. Additionally, the tests with the direct connection between the physical servers have been successfully carried out.

Note, that the traffic from the interfaces *eth0* on the *debian1* and *debian2* computers remains in the native VLAN 1. Hence, it is possible access to the Internet (in the both directions) without any additional effort after the IPv4 addresses assignment.

Concluding, the VLANs and the trunk connections can be configured and work correctly with the IPv6 protocol in the Xen 4.0 virtualization environment on Debian 6.

## 4   IPv6 Static Routing

The routing experiments in the Xen environment with VLANs have been performed with the use of the virtual machines *debian11*, *debian12*, *debian21* and *debian22*. The *debian11* and *debian21* virtual machines have been configured as routers. The aim was to obtain the connection between *debian12* and *debian22*. The addresses and the gateways of machines have been configured as shown in Table 1. The following line has been added to the configuration file */etc/sysctl.conf* on *debian11* to enable forwarding in the routing process.

```
net.ipv6.conf.all.forwarding=1
```

The file */etc/network/interfaces* of *debian11* contain:

```
auto lo
iface lo inet loopback
allow-hotplug eth0
iface eth0 inet6 static
  address 2001:db8:0::11
  netmask 64
  post-up route -6 add 2001:db8:2::/64 gw 2001:db8:0::21
allow-hotplug eth1
iface eth1 inet6 static
  address 2001:db8:1::11
  netmask 64
```

The *debian21* virtual machine has been configured analogously. The *debian12* and *debian22* machines have been configured as the network clients with gateways only. The results of the tests with the *ping6* and the *traceroute6* tools confirmed the correctness of the static routing processes.

## 5   Tunnelling at the IPv6 Logical Addressing Layer

The experiment aim was to transfer packets between *cloud3* and *cloud4* via the IPV6-in-IPv6 tunnel (Fig. 1). The configuration for the IPv6 static routing allowed the connection between *debian12* and *debian22*. The communication between *cloud3* and *cloud4* (*debian13* and *debian23*) is impossible. However, the configuration of the tunnel and the routing on *debian12* and *debian22* should allow to transmit data between *debian13* and *debian23* without the configuration of the routing to *cloud3* and *cloud4* on *debian21* and *debian11*, respectively. This property allows to build virtual IPv6 networks with complex structure which are independent of the underlying IPv6 networks.

The IPv6-in-IPv6 tunnel has been configured as follows (the configuration for *debian12*). The script */etc/network/tunnels/mytun* has been created with the following content:

```
#!/bin/bash
ip -6 tunnel add mytun mode ip6ip6 \
   remote 2001:db8:2::22 local 2001:db8:1::12 dev eth0
ip link set dev mytun up
ip -6 addr add 2001:db8:100::12/64 dev mytun
ip -6 route add 2001:db8:4::/64 via 2001:db8:100::22 dev mytun
```

The script execution has been associated with the initialization of the *eth0* interface as presented in the configuration file */etc/network/intefaces*.

```
allow-hotplug eth0
iface eth0 inet6 static
   address 2001:db8:1::12
   netmask 64
   gateway 2001:db8:1::11
   post-up sleep 5; /etc/network/tunnels/mytun
```

The above static configuration can be performed from command line, which allows dynamical changes of the virtual networks topology. The tests with *ifconfig*, *ip*, *route*, *ping6* and *traceroute6* confirmed the configuration correctness and the proper behavior of the virtual networks. Note that, the packets inside the tunnel have been transmitted across the routers that have not been aware of the addresses of the packets encapsulated in the IPv6-in-IPv6 tunnel.

# 6   Dynamic Routing with OSPFv3

During the experiments with OSPFv3 the following assumption have been taken:

- *debian11* and *debian21* are routers which carried the tunnels (a kind of transport infrastructure);
- *debian12* and *debian22* are routers on which the virtual network consists of the IPv6-in-IPv6 tunnels for applications are configured;
- *debian13* and *debian23* are considered as the end user and application machines.

The dynamic routing with the OSPFv3 protocol has been configured on the *debian11* and *debian21* routers. The automatic configuration for the tunnel transport infrastructure is acquired. Similarly to the static routing, the routers have not been aware of cloud3 and cloud4. The routers have been configured as follows.

The OSPFv3 protocol has been run using the *quagga* package [6]. The *zebra* and the *ospf6d* daemons have been enabled and configured. For the *ospf6* daemon, the *ospf6.conf* configuration file is presented bellow (for *debian11*).

```
interface eth0
 ipv6 ospf6 instance-id 0
!
interface eth1
 ipv6 ospf6 instance-id 0
!
router ospf6
 router-id 255.1.1.11
 interface eth0 area 0.0.0.0
 interface eth1 area 0.0.0.0
```

The router-id for *debian21* has been configured as 255.1.1.21.

The *debian12* and *debian22* routers have been configured in a similar manner. The *ospf6.conf* file for *debian12* was in the following form.

```
interface eth1
 ipv6 ospf6 instance-id 0
!
interface mytun
 ipv6 ospf6 instance-id 0
!
router ospf6
 router-id 255.1.1.12
 interface eth1 area 0.0.0.0
 interface mytun area 0.0.0.0
```

The router-id for *debian22* has been configured as 255.1.1.22. The crucial in the *debian12* and *debian22* configuration is that the *eth0* interfaces connected to the transport routers (*debian21* and *debian22*) are not enabled in the OSPFv3 configuration. This allowed to isolate the configuration of the routers responsible for routing in virtual networks (*debian12* and *debian22*) from the configuration of the routers responsible for tunnel transport (*debian11* and *debian21*).

The tests of the configured environment confirmed the configuration correctness and the proper behavior. The area of the applications and the virtual networks worked independently and do not exchange any information about the network configuration. However, *debian11* and *debian12* could route packets to any network and the routing protocol worked with one process and one routing table. Therefore, the address spaces have to be planned carefully.

## 7    802.1ad (QinQ) Standard – VLAN in VLAN Tunneling

The Cisco switch in Fig. 1 has been replaced by the structure illustrated in Fig. 2 for the 802.1ad (QinQ) standard implementation.

Two Juniper EX 4200 switches (*S_J1*, *S_J2*) have been configured for carrying the VLAN in VLAN traffic between them. The ports between the switches have been configured in the trunk mode, while the ports connected to *debian1*

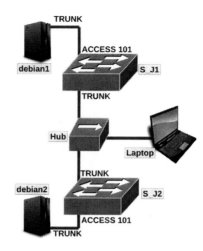

**Fig. 2.** The 802.1ad (QinQ) test environment

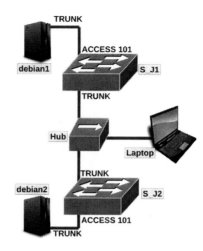

**Fig. 3.** The packet transmitted inside the IPv6-in-IPv6 tunnel and the VLAN in VLAN encapsulation, simultanously

and *debian2* as the members of VLAN 101. Hence, packets received from the *debian1* and *debian2* have been encapsulated in VLAN 101 between the switches. The configuration of the *debian1* and *debian2* computers remained unchanged. A hub and a laptop computer with the *wireshark* sniffer have been used to monitor traffic on the trunk ports of the switches. The *ping6* command has been executed between *debian13* and *debian23*. The sample packets are shown in Fig. 3.

All the levels of encapsulation may be observed, ie IPv6-in-IPv6 and VLAN 10 in VLAN 101.

It is worth emphasizing that not all switches support 802.1ad (QinQ). The Cisco switch used in the basic environment does not support this technology. The network equipment should be selected very carefully for 802.1ad (QinQ).

## 8   Conclusions

The process of creation of the complex computer networks based on IPv6 that use virtual and real resources is presented. The experiments performed on the designed and implemented environment confirm that the Xen virtualizer can be used to building such networks. The paper demonstrates how to build the environment that provides virtualization at Layer 2 and 3. For Layer 2, the way in which VLANs should be configured within the Xen virtual environment and the method in which the real network devices are configured for the 802.1ad standard is proposed. For Layer 3, the configuration of the IPv6-in-IPv6 tunnels with the dynamic routing using the OSPFv3 protocol is presented. The applied tunnel configuration does not provide the secure way for the data exchange. Therefore, further work will be focused on the protocols that ensure the secure transmission, for example the IPsec protocol.

**Acknowledgement.** The research presented in this paper has been partially supported by the European Union within the European Regional Development Found program no. POIG.01.01.02-00-045/09-00.

## References

1. Wolf, C., Halter, E.: Virtualization From the Desktop to the Enterprise. Apress (2005)
2. Tulloch, M., The Microsoft Virtualization Teams: Understanding Microsoft Virtualization Solutions from the Desktop to the Datacenter, 2nd edn. Microsoft Press (2010)
3. The Xen.org Website (February 2012), http://www.xen.org
4. Barham, P., Dragovic, B., Fraser, K., Hand, S., Harris, T., Ho, A., Neugebauer, R., Pratt, I., Warfield, A.: Xen and the art of virtualization. SIGOPS Oper. Syst. Rev. 37(5), 164–177 (2003)
5. Hagen, W.: Professional Xen Virtualization. Wiley Publishing, Inc. (2008)
6. Quagga Routing Suite (February 2012), http://www.quagga.net

# Multi-agent Based Approach
# of Botnet Detection in Computer Systems

Oleg Savenko, Sergiy Lysenko, and Andriy Kryschuk

Institute of Telecommunication and Computer Systems,
Department of System Programming, Khmelnytskyi National University,
Instytutska, 11, Khmelnytskyi, Ukraine
kism@beta.tup.km.ua, sirogyk@ukr.net, rtandrey@rambler.ru
http://spr.khnu.km.ua

**Abstract.** A new approach for the botnet detection based on multi-agent system is proposed. For increasing of the efficiency of botnet detection multi-agent systems were involve that allowed to make antivirus diagnosis via agents' communication within corporate network. The structure and main principles of antiviral agents' functioning within multi-agent system are developed. The principles of communication between the agent's units before and after attack on the computer system were developed. A new technique for sensor diagnosis in monitor mode which uses fuzzy logic was developed. A new technique for sensor diagnosis in scanner mode with generation of detectors using the modified negative selection algorithm was developed.

**Keywords:** botnet, Trojan, worm-virus, antivirus detection, multi-agent system, agent, sensor, fuzzy logic.

## 1 Introduction

The analysis of the situation of development of the malware shows dynamic growth of its quantity. The most numerous classes of malware during last 10 years are Trojans and worm-viruses which spread and penetrate into computer system (CS) for the purpose of information plunder, DDoS attacks, anonymous access to network, spy actions, spamming that represents real danger [1] (Fig. 1). Despite the regular refinement of methods of the search, detecting and removal

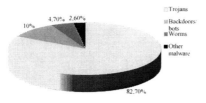

**Fig. 1.** Malware rate in 2011

A. Kwiecień, P. Gaj, and P. Stera (Eds.): CN 2012, CCIS 291, pp. 171–180, 2012.

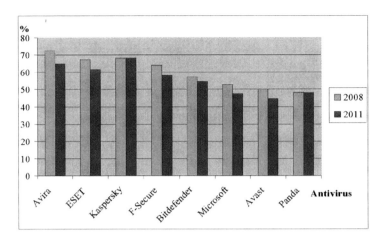

**Fig. 2.** Worm-viruses' detection in 2008 vs 2011 years

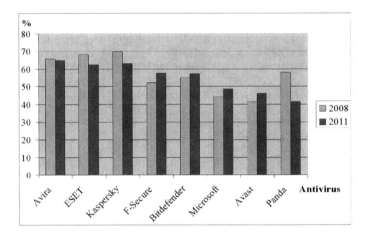

**Fig. 3.** Trojans' detection in 2008 and 2011 years

Trojans and worm-viruses of different function, regular updates of anti-virus bases, the numerous facts of plunder of the confidential information are observed and the various destructive operations are performed which lead to serious negative consequences. Common techniques used in modern antivirus software of Trojans' and worm-viruses' detection are signature-based one, code emulators, encryption, statistical analysis, heuristic analysis and behavioural blocking [2]. However, the accuracy of detection of new malware is low, and in recent years it has constantly decreased [3] (Figs. 2 and 3). One of the main reasons for the lack of detection accuracy is cooperating of Trojans with worm-viruses.

Over the past 3–5 years there is a clear dynamics of conception of a new malware class – botnet (Fig. 4). Botnet today represents a real threat to computer systems users; the accuracy of its detection is low because of its complicity.

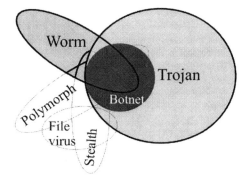

**Fig. 4.** Place of botnet among all malware

That is why the actual problem of safety of various CS is a development of new more perfect approach of antivirus detection. One of possible way to increase the detection efficiency is a construction of virus multi-agent system in CS for new botnet detection. For this purpose it is necessary to develop the principles of such system functioning; to describe the communication and functions' features of agents; to formalize sensors' and effectors' properties.

## 2   Multi-agent System of Botnet Detection

To increase the efficiency of botnet detection we involve multi-agent systems that will allow us to make antivirus diagnosis via agents' communication within corporate network [4]. Usage of multi-agent systems for botnet detection requires a generation of agents set with some structure and functionality [5].

Each agent should implement some behaviour and should include a set of sensors (components that directly is effected by the CS), a set of effectors (components of that effect the CS) and CPU – information processing unit and memory [6]. The scheme of antiviral agent multi-agent system operation is shown in Fig. 5. Let us present agent as a tuple:

$$A = \langle P, R, K, S_1, S_2, S_3, S_4, S_5, S_6, \ldots, S_n \rangle \tag{1}$$

where $P$ – processor, which provides integration and processing data, processing optimal response to the incoming information about the computer system state, decision on the steps to be done; $R$ – rules, that change agent behaviour according to incoming information; $K$ – agent knowledge – part of rules and knowledge, that could be changed during its functioning; $S_1$ – communication sensor, communicates with other agents via network protocols; $S_2$ – agent of signature-based analysis; virus detection is performed by searching signatures in database [7]; all signatures are detectors generated using the modified negative selection algorithm [8,9]; antivirus system alarms if computer is infected; $S_3$ – checksum sensor; $S_4$ – sensor of heuristics analysis; detection is performed in

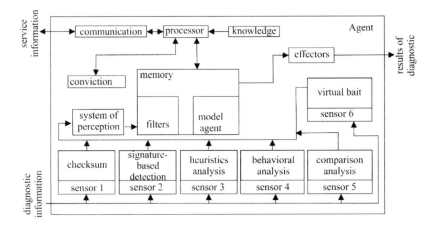

**Fig. 5.** The scheme of antiviral agent multi-agent system operation

monitor mode with the use of fuzzy logic; sensor makes a conclusion about the danger degree of computer system infection with a new botnet [10]; $S_5$ – sensor of comparative analysis through application programming interface API and driver disk subsystem via IOS. If data on file received the first way differ from those obtained by the second way, file is infected; $S_6$ – sensor – "virtual bait"; it is used for modelling of possible attacks or unauthorized access and it allows to learn the strategy of attacker and to identify a list of tools and actions intruder can do on infected CS. If a remote administration of network is not carried out, all incoming ssh-traffic is redirected to this sensor.

The processor processes the input data and determines the level of risk of specified object in the computer system. There is a knowledge base of trusted software.

Conviction unit provides knowledge for agent in unusual situations. This will reduce the number of false positives in the new botnet diagnosis of CS. The filters system for each sensor proposed to establish the risk factors for the evaluation of objects. Exceeding the limit values of the coefficients including the experience of all agents indicates the computer system infection with botnet.

Diagnostic information according to their functional properties each sensor is submitted. Work results of the checksums and signature analysis sensors may not require full engagement of the agent functioning for notification of the infection with botnet, but in conjunction with results of other sensors and communication with other agents this sensors may assert this signal detection of the botnet.

Unit of perception holds summary information to the general form for further work. Then the information goes to the input of filters. Filters reject data generated by trusted programs or units.

Depending on the level of danger detected attacks the coefficients are defined by filters. The data from the filters are to be processed by agent processor which determines whether the computer system is infected. Because of lack of data, the agent communicates with other agents for similar influence of programs' actions.

The availability or absence of such information from other agents affects the final agent decision on a particular file or process. When comparing the results obtained with the conviction unit data changes of coefficients and trusted programs are held. Communications unit is responsible for encryption and decryption of interagents' information.

Agent results are transmitted to the effectors, as a means of influence on the computer system. If malware is detected agent through effectors blocks the process or processes that are responsible for performance of some malware and then notifies the user about the infection.

Agent model ensures the integrity of the agent's structure. It is realized by implementation of system checkpoints to provide the serviceability of this agent. Also after each checking all agent critical elements are stored for later restoring in case of virus attack on antivirus multi-agent system or possible failures in the computer system.

Each agent can activate the recheck the selected number of sensors to refine the results. In situation when agent cannot communicate with other agent it is as autonomous unit and is able to detect different malware relying on knowledge of the latest updates and corrections in the trusted software baseIt is advisable to keep all the given values.

## 2.1  Sensor of Botnet Detection in Monitor Mode

A new technique for sensor diagnosis in monitor mode which uses fuzzy logic was developed. It is based on behavioural model of malware [10]. This sensor enables to make a conclusion about the degree of danger of computer system infection by malware. For this purpose we construct the input and output linguistic variables with names: "suspicion degree of software object" – for the input linguistic variable, and "danger degree of the infection" – for output one.

The task of determination of membership function for input variable we will consider as the task of the ranking for each of mechanisms (functions) $m_i$ of penetration ports $p_j$ with the set of indications of danger $Z$ and a choice of the most possible $p_j$ with activation of some function $m_i$. Then we generate a matrix of advantage $M_{\text{adv}} = |q_{ij}|$. Elements of given matrix are positive numbers: $q_{ij} = q_i/q_j$, $0 < q_{ij} < \infty$; $q_{ji} = 1/q_{ij}$, $q_{ii} = 1$ $i, j = \overline{1, l}$, $l$ – amount of possible results. Elements $q_{ij}$ of matrix $M_{\text{adv}}$ are defined by calculation of values of pair advantages to each indication separately taking into account their scales $Z = \{z_k\}$; $k = \overline{1, r}$; with usage of such formula

$$q_{ij} = \left( \sum_{k=1}^{r} q_{ij} \cdot p_k \right) / \left( \sum_{k=1}^{r} q_{jk} \cdot p_k \right) . \tag{2}$$

Eigenvector $\Pi = (\pi_1, \ldots, \pi_m)$ is defied by using a matrix of advantage. This eigenvector answers maximum positive radical $\lambda$ of characteristic polynomial $|M_{\text{adv}} - \lambda \cdot E| = 0$. $S \cdot \Pi = \lambda \cdot \Pi$, where is an identity matrix.

Elements of vector $\Pi(\sum \pi_i = 1)$ are identified with an estimation of experts who consider the accepted indications of danger. The same procedure is performed for all $m_i$. As a result we receive a matrix of relationship $V_p = |m_i, p_j|$, in which each pair (relationship) $m_i$, $p_j$ value $0 < \pi < 1$ responds.

Using matrix $V_p = |m_i, p_j|$, we build matrix $V_p^* = |m_i, p_j|$ in which the relationship $(m_i, p_j)$ is used and the elements of this relationship have value $\pi_{\max}(0 \leq \pi_{\max} \leq 1)$. Using matrix $V_p^* = |m_i, p_j|$, we build normalized curve for membership function $\mu_{Xp}(R)$ of an input variable.

Formation of function membership and at the stages of activation $\mu_{Xa}(R)$ and executing of the destructive actions $\mu_{Xe}(R)$ are similar.

As a part of the solution of the problem the FIS using Mamdani algorithm was realized (Figs. 6 and 7). The results of fuzzy inference system 0.804 are

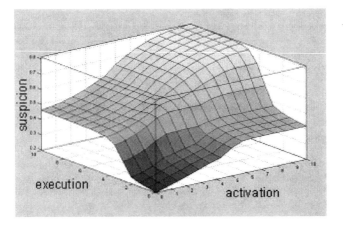

**Fig. 6.** Results of the fuzzy inference system implementation

interpreted as the degree of CS infection with malware. If the resulting number exceeds some adopted threshold of danger antivirus system will block actions of the aqueous object. The sensor also transmits information about suspicious software to other agents.

## 2.2   Sensor of Botnet Detection in Scanner Mode

The scanner mode detection involves the following steps: forming a set of files to be scanned: system libraries, executables system services and device drivers, which can be taken as the samples; generate protected sequences and detectors depending on operating system; comparison of the protected sequences with detectors at the stage of virus scanning; notification about the substitution when the protected sequences match with detector; check the suspicion of software actions.

Thus protected sequences and detectors have format for GNU / Linux operating system:

$$D_i^L = \langle m_1 \ldots m_i \ldots m_{x1}, u_1 \ldots u_i \ldots u_{x2}, g_1 \ldots g_i \ldots g_{x3}, s_1 \ldots s_i \ldots s_{x4}, \\ t_1 \ldots t_i \ldots t_{x5}, C_1 \ldots C_i \ldots C_{x6} \rangle \tag{3}$$

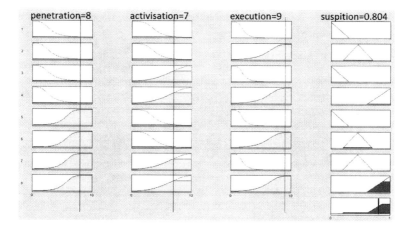

**Fig. 7.** Membership function of fuzzy set "suspicion degree"

where $m_1 \ldots m_i \ldots m_{x1}$ – file mode (type, permissions); $u_1 \ldots u_i \ldots u_{x2}$ – identifier of the file owner; $g_1 \ldots g_i \ldots g_{x3}$ – identifier of the group owner; $s_1 \ldots s_i \ldots s_{x4}$ – file size; $t_1 \ldots t_i \ldots t_{x5}$ – time of last file modification; $C_1 \ldots C_i \ldots C_{x6}$ – CRC of the file, $i = \overline{1, n}$ , $n$ – number of detectors.

Protected sequences and detectors have format for MS Windows operating system:

$$D_i^{W} = \langle s_1 \ldots s_i \ldots s_{z1}, t_1 \ldots t_i \ldots t_{z2}, a_1 \ldots a_i \ldots a_{z3}, C_1 \ldots C_i \ldots C_{z4} \rangle \qquad (4)$$

where $s_1 \ldots s_i \ldots s_{z1}$ – file size; $t_1 \ldots t_i \ldots t_{z2}$ – time of last file modification; $a_1 \ldots a_i \ldots a_{z3}$ – file attribute (read-only, hidden, system, archived); $C_1 \ldots C_i \ldots C_{z4}$ – CRC of the file, $i = \overline{1, n}$, $n$ – number of detectors.

Generation of detectors is performed using the modified negative selection algorithm [7,9,10].

### 2.3 Agents' Functioning

Let a communication agent message present as a tuple:

$$\langle g, h, Com_x, Com_v, Mes, t \rangle \qquad (5)$$

where $g$ indicates whether it is a report, order or fetch of communication message; $h$ – type of the agent message; $Com_v$ – message receiver; $Com_x$ – message sender, $Mes$ – agent message content; $t$ – sending time.

Thus the communication between the units within its sensors before attack or intrusion can be represented:

$$\begin{aligned}
\langle R, N, Se, P, Inf_{int}, t_0 \rangle &\Rightarrow \langle f, Int, P, M, Inf_{int}, t_1 \rangle \cap \\
\langle O, C, P, Se, Inf_M, t_2 \rangle &\cup \langle O, S, P, Se, Inf_M, t_2 \rangle \cap \\
\langle O, R, P, E, Inf_{int,Sh}, t_2 \rangle &\cap \langle O, I, P, Sh, Inf_{int}, t_2 \rangle
\end{aligned} \qquad (6)$$

where $R$ – report, $O$ – order, $F$ – fetch of the communication messages; $N$ – new attack, $Int$ – intrusion, $C$ – continue, $S$ – stop, $Red$ – redirect, $I$ – initialization as a type of the massage; $P$ – agent processor; $Se$ – sensors $S_1 \ldots S_5$; $Sh$ – virtual bait; $E$ – effectors; $M$ – are respectively the sender and receiver of the message; $Inf$ – the content of the message; $t$ – time of the message sending.

The communication (interactions) between the units within its sensors after attack or intrusion can be represented:

$$\begin{aligned}
\text{AttackApproved} &\Rightarrow \langle R, At, P, M, Inf_{int}, t_3 \rangle \cap \\
\langle R, At, P, E, Inf_{int}, t_3 \rangle &\cap \langle O, S, P, Sh, Inf_{int}, t_3 \rangle \cap \\
&\langle R, At, P, Se, Inf_{int}, t_3 \rangle
\end{aligned} \tag{7}$$

$$\begin{aligned}
\text{AttackDisapproved} &\Rightarrow \langle O, S, P, Sh, Inf_{int, Se}, t_3 \rangle \cap \\
\langle O, Red, P, E, Inf_{int, Se}, t_3 \rangle &\cap \langle O, C, P, Se, Inf_{int, Sh}, t_3 \rangle
\end{aligned} \tag{8}$$

where $At$ means – attack to computer system.

Let us formalize the function $F$ which identifies the worth of agent $A_1$ at time $t$ and associates a real number to each of agents as the worth of that agent:

$$F : 2^{Ag} \times T \to R, \ F(A_1, t) = \sum_{k=1}^{r} \frac{1}{(T_a - t)} \cdot N_{lk}^2 + \frac{1}{t} \cdot \frac{\Theta_{lk}}{N_{lk} + \varepsilon}, \tag{9}$$
$$F(A_1 \cup A_j) \geq F(A_1, t) + F(A_k, t), \ l \neq k$$

where $Ag$ – a set of agent units which are formed by combination of units with different types; $Ta$ – the time of performing diagnosis actions; $d$ – the number of agent components types; $N_{lk}$ – the number of sensors of type $k$ in agent $Al$ and $\Theta_{lk}$ – the sensors weight of type $k$ within the agent no matter of their amount.

A good incentive for agents at the initial moments of reporting intrusion can be provided by sensors Se in the system in the sense that they will form better coalitions and thus collaborate. As we can see in (6) no agent in AMAS can get more advantage by changing its actions. Also the function F does not increase by changing the agents set.

## 2.4    Experimental Research

Software for realisation of antivirus multi-agent system on proposed techniques was developed.

Interface results window of botnet diagnosing of computer system is shown in Fig. 8. For the experimental determination of the efficiency of developed software 217 programs with the botnets' properties were generated and launched on different amount of workstations (Table 1). Accuracy of botnet detection of the developed software in comparison with known shows growth of accuracy by 3%–5% in comparison with known antivirus software.

Also we performed false detection experiments and it is about 3%–7%. But with the growth of agents amount false detection is reducing to 2%–4%.

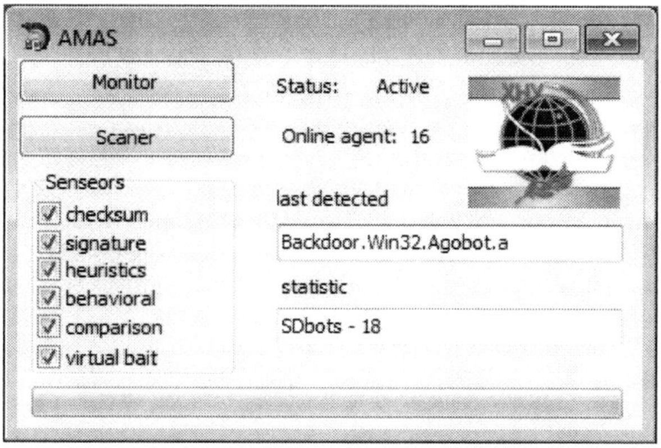

**Fig. 8.** Software of botnet detection

**Table 1.** Accuracy of botnet detection with developed software

| Botnets(number) | Workstations (agents) | | | |
|---|---|---|---|---|
| | 16 | 24 | 32 | 40 |
| SDbot (80) | 74% | 75% | 78% | 81% |
| Rbot (49) | 61% | 63% | 64% | 67% |
| Agobot (54) | 60% | 60% | 62% | 63% |
| Spybot (18) | 64% | 67% | 70% | 74% |
| Mytob (16) | 53% | 56% | 59% | 62% |
| **Accuracy [%]** | 62.8% | 64.6% | 67.8% | 69.8% |

# 3    Conclusions

A new approach for the botnet detection based on multi-agent system is proposed.

The structure and main principles of antiviral agents' functioning within multi-agent system is developed.

A new technique for sensor diagnosis in monitor mode which uses fuzzy logic was developed. The principles of communication between the agent's units before and after attack on the computer system were developed. Software for realisation of antivirus multi-agent system on proposed techniques was developed. It shows growth of accuracy by 3%–5% in comparison with known antivirus software.

# References

1. Goshko, S.: Encyclopedia of protection against viruses. SOLON-Press, Moscow (2005) (in Russian)
2. Savenko, O., Lysenko, S., Kryshchuk, A.: Research of the antivirus technologies for malware detection. In: XII Conference: Modern Informations & Electronic Technologies – 2011, Odessa, pp. 95–96 (2011)

3. AV Comparatives laboratories, http://www.av-comparatives.org
4. Wooldridge, M.: An Introduction To Multiagent Systems. John Wiley & Sons Ltd. (2002)
5. Shoham, Y., Leyton-Brown, K.: Multiagent Systems Algorithmic, Game-Theoretic, and Logical Foundations. Cambridge University Press (2009)
6. Alkhateeb, F., Maghayreh, E., Doush, I.: Multi-Agent Systems – Modeling, Control, Programming, Simulations and Applications. InTech (2011)
7. Savenko, O., Lysenko, S.: Developing process of detection of Trojan programs using artificial immune systems, pp. 183–188. Bulletin of the Khmelnytsky National University, Khmelnytsky (2008) (in Ukrainian)
8. Forrest, A., Perelson, L., Cherukuri, R.: Self-nonself discrimination in a computer. In: Proceedings of the IEEE Symposium on Research in Security and Privacy. IEEE Press, Oakland (1994)
9. Castro, L., Timmis, T.: Artificial Immune Systems: A New Computational Approach. Springer, London (2001)
10. Bernikov, A., Grafov, R., Savenko, O., Lysenko, S.: Malware search in the distributed simulators using the technology of fuzzy logic. In: Information Technologies, Moscow, pp. 42–47 (2011) (in Russian)

# Preventing TMTO Attack in AES-CCMP in IEEE 802.11i[*]

Iman Saberi[1], Bahareh Shojaie[1], Mazleena Salleh[1], Mahan Niknafskermani[1], and Mohammad Javad Rostami[2]

[1] Faculty of Computer Science & Information Systems,
Universiti Teknologi Malaysia, Johor Bahru, Malaysia
{isaberi,bahar.shojaie,mahan.nk}@gmail.com, mazleena@utm.my
[2] Bahonar University, Kerman, Iran
mjrostami@uk.ac.ir

**Abstract.** This study is conducted to establish an alternative, creative technique for the structure of Advanced Encryption Standard-Counter Mode with Cipher Block Chaining Message Authentication Code Protocol (AES-CCMP) key in IEEE 802.11i. the structure of proposed method increase the length of AES-CCMP key from 128 bits to 256 bits to eliminate Time-Memory Trade-Off (TMTO) attacks by using three proposed solutions including Random Nonce$_{Key}$, Four Way Handshake alteration and Pseudo Random Function (PRF). Besides, two proposed and classic methods are compared in terms of TMTO attack probability, avalanche effect, changes in neighbor blocks, memory usage and execution time. According to the results, the proposed method is completely resistant to TMTO attack. In addition, avalanche effect and change in neighbor blocks of proposed method are so near to optimized state and also, two classic and proposed methods are approximately the same in case of memory usage and execution time.

**Keywords:** TMTO, AES-256, AES-CCMP, 802.11i.

## 1 Introduction

802.11i uses encryption method of AES-128 in counter mode for providing confidentiality and Cipher Block Chaining Message Authentication Code (CBC-MAC) for creating Message Integrity Code (MIC) for providing integrity and preventing message injection and modification of man in the middle attack.

Encryption phase of AES-CCMP starts with creating Initial Counter (IC) by joining Nonce, Flag and Counter (CTR) fields [1]. Also, Nonce is created from Packet Number (PN), Priority and MAC Address 2 (A2). In encryption phase, MIC and original data contribute [2].

---

[*] The work is funded by RMC, Universiti Technologi Malaysia (UTM) under the vote number C.C. NO. Q.J130000.7128.00J14.

A. Kwiecień, P. Gaj, and P. Stera (Eds.): CN 2012, CCIS 291, pp. 181–190, 2012.

In counter mode, IC is encrypted and this encrypted 128 bits is XOR with plaintext, then the value of CTR in each block is increased one unit [3,4]. So, IC value is updated for next block. This procedure is repeated until the cipher text is completely generated. Thus, security of counter mode is dependent to the Temporal Key (TK), and IC structure.

The predictable structure of IC in AES-CCMP and probability (almost one) of creating IC by unauthorized person, also reducing the effective key length in pre computational attack increase the probability of Time-Memory Trade-Off (TMTO) attack in AES-CCMP [5]. TMTO decrease launch time of attack by using memory and by creating trade off between memory and time is done in a short time. So, enhancing IC structure and TK length in AES-CCMP is crucial.

The scope of this study is encryption method of AES-CCMP in 802.11i to create an innovatory method of generating key to increase key length and build a robust secure method in order to prevent TMTO attack.

The presentation of the paper is as follows: Section 1 introduces the AES-CCMP structure and defines the scope of this study, while Sect. 2 identifies TMTO attack properties and three proposed methods of creating 256 bit key in AES-CCMP. Section 3 describes the results of proposed method evaluation of probability of TMTO attack, avalanche effect, neighbor block changes, memory usage and execution time is compared with classic method. At last, Section 4 concludes the paper based on the innovatory method and results of Sect. 3.

## 2   Discussion

Based on the weaknesses of Nonce structure and key length in AES-CCMP structure, one of the major attacks in mentioned structure is TMTO attack. In this attack, attacker tries to make a balance between time and memory usage. In general, two solutions for finding $n$ bit key are existent:

1. Exhaustive key search: In this method, the usage of memory is zero. But, the role of time is significant, and $2^{n-1}$ times the attack should be done (time=$2^{n-1}$, memory=0). Such as Brute force attack.
2. Pre computation: In this method, if $C = E(P, K)$, then $2^n$ possible keys are created. So, the memory usage is maximum value and it is equal to $2^n$. But the time needed for attack is decreased to 1 (memory=$2^n$, time=1).

TMTO attack lies between the first and second mode and TMTO approximately requires $T = t^2$ times ($N^{1/3}$ operations per each table), $M = mt$ word of memory ($N^{1/3}$ tables, each table with $m = N^{1/3}$ words) [6]. Also, the value of $N$ is equal to $mt^2$ ($N = 2^n$ (possible solutions of key)). Letting $m = t = N^{1/3}$ results in $M = T = N^{2/3}$. So, the effective length of n bit key is reduced to 1/3 which is equal to $2n/3$. Considering the recommendation of 1996 ad-hock network, minimum length of key for symmetric cipher methods to provide proper security is equal to 75 bits key length in that time [7]. Also, the mentioned report recommends adding 14 bits to keep it secure for next 20 years at least.

According to Moore's law [8], for 16 years (1996 to 2012) and five more years for validity of new method (21 years), it requires adding 14 bits to minimum effective key length. In these days, recommended minimum key length is equal to 103 bits (75+14+14).

Considering the decreasing effective key length in TMTO from n to $2n/3$, AES-CCMP in 802.11i uses 128 bits cipher key. So, the effective key length in AES-CCMP is approximately equal to 85 bits ($2*128/3 \approx 85$). However, the least recommended key length is equal to 103 bits in order to resistant TMTO attacks. Also, according to weak structure of Nonce and IC in first block, the probability of predicting these values are near one. So, attacker can easily calculate the input of first block. Thus, the total workload for a successful TMTO attack in AES-CCMP is only $2^{85}$ times [7,9]. For solving mentioned issue, two general solutions are suggested [7,10]:

1. Increasing key length: For $n$ bit key, the attacker needs $2^{2n/3}$ time trial and error, so the probability of this attack can be zero for 256 bits key. Because creating and comparing $2^{171}$ states is impossible in practice.
2. Adding the random bytes in Nonce structure: By adding random bytes to Nonce structure, attacker should generate all the possible values of random Nonce for all the keys, that is so long for calculating these values and the attacker cannot do it.

This paper is focused on the first mentioned solution which increases key length. One of the existing ways is using AES-256 instead of AES-128 in AES-CCMP structure. In this method, in encryption structure of AES-CCMP, which shows in Fig. 1, instead of using AES-128 modules, AES-256 modules are used.

For increasing key length from 128 bits to 256 bits in AES-CCMP, in this research, three general solutions are proposed.

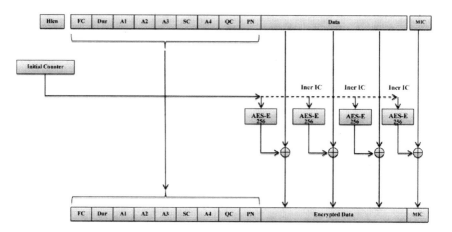

**Fig. 1.** Suggested structure of AES-CCMP by using AES-256

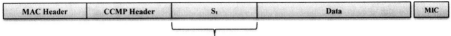

Nonce $_{key}$ XOR TK or Encrypted Nonce $_{key}$ by using TK

**Fig. 2.** Schema of Transmission of Nonce$_{Key}$

## 2.1 128 Bit TK and 128 Bit Random Nonce$_{Key}$

This method is based on 256 bit key which is named K$_{Tot}$, and first 128 bits of K$_{Tot}$ are created from TK and then next 128 bits of this key are generated by Nonce$_{Key}$ which is 128 bit random key created by the sender. For the awareness of receiver from 128 bits of Nonce$_{Key}$, two methods are suggested:

1. The first method is 128 bits of TK XOR with 128 bit Nonce$_{Key}$ and the result in the second block of each encrypted message ($S_1$) is located and after $S_0$, 16 octets $S_1$ is transmitted clear text. As the value of TK is not known to the attacker, so he cannot calculate the value of Nonce$_{Key}$. But the receiver uses TK to XOR with $S_1$ to extract Nonce$_{Key}$ easily.
2. In the second method, the sender encrypts 128 bits of Nonce$_{Key}$ with the TK and AES-128 algorithm and the result is transmitted in 16 octets $S_1$. And the receiver after receiving $S_1$ decrypts it with TK and extracts Nonce$_{Key}$.

After receiver extracts Nonce$_{Key}$ in each method, TK and Nonce$_{Key}$ are bind and K$_{Tot}$ is generated and for decrypting the rest of packets, the new key is used. Figure 2 shows the schema of new method.

## 2.2 256 Bit TK in Four Way Handshake Process

802.11i uses different keys in different layers, as Fig. 3 shows in the highest level the master key is present that is called Pairwise Master Key (PMK), then in four way handshake, between access point and supplicant, PTK is created which length in CCMP is 384 bits (in TKIP the length is 512 bits) then in four way handshake, Pairwise Transient Key (PTK) is spit to KEK, KCK, and TK.

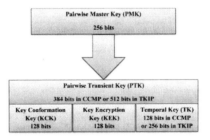

**Fig. 3.** Key generation mechanism in 802.11i

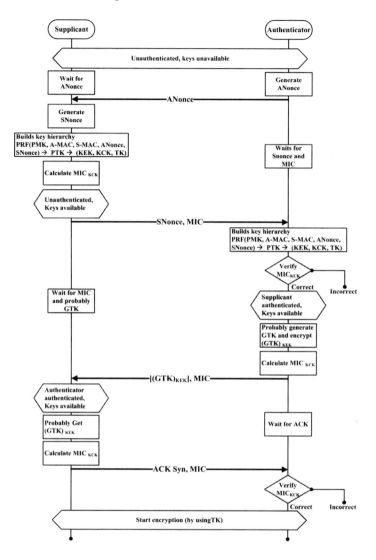

**Fig. 4.** Key distribution in four way handshake

EAPOL-key message is used to transfer information among supplicant and authenticator in four way handshake. KCK is used for calculating MIC of EAPOL-key message. KEK is used for encrypting Group Temporal Key (GTK), which GTK is used for multicast traffic between access point and mobile terminals, and for encrypting unicast user data traffic, TK is employed. The structure of encryption key distribution in four way handshake shows in Fig. 4.

As shown in Fig. 4, in the first place, supplicant and authenticator do not authenticate each other and the only shared key between these two entities is PMK. The four ways handshake process is started by sending Authenticator

Nonce (A-Nonce) to the supplicant, which is a random number. So, the first message of the first step in four way handshake is created.

Once supplicant receive the message, Supplicant Nonce (S-Nonce) is created, and then inputs of Pseudo Random Function (PRF) is provided to create PTK (384 bits for CCMP, and 512 bits for TKIP). These pieces which are the inputs of PRF, includes: PMK, Supplicant MAC Address(S-MAC), Authenticator MAC Address (A-MAC), S-Nonce and A-Nonce [11]. Then, one PTK (384 bits or 512 bits) is generated after calling PRF function, and then PTK split to three distinct phases KEK (128 bits), KCK (128 bits) and TK (128 bits for CCMP or 256 bits for TKIP). So, key hierarchy is created for supplicant and authenticator, but it is still not usable, because it is not authenticated by authenticator.

In the second message, supplicant sends S-Nonce with MIC of content of the second message that is calculated by using KCK as key, and SHA or MD5 algorithm. Then, the authenticator collects the pieces required for PRF (PMK, S-MAC, A-MAC, S-Nonce and A-Nonce) and PTK is created. And then, it extracts KEK, KCK, and TK. Then, the received S-Nonce is hashed by using extracted KCK and creates new MIC. Then authenticator compares new and received MIC from supplicant. If two mentioned MICs are the same, it ensures received S-Nonce and created key hierarchy are correct.

In the third message, authenticator sends the finished message including encrypted GTK (by using KEK) and MIC (by using KCK) of the contents of the third message to the supplicant. After this step, three derived keys from PTK are authenticated by supplicant and authenticator [11].

After receiving the third message, supplicant cerates MIC of third message by using KCK, and compare it to the received MIC from the authenticator. So, authenticator is authenticated and confirms the correctness of key hierarchy. Then, encrypted GTK is extracted from message and decrypted by using KEK.

The supplicant creates the fourth message for synchronizing and announcing success of four way handshake steps. For confirming authenticator that this message is from real supplicant, it sends MIC of fourth message of four way handshake. After receiving fourth message, authenticator calculates MIC of last message and compares it with received MIC, then for starting encryption phase, by using AES-CCMP and TK, authenticator and supplicant are both ready.

In CCMP, bits from 256 to 383 of PTK are used as TK. But in TKIP method, because of difference in structure of PRF output, bits from 256 to 511 of PTK are used as TK. As mentioned before, length of 128 bits for TK is not suitable. So, by using PRF method of TKIP, in second and third steps of four way handshake, 512 bit PTK is created, and last 256 bits are used for TK. As a result, from the beginning, supplicant and authenticator negotiate a 256 bit TK.

### 2.3    128 Bit TK and 128 Bit Generated by PRF

For creating the second 128 bits of $K_{Tot}$, in this method, PRF is used and by using TK as a main input of PRF, 128 right output bits of PRF is used for increasing key length. Structure of new $Nonce_{Key}$ is: $Nonce_{Key} = PRF(TK, A_2, A_1, 128)$. Pseudo code of mentioned PRF shows below:

$R \leftarrow SHA - 1(\text{TK}, A_2, A_1, 0)$
For $(i = 1$ to $((\text{int}) \text{ TK } (b126, b127) \bmod 3) + 1)$ do
$R \leftarrow SHA - 1(\text{TK}, R, i)$
Return $R$ 128 right bits
($A_2$: Source Address, $A_1$: Destination Address, $b$: bit)

So, by using Hash MAC (HMAC), final output is calculated, and 128 bits used to create Nonce$_{\text{Key}}$, so all of Secure Hash Algorithm (SHA) can be used, the most optimized one is SHA-1. As a result, in this method, there is no need to change four way handshake, and after calculating Nonce$_{\text{Key}}$ by using proposed PRF, result is bind by TK and 256 bit K$_{\text{Tot}}$ is created and used for encrypting.

## 3   Results

The security structure of AES-CCMP is dependent on TK length and IC which is created with joining the Nonce, Flags and CTR; also Nonce is created with joining PN, $A_2$ and Priority. The innovatory algorithm is implemented on .Net Framework 2010 in C Sharp language and it is executed on the platform with the property of 64 bit Operating System, 2.30 GHz CPU and 3.00 GB RAM.

In the following sections the probability of TMTO attack, avalanche effect, neighbor block, memory usage and execution time in proposed method are discussed and compared with the classic method.

### 3.1   Evaluating TMTO Attack

In TMTO attacks, the effective key length from $n$ bit is reduced to $2n/3$ bits. As a result, the effective key length in classic AES-CCMP from 128 is reduced to 85 bits and it becomes vulnerable to TMTO attack. On the other hand, as described in Sect. 2, today the minimum effective key length is equal to 103 bits. In proposed method, the effective key length is increased from 85 to 171 bits as a result of increasing cipher key length from 128 to 256 bits. And the proposed structure of AES-CCMP is resistant completely to TMTO attack.

As mentioned, waiting for $2^{85}$ different sessions for specific station is long duration [9], but not impossible for the attacker to launch a successful attack. In a new proposed AES-CCMP, this problem by increasing key length from 128 to 256 bits is completely solved. And $2^{171}$ trial and error for specific station is absolutely impossible.

### 3.2   Avalanche Effect

The concept of avalanche effect means; if one bit is altered in plain text, generated cipher text is totally different from previous cipher text. As Figure 5 shows, avalanche effect of IC samples which are selected random in classic AES-CCMP and proposed AES-CCMP. As the length of data block in AES-CCMP is 128 bits, the best optimized avalanche effect that can be observed is 64 bits [12].

Figure 5 demonstrates the average of avalanche effect in IC samples, which are different only in one bit with each other, in two types of AES-128 and AES-256 in counter mode. As the results show, the mentioned average for classic AES-CCMP is equal to 64.23 bit changes and for proposed AES-CCMP is equal to 63.39 bit changes. So, mentioned statistics clarify the avalanche effect in two techniques of proposed and classic method is so near and approximately equal to optimized avalanche effect in AES-CCMP which is 64 bits change.

### 3.3    Output Neighbor Blocks

In this comparison, sample ICs which increase 1 unit compared to the previous one, are the inputs of AES-128, AES-256 in counter mode, and then, outputs of AES modules, are compared neighbor to neighbor. And in this comparison, the changed bits compared to the adjacent block are calculated and the result shows in Fig. 6.

As Figure 6 shows, the changed bits of neighbor blocks in the output of AES-128 in counter mode is in average 63.51 changed bits while mentioned statistics for AES-256 output is equal to 63.85 changed bits.

The goal of using AES, in counter mode is creating random keys compared to last generated keys, considering criteria of random numbers and binomial distribution, the optimized mode of changing two $n$ bit neighbor blocks is when $n/2$ bits are changed. In this mode, the probability of guessing second block from first block in the least possible value equals to (1).

$$P = \binom{n}{\frac{n}{2}} \times \left(\frac{1}{2}\right)^{\frac{n}{2}} \times \left(\frac{1}{2}\right)^{n-\frac{n}{2}} . \tag{1}$$

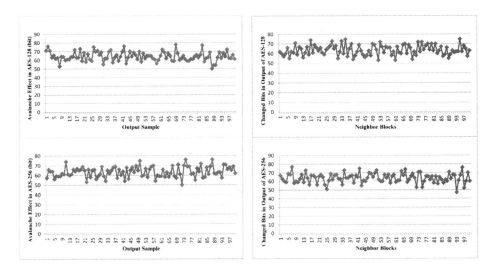

**Fig. 5.** Comparing Avalanche effect        **Fig. 6.** Comparing output neighbor blocks

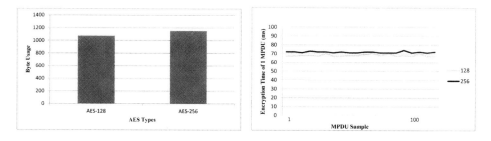

**Fig. 7.** Compareing memory usage for one MPDU

**Fig. 8.** Execution time evaluation

So, considering (1), the optimized mode for 128 bit block is equal to 64 bit change between two neighbor blocks. The probability of guessing the second block from the first block is equal to the least possible value which shows in (2).

$$P = \binom{128}{64} \times \left(\frac{1}{2}\right)^{128} . \tag{2}$$

### 3.4 Memory Usage

One of the criteria used in AES-CCMP is the memory consumed, because AES-CCMP is designed for 802.11i which face the limitations of wireless technology. Figure 7 shows memory comparison between proposed and classic method in AES-CCMP encryption phase for one block of MAC Protocol Data Unit (MPDU).

As Figure 7 displays, memory usage in encryption phase for one block of MPDU between proposed and classic method of AES-CCMP is slightly different, about 7.407% increase in memory usage that can be ignored as the proposed method is more robust in the field of security than the classic method.

### 3.5 Execution Time

One of the important topics in AES-CCMP is execution time. As in the wireless systems, often there are some limitations in system resource and time. Figure 8 shows the difference between execution time of classic and proposed methods.

As Figure 8 shows random MPDU block samples are being examined in encryption phase of two mentioned algorithms. Considering results proposed AES-CCMP have higher execution time. Which is 71.7041 ms in average (5.441% increasing execution time than the classic method) which can be ignored.

## 4 Conclusion

This research investigates a proposed method in AES-CCMP key structure for eliminating TMTO attack. The weakness of AES-CCMP security structure is the result of two gaps including predictable IC and the short effective key length. As a result, the proposed method focused on the second gap and three solutions are proposed for increasing key length including 128 bit Random Nonce$_{Key}$, changing four way handshake structure and 128 bit PRF.

According to results, proposed method is entirely resistant to TMTO attacks. avalanche effect and changes in output neighbor blocks of proposed method are completely close to the optimized goal. Also, proposed method memory usage and execution time are approximately same as classic method.

# References

1. Samiah, A., Aziz, A., Ikram, N.: An Efficient Software Implementation of AES-CCM for IEEE 802.11i Wireless Standard. In: 31st Annual International Computer Software and Applications Conference, COMPSAC 2007, Beijing (2007)
2. Razvi Doomun, M., Sunjiv Soyjaudah, K.M.: Resource Saving AES-CCMP Design with Hybrid Counter Mode Block Chaining – MAC. IJCSNS International Journal of Computer Science and Network Security 8, 1–13 (2008)
3. RFC 3610: Counter with CBC-MAC (CCM), pp. 1–26 (2003)
4. Algredo-Badillo, I., Feregrino-Uribe, C., Cumplido, R., Morales-Sandoval, M.: Efficient hardware architecture for the AES-CCM protocol of the IEEE 802.11i standard. Computers & amp; Electrical Engineering 36, 565–577 (2010)
5. Khan, M.A., Cheema, A.R., Hasan, A.: Improved Nonce Construction Scheme for AES CCMP to Evade Initial Counter Prediction. In: Ninth ACIS International Conference on Software Engineering, Artificial Intelligence, Networking, and Parallel/Distributed Computing, Phuket (2008)
6. Hellman, M.E.: A cryptanalytic time-memory trade-off. IEEE Transactions on Information Theory IT-26, 6 (1980)
7. Junaid, M., Mufti, M., Ilyas, M.U.: Vulnerabilities of IEEE 802.11i Wireless LAN CCMP Protocol. Transactions on Engineering, Computing and Technology (World Enformatika Society) 11, 6 (2006)
8. Moore's law,
   ftp://download.intel.com/museum/Moores_Law/
   Printed_Materials/Moores_Law_2pg.pdf1965
9. He, C., Mitchell, J.C.: Security analysis and improvements for IEEE 802.11i. In: Proceedings of the 12th Annual Network and Distributed System Security Symposium, p. 19 (2005)
10. Saberi, I., Shojaie, B., Salleh, M., Niknafskermani, M.: Enhanced AES-CCMP Key Structure in IEEE 802.11i. In: International Conference on Computer Science and Network Technology, ICCSNT 2011, Harbin, China (2011)
11. Nguyen, T.M.T., Sfaxi, M.A., Ghernaouti-Helie, S.: 802.11i Encryption Key Distribution Using Quantum Cryptography. Journal of Networks 1, 9–20 (2006)
12. Saberi, I., Shojaie, B., Salleh, M.: Enhanced Key Expansion for AES-256 by Using Even-Odd Method. In: 2nd International Conference on Research and Innovation in Information Systems, ICRIIS 2011, Kuala Lumpur (2011)

# Reverse Engineering of Microprocessor Program Code

Andrzej Kwiecień, Michał Maćkowski, and Krzysztof Skoroniak

Silesian University of Technology, Institute of Computer Science,
Akademicka 16, 44-100 Gliwice, Poland
{akwiecien,michal.mackowski,krzysztof.skoroniak}@polsl.pl
http://www.polsl.pl/

**Abstract.** This paper has an experimental character. Theoretical backgrounds presented here allow creating a research method. The research focus on analysis of microprocessor voltage supply changes. Such analysis based on the presented research assumptions allowed for a rather high efficiency of decoding program without interference in the internal structure of microprocessor. The obtained results show, that there is a possibility of uncontrolled access to program codes. Thus, it is necessary to search for and develop appropriate methods used for protecting program.

**Keywords:** reverse engineering, program code, microcontroller, conducted emission, electromagnetic disturbances, Hamming distance, embedded system.

## 1 Introduction

The current continual trend towards more and more miniaturization and integration has led to existence a very large scale integration circuits. In recent years, SOC (*System On Chip*) which merge into one chip analog, digital, mixed-signal, and often radio-frequency functions, has become very popular solutions – especially in embedded systems. This may indicate a present level of advancement of production technology of integrated units. The high integration scale and continuous increase of microprocessor circuit frequency cause the current peaks to be generated with higher amplitudes and shorter rise times on power supply and I/O lines of the electronic circuits. Such impulses are generated by thousands/millions of transistors inside the integrated unit, which are switching simultaneously. Propagation of such currents through wires and paths on a PCB (*Printed Circuit Board*) to other electronic systems may cause the problems with their normal functioning. On the other hand, the total current drawn by all the gates during execution of a single instruction, may indicate what instruction is currently executed.

During the analysis of microprocessor program code based on the measurement of power supply changes, the system is treated as "a black box" executing a sequence of instructions stored in a program memory. In idealistic approach

A. Kwiecień, P. Gaj, and P. Stera (Eds.): CN 2012, CCIS 291, pp. 191–197, 2012.

it can be assumed that the operation of the implemented algorithm consists of processing some input data and returning the result without any interaction between the device and environment. In fact, any device powered by electric energy and processing digital signals affects the environment and other devices through the emission of electromagnetic disturbances in the way of conducted and radiated emission [1,2,3,4]. Changes in power consumption of the device and emitting the electromagnetic field can be described as side effects of realization of the algorithm, which make up affiliate channel providing additional knowledge about the algorithm.

The paper deals with an aspect of this problem resulting from the fact, that for example an author of software of embedded system is not aware of the possibility to recognize, to a certain extent, a program code without direct interference into microcontroller program memory. Microprocessor units currently used in network devices as network controllers can be also considered as advanced chips being responsible for data processing and reconstruction of transmitting frames. Such units can also be source of electromagnetic disturbances. The authors in previous papers [5,6] presented the research referring to analysis of the influence of data bus, instruction operand, its result and address in memory to the supply power line during subsequence instruction cycles and recognition of instruction that are realized with the arguments with the value of zero.

In this paper, the authors analyze 8-bit microprocessor program code based on the power supply changes, and focus on recognition of instructions that operate on arguments with the value of any kind.

## 2    Test Bench and the Research Procedure

Test bench consisted of Microchip microprocessor with PIC16F84A signature, connected to the power supply and an external square-wave generator with a frequency of 250 kHz. To supply the microprocessor Agilent stabilized power supply was used. Oscilloscope probe was connected to microcontroller supply lines to monitor voltage drop during realization of following instructions. The test bench, for the period of research was placed in shielded cell – GTEM (*Gigahertz Transverse ElectroMagnetic*) which provided total separation of measuring area from external electromagnetic influences. The exact description of the test bench, methods used to measure voltage disturbances and ways to analyze the obtained results in the time and frequency domain, have been presented in the previous papers of the authors [6,7].

In papers [5,8] the authors developed a method for recognizing instructions that operate on arguments with the value of zero, based on the analysis of voltage disturbances. The first step to do this is to measure the microprocessor voltage supply waveform while running the entire program. The next step is to cut the part of time waveform referring to the instruction being tested. Then the minimum and maximum value of the voltage for the first three machine cycles is saved – a total of six values are saved. In this way the sample database was created, in which each microprocessor instruction is characterized by 6 points – three maximum and three minimum values of voltage, measured in particular machine cycles Q1, Q2, and Q3. Based on the obtained results, it was proved that the method developed to recognize instructions operating on arguments with the value of zero, is effective in 91%.

In this paper the authors extended the scope of the research already discussed in [5] by taking into consideration both instruction and the instruction operand. In case of compiling the database of samples used for recognizing instructions that operate on arguments with the value of any kind, it is required to take into consideration not only the instruction code but also instruction argument and its result. Hence, the procedure of compiling the samples is a very time consuming process and requires a careful synthesis of each instruction, which is based not only on technical specification, but also on changes in voltage supply. Therefore, in the process of compiling the database of samples the study focuses on ten instructions of the microprocessor instruction list: ADDLW, ANDWF, BSF, CLRF, COMF, INCF, MOVF, MOVLW, NOP, and XORLW.

In previous research the authors showed that the voltage waveform during realization of the first and third machine cycle is not directly affected by instruction argument and the result of operation. It appears that the voltage waveform during realization of instruction cycle, is affected not only by the operation code, but also by the differences between the state of data bus and instruction argument (*machine cycle Q1*), and between instruction argument and the result of operation (*machine cycle Q3*). These differences can be measured by using Hamming distance parameter. In this case a schema based on Hamming distance calculated for instruction arguments, can be used to create database of samples. This information allows for a significant simplification of the construction of database used in the process of instruction recognition.

Figure 1 presents the schema of database construction and the process of recognizing instructions that operate on arguments with the value of any kind. The database consists of samples which describe 10 instructions, where samples from 1 to N describe instruction 1, next M samples describe instruction 2, etc. In this way, it was possible to create a database for the previously mentioned instructions, consisting of 1935 samples and used then in the process of recognition of microprocessor program code.

Having compiled the database, three test programs consisting of 100 instructions were generated. They were used to determine the effectiveness of method for instruction recognition. The programs were created by using the same 10 instructions, which were previously used in the process of compiling database. Both, the order of instructions in the test programs, as well as instruction arguments were selected at random.

## 3   The Research Results

According to research procedure, each instruction included in test program was compared to 1935 samples compiled in database.

Figure 2 presents the numbers of correctly recognized instructions in subsequent instructions proposition, which were received as a result of a method

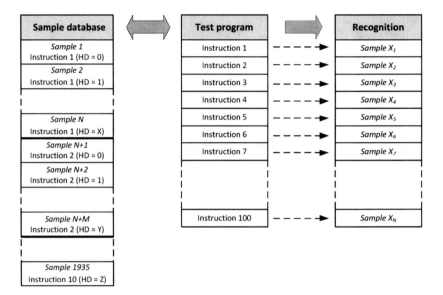

**Fig. 1.** The process of recognizing instructions

used for all three test programs. Further analysis of the bar chart reveals that the largest number of instructions was correctly recognized in the first three propositions. Table 1 presents the number of correctly recognized instructions in following instructions suggestions and their average values for test programs 1, 2 and 3. Based on the obtained research results it can be stated that for the first three propositions it was possible to recognize on average 33.67%, 22.00%, and 16.67% of instructions in the test programs. For the next columns the average number of correctly recognized instructions was less than 10%.

Table 2 presents the effectiveness of the method used for instruction recognition. Efficiency is the ratio of the numbers of instructions correctly recognized in one of the first three instructions propositions, to the total number of samples in test program. For the test program 1, in 72% of cases the correctly recognized instruction was among the suggestions of the first three the most similar instructions, returned as a result of the method used in the research. For the other test programs the achieved effectiveness of instruction recognition is about 74 and 71%.

**Table 1.** Number of correctly recognized instructions in the subsequent instructions propositions and their average values for test programs 1, 2 and 3

| | Number of correctly recognized instructions | | | | | | |
|---|---|---|---|---|---|---|---|
| | Instr. 1 | Instr. 2 | Instr. 3 | Instr. 4 | Instr. 5 | Instr. 6 | Instr. 7 |
| Test program 1 | 35 | 22 | 15 | 9 | 9 | 10 | 0 |
| Test program 2 | 37 | 20 | 17 | 9 | 11 | 6 | 0 |
| Test program 3 | 29 | 24 | 18 | 11 | 8 | 9 | 1 |
| Average value | **33.67** | **22.00** | **16.67** | 9.67 | 9.34 | 8.34 | 0.34 |

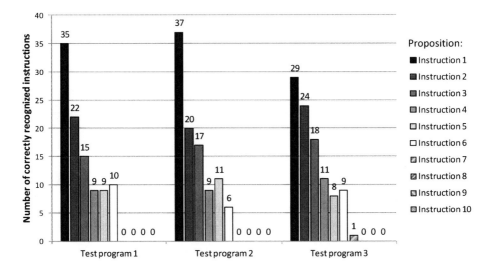

**Fig. 2.** Number of correctly recognized instructions in the subsequent instructions propositions, received as a result of method used for test programs 1, 2 and 3

**Table 2.** Comparison of the method effectiveness used for instructions recognition, that operate on arguments with the value of any kind, for test programs 1, 2 and 3

| | Number of instructions in the first three instructions suggestions | | The effectiveness of method used for recognizing instructions |
|---|---|---|---|
| | correctly recognized | incorrectly recognized | |
| Test program 1 | 72 | 28 | 72% |
| Test program 2 | 74 | 26 | 74% |
| Test program 3 | 71 | 29 | 71% |

Decrease of effectiveness of instructions recognition, presented in this paper, compared to the effectiveness of instructions recognition that operate on arguments with the value of zero, is caused in this case by the influence of data processing on the voltage supply waveform when a particular instruction is executed. The result is that various instructions of microprocessor unit for specific values, such as: state of data bus, instruction argument and result of operation, can have the same or very similar voltage waveforms measured in microprocessor power circuit. Based on the database of samples, the research method returns propositions of instructions which were recognized correctly together with the value of similarity for each instruction in the test program. The study assumes that the effectiveness of method for recognizing microprocessor program code will be determined based on the numbers of recognized instructions in the first three instructions suggestions.

## 4    Conclusion

The paper concerns the problem of microprocessor systems security, and in particular the threats to the programs stored in memory of such systems, and information they process.

The authors proved the possibility of partial recognition of the instruction currently executed based on the changes of microprocessor voltage supply. Moreover, the authors indicate that it is possible to determine to a certain extent Hamming distance between the state of data bus and instruction argument, and between instruction argument and result of operation. Such possibility in this case refers to the recognition of numbers of bits changes on instruction argument, in consequence of instruction realization.

As a result of conducted research, a database consisted of 1935 samples was compiled, and then was compared to three test programs. Each test program consisted of 100 instructions and arguments which both were selected at random. It was assumed that if the instruction after having been compared to the samples database is in one of the first three places, then it was correctly recognized. With such assumptions, the effectiveness of the presented method for the following test programs is: 72%, 74% and 71%. Results of research can be generalized to other microprocessor systems executing program code stored in the memory.

Results of research can be also generalized to other microprocessor units with very similar internal architecture which execute program code stored in the memory. In case of microcontrollers with completely different datapaths, such as multiple cycle and pipeline, to fetch and execute instructions it is necessary to conduct further research.

This study and the results presented here should be considered also as an attempt to draw attention to the threats resulting from the phenomena of emanation emission, or any kind of unintended signals, which in case of being captured and analyzed reveal the information processed by the device. Moreover, several questions arise. The first is, whether obtained results have only to draw attention to the risk of decoding programs with the use of presented method? Another refers to the necessity of improving the method described above (*or develop a new one*) to get more effectiveness, and thereby indicate threats in a more precise way, and at the same time to build protection against unauthorized access to the source version (*following assembler instructions*) of a software. Finding answers to these questions is so far an open issue.

## References

1. Bao, F., Deng, R.H., Han, Y., Jeng, A., Narasimhalu, A.D., Ngair, T.: Breaking Public Key Cryptosystems on Tamper Resistant Devices in the Presence of Transient Faults. In: Christianson, B., Lomas, M. (eds.) Security Protocols 1997. LNCS, vol. 1361, pp. 115–124. Springer, Heidelberg (1998)
2. Biham, E., Shamir, A.: Differential Fault Analysis of Secret Key Cryptosystems. In: Kaliski Jr., B.S. (ed.) CRYPTO 1997. LNCS, vol. 1294, pp. 513–525. Springer, Heidelberg (1997)

3. Maćkowski, M.: The Influence of Electromagnetic Disturbances on Data Transmission in USB Standard. In: Kwiecień, A., Gaj, P., Stera, P. (eds.) CN 2009. CCIS, vol. 39, pp. 95–102. Springer, Heidelberg (2009)
4. Mangrad, S., Oswald, E., Popp, T.: Power Analysis Attacks – Revaling the Secrets of Smart Cards. Springer (2007)
5. Kwiecień, A., Maćkowski, M., Skoroniak, K.: Instruction Prediction in Microprocessor Unit. In: Kwiecień, A., Gaj, P., Stera, P. (eds.) CN 2011. CCIS, vol. 160, pp. 427–433. Springer, Heidelberg (2011)
6. Kwiecień, A., Maćkowski, M., Skoroniak, K.: The Analysis of Microprocessor Instruction Cycle. In: Kwiecień, A., Gaj, P., Stera, P. (eds.) CN 2011. CCIS, vol. 160, pp. 417–426. Springer, Heidelberg (2011)
7. Maćkowski, M., Skoroniak, K.: Instruction Prediction in Microprocessor Unit Based on Power Supply Line. In: Kwiecień, A., Gaj, P., Stera, P. (eds.) CN 2010. CCIS, vol. 79, pp. 173–182. Springer, Heidelberg (2010)
8. Maćkowski, M., Skoroniak, K.: Electromagnetic Emission Measurement of Microprocessor Units. In: Kwiecień, A., Gaj, P., Stera, P. (eds.) CN 2009. CCIS, vol. 39, pp. 103–110. Springer, Heidelberg (2009)

# Network Malware Activity – A View from Honeypot Systems

Mirosław Skrzewski

Politechnika Śląska, Instytut Informatyki, Akademicka 16, 44-100 Gliwice, Polska
miroslaw.skrzewski@polsl.pl

**Abstract.** There is little detailed information about the lifetime of network malware and malware infected systems. There is a lot of information about the emergence of some new versions of malware, used security vulnerabilities, methods and the range of infection but they are generally related to the start of life cycle of malware. Much less information is available about what happens next – how long the malware actively is distributed on the network, how long operate infected hosts and how the amount of infected systems fluctuates with time.

The paper presents the results of nearly two years' operation of honeypot systems, installed on unprotected research network, throwing some lights on the longtime malware activity.

**Keywords:** malware activity monitoring, malware survivability, honeypot systems.

## 1  Introduction

There is little detailed information about the lifetime of network malware and malware infected systems. A lot of information is available about the emergence of some new versions of malware, its exploited security vulnerabilities, methods and the range of infection but they are generally related to the start of life cycle of malware. Much less information is available about what happens next – how long the malware actively is distributed on the network, how long operate infected hosts and how fast changes with time the population of infected systems.

For gathering such information on the presence and activity of malware programs the honeypot systems became one of the main tools, used both in academia research as well by security system vendors [1]. Their usage is focused on collection of samples of current internet threats. Captured samples are analysed to derive signatures for threats detection [2], to discover new ways exploits use for system infections, to prepare methods of system protection or malware removal. Statistics of this honeypot activity, especially numbers of captured samples in some time (day, month) [3,4] are often reported as current threats level or as a measure of new malware activity.

But the honeypot systems could deliver also a lot of additional information on the behavior of malware population and the longevity of its operation. The honeypot operation is monitored in details and the records are stored in local

A. Kwiecień, P. Gaj, and P. Stera (Eds.): CN 2012, CCIS 291, pp. 198–206, 2012.

database with attached timestamps of each operation. The records contains the evidence of each external contact with the honeypot, both succesfull (malware delivered) and unsuccessful.

The data derived from honeypots operated on research network are the basis of presented analysis of malware landscape, visible through class C LAN window. The paper is organized as follow: Section 2 describes used honeypot systems, Sect. 3 presents and analyzes the results of honeypots operation. The following sections present the malware distribution models, statistics of connections to honeypot systems, discuss the distribution of malware programs, and the activity of infected systems.

## 2   Selection of Honeypot Systems

During the malware monitoring on experimental network a set of different honeypots systems were installed and tested. The systems represented all classes of honeypots – low-interaction, middle- and high-interaction systems, Linux and Windows based. Most of them deliver very few information about malware – low-interaction honeypot records only connection attempts to simulated services, without any further action. High-interaction honeypot, Windows based remained uninfected through months of operation, despite default, post-install configuration and switched off actualization services. Both were disassembled after few months.

The middle-interaction honeypots, based on full emulation of network services worked much better. Nearly immediately after installation they started gathering malware samples. Two types of honeypots were installed: Windows XP based system from www.iDefense.com [5] described on its site as emulation honeypot or multipot and Linux based dionaea [6], representing different generations of trap systems.

Multipot was based on emulation of few security holes and communication ports, typically remaining open after successful infection of some, at some time popular malware (MyDoom [7] – 3 127, Bagle [8] – 2 745, 12 345, Sub7 [9] – 27 347, Kuang2 – 17 300, Veritas – 6 101, 10 000, Optix – 2 060, 500) and attempt to capture specialised malware programs, re-infecting systems previous attacked by other malware. For port 445, connected with smb and netbios protocols, was prepared special emulated service, allowing for capturing of different types of new malware attacking this port.

Dionaea honeypot represents another, more universal concept of initiating generic or specialised service on-demand on any requested by connection attempt port, based on virtual machine type of emulation (packet emu) and some shellcode processing services. For unknown shellcodes types the mirroring operation, resending back to attacking system all intruder operation was prepared. Running system services include TCP ports 21, 25, 42, 80, 135, 443, 445, 1433.

Both honeypots have used file-database to register the course of the monitored events. Multipot uses the access database (.mdb file) and Dionaea uses sqlite. Captured malware were stored in separated folders, under differently generated file names. Multipot assigns some random name (with .dat extension) to each captured copy of malware or shellcode, and saved them in separated folders for each serviced port. The same malware may be stored multiple times under different names.

Dionaea for each captured malware calculates its MD5 hash and uses it as file name, thus preventing storage of multiple copies of the same malware.

## 3    Results of Honeypots Operation

Both honeypots operated quite a long time. Multipot began its operation in January 2010, dionaea started its operation about year later, in the middle of March 2011. Their records documents nearly two years and respectively over nine month of continuous malware monitoring operation. General summary of results of their operation for the end of 2011 year looked as follows (Table 1).

Multipot has recorded 1986 files of different size from 1106 to 332 800 B, delivered by 1190 remote systems. Most of recorded files (1760) represents initial attack attempts (classified as shellcode) and have size of up to 5237 B and 28 different MD5 sums. Remaining 226 files (malware samples) have sizes starting from 94 721 B, and in this group were recorded 57 different MD5 sum values, 6 of them appeared in records more then 10 times. Most active malware (Win32/Agobot) was recorded 30 times, all other samples were classified as different versions of Win/Rbot malware.

There were registered 193 different remote host IP addresses related to malware download, 14 of them multiple times. Top-active IP were listed 5 times, and have distributed the same versions of malware. All these addresses belong to 83.x.x.x/8 network.

Dionaea recorded 169 166 connection attempt events, registered as reject (49 672), accept (114 425), connect (3081) and listen (1988). As result were

**Table 1.** Honeypot operation summary

| Property | Multipot | Dionaea |
|---|---|---|
| operation start date | Jan. 10, 2010 | Mar. 23, 2011 |
| last recorded malware | Aug. 30, 2011 | Dec. 30, 2011 |
| connection attempts | not recorded | 169 166 |
| downloaded files | 1986 | 3675 |
| remotr IP's | 1190 | 1189 |
| malware samples | 226 | 3675 |
| shellcode samples | 1760 | not recorded |
| different MD5 sum | 57 | 537 |
| different IP addresses | 193 | 1189 |
| multiple times registered IP | 14 | 181 |
| top-frequent IP registration | 5 | 616 |
| top-frequent MD5 registration | 30 | 693 |
| longest activity time MD5 (days) | 258 | 269 |
| longest visible IP (days) | 131 | 280 |

captured 3675 malware samples with 537 different MD5 sums from 1189 IP addresses. Of them 181 addresses were recorded more then once, and five most active more then 150 times. Majority of recorded addresses belong to 83.x.x.x/8 network.

## 3.1 Malware Distribution Models

Analyzing details of malware capture records from dionaea honeypot you can easily identify two models of the distribution of malware. Both are based on multistep infection process [10,11]. After successful initial infection of victim systems attacking malware components (attack code) initiates download of main part of malware to attacked system, creating connection to distribution server.

In first model such server was on demand launched on attacker system – eg. http, ftp or tftp server started just before start of download on an arbitrary selected random port. In the second, the attack code directs victim systems to some other external malware distribution server (often via http connection). Both models are present in honeypot records. Description of each infection event includes malware download links, pointing to such servers.

Part of malware e.g. Win32/Rbot, Win32/Agent.NJF, Win32/Pepex uses attacker system as distribution server (self server model). Some others e.g. Win32/AutoRun.IRCBot.xx, Win32/Injector.vvv, Win32/Kryptik.yyy, IRC/SdBot downloads components from external distribution servers (foreign server model), and many different attacking systems directs victims to the same servers for download. Both models looks very stable, and the same external servers were used through whole observation time by various (different MD5 sums) versions of given malware family.

Quite a big part of malware records shows the evidence of honeypot troubles with processing of malware attack code. When suitable processing module is missing, dioneae switches to emulation mode, sending back to attacking system its actions. Often this method results in the download of malware to the honeypot, but in the records misses details of download url, and the description of protocol and url is replaced with the "emulate" message e.g. for some versions of Win32/Injector.xxx, Win32/AutoRun.IRCBot.xx or Win32/Allaple worms.

## 3.2 Honeypot Connection Statistics

External connections to honeypot system sometimes only are successful and allows for transfer of malware code to honeypot (connection of 'accept' type). In most of the cases they are rejected by the honeypot or handled another way. Statistics, which external systems attempted to connect with honeypot and on which ports show that accepted connection came from 2494, rejected ones from 10 506 different IP addresses. The Table 2 show statistics for most active honeypot ports, both from accepted and rejected connection classes, with numbers of connections to each port, ordered by port activity interval.

Most of the ports activity presented in Table 2 has no relationship with malware downloads. Nearly all malware were captured through connections on ports

**Table 2.** Most active dionaea honeypot ports

| Accepted connections | | | Rejected connections | | |
|---|---|---|---|---|---|
| interval | port | count | interval | port | count |
| 281.9 | 445 | 18450 | 281.9 | 139 | 34622 |
| 281.8 | 135 | 4769 | 281.5 | 23 | 8356 |
| 281.5 | 9988 | 180 | 281.5 | 3389 | 1352 |
| 279.2 | 80 | 2067 | 280.9 | 4899 | 1015 |
| 278.1 | 1433 | 87338 | 281.2 | 8080 | 733 |
| 271.9 | 21 | 94 | 280.0 | 9988 | 646 |
| 211.0 | 63869 | 3 | 280.0 | 25 | 596 |
| 209.4 | 63404 | 3 | 280.8 | 5900 | 585 |
| 205.5 | 63815 | 4 | 280.5 | 3128 | 400 |

445 (3371 samples) and 135 (261), there were also gathered some samples via ports 1957 (35 copies of Win32/Rbot) and 1130 (7 copy of Win32/Hatob.E).

Intervals of ports activity of 281–270 days with big numbers of connection attempts mean, that from the very start of honeypot operation all these ports were constantly or regularly probed from external systems, in search for some usable security vulnerabilities. Long interval value alone, without the reasonable number of connections means that some ports were probed very rare, from time to time e.g. once per 4–5 month on ports over 63 000.

### 3.3   Malware Activity Analysis

Malware infection events were analyzed to select most frequent versions of malware (MD5 sums), and also the longest visible (active) samples on the network. For selected malware samples were also analyzed the path of infection – which infection propagation model they use and where are situated distribution servers.

To classify captured samples of malware and to get their names, all samples were scanned with antyvirus program NOD32 [12], and assigned to each MD5 sum names were exported to sqlite database. AV program allocated 129 different names to dionaea captured malware files, divided on more then ten malware families.

The process of assigning names was not very selective – many MD5 sum received the same name with the addition "a version of", "probably a variant of" (there were 537 different MD5 sums). The most numerous was Win32/Injector.XX – 84 variants, then Win32/Kryptik.XXX – 11 variants and Win32/AutoRun.XXX – 7. Multipot captured samples (57 total) in big majority represents multiple variants of one malware family – Win32/Rbot (49 different MD5 sums), there were also recorded single variants of Win32/Injector.XX, Win32/Agent.ORE and Win32/Agobot.

Table 3 presents the summary information of the most active malware: the day of first capture, interval of malware activity, name assigned by NOD32 AV, number of recorded malware appearances and IP addresses of infecting host and of download server. For foreign server model both addresses differs, as is the case for majority of malware in the table.

**Table 3.** Most often visible malware versions

| first day | interval | malware name | count | remote host | download url |
|---|---|---|---|---|---|
| 2011-03-23 | 281.2 | Win32/AutoRun.IRCBot.FC | 1247 | 83.222.59.21 | 83.222.59.21 |
| 2011-04-04 | 87.1 | Win32/AutoRun.IRCBot.HY | 537 | 83.229.8.189 | 60.10.179.100 |
| 2011-03-23 | 280.2 | Win32/Rbot | 284 | 83.222.94.239 | 83.222.94.239 |
| 2011-06-25 | 179.1 | Win32/AutoRun.AFQ | 228 | 83.96.75.81 | 146.185.246.45 |
| 2011-03-26 | 266.1 | Win32/AutoRun.KS | 123 | 83.149.44.249 | 146.185.246.52 |
| 2011-03-26 | 272.3 | Win32/Allaple | 89 | 145.236.8.72 | emulate |
| 2011-03-31 | 237.9 | Win32/Injector.ITQ | 82 | 83.230.250.194 | 46.45.164.163 |
| 2011-08-27 | 119.0 | Win32/AutoRun.Delf.AG | 68 | 83.230.12.214 | 83.230.12.214 |
| 2011-04-05 | 216.3 | IRC/SdBot | 50 | 83.149.44.114 | 146.185.246.71 |
| 2011-09-11 | 7.4 | Win32/Injector.JGG | 43 | 83.149.44.62 | 109.201.134.138 |

Malware samples visibility varies significantly. Most of recorded events (318 per 537) represents single infection contact or casual, less then one day duration attempts, and they represents 59 of 129 assigned malware names. Very long presence (more then 100 days of visibility) exhibits 30 malware species (MD5 sums) representing 22 names, five of them were active through all the time of the honeypot operation (about 280 days), however the frequency of contacts varies significantly, from one per 55 hours to one per 120 days. Most of them use self server model of distribution. In the middle of the stake are most active malware cases, with up to 693 infections in 82 days from single infecting system.

### 3.4   Infected Host Activity Analysis

Malware infection events were also analysed to select IP addresses of the hosts most frequently recorded as the sources of malware distribution and with the longest time records of spreading malware infection. For selected top adresses there were analyzed detailed host history, to find, what and how many malware versions were distributed, how long one malware type was served by given host, and what infection model were used.

There was also analyzed hosts population – how long given IP appears in infection records, and how many hosts attempts to infect analyzed honeypot and deliver malware samples in given time. The Table 4 present the longest visible infecting hosts, and Table 5 the most active (highest numbers of infection attempts) hosts. It is clearly evident, that both groups of hosts represents the opposite ends of malware activity picture. For selected most active (both in visibility and connection attempts) hosts there were analyzed detailed host profile, to find which and how many malware versions were distributed, how long one malware type were served by given host, what infection model were used and where are located systems involved in malware distribution. Examples of systems location for selected infection path contain Table 6.

Table 7 presents the information on the malware profiles of selected active hosts. In most cases a lot of different versions (MD5 sums) of malware were distributed sequentially in time. Some from analysed hosts distributed all the time at least two different versions of malware simultaneously.

**Table 4.** Most active hosts, by days of visibility

| first day | last day | interval | port | remote host | count |
|---|---|---|---|---|---|
| 2011-03-24 | 2011-12-30 | 281.0 | 445 | 83.230.250.194 | 658 |
| 2011-03-25 | 2011-12-27 | 277.6 | 445 | 83.230.244.205 | 13 |
| 2011-04-02 | 2011-12-02 | 243.9 | 445 | 83.149.38.2 | 3 |
| 2011-04-10 | 2011-12-06 | 239.6 | 445 | 83.149.38.21 | 4 |
| 2011-05-23 | 2011-12-30 | 220.5 | 445 | 83.149.38.70 | 10 |
| 2011-05-05 | 2011-12-08 | 217.3 | 445 | 83.149.38.28 | 4 |
| 2011-03-23 | 2011-10-25 | 215.4 | 445 | 83.230.180.92 | 1386 |
| 2011-04-21 | 2011-11-09 | 201.6 | 1433 | 222.133.189.12 | 5 |
| 2011-06-02 | 2011-12-21 | 201.4 | 445 | 88.101.106.226 | 55 |

**Table 5.** Most active hosts, by number of connections

| first day | last day | interval | port | remote host | count |
|---|---|---|---|---|---|
| 2011-05-30 | 2011-05-31 | 0.3 | 1433 | 83.103.179.130 | 25574 |
| 2011-08-17 | 2011-08-17 | 0.1 | 63963 | 83.38.161.253 | 5377 |
| 2011-08-30 | 2011-08-31 | 1.4 | 63177 | 83.246.204.245 | 5364 |
| 2011-08-27 | 2011-08-27 | 0.0 | 63768 | 83.238.230.56 | 5356 |
| 2011-05-08 | 2011-05-09 | 0.9 | 135 | 83.230.3.239 | 5352 |
| 2011-05-21 | 2011-05-21 | 0.2 | 63940 | 83.238.229.143 | 5350 |
| 2011-05-15 | 2011-05-15 | 0.1 | 63928 | 83.119.156.70 | 5349 |
| 2011-04-22 | 2011-04-22 | 0.1 | 63440 | 83.191.129.241 | 5345 |
| 2011-08-25 | 2011-08-25 | 0.3 | 1433 | 83.213.82.137 | 5339 |

**Table 6.** Most active hosts, by number of connections

| days | count | malware name | remote host host location | download url server location |
|---|---|---|---|---|
| 281 | 319 | Win32/Injector.MER | 83.230.250.194 Madrid, Spain | 46.45.164.165 Mustafa, Turcja |
| 215 | 616 | Win32/AutoRun.IRCBot. | 83.230.180.92 Madrid, Spain | 119.188.6.227 Jinan, China |
| 194 | 185 | Win32/Injector.KAO | 83.230.10.249 Gliwice, Poland | 146.185.246.28 Sedova, Russia |
| 165 | 447 | Win32/AutoRun.IRCBot. | 83.229.8.189 Dijon, France | 119.188.6.227 Jinan, China |
| 53 | 49 | Win32/Kryptik.WUJ | 83.229.125.104 Hertfordshire, UK | 146.185.246.52 Sedova, Russia |
| 118 | 84 | Win32/AutoRun.Delf.AG | 83.230.192.106 Madrid, Spain | 83.230.192.106 |
| 186 | 35 | Win32/AutoRun.IRCBot.FC | 83.230.161.89 Madrid, Spain | 46.45.164.228 Mustafa, Turcja |
| 168 | 25 | Win32/Rbot | 83.2.141.248 Czestochowa, PL | 83.2.141.248 |
| 191 | 21 | Win32/Allaple | 201.202.252.194 Alajuela, Costa R. | 201.202.252.194 |
| 162 | 17 | Win32/Rbot | 83.143.47.7 Warszawa, PL | 83.143.47.7 |

**Table 7.** Details of distribution process for selected remote hosts

| days | count | remote host | host location | MD5 | servers |
|------|-------|-------------|---------------|-----|---------|
| 281 | 319 | 83.230.250.194 | Madrid, Spain | 28 | 8 |
| 215 | 616 | 83.230.180.92 | Madrid, Spain | 3 | 2 |
| 194 | 185 | 83.230.10.249 | Gliwice, Poland | 27 | 27 |
| 165 | 447 | 83.229.8.189 | Dijon, France | 2 | 2 |
| 53 | 49 | 83.229.125.104 | Hertfordshire, UK | 14 | 12 |
| 118 | 84 | 83.230.192.106 | Madrid, Spain | 3, p | 2 |
| 186 | 35 | 83.230.161.89 | Madrid, Spain | 5, p | 2 |
| 168 | 25 | 83.2.141.248 | Czestochowa, PL | 1 | 1 |
| 191 | 21 | 201.202.252.194 | Alajuela, Costa R. | 1 | 1 |
| 162 | 17 | 83.143.47.7 | Warszawa, PL | 1 | 1 |

One of the most interesting question in analysis was how to assess, what role play registered hosts in malware distribution process – are they victims or attackers, remain without definitive answer. Some hints may be derived from the time, the IP address has been recorded as connecting to honeypot system. Dionaea was contacted from 2494 IP addresses, from 1975 IP more then once, 348 addresses were visible by more then one day and 252 more then seven days. Malware samples were downloaded from 1189 addresses and 181 addresses were the sources of more then one sample.

We can probably assume that the systems seen once or for a short period of time are victims of accidental infection, and have been cleaned by the user. Longer visible systems can represent elements of the network of bots, or poorly managed home systems of inexperienced users. Systems distributing the same malware for a long time probably belong to this second group, the systems often sending new versions of malware can be part of the first group, but these are only our guesses.

## 4   Conclusion

The information about the life time of malware programs are rare and rather incomplete. There are a lot of information about the emergence of some new versions of malware, methods and the range of infection but they describe the begining of life cycle of malware. Much less information is available of how long the malware is active on the network and how fast the population of infected systems changes with time.

Different sources present conflicting data on this subject. [3] presents plots of malware outbreak – intensive distribution of multiple (up to few 1000 or more) new version a day lasting 3–5 weeks, and later the number of recorded new samples fall nearly to zero. [13] talks about bigger numbers, up to 63 000 new malware variants a day, but most of them is active for less then a day. In contrary [14,15] presents examples that old malware never die, and even after few years some malware copies are still active and attempt to infect vulnerable systems, so antivirus systems do not retired any malware signatures.

Malware activity picture from the analysis of honeypot operation confirms rather the last point of view. There are a lot of malware samples arriving more or less frequent all the time of honeypots operation. Multipot stopped capturing samples of Win32/Rbot family in the end of August 2011, but the dionaea keeps gathering copies of this malware all the time, to the end of 2011 year. A lot of other versions of malware appears very rare, once per few months, so lack of any new copy in given time instant can not be treated as the proof of the end of life of some malware program.

# References

1. Trustwave's SpiderLabs: Monthly Web Honeypot Status Report,
   http://blog.spiderlabs.com/honeypot-alert/
2. IBM X-Force® 2011 Mid-year Trend and Risk Report (2011),
   http://www-935.ibm.com/services/us/iss/xforce/trendreports/
3. Q1 2007 Malware Outbreak Trends (2007), http://www.commtouch.com
4. Annual Report PandaLabs (2011),
   http://press.pandasecurity.com/wp-content/uploads/
   2012/01/Annual-Report-PandaLabs-2011.pdf
5. iDEFENSE Labs Releases Multipot, http://seclists.org/honeypots/2005/q3/17
6. Dionaea catches bugs, http://dionaea.carnivore.it/
7. http://us.mcafee.com/virusInfo/default.asp?id=helpCenter&hcName=mydoom
8. http://www.f-secure.com/v-descs/bagle_at.shtml
9. http://netsecurity.about.com/od/hackertools/p/aapr092004.html
10. Roberts, R.: Malware Development Lifecycle. In: Virus Bulletin Conference (October 2008), http://www.microsoft.com/download/en/details.aspx?id=18919
11. Ollmann, G.: Behind Today's Crimeware Installation Lifecycle: How Advanced Malware Morphs to Remain Stealthy and Persistent,
    http://www.damballa.com/downloads/r_pubs/
    WP_Advanced_Malware_Install_LifeCycle.pdf
12. ESET NOD32 Antivirus,
    http://www.eset.pl/Dla_domu_i_firmy/Produkty/ESET_NOD32_Antivirus
13. Barrett, L.: Malware Volume Surges, But Lifespan Is short,
    http://www.esecurityplanet.com/trends/article.php/3914966/
    Malware-Volume-Surges-But-Lifespan-Is-Short.html
14. Fisher, D.: The Infections That Will Not Die: Conficker and Auto Run,
    http://threatpost.com/en_us/blogs/
    infections-will-not-die-conficker-and-autorun-011712
15. Ollmann, G.: Old threats never die,
    http://www.iss.net/documents/whitepapers/old_threats_never_die_wp.pdf

# The Method of Information Security Based on Micrographics

Ivanna Dronjuk, Mariya Nazarkevych, Nikola Medykovski,
and Olena Gorodetska

Automated Control Systems Department, Institute of Computer Science
Lviv National Polytechnic University
nazarkevich@mail.ru
http://www.lp.edu.ua

**Abstract.** Printed and electronic data storage were considered. Existing protective methods of files and documents were analyzed. The method of micrographics was suggested for protection. Micrographics as background nets are built on the basis of mathematical formulas of Ateb-functions theory. The investigation was conducted, and selection of the line width for the protective image generation was rationalized. Technical data of the paper and printing inks were taken into account, and selection of the printing method was substantiated. The file format for the protected information and the encryption method for electronic documents were considered. The algorithms of micrographics generation for information protection were presented. The work of the method was instantiated.

**Keywords:** information security, Ateb-functions, micrographics, computer networks.

## 1 Introduction

The security of information is an actual issue of nowadays because of it belonging to category of electronic instruments of information dissemination. GuardSoft, Jura JSP are software companies in the field of security design and security technologies [1].

Their software products are designed for creation of guilloche elements used for protection of documents and securities. Original mathematical algorithms for guilloche elements creation are used to protect and based on Bezier curves. The elements of design cannot be faithfully copied even with the help of this program without knowing the precise numerical values. Guilloche elements are specified geometrically and are controlled by a number of parameters. The results are exported to PostScript file.

The authors developed a new protection method, which ideologically is similar to the methods in question, but mathematical discription is based on the theory of Ateb functions. The advantage of this method is a possibility of a precision documents identification.

A. Kwiecień, P. Gaj, and P. Stera (Eds.): CN 2012, CCIS 291, pp. 207–215, 2012.

In the article the method of information protection, which is spread in computer networks as files, is presented. This method can be used for security of printed matter as well.

The idea of method consists in adding of specific protective elements as micrographics in file. It is grounded on the hidden image effect which is based on high resolution of lines capacity. Visually, the micrographics is perceived as indissoluble line, but consists of strokes, which can be considered via significant increasing.

This method can be employed for protecting financial documents such as banknotes and securities in general. The protection method creates hidden messages or images, which become visible on counterfeited documents.

## 2  The Theory of Ateb-functions for Creation Micrographics

One of possible ways of information security is the addition of special security elements into the file. One of those is background nets. Let us consider the already known methods of background nets construction which are implemented in specialized software. In particular, Securedraw software, by JSP Jura Company which specializes in the sphere of software for creating protected design of documents [1]. The suggested software for designing lines and curves is based on Bezier curves.

Method of curve constructing through Bezier does not have a distinct binding to co-ordinate plane. In this method background nets will be built by means of graphical element which is based on Bezier curves apparatus. For creation the background net one graphical element is not enough. Therefore, several segments of Bezier curve which are interlinked are used for construction of graphical element. Thus, the resulting graphical element is a combination of several segments. Consequently there is no single analytical expression for the construction of the whole graphical element. It needs to be calculated partly. This method has no clear attachment to the co-ordinate plane, it is impossible to define whether the set point belongs to the graphical element. The Bezier curve which is built based on 3 points does not necessarily contain these points. This method has low construction efficiency of protective nets.

We suggest a different method of creating background nets, which is based on Ateb-functions. In contrast known methods based on Bezier curves it is presented the construction of background nets with the help of Ateb-function theory. These functions are the solutions of the nonlinear differential equations system of the first order [2,3]. Ateb-functions tabbing allows to get the graphic images of the exact solutions of nonlinear differential equations. Therefore background nets which are the exact solutions of nonlinear differential equations can be used protection technology. In our case the net will be built based on the method known only by the developer. When changing parameters of nonlinear function the net will have a different form.

We take as the basis for constructing background nets aperiodic Ateb-functions. Aperiodic Ateb-functions $sha(n, m, \omega^*), cha(m, n, \omega^*)$ are the inverses of the proper integrals

$$\omega^* = \frac{n+1}{2} \int_0^V \left(1 + \bar{V}^{n+1}\right)^{-\frac{m}{m+1}} d\bar{V} , \qquad V = sha(n, m, \omega^*) \tag{1}$$

$$\omega^* = \frac{m+1}{2} \int_1^U \left(\bar{U}^{m+1} - 1\right)^{-\frac{n}{n+1}} d\bar{U} , \qquad U = cha(m, n, \omega^*) \tag{2}$$

in which $\omega^*$ is an independent variable, and m and n are parameters which are determined by the formulas

$$n = \frac{2v_1^* + 1}{2v_1^{**} + 1}, \qquad m = \frac{2v_2^* + 1}{2v_2^{**} + 1} \qquad (v_1^*, v_2^*, v_1^{**}, v_2^{**} = 0, 1, 2, 3 \ldots) . \tag{3}$$

The change of parameters m and n changes a domain of Ateb-functions. The change in question influences the form of Ateb-function curve. Aperiodic Ateb-functions satisfy the correlation

$$cha^{m+1}(m, n, \omega^*) - sha^{n+1}(n, m, \omega^*) = 1 . \tag{4}$$

As it follows from correlations (1) and (2), functions $sha(n, m, \omega^*)$ and $cha(m, n, \omega^*)$ in relation to the parameter $\omega^*$ for different values m and n, which are determined by formulas (3).

Taking (4) into account, we arrive at the conclusion, that only one of Ateb-functions should be tabulated. We will tab $sha(n, m, \omega^*)$ or $cha(m, n, \omega^*)$ function. The algorithm of construction is showed in [3]. The tabulated function is shown in Fig. 1–Fig. 4. The values of the tabulated Ateb-functions are arranged in the array of values of PostScript-program. On the basis of this array we build a graphic image which is built on the basis of exact mathematical formulas. It will have an irregular structure and depend on the parameters of Ateb-function. Complexity of calculations consists in the simultaneous use of rough integral calculation and search of a zero value of functions set implicitly methods. When calculating the double precision data type was implemented.

## 3    Information Protection Based on Ateb-Functions

By means of the suggested method with the help of certain set of mathematical equations graphical elements of the background net are constructed. For creation of the protective net the graphical element is multiplied through parallel transference, rotation, clenching and stretching, copying, making different combinations. When constructing protective net color, thickness and type of lines can be customized.

On basis of analysis Sheberstov-Murrey-Davis formula with the corrective Yul-Nilsen coefficient [4,5] we considered that the designed method of information protection has to be realized with the stroke thickness 40–90 $\mu$m, which the

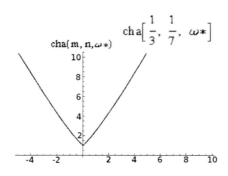

**Fig. 1.** The aperiodic Ateb-function $cha(m, n, \omega^*)$ at $m = 1/3$, $n = 1/7$

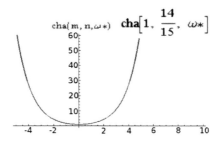

**Fig. 2.** The aperiodic Ateb-function $cha(m, n, \omega^*)$ at $m = 1$, $n = 14/15$

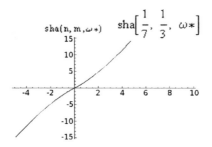

**Fig. 3.** The aperiodic Ateb-function $sha(n, m, \omega^*)$ at $m = 1/3$, $n = 1/7$

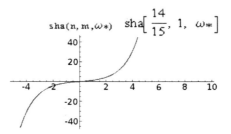

**Fig. 4.** The aperiodic Ateb-function $sha(n, m, \omega^*)$ at $m = 1$ $n = 14/15$

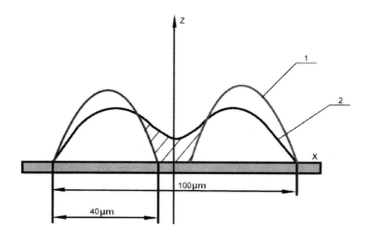

**Fig. 5.** Toner participles distribution on copy formation at copier where $Z$ – the number of toner participles, $X$ – (current value) stroke width

offset machine can print with high quality. It should be mentioned that colorful impress is gained by the colors unification of Cyan (blue), Magenta (purple), Yellow (yellow) and Key color (black) – CMYK. During the calculation of stroke thickness separately for each color can distinguish that the yellow color is printing worst, then purple, blue and black. Therefore, for better documents protection the yellow color should be used, as the optical density of it is the smallest.

Let us examine the image formation on the copies via office equipment, in particular printer and copiers. In them the principle of image creation is based on electronic photography. The receiving of copy is reached by the transfer of toner charged particles on printing material or paper. The created representation by toner is carried on paper and fixed by toner adhesion to paper, caused by heat effect in result.

The toner consists of niggling elementary particles, the average radius of which is 0.1–0.2 $\mu$m. The thinnest stroke size, which can be created on the basis of electronic photography, is 40 $\mu$m (Fig. 5, curve 1). Although, the analysis of wider strokes on obtained copy concludes that the boundary effect is presented. It is illustrated at Fig. 5, curve 2. As seen, the gaps between the strokes are absent, as between thin lines, and the hollow is formed, representation of which is at Fig. 5 by shaded area.

As outcome, after analysis the line width should be within 40–90 $\mu$m for effective model protection. With line width more than 90 $\mu$m the copier will reproduce micrographics without any distortion, and the needed security level will not be reached. With line width less than 40 $\mu$m – the offset printing will not reconstitute the lines.

Hereby, the designed protection method for printing information is based upon formation of niggling strokes that the offset machine prints with high quality, comparing with copier that can reproduce the information or reproduce it with errors. At Fig. 5, curve 2 illustrates the stroke reproduction with width 100 $\mu$m and boundary effect. As a result, on the stroke midst the toner is settled, and splitting of strokes is non-available. That is why the guilloche composition is hard to reproduce at copier, as small lines width and permanent change of each line curvature cause the interferences before reproduction.

Figure 6 shows the information with the protection net. For demonstration the work of the method in this paper the net has low density, thick line and it is done in black colour. For the effective document protection the net is supposed to have high density, thin lines and to be light coloured. The worked out method allows to build protective information file in vector format (PDF), which widely used in computer networks.

The suggested method is based on Post-script file structure analysis and implementation of Ateb function theory as well as Post Script language programming. It is rational to use it in cases when it is essential to presecve the exact formatting of the document. Thus, a lot of companies store the documentation of their products in PDF. The additional benefit of PDF-files is possibility of effective compression which simplifies their transfer in the net.

**Fig. 6.** Example of background net on the basis of Ateb-function

## 4    Information Protection Based on Micrographics

After description of the method micrographics lines for security information, let us instantiate the development of information protection method for electronic storages based on letter micrographics.

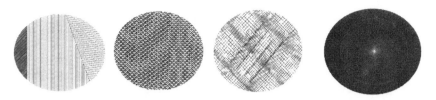

**Fig. 7.** Well-known approaches of information protection by micrographic (thin parallel lines, periodical protective nets, microprint, negative guilloche elements)

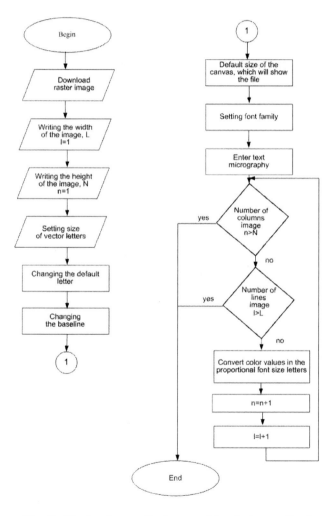

**Fig. 8.** Block schema of micrographics design method

**Fig. 9.** Output image

**Fig. 10.** Output image in large scale

For information protection in files, which are spread in computer networks, the creation of invisible micrographics elements in defined parts of output files is proposed. The authorship of that file can be proved with their assistance. The detection of such protective elements from file is impossible. The well-known approaches of information security by micrographics are viewed at Fig. 7.

The author propose to apply the method for micrographics design, which can be involved in file with textual, graphical or table information. The block schema of developed algorithm is represented at Fig. 8.

On the basis of this algorithm the appropriate software is developed. The application downloads the input document and after processing creates file with protection elements, which at first sight does not change document from input one, endowed by security elements. The example of method realization is illustrated at Figs. 9 and 10. At this figure the defensive element is presented in visible manner, but the parameters of such element (color, size, location) can be changed that can cause its invisibility. The advantage of the method is the arbitrary format of input file, and the output file is created in the .pdf format. Due to using PostScript language the output image is built in vector format.

## 5    Conclusion

The proposed method of information security can be used for preventing of printing information fakes, and information security in electronic data storages that are presented in different computer networks, in particular in Internet. Information protection is realized in the method of constructing background nets on the basis of graphical elemements obtained with the help of Ateb-functions. The advantages of this method is an analytical expression of the graphical curve element of the net and the possibility to check whether the point belongs to the graphical element. Another way for information protection is micrographics wich based on creation vector images with hidden word. The suggested method allows to build a model with protective elements in wector format, which provides high quality of pre-printing package preparation. The advantages of such method are the slight size change of protected file in comparison with output file and the possibility of security realization in runtime.

**Acknowledgments.** The work is performed under the contract DZ 465-2011 "Creation and development of information technology for protection of securities based on new digital processing methods of graphic information".

## References

1. www.jura.at/en/index.html
2. Grytsyk, V.V., Dronyuk, I.M., Nazarkevych, M.A.: Information Technologies for Document Protection by Means of Ateb-Functions. Part I. Construction of Database of Ateb-Functions for Protection of Documents. Journal of Automation and Information Sciences 41(i3.70), 56–70 (2009)
3. Grytsyk, V.V., Dronyuk, I.M., Nazarkevych, M.A.: Information Technologies for Document Protection by Means of Ateb-Functions. Part II. Construction of Database of Ateb-Functions for Protection of Documents. Journal of Automation and Information Sciences 41(i6.50), 70–79 (2009)
4. Murrey, C.H., Davis, F.J.: Appl. Phot. 13(2), 243–248 (1943)
5. Ulichney, R.: Digital Halftoning. MIT Press, Cambridge (1987)

# IP Storage Security Analysis

Tomasz Bilski

Poznań University of Technology
tomasz.bilski@put.poznan.pl

**Abstract.** The paper presents analysis of different methods used to protect IP storage systems. The problem is, if the existing tools and methods for securing common Internet services like WWW, VoIP are suitable to IP storage system protection? The necessary security services such as: authentication, integrity and encryption may be provided in many different ways. It is necessary to select optimal method for a given application. The main part of the paper is related to evaluation of various IP storage systems that utilize iSCSI protocol for transfer and IPSec protocol for protection. Several methods and implementations of security systems with different crypto module locations (client, server, storage medium) and with different implementations (software or hardware) are examined. The following attributes of the systems are analyzed: security level, performance/throughput overhead, cost, flexibility.

**Keywords:** IP, IP storage, SCSI, Fibre Channel.

## 1   Introduction

IP (Internet Protocol) storage system is a system in which storage media are accessed by users with a use of IP network (usually public). It is assumed that data are accessed without needing to alter storage applications. An important feature is block-level data transport. IP networks are used to overcome the distance limitations of such storage buses as FC (Fibre Channel) and SCSI (Small Computer System Interface) with low cost of design, implementation and management. Main applications of IP storage systems are: data protection (backup and recovery), tier-2 storage level, replication, storage consolidation for systems with thousands of servers. For Storage Service Providers, IP storage promises to enable efficient and secure sharing of storage resources among multiple customers. IP storage solutions are available from a number of vendors to extend and complement existing networked-storage environments. It may seem that data security problems may be solved by standard methods for network protection. Nevertheless, both features of IP storage system traffic as well as security requirements of the systems are not identical to other Internet service features and requirements. IP storage security problems cannot be solved with the same methods and tools that are used to protect email or WWW (World Wide Web) services. Furthermore, threats related to Internet are relatively new to storage experts. For example, FC networks used in high end storage systems are not as

A. Kwiecień, P. Gaj, and P. Stera (Eds.): CN 2012, CCIS 291, pp. 216–228, 2012.

open as Internet and are relatively immune to DoS (Denial of Service) attacks, intrusions and malicious software. Security problem in contemporary networks is especially hard to solve due to IPv4 to IPv6 transition phase.

## 2 Traffic Characteristics

Packet flows in IP storage systems are different from packet flows related to Internet services like WWW, email, VoIP (Voice over IP). The main distinctive features of IP storage networking traffic are: gigabyte level of volume of transmitted data, relatively great number of large packets, long-lived sessions between endpoints, intensive handshaking processes related to iSCSI (Internet SCSI) session creation. Sessions between IP storage end-points are typically long-lived to avoid latencies due to TCP (Transmission Control Protocol) slow start algorithm inefficiencies. Slow start is just one of many obstacles related to TCP in IP storage. It must be said that TCP/IP protocols are not well suited for given above traffic characteristics. The protocols were designed many years ago for completely different purposes – telnet, email, FTP (File Transfer Protocol).

The solution is hardware TCP/IP implementation. Some IP storage endpoints use TOE (TCP/IP Offload Engines) for low-latency processing. The general-purpose packet handling capability provided by common network processors is missing. From the security point of view it is a drawback because IPSec acceleration hardware must incorporate the full range of IPSec functions, including crypto processing, security association handling, security policy management and link layer adaptations. The uniqueness of the transfer characteristics has significant impact on the performance as well as the security tools and methods. In general, it may be said that the existing tools and methods for securing common services (WWW, email, ... ) are not attuned to IP storage. Existing IPSec solutions will not provide the throughput required for storage systems, while new high-speed IPSec processors that are designed for VPN (Virtual Private Network) traffic will not work efficiently with IP storage traffic. Furthermore, security solutions for Internet communication do not protect data stored on remote media.

## 3 Threats and Security Requirements

The security requirements of IP storage system involve three common aspects and two sets of threats. The security aspects are well known: confidentiality, integrity and availability. The first set of threats is related to communication channel, the second – to storage media. Three security features should be preserved while data are transmitted and stored. Data integrity and availability are always required. Confidentiality is obligatory in the case of sensitive data. From the security point of view IP storage system is extraordinary. Private data sent across the public TCP/IP networks are exposed to all (passive and active) attacks. The same threats are related to such services as email and VoIP but in the case of these services confidential information is transmitted from sender to receiver just once and exact time of transmission is usually hard to predict. On the other hand in IP storage system a single piece of sensitive data may be transmitted many times from client to storage server and back. If remote storage is used as backup then transmission point in time may be predicted – some backup services transmit data on a regular basis. So, eavesdropping and session hijacking risk levels are higher. Another IP storage specific security issue is related to data sharing scheme. In IP storage systems we may encounter two situations: private and shareable data. If data are private then protection methods may be unique to particular data owner. But, if data should be shareable between many clients then security solutions must work in such a way that each client is able to access (read) data that were written and eventually secured (encrypted) by another client.

## 4     Protocols

### 4.1     General Overview

Communication client – remote storage device may be conducted in two modes of data organization and operation: file-based and block-based. In the first mode of operation the smallest objects that are recognized on TCP/IP application layer are files – logical units of data on disk. File-based organization (e.g. NFS (Network File System), CIFS (Common Internet File System)) and protocols (e.g. FTP) are widely used in NAS (Network Attached Storage) systems and for some Internet services, including WWW. The drawback of this approach is relatively high complexity of the communication channel and in effect relatively small efficiency. The second mode of operation is based on blocks which are related to physical units of data on disks. Block-based approach and FC protocol are used in SAN (Storage Area Network) systems. This mode of operation is also dominant in IP storage. In general two groups of block-based application layer protocols may be used in IP storage systems: based on FC and based on SCSI (Table 1).

**Fibre Channel.** FC is a stack of protocols optimized for networks dedicated to connecting servers to storage arrays. There are two versions of application layer protocols based on FC: FCIP (FC over IP) and iFCP (Internet FC Protocol). FCIP is used to connect distant SAN networks with a use of Internet channels (Fig. 1). It is almost exclusively used for bridging FC fabrics at multiple sites to enable SAN-to-SAN replication and backup over long distances. FC frames from

**Table 1.** Stacks of protocols in IP storage systems

| Application layer | mFCP | iFCP | FCIP | iSCSI |
|---|---|---|---|---|
| Transport layer | UDP | TCP | TCP | TCP |
| Network layer | IP | IP | IP | IP |
| Link layer | Ethernet | Ethernet | Ethernet | Ethernet |

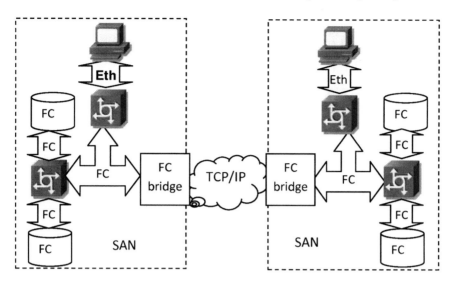

**Fig. 1.** Model of IP storage network with FCIP protocol

host or storage device (with FC interface) in one SAN are encapsulated within TCP/IP units of data and transferred to another SAN. The disadvantage of this solution is the requirement (and cost) to support two types of networks (SAN and Internet) with two different protocol stacks. Furthermore, it is necessary to utilize relatively expensive gateways (FC-to-FCIP bridges) to remote SANs.

The second protocol iFCP is used together with FC storage devices but SAN is not mandatory (Fig. 2). The iFCP protocol gives native TCP/IP connections between end devices. The main advantage of the solution is higher level of flexibility and lower cost – existing FC SAN applications can be used in iFCP systems without modifications. iFCP may be also used to link FC SANs with iSCSI networks. In transport layer for iFCP TCP is used. UDP replaces TCP in protocol named mFCP (Metro FC Protocol). An additional drawback of all IP storage systems using FC is related to management issues. FC infrastructure management is relatively difficult and lack of experienced specialists is a concern.

**SCSI.** Another group of protocols are based on SCSI bus. Internet SCSI (iSCSI) is application protocol which enables block-level operations on distant storage devices running over a TCP/IP transport. At the source, SCSI commands, status and data blocks are placed in iSCSI messages which are transferred by TCP/IP (Fig. 3). At the destination, command or data are retrieved from iSCSI message and moved to storage device or host. iSCSI devices are identified with a use of names and addresses. Names are defined according to iSNS (Internet Storage Name Service), addresses are built according to URL (Uniform Resource Locator) specification. iSCSI protocol is commonly used in LANs (Local Area Networks) and WANs (Wide Area Networks). In LANs, to minimize the impact of disruptive IP traffic, iSCSI traffic may be isolated from nonstorage traffic via

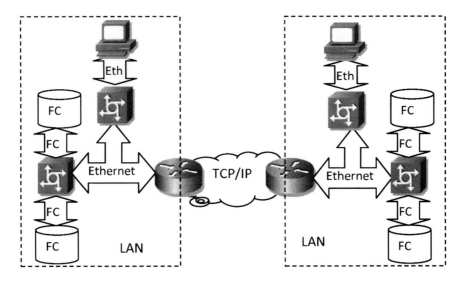

**Fig. 2.** Model of IP storage network with iFCP protocol

dedicated iSCSI networks that have no physical connection to the rest of the network, or by Ethernet isolation techniques like Virtual LANs (VLANs). It is clear that the methods may not be used in public WANs. Typically, SCSI does not involve security as it connects the components inside a computer so iSCSI operates as a cleartext protocol without protection for data in motion during SCSI transactions. An attacker listening in on iSCSI traffic can: violate confidentiality and integrity of the files being transferred, corrupt filesystems being accessed by initiators, exposing servers to software flaws in poorly-tested filesystem code. These problems do not occur only in systems with iSCSI, but rather apply to any storage protocol without cryptographic protection.

### 4.2    FC – iSCSI Comparison

General differences between three IP storage protocols are related to the communication channel structure (Table 2).

From performance point of view systems based on FC are better, the latencies are smaller. Moreover, CPU utilization in iSCSI systems is higher – most iSCSI implementations use software initiators (initiators work as iSCSI clients and are implemented in operating systems e.g. MS Windows Vista/7) using the server's

**Table 2.** Basic differences between IP storage protocols

| Protocol | Terminal/storage interface | Switching | Medium |
|---|---|---|---|
| iSCSI | IP/Ethernet | IP/Ethernet | IP/Ethernet |
| iFCP | FC | IP/Ethernet | IP/Ethernet |
| FCIP | FC | FC | IP/Ethernet |

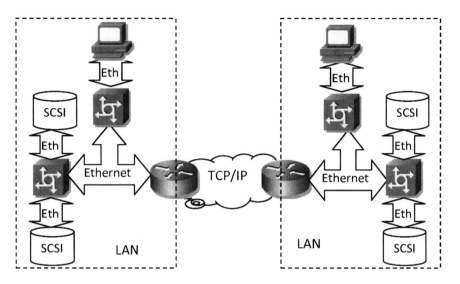

**Fig. 3.** Model of IP storage network with iSCSI protocol

processor for such tasks as creating, sending, and interpreting storage commands. It may be assumed that TCP and iSCSI overhead of 1 Gbit/s connections is at the level of 10% [1,2]. Hardware iSCSI HBA (host bus adapters) do exist but they are not used frequently.

iSCSI advantage is its relative simplicity. Much of the skill set required to implement iSCSI overlaps with that of general network operation. This makes iSCSI attractive to smaller enterprises with limited IT staff and explains its popularity in that market segment. iSCSI performance problems may be solved in some ways: multiple connections, optimized (in hardware) iSCSI HBA, also called iSCSI NIC (Network Interface Card) or Storage NIC (SNIC), with realization of a TCP/IP off-load engine (TOE) to minimize processing overhead [3], Tx/Rx IP, TCP and UDP (User Datagram Protocol) checksum offloading capabilities, Tx TCP segmentation offload (e.g. Intel® 82576 Gigabit Ethernet Controller).

Some differences are related to security. While FC uses FC switches for zoning, arrays for LUN (Logical Unit Number) presentation and host identification through worldwide names, iSCSI secures storage access through a combination of the physical and virtual isolation (in LAN) of the iSCSI network, as well as access restriction by IP addresses, initiator-target names and internal/external CHAP (Challenge Handshake Authentication Protocol) authentication.

It may be expected that in numerous occasions both protocols will function side by side. Common architecture of the system will consist of FC SANs connected with a use of iSCSI. Such complex storage architecture requires IP and FC knowledge as well as the ability to manage and troubleshoot a multiprotocol storage environment. Storage architects must be familiar with iSCSI design options as they affect performance, security and availability.

# 5    Survey of Security Solutions

## 5.1    General Overview

In order to fulfill the security requirements we have to provide the following
services to IP storage system: authentication, access control, integrity control
and encryption. The above services may be provided in many different ways.
In a given system some of the security services may be redundant, e.g. two
independent crypto services – one for transmission, the other for data storage.
Encryption/decryption module or modules may be located in various places
between client and remote storage server. Encryption/decryption modules may
have software or hardware implementations.

## 5.2    Confidentiality and Integrity Protection

In the case of sensitive data we have to provide confidentiality protection. Data
encryption is obvious choice. The question is how to implement encryption modules? There are many general solutions for networking storage confidentiality.
They may be classified according to cryptographic module location. Data encryption may be performed by the following components: operating system in
the data source system, application (e.g. database), network hardware device
and remote storage system. From this point of view different objects may be en-
crypted/decrypted: file, database table, IP packet, disk sector. Each method has
its weaknesses and drawbacks. Evaluating and selecting the solution for a given
information system we have to analyze many intricate criterions: security level,
performance, key management and cost [4].

Transmitted data are usually protected with a use of security mechanism at
transport layer (Transport Layer Security TLS) or internetwork layer (IPSec).
Both protocols are widely adopted for a lot of Internet services. IPSec is more
often used in IP storage systems based on iSCSI.

By the way, it may be noted that for FC networks FC-SP (FC-Security Protocol) was specified. It provides different security mechanisms for various portions
of FC traffic. The ESP (Encapsulating Security Payload) header protocol provides security to the data traffic and the authentication protocol secures the
control traffic. FC-SP uses an authentication mechanism which is a combination
of Diffie-Hellman algorithm and CHAP algorithm. Both protocols belong to the
FC2 layer of the FC stack of protocols and are responsible for authentication of
the fiber channel frames.

Complete stack of FC protocols (including FC-SP) may be used in SAN environment but not in global, public networks such as Internet – IPSec is necessary.
Decision to use IPSec is initial step of designing security system for IP storage.
We have to select several options: location of crypto modules, mode of crypto
implementation (software or hardware), mode of IPSec operation (transport or
tunnel mode).

IP storage system is a set of clients, communication channels, servers and
storage media. Each of the elements may be integrated with encryption module.

Data in IP storage communication channel may be protected with a use of IPSec – typically transport mode is used. Data in storage media may be protected with a use of encryption system integrated with hard disk. IPSec on client side, IPSec on server side as well as hard disk encryption may have software or hardware implementations. In addition IPSec may be implemented by deploying a VPN devices in front of clients and remote storage resources. In the following part we examine some different implementations of IPSec for IP storage systems.

**IPSec Issues in IP Storage.** In the case of mass volume transmissions encryption/decryption processes (especially implemented in software) trigger significant performance degradation. It was shown [5] that on a typical IPSec implementation (Linux kernel 2.6), point-to-point throughput reduction may reach up to 75%. Furthermore, in software implementations of cryptography algorithms the computation time may vary depending on the current CPU load. IPSec degrades not only throughput. The CPU usage during IPSec (with AES (Advanced Encryption Standard)) processing may reach 100% [5,6]. Such CPU load is unacceptable in many cases.

Some different solutions to performance problem were presented: hardware cryptography accelerators, double implementation of IPSec with packet scheduling, lazy crypto approach. Additional delays related to crypto processes are just one of the drawbacks. Further performance degradation is an effect of packet size growth resulting from addition of AH (Authentication Header) and ESP headers. ESP and AH headers contain tens of bytes that are added to each IP datagram.

In general, the performance problem may be solved in two ways: by accelerating some processes (e.g. with a use of hardware instead of software crypto implementations) or by removing the redundant processes in such a way that security is not compromised (the option is used in 'lazy crypto approach').

Another problem of IPSec is incompatibility between IPSec and NAT (Network Address Translation). The problem is widespread and exists in many Internet services so it is not discussed here.

**Hardware IPSec Implementations.** Obvious solution to performance problem is hardware (e.g. NSP (Network Security Processor), ASIC (Application Specific Integrated Circuit), FPGA (Field Programmable Gate Array)) cryptography implementation. For symmetric-key algorithms, performance increase of 4000% over a modern CPU at similar clockspeeds may be achieved [7]. At the same time, the cryptography processor may decrease the overhead of the general-purpose processor.

Accelerators may be placed off-line (outside the main data path) or in-line (on the main data path). The first architecture (known also as look-aside) assumes IPSec encryption accelerators act as co-processors to network processors. The intelligence for managing the IPSec function and full IPSec protocol support, including packet processing, link layer adaptations and security association handling, resides within the network processor. The look-aside approach places a significant burden on the network processor and requires a high bandwidth sideband interface for data transfer between the network processor and IPSec accelerator. The second architecture is called flow-through. All the network-related data flows through the accelerator – the advantage is that the other host machine processing units may be completely unaware of IPSec.

Further performance improvement may be achieved with some dedicated packet grouping and scheduling algorithms. Dedicated algorithms may minimize the time needed to transfer data and to set up the accelerators [8].

Some hardware IPSec implementations are specially dedicated to the requirements of IP storage (including iSCSI, FCIP). They are based on architecture that handles the entire IPSec protocol in a single chip. Both the IPSec encryption data path functions, inbound and outbound policy processing, SA (Security Association) lookup and context handling and the IKE (Internet Key Exchange) handshake are handled completely on chip and in-line for Gbit networks (examples are: Hifn HiPP III 4300 Storage Security Processor and Intel 82576 Gigabit Ethernet Controller).

Fast, ASIC IPSec implementations are expensive, e.g. 1–10 Gbit/s crypto ASIC may cost from about \$50 to more than \$500. Gbit Ethernet adapters top out at about \$200, while FC adapters cost as much as \$1,000. In both cases IPSec ASIC significantly increases total cost of network hardware. Crypto ASIC chips should be integrated with many devices: HBAs, target bus adaptors, SAN switches and storage servers. Another drawback is lack of the flexibility, especially in the case of NSP and ASIC. Both platforms are not re-programmable. FPGA offers some level of flexibility [9] but throughput of FPGA is smaller [10,11].

**Double Implementation of IPSec.** Double IPSec implementation [12] is encryption system that uses two parallel implementations of IPSec (one software and one hardware). The scheduling algorithm is used to schedule packets to be processed either by the CPU with a software implementation or by the accelerator. Similar scheduling algorithm for distributing IPSec packet processing over the CPU with a software implementation of the cryptographic algorithm and multiple cryptographic accelerators is described by [13]. In both cases the goal is to minimize the average waiting time of the packets giving the best throughput.

The scheduling algorithms utilize IPSec processing features: approximate hardware processing time (of symmetric cryptography algorithm) is known in advance and no data dependency exists between different packets – according to IPSec specification each packet must carry any data required for its processing.

The advantage of double (software and hardware) IPSec implementation is the lack of tradeoff between performance and flexibility. Hardware implementation provides performance while software provides flexibility. Flexibility is necessary, for example, for protocol updates. It must be noted that achieving simultaneously some flexibility and relatively high performance is possible with a use of modern FPGA platforms. A quite new feature of FPGA platforms, is partial reconfiguration in which part of the chip is reconfigured while the remaining portion is operational.

**IPSec Implemented in VPN Device.** VPNs are another solution to IP storage security problem. Clients and storage systems are connected to the Internet by security gateways which implement IPSec in tunnel mode. It is clear that VPN protects data in transit and does not protect data in rest on storage media.

**Lazy Crypto Approach.** In [14] data storage and transmission convergence concept had been proposed. The idea of the concept is to solve network and storage individual problems in uniform way. The concept implementation may be found in [5]. Chaitanya et. al propose a solution to IPSec performance problem called lazy crypto approach. In the method data encryption/decryption process at storage site (medium) is eliminated. Data written to remote disk are encrypted before transmission to the disk at the client site with a use of modified IPSec procedure. Encrypted data are transferred and written to remote disk. During read process encrypted data from disk are read and transferred to client. Decryption process is done by client after receiving data from remote storage site.

## 5.3  Authentication

In the case of private networks administrators most commonly run iSCSI only over logically-isolated back channel networks – only the management ports of storage arrays are exposed to the general-purpose network and the iSCSI protocol is run over dedicated network segments or VLANs. Administrator may choose to use physically separate switches (unauthorized users aren't physically provisioned for iSCSI) dedicated to iSCSI VLANs only, to further prevent the possibility of an incorrectly connected cable plugged into the wrong port bridging the logical barrier between VLAN and public LAN. Simple initiator-target name authentication will typically suffice. If the iSCSI network is physically connected to the LAN, the stronger CHAP authentication is deployed, eliminating the threat of spoofed IP addresses accessing iSCSI LUNs. If large number of iSCSI devices is used, central authentication via a Radius server solves the user credentials in iSCSI targets management problem.

Authentication in public networks may be performed with a use of mechanisms existing in iSCSI specification. iSCSI initiators and targets prove their identity to each other using the CHAP protocol and SRP (Secure Remote Password) protocol, which include a mechanism to prevent cleartext passwords from appearing on the wire. An initiator must establish a session with the target (remote storage server) before obtaining access to storage data. The initial iSCSI authentication phase required for providing an application-level access control scheme may be carried on a secure IPSec-based TCP connection between the iSCSI initiator and target.

By itself, the CHAP protocol is vulnerable to dictionary attacks and spoofing attacks. If followed carefully, the rules for using CHAP within iSCSI prevent most of these attacks. It must be added that iSCSI does not account for per packet authentication and integrity of data during the full featured data transmission phase. In addition, privacy of sensitive data sent across the vulnerable IP networks is also not taken into consideration.

Authentication process in FC may use several mechanisms: FCAP (FC Authentication Protocol), FCPAP (FC Password Authentication Protocol), DH-CHAP (Diffie Hellman – CHAP), FC-SP (FC – Security Protocol). The last one is a security framework for authentication of FC devices, secure key exchange and secure communication between FC devices.

## 5.4 Access Control

IP storage access control requirement may be fulfilled with masking and zoning. The services are integrated with FC but may be used also in iSCSI networks. The access control is implemented either in the hardware drivers on the initiators or incorporated in FC switches. Masking means that storage devices identified by LUN are available only to authorized hosts identified by WWNs (World Wide Name) – WWN addresses HBA. The access control function is provided by HBA so successful attack on HBA is in effect equal to access control system abuse.

Functions of zoning service are similar. Zoning is used to block access to a zone from any device outside the given zone. Zoning may be implemented in the form of hard or soft. In SANs zoning is done by FC switches and is linked to ports while in Internet it is based on WWNs. Zoning is a method where restrictions are placed on the communication between certain sets of WWNs. Zoning and masking were not created to be security tools. They are often used as security tools but do not always contain any enforcement capabilities such as strong authentication. Both security services (masking and zoning) which are based on WWNs are vulnerable to WWNs spoofing since WWN of a given device is configurable parameter and may be changed easily (changing WWN is a feature provided by HBA vendors) by malicious user in order to obtain authorized access to data. Data at rest encryption is another solution to access control. In the case of encrypted data even physical access to blocks of data (without decryption key) is not equal to the access to data values.

## 5.5 Availability

Availability needs to be architected at the server, network and storage device level. At the LAN level, redundancy is achieved by deploying switches in pairs and using Ethernet failover techniques like spanning tree and dynamic routing. At the server level, high availability is achieved by dual-connecting servers to Ethernet switches (such mode of operation is supported by e.g. Microsoft 2.0 release of its iSCSI Initiator). Availability in WAN is achieved with common routing services. In IPv6 networks availability may be supported with a use of multihoming.

Redundancy options in iSCSI storage devices vary by vendor and product type. iSCSI gateway appliances, intelligent storage switches and server-based iSCSI devices are typically available in cluster configurations in which two devices run in active-active or active-passive mode. Some storage arrays provide redundancy with a use of dual-controller architecture or architectures based on chassis with multiple iSCSI blades.

# 6    Conclusion

IP storage security is a complex problem. The problem may not be solved with a use of common Internet security tools and methods without some modifications. A general, observable trend is to use IPSec and integrate its functionality directly into the iSCSI host and storage devices. But storage vendors may be resistant to building IPSec into their products when they have no guarantee that it will be used. Issues of cost, security management, interoperability will need to be addressed in order to make these solutions practical and scalable. Because adding the cost (several hundred dollars) of fast IPSec ASICs to storage devices may not be realistic, vendors are likely to choose another implementation. Given the contention surrounding IPSec for IP storage, users might end up having to trade interoperability for security.

**Acknowledgements.** The research project is scheduled for years 2010–2013 and supported by scientific grant from the Polish Ministry of Education and Science.

# References

1. Aiken, S., et al.: A Performance Analysis of the iSCSI Protocol. In: Proc. 20th IEEE Symposium on Mass Storage Systems and Technologies, pp. 123–134 (2003)
2. Xinidis, D., Flouris, M., Bilas, A.: Performance Evaluation of Commodity iSCSI-based Storage Systems. In: Proceedings of the 22nd IEEE/13th NASA Goddard Conference on Mass Storage Systems and Technologies, MSST 2005 (2005),
   http://storageconference.org/2005/papers/27_xinidisd_iscsi.pdf
3. Todorov, D., et al.: Storage Networks on IP infrastructure. In: International Workshop NGNT 50 Storage Networks on IP Infrastructure, pp. 50–56 (2008)
4. Bilski, T.: Network Backup Confidentiality Evaluation. In: Pejas, J., El Fray, I., Saeed, K. (eds.) ACS 2006 13th International Multi-Conference on Advanced Computer Systems, Miedzyzdroje, October 18-20, vol. I, pp. 321–330 (2006)
5. Chaitanya, S., Butler, K., Sivasubramaniam, A., McDaniel, P., Vilayannuret, M.: Design, Implementation and Evaluation of Security in iSCSI Based Network Storage Systems. In: StorageSS 2006, October 30, pp. 17–28. ACM (2006),
   http://www.patrickmcdaniel.org/pubs/storss06.pdf
6. Ferrante, A., Piuri, V., Owen, J.: IPSec Hardware Resource Requirements Evaluation. In: Proceedings of 1st Euro-NGI Conference on Next Generation Internet Networks – Traffic Engineering, Rome, pp. 240–246 (2005)
7. Birman, M.: Accelerating Cryptography in Hardware. In: HOT CHIPS Symposium on High Performance Chips (1998),
   http://www.hotchips.org/archives/hc10/2-Mon/HC10.S4/HC10.4.2.pdf
8. Taddeo, A.V., Ferrante, A.: Scheduling Small Packets in IPSec Multi-accelerator Based Systems. Journal of Communications (JCM) 2(2), 53–60 (2007)
9. Lu, J., Lockwood, J.: IPSec Implementation on Xilinx Virtex-II Pro FPGA and Its Application. In: Proceedings of the 19th IEEE International Parallel and Distributed Processing Symposium, IPDPS 2005 – Workshop 3, vol. 04 (2005)

10. Saravanan, P., et al.: High-Throughput ASIC Implementation of Configurable Advanced Encryption Standard (AES) Processor. In: IJCA Special Issue on Network Security and Cryptography, NSC (2011),
    http://research.ijcaonline.org/nsc/number3/SPE028T.pdf
11. Wang, H., Bai, G., Chen, H.: A Gbps IPSec SSL Security Processor Design and Implementation in an FPGA Prototyping Platform. Journal of Signal Processing Systems 58, 311–324 (2010)
12. Amiri, R.: IPSec Available (2007),
    http://www.slideshare.net/Franklin72/ipsec-protocol
13. Castanier, A., Ferrante, A., Piuri, V.: Packet Scheduling Algorithm for IPSec Multi-Accelerator Based Systems. In: Proceedings of the 15th IEEE International Conference on Application-Specific Systems, Architectures and Processors, ASAP (2004)
14. Bilski, T.: Data storage and transmission convergence concept. In: Kozan, E. (ed.) Proceedings of the Fifth Asia Pacific Industrial Engineering and Management Systems Conference, pp. 14.8.1–14.8.16. Queensland University of Technology (2004)

# Usage of Pseudo-estimator LAD and SARIMA Models for Network Traffic Prediction: Case Studies

Maciej Szmit[1,2] and Anna Szmit[2]

[1] Orange Labs, ul. Obrzezna 7, 02-691 Warszawa, Polska
maciej.szmit@gmail.com
[2] Technical University of Lodz, ul. Zeromskiego 116, 90-924 Lodz, Polska

**Abstract.** This article focuses on the application of SARIMA models and Last Absolute Deviation pseudo-estimator in Auto Regression models of network traffic for various types of network protocols in sample computer networks. The models are used to build predicted patterns of traffic.

**Keywords:** intruder detection systems, anomaly detection, autoregression models, ARIMA models.

## 1   Problem Description

An important aspect of security management in computer systems is to detect abnormal situations, which may be related with potentially dangerous activities, such as attacks of intruders or malicious software operations. One of phenomena which is often monitoring traffic in computer networks. Detection of abnormal behavior of computer networks (so called Network Behavior Anomaly Detection, NBAD) is one of the three groups of methods, including misuse detection systems and integrity verification, used in Intruder Detection Systems.

The behavioural analysis of network traffic is a promising path in the development of anomaly detection systems. In large widespread networks, teams responsible for their functioning analyse traffic to detect problems with infrastructure of the backbone network (especially carrying out BGP traffic monitoring) as well as to detect activity of malicious software (primarily to detect and remove botnets and blocking DDoS). In smaller networks, traffic analysis and monitoring is carried out sporadically and used mainly to monitor performance and possibly diagnose sources of malfunctions. As it seems, using methods of traffic anomalies detection can be useful also in the case of smaller networks, in particular to detect certain types of attacks, which can easily be identified as the network traffic characteristics changes.

In our works (see e.g. [1–5]) we develop a software dedicated to Snort IDS, designed to enhance its possibilities to monitor, analyze and detect network traffic anomalies using NBAD approach. The software consists of three elements:

A. Kwiecień, P. Gaj, and P. Stera (Eds.): CN 2012, CCIS 291, pp. 229–236, 2012.

- AD preprocessor, which periodically, with a given interval, logs information about a set of parameters (as a number of TCP/UDP packets sent/received from outside/inside the current IPv4 subnet, www download/upload speed, a number of UDP 53 (DNS) datagrams etc.) of the network traffic. The second function of the preprocessor is reading a predicted pattern of the network traffic from the "profile" file and generates alert when the current value exceeds "minimum" to "maximum" range for the current moment.
- Profile generator, which – among other things – generates file containing predicted pattern (expected future values of parameters) of the network traffic based on the statistical model for a given future time period.
- Evaluator used to test the quality of the models.

Our current research is to compare fits of different class of models used to used to explain the variability different kinds of network traffic and building its predicted pattern. The models was tested on several time series collected in five different computer networks:

- W – amateur campus network consisting of circa 25 workstations. Snort has worked on the router which acts also as the gateway to the Internet and as the few servers as ftp, www, etc.
- T1 – Campus network provided by a mid-size Internet Access Provider – (about 400 clients).
- T2 – A network in a block of flats; one of the subnetworks mentioned in the examples T2 containing about 20 clients.
- M – Home network connected to the campus amateur network (with maximum speed of inbound traffic set on the bandwidth manager to 4 Mbps. Home network consists of five computers protected by corporate firewall and two intranet servers (ftp and PrintServer).
- I – Local Area Network in small company (about 40 personal computers, two intranet servers, detailed description of the networks can be find in our previous works, see: [3, 4]).

For each type of traffic in each network we get, a time series of length from a few thousand to over 20 000 observations (experiments on different networks were independent for different time intervals). Then the time series were subjected to statistical analysis and modeling using the naive method, moving average, Holt-Winters [6] and Taylor model and linear regression models with 0-1 variables. In the article we discuss on the one hand the result of using a few of the family of Auto Regression-based models (ARMA/ARIMA/SARIMA) and on the other hand, an alternative method for estimating the parameters of autoregressive model – Least Absolute Deviations. We analysed three types of packages in the seven time series (one from each W, T1, T2 and two from M and I networks). For each method we compare level of the model fit. We decided to use quotient $MAE/M$ as a measure of fit rather than MSE (Mean Squared Error) based measure because there were a lot of so-called outliers are noted in the analysed samples and MSE-based measure can be oversensitive in those cases).

## 2   AR Models

Linear autoregression models are a type of linear regression models, where explanatory variable are lagged values of this variable (values of the time series in the past):

$$\hat{y}_t = a_0 + a_1 y_{t-1} + a_2 y_{t-2} + a_k y_{t-k} \tag{1}$$

where:

$\hat{y}_t$ is the theoretical value of the response variable in moment $t$,
$y_t$ is the actual (measured) value of the variable,
$k$ is the number of explanatory (independent) variables.

Two models with autoregression were built for the analysed series. In the first case, the value of the response variable lagged by one measurement (ten minutes) was used as the explanatory variable. In the second case, three explanatory variables were used: the value of the variable lagged by one measurement, by 24 hours and by a week. The goodness of fit was subsequently calculated for the models, measured by the $\frac{MAE}{M}$, where $MAE$ denotes Mean Absolute Error:

$$MAE = \frac{1}{n} \sum_{t=1}^{n} |y_t - \hat{y}_t| = \frac{1}{n} \sum_{t=1}^{n} |e_t| \tag{2}$$

and $M$ denotes Arithmetic Mean.

The obtained results (for the series described in articles [2] and [3]) are presented in the Table 1.

As results from data in the table above, adding more explanatory variables does not significantly improve the quality of the model. The modelled phenomena depend more on the value of the last measurement than on a value noted a day or a week ago. Therefore, the present traffic volume is more important than the potential daily or weekly periodicity. Only in the case of ICMP traffic in I5 network adding more explanatory variables resulted in considerable improvement (by over one fourth). The AR model obtained the best (among all of the models being analysed) fit, measured with the value of the $MAE/M$ error for TCP series in T2, T3 and II networks and for UDP traffic in T3 network.

## 3   ARIMA Models

The ARIMA model, i.e. the integrated autoregression and moving average process, popularised by Box and Jenkins (see [7]) is one of the classic tools used in the analysis of time series. The model presents the series $y_t$ as a result of the so-called white noise filtering. It assumes that the current value (or its certain increment) in the series $y_t$ depends directly on $p$ previous values of this series and on $q$ previous values of the white noise series $\xi_t$.

The ARIMA model can be treated as a complex phenomenon, composed of the following processes: Autoregression (AR), Moving Average (MA) and summation ("integration") component. The values of the series $y_t$ in ARIMA can be represented as:

– in $y_t$ is a stationary series where the average equals $\mu$ (ARMA$(p,q)$ process):

$$y_t = \mu + \phi_1 y_{t-1} + \phi_2 y_{t-2} + \phi_p y_{t-p} + a_t - \theta_1 a_{t-1} - \theta_2 a_{t-2} - \cdots - \theta_q a_{t-q} \ , \quad (3)$$

– if $y_t$ is an ARIMA$(p,1,q)$ process:

$$y_t = y_{t-1} + \phi_1(y_{t-1} - y_{t-2}) + \phi_2(y_{t-2} - y_{t-3}) + \cdots \\ + \phi_p(y_{t-p} - y_{t-p-1}) + a_t - \theta_1 a_{t-1} - \theta_2 a_{t-2} - \cdots - \theta_q a_{t-q} \quad (4)$$

where:

$\phi_1, \phi_2, \cdots, \theta_1, \theta_2, \cdots$ parameters of the model,
$a_t, a_{t-1}, \cdots$ residuals of the model.

**Table 1.** Errors of fit of models with autoregression (source: own research): (1) having one variable lagged by one measurement or (2) having variables lagged by one measurement, 24 hours and a week

| Network | Protocol | $MAE/M$ for AR (1 measurement) | $MAE/M$ for AR (1 measurement, 24 hours, a week)[a] |
|---|---|---|---|
| W1 | TCP | 53.52% | 54.03% |
| W1 | UDP | 34.06% | 33.07% |
| W1 | ICMP | 32.68% | 32.28% |
| T2 | TCP | 4.22% | 4.15% |
| T2 | UDP | 16.56% | 16.64% |
| T2 | ICMP | 8.71% | 8.59% |
| T3 | TCP | 4.06% | 4.07% |
| T3 | UDP | 15.10% | 15.12% |
| T3 | ICMP | 8.82% | 8.75% |
| M4 | TCP | 58.48% | 60.06% |
| M4 | UDP | 32.12% | 30.36% |
| M4 | ICMP | 16.03% | 15.76% |
| I5 | TCP | 48.46% | 51.45% |
| I5 | UDP | 62.28% | 63.29% |
| I5 | ICMP | 149.86% | 112.29% |
| MM | TCP | 77.26% | 74.67% |
| MM | UDP | 35.60% | 33.79% |
| MM | ICMP | 11.45% | 11.40% |
| II | TCP | 43.73% | 39.70% |
| II | UDP | 64.53% | 54.78% |
| II | ICMP | 135.20% | 137.09% |

[a] In view of the necessity to take into account the variable lagged by a week, the measurement error was possible to calculate only starting from the second week of measuring. In the case of the first model (taking into account only the variable lagged by one measurement) – starting from the second measurement. Therefore, certain results are present in the table where the error displayed by the model with many variables is higher than in the case of the model with one variable, which could not occur if errors from the same number of forecasts were to be compared (adding more variables to the additive model cannot deteriorate its quality).

The generalised ARIMA model can be applied to periodic processes with a fixed duration of the $s$ period. Then, the series $y_t$ is treated as (multiplicative) compounding of two processes: ARIMA$(p, d, q)$, in the case of lags by one period, and ARIMA$(P, D, Q)$ in the case of lags by $s$ periods. We are then dealing with the ARIMA$(p, d, q) \times (P, D, Q)_s$ process, sometimes represented as SARIMA (see e.g. [8]).

For example, in the case of ARIMA$(1, 1, 1)(1, 1, 1)s$, is:

$$
\begin{aligned}
y_t = {} & y_{t-1} + y_{t-s} - y_{t-s-1} + \phi_1(y_{t-1} - y_{t-2} - y_{t-s-1} + y_{t-s-2}) + \\
& + \Phi_1(y_{t-s} - y_{t-s-1} - y_{t-2s} + y_{t-2s-1}) - \\
& + \phi_1 \Phi_1(y_{t-s-1} - y_{t-s-2} - y_{t-2s-1} + y_{t-2s-2}) + \\
& + a_t - \theta_1 a_{t-1} - \Theta_1 a_{t-s} + \theta_1 \Theta_1 a_{t-s-1} \ .
\end{aligned}
\tag{5}
$$

The Table 2 summarises the results obtained using the particular models for selected series.

**Table 2.** Errors of fit of ARMA/ARIMA/SARIMA models (source: own research)

| Network | Protocol | Model used | $MAE/M$ |
|---------|----------|------------|---------|
| W1 | TCP | (7,0,0)(0,0,2) | 50.22% |
| W1 | UDP | (2,1,0)(0,0,1) | 31.00% |
| W1 | ICMP | (1,0,1)(1,0,1) | 31.57% |
| T2 | TCP | (5,0,0)(0,0,4) | 4.20% |
| T2 | UDP | (0,1,3)(3,0,0) | 16.20% |
| T2 | ICMP | (4,1,0) | 8.60% |
| T3 | TCP | (3,1,1)(0,1,2) | 4.25% |
| T3 | UDP | (1,0,1)(1,0,1) | 14.99% |
| T3 | ICMP | (1,0,1)(1,0,1) | 8.84% |
| MM | TCP | (3,0,0) | 65.86% |
| MM | UDP | (4,0,0)(0,0,3) | 29.87% |
| MM | ICMP | (0,1,5) | 10.71% |
| II | TCP | (1,0,1)(1,0,1) | 38.62% |
| II | UDP | (6,0,0)(1,0,0) | 48.31% |
| II | ICMP | (4,0,0)(0,0,1) | 102.21% |

## 4    AR Models with LAD Estimator

The AR models presented above are characterised by a relatively low fit. This results from the $MAE/M$ fit method used, which is natural in the case of minimising the total (or average) absolute error, whereas the model parameter values estimation was performed, as is usually the case, using the least squares method estimator. LSM (Least Squares Method) estimation is performed to retrieve parameters minimising the sum of squared residuals (and thus the MSE mean-square error and the RMSE root mean squared error). In the cases where the so-called outliers are noted in the analysed sample, MSE minimisation will lead to determination of parameters which yield the lowest possible value of

$MAE/M$. Considering the characteristics of each of the series, one can expect this type of phenomena to occur relatively frequently while analysing network traffic. Therefore, it seems that in traffic modelling robust statistics (see e.g. [9]) methods of estimation should be favoured, in this case the simplest method of minimising the residuals module (Least Absolute Deviations – LAD, Least Absolute Errors – LAE, Least Absolute Value – LAV) based on a function returning the absolute value of the sum of residuals (Sum of Absolute Errors – SAE):

**Table 3.** Errors of fit of models with autoregression (source: own research): (1) having one variable lagged by one measurement or (2) having variables lagged by one measurement, 24 hours and a week while estimating the parameters of the model using the LAD estimator

| Network | Protocol | $MAE/M$ for AR (1 measurement) | $MAE/M$ for AR (1 measurement, 24 hours, a week)[a] |
|---------|----------|-------------------|-------------------|
| W1 | TCP | 45.16% | 45.07% |
| W1 | UDP | 30.35% | 29.70% |
| W1 | ICMP | 31.71% | 31.32% |
| T2 | TCP | 4.21% | 4.15% |
| T2 | UDP | 15.54% | 15.58% |
| T2 | ICMP | 8.60% | 8.47% |
| T3 | TCP | 4.06% | 4.07% |
| T3 | UDP | 15.02% | 15.09% |
| T3 | ICMP | 8.71% | 8.62% |
| M4 | TCP | 50.25% | 51.33% |
| M4 | UDP | 30.28% | 29.31% |
| M4 | ICMP | 14.48% | 14.35% |
| I5 | TCP | 39.66% | 42.71% |
| I5 | UDP | 49.47% | 51.26% |
| I5 | ICMP | 100.03% | 75.78% |
| MM | TCP | 64.72% | 62.40% |
| MM | UDP | 29.68% | 29.25% |
| MM | ICMP | 11.13% | 11.08% |
| II | TCP | 36.93% | 35.45% |
| II | UDP | 50.83% | 47.99% |
| II | ICMP | 94.57% | 97.25% |

[a] In view of the necessity to take into account the variable lagged by a week, the measurement error was possible to calculate only starting from the second week of measuring. In the case of the first model (taking into account only the variable lagged by one measurement) – starting from the second measurement. Therefore, certain results are present in the table where the error displayed by the model with many variables is lower than in the case of the model with one variable, which could not occur if errors from the same number of forecasts were to be compared (adding more variables to the additive model cannot deteriorate its quality).

$$SAE = \sum_{t=1}^{n} |y_t - \hat{y}_t| = \sum_{t=1}^{n} |e_t| \; . \tag{6}$$

The Table 3 summarises the results of $MAE/M$ fit obtained for AR models, whose parameters were calculated by minimising SAE.

As can be noted, owing to employment of a different method of estimating parameters, the goodness of fit was improved by decimals of a percentage point in the case of models with a good fit (e.g. TCP traffic in T2 and T3 networks) or full percentage points in the case of models with a slightly worse fit (e.g. ICMP or UDP in T2). When the AR model with coefficients estimated using the least squares method resulted in a very bad fit (e.g. UDP in I1 network or TCP in I5), owing to the use of the LAD estimator, an improvement by several percentage points can be achieved. Nevertheless, if the $MAE/M$ error reaches several tens per cent, one can expect that in the case of such a network the AR model will have no practical application. As is intuitive, volumes of certain types of traffic in certain networks being analysed were in fact difficult to model using any model at all.

## 5   Conclusions

1. Proportions of traffic are different in various networks. Although TCP segments are the most frequent, the dominance of this protocol can be less or more pronounced. Traffic in the case of various protocols, even in the same network, can have totally different characteristics and, as a consequence, relatively optimal fits for various protocols can yield different modelling methods. Specific phenomena (f.e. particular type of attack) can be observed while monitoring selected protocol. While analysing network traffic, it is necessary to at least differentiate between protocols used in communication.
2. Series describing traffic volume, at least those collected in the analyses performed for the purpose of this paper, are characterised by a relatively high number of outliers (with the exception of the network T2 and T3). Therefore, it seems appropriate to use methods of robust statistics to estimate parameters of particular models. Assuming the fit factor used, the application of these methods yields better results than the use of even the most complex ARIMA models. In the case of series without outliers use LAD estimation did not lead to significant improvements.
3. Perhaps, in view of the above, one should consider the possibility of using robust statistics methods to estimate parameters of models which are more complex than AR, which might be the object of focus in further research. It is also necessary to remember that the decisive criterion for anomaly detection is not the quality of the "point" fit of the model to empirical data. From the perspective of anomaly identification, it is important that the anomaly detection method correctly defines the limits outside which the given type of traffic is classified as an anomaly.

# References

1. Skowronski, M., Wezyk, R., Szmit, M.: Detekcja anomalii ruchu sieciowego w programie Snort. Hakin9 3, 64–68 (2007)
2. Skowronski, M., Wezyk, R., Szmit, M.: Preprocesory detekcji anomalii dla programu Snort. In: Sieci Komputerowe. Aplikacje i zastosowania, vol. 2, pp. 333–338. Wydawnictwa Komunikacji i Lacznosci, Gliwice (2007)
3. Szmit, M., Szmit, A.: Use of Holt-Winters Method in the Analysis of Network Traffic: Case Study. In: Kwiecień, A., Gaj, P., Stera, P. (eds.) CN 2011. CCIS, vol. 160, pp. 224–231. Springer, Heidelberg (2011)
4. Szmit, M.: Wyuziti nula-jednickovych modelu pro behavioralni analyzu sitoveho provozu. In: Internet, Competitiveness and Organizational Security, TBU Zlin 2011, pp. 266–299 (2011)
5. Szmit, M., Wezyk, R., Skowronski, M., Szmit, A.: Traffic Anomaly Detection with Snort. In: Information Systems Architecture and Technology. Information Systems and Computer Communication Networks. Wydawnictwo Politechniki Wroclawskiej, Wroclaw (2007)
6. Brutlag, J.D.: Aberrant Behavior Detection in Time Series for Network Monitoring. In: 14th System Administration Conference Proceedings, New Orleans, pp. 139–146 (2000)
7. Box, G.E.P., Jenkins, G.M.: Analiza szeregow czasowych. Prognozowanie i sterowanie. PWN, Warszawa (1983)
8. Klevecka, I.: Forecasting Network Traffic: a Comparison of Neural Networks and Linear Models. In: Proceedings of the 9th International Conference "Reliability and Statistics in Transportation and Communication", RelStat 2009, Riga, October 21-24 (2009)
9. Maronna, R., Martin, R.D., Yohai, V.: Robust Statistics – Theory and Methods. Wiley (2006)

# Anonymization of Web Client Traffic Efficiency Study

Tomas Sochor

University of Ostrava, Ostrava, Czech Republic
`tomas.sochor@osu.cz`
`http://www1.osu.cz/home/sochor/en/`

**Abstract.** Anonymization in the sense of hiding the originator of the web request could be sometimes useful and it can be done using various commercial and public domain tools. The adverse aspect of anonymization is that the anonymized traffic is significantly slower than normal one. The study focused to describe the slowing down and to quantify its rate is presented in this article. The best known free anonymization tools were involved in the study, namely TOR, JAP and I2P. The set of files was formed and their download transmission speed was measured with and without use of anonymization. Also the set of webpages was formed where both latency and transmission speed were measured. All the measurements were done at the application layer. The final comparison showed that the TOR remains to be the best tool for anonymization despite the fact that JAP excelled in latency. The price paid for anonymization also remains to be quite high because it was confirmed that at least 90% transmission speed decrease is inseparable from using free anonymization tools.

**Keywords:** traffic anonymization, TOR, JAP, I2P, latency, round-trip time, transmission speed, WWW.

## 1 Introduction – WWW Traffic Anonymization

There are numerous situations when a user desire to hide details about his/her activity in the Internet. In such cases various tools and methods for anonymization are used. TOR (The Onion Routing) belongs among the most famous free tools but recently also other tools (JAP, I2P) emerged. After preliminary study [1] devoted to TOR only the more thoroughgoing study involving other tools was performed.The study focused primarily to the comparison of latency and transmission speed on the application level between normal and anonymized WWW traffic.

There are numerous tools for anonymization of Internet traffic. Most of them including TOR rely on concealing the IP address of the originator. The operation of public anonymization tools is possible thanks to the fact that many people throughout the world allow using their computers as anonymization nodes (in case of TOR called *onion routers*). More detailed information about TOR operation can be found in [2] while I2P is described in [3] and JAP in [4]. The

A. Kwiecień, P. Gaj, and P. Stera (Eds.): CN 2012, CCIS 291, pp. 237–246, 2012.

main difference between measured tools is that TOR in based on onion routing while the others on cascade mix but the practical difference between these two approaches seems to be somehow unclear and it was not investigated in this study in details.

An obvious drawback associated with the use of any anonymization tool is the increase in latency and/or decrease of transmission speed. Significant worsening of transmission quality expressed in latency and transmission speed is obviously not only because of adding more nodes into the path between the client and target but also due to encryption of data traffic between intermediate nodes. The main aim of the study described here was to quantify such decrease to allow the assessment of efficiency of using the anonymization tools studied.

## 2    Measurement Methods

### 2.1    Model Traffic Selection

Similarly as in the case of the previous work [1] we decided to focus our measurements to the WWW traffic. The WWW traffic was chosen due to three main reasons. First one was the significant and even growing share of WWW in total Internet traffic (between 16% and 34% in 2008-9, see e.g. [5]). Another reason for this choice was the nature of WWW traffic comparing with e-mail also making significant share of Internet traffic; WWW communication is much more likely to require anonymization. And third but not last reason was the previous experience with application-level measurements in the web browser.

As the plan for the experiment was to measure latency and transmission speed without anonymization and compare it with the results obtained when anonymization is used, the important part of the study was proper selection of two test sets. The first test set was the set of WWW pages and the other was the set of files available via WWW. The sets were built rather in subjective-casual way with trial&error refinement. The selection process was relatively lengthy however because the first idea to reassume the measurements as described in [1] with the same test sets proved to be inapplicable due to great changes in the original test sets. The WWW test set had to be recreated completely because almost all web pages in the original test set were modified significantly or even removed. The WWW file test set allowed modification so that it is comparable with the original test set with certain limitations (consisting especially in storing some files no more available in the web on the own test server). The most important factors for selecting pages and files into the sets were their stability during the testing period (both in availability and in the size of page or file). The geographical (or topological) criterion was not considered because our previous work [1] did not confirm the importance of the "network distance" to the latency increase.

The resulting lists of 12 WWW pages (only 11 with complete set of measurements) and 9 files (only with 8 the complete set) forming both test sets are listed in the Tables 1 and 2 below. As it can be seen in the tables, items 11 and 12 in the WWW page set and 2 a 9 in the WWW file set were prepared on the test server.

**Table 1.** Details of WWW pages test set (the crossed item 4 did not allow to perform the whole measurement set). Also referred to as Testset 1.

| Item No. | URL | Size [kB] | Images | HTML Files | Scripts |
|---|---|---|---|---|---|
| 1 | `http://www.ispforum.cz/` | 106.3 | 63 | 3 | 2 |
| 2 | `http://www.sambaby.com/` | 883.9 | 36 | 12 | 6 |
| 3 | `https://www1.osu.cz/home/sochor/` | 5.1 | 9 | 9 | 0 |
| 4 | `http://moodle.osu.cz` | 261.1 | 20 | 20 | 8 |
| 5 | `http://www.vscr.cz/` | 580.3 | 31 | 15 | 11 |
| 6 | `http://www.instalateri-ostrava.cz/` | 354.9 | 37 | 3 | 3 |
| 7 | `http://www.css-poruba.cz/` | 104.8 | 18 | 1 | 3 |
| 8 | `http://pocasi.divoch.cz/` | 55.2 | 19 | 19 | 4 |
| 9 | `http://www.corecom.cz/` | 4.9 | 6 | 6 | 0 |
| 10 | `http://www.sopza.sk` | 49.5 | 60 | 4 | 5 |
| 11 | `http://89.176.170.nnn/1/` (own TestPage 1) | 81.9 | 0 | 0 | 0 |
| 12 | `http://89.176.170.nnn/2/` (own TestPage 1) | 265.1 | 40 | 20 | 7 |

**Table 2.** Details of WWW files test set. Also referred to as Testset 2.

| Item No. | File description / URL | Size [kB] |
|---|---|---|
| 1 | K-Lite Codec Pack 7.6.0 `http://www.slunecnice.cz/sw/k-lite-codec-pack/full/stahnout/` | 15,590.8 |
| 2 | K-Lite Codec Pack 7.6.0 (Testfile 1) / `http://89.176.170.nnn/file1` | 15,590.8 |
| 3 | nVidia Forceware 81.98 / `http://http.download.nvidia.com/Windows/81.98/81.98_forceware_win9x_international.exe` | 40,401.9 |
| 4 | Adobe Reader 9.3 `http://oldapps.com/adobe_reader.php?old_adobe=21?download` | 26,666.1 |
| 5 | Adobe Reader 9.3 / `http://ardownload.adobe.com/pub/adobe/reader/win/9.x/9.3/enu/AdbeRdr930_en_US.exe` | 27,386.3 |
| 6 | Nero 9.4 / `http://www.instaluj.cz/nero-burn-lite/starsi-verze/nero-lite-9-4-12-708b/` | 31,177.7 |
| 7 | Nero 9.4 – `http://www.filehippo.com/download-ner-lite/download` | 32,747.8 |
| 8 | ATI Catalyst 10.3 `http://www.porse.cz/Ati-catalyst-10.3/download.html` | 55,330.8 |
| 9 | ATI Catalyst 10.3 (Testfile 2) – `http://89.176.170.nnn/file2` | 55,330.8 |

## 2.2 Latency and Transmission Speed Measurement

Parameters like latency (to be more precise *round-trip latency* because all measurements were made locally) are quite difficult to measure exactly because its *soft* (i.e. insufficiently exact) definition. The key problem is represented by the issue how to define the start and especially the end of communication. This is particularly complicated in case of WWW communication where the traffic consists of downloading multiple files consecutively initiated by a WWW client. Therefore our study assumed certain additional limitations. The main limitation was only to the single web client software (Mozilla Firefox, see details in section *Measurement of Webpages*) so as to eliminate extrinsic fluctuations in measured data due to different implementation of client functions.

All the measurements were made solely on the application layer despite the fact that in some cases the L3 measurement tools could be more precise. The limitation of the measurement to L7 is given to the nature of comparison that was pursued after. The main obstacle making the L3 measurement useless was the fact that all anonymization tools studied here perform the anonymization on the application layer so the measurement of L3 parameters could hardly lead to any result.

The description of the measurement techniques is divided into two parts, each one focusing to the corresponding Testset mentioned above.

**Measurement of Webpages.** All measurements must be made on the application layer. i.e. inside the client application. To eliminate variation given to the processing of HTTP requests and responses in the client software the single client was chosen and used for all experiments. The client chosen was Mozilla Firefox version 3.6.17. It was not the newest version available at the moment of measurement commenced (beginning 2011) but this was the version still being supported (patched). The most important factor for choosing the version 3.6.17 was the wide choice of plugins. Some plugins we planned to use for our experiments were unable to run on newer versions of Mozilla Firefox. To eliminate fluctuations in results due to caching of previously downloaded data the use of client cache was eliminated. First experiments showed that the web page cache deactivation was not sufficient so that to obtain repeatable results. Therefore more deep erasing of the cache after every measurement was performed using the Empty Cache Button plugin (see [6]).

The measurement consisted in measuring the time necessary for complete the request and get the first response (TtFB – Time to First Byte) and the time necessary for complete the whole request up to the end of downloading of all components of the webpage (TTC – Time To Completion). The downloaded webpages were displayed. The measurement itself was made by the following Firefox plugins. Fasterfox plugin (see [7]) was used to measure the total time of the webpage download (as well as for check the results produced by other tools), The main component used was Lori (Life of Request Info – see [8]) plugin that was able to measure TtFB as well as TTC precisely also when anonymization programs were used. Lori provides also very convenient passing of the measured data for further processing via its Copy Stat feature. Other plugins like HTTP Analyzer and Firebug were also tested but they tend to give unsatisfactory results when anonymization was used All other applications on the computer used for measurements were inactive (excluding antivirus system proven to have no influence on the results) and it was checked regularly using Wireshark program (see [9]).

Using the TtFB and TTC times and described above both latency and transmission speed could be easily calculated as follows:

$$latency = TtFB \tag{1}$$

$$transmission\ speed = (page\ or\ file\ size)/(TTC - TtFB) \ . \tag{2}$$

The TtFB and TTC times of WWW pages and were measured in 15 sets. Each set of measurements comprising gradual displaying of each of the webpage from the set described in the Table 1 was performed in different weekday and time of the day to avoid systematic circadian and hebdomadal errors. The data were than preprocessed by computing maximum, minimum and average values and subsequent excluding the extremal data (such data were excluded where the single measured time, i.e. TTC or TtFB, was higher than 200% of the average and at the sane time the difference of the value normalized by the standard deviation was higher than 300%) to avoid extremal fluctuation to affect results. This was used for 4 TTC and 2 TtFB measurements without anonymization, 3 TTC measurements with TOR, 2 TtFB measurements with JAP and 3 TTC and 5 TtFB measurements with I2P.

**Measurement of File Download.** For file download measurement the same client as above was used, just instead of Lori not supporting the measurement of web file downloaded another plugin called Download Statusbar (see [10]) was used. The measurement process of file download time was similar in the latter part of measurement, namely transmission speed. The plugin we used did not provide the data necessary for latency measurement but the measured set of WWW pages was sufficient to make conclusions regarding latency.

The measurements of file download were made in the following five modes of operation:

1. WWW client without anonymization,
2. WWW client with TOR in default setting,
3. WWW client with TOR in bridge mode with the bridge in the same LAN,
4. WWW client with TOR in bridge mode with the public bridge,
5. WWW client with JAP.

I2P measurements were unable to perform because I2P does not support file download. The similar situation was in the free version of JAP so the decision was made to allow an exception from free tools here and the last mentioned set of measurements was made using JAP with prepaid file download service. File download was measured in 10 sets (for cases 1–3 above) and in 5 sets (for remaining cases 4 a 5). The lower number of measured sets in case 4 was due to problems with stability of the TOR configured circuit while in case 5 by the lack of money available for prepaid file download credit. Regarding the purpose of those two sets the number of 5 sets was considered to be sufficient. Like in the case of webpages each set consisted of subsequent downloading of all files from the set described in the Table 2. Measured data were preprocessed in a similar way to the previous case.

# 3 Results

The results obtained from measurements and processed as described in the Sect. 2 are presented here. The results are split into two separate parts. The first part focuses to latency that has been measured for the webpage test set (see Table 1). The other part of results summarizes the transmission speed measured in various configurations of anonymization tools as well as without them.

## 3.1 Latency

The main results for latency are shown in the following Fig. 1 and 2 and Table 3. The var. coeff. means variational coefficient, ie. the ratio between standard deviation and the average.

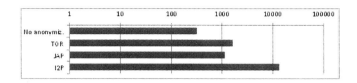

**Fig. 1.** Latency distribution along the webpages in the Testset 1 in logarithmic scale (for every page from above: no anonymization, TOR, JAP, I2P)

It can be easily seen that the latency increase is significant in cases of all anonymization tool studied. The most favorable results were obtained in the case of JAP (increase factor 3.1 for TOR and 2.2 for JAP) while the behavior of I2P was completely unsatisfactorily (increase factor almost 28).

## 3.2 Transmission Speed

The main results for transmission speed are shown in the Fig. 3–6 and in the Tables 3 and 4. The situation is similar in both cases, namely Testset 1 (web pages) and Testset 2 (WWW files). As one can see, the results show significant decrease of transmission speed. The decrease for web pages (Testset 1) was to 37% for TOR, to 23% for I2P and to less than 2% for JAP. The decrease for files (Testset 2) was much more significant, namely to less than 5% for TOR default configuration, 8 to 9% for TOR bridge mode (with slightly better result for public bridge) and 20.5% for JAP. Measurement for I2P for Testset 2 has not been performed because of missing file download support. It should also be noted that JAP measurement for Testset 2 that are much better than in case of Testset 2 is made using *prepaid* version which declares much higher speed (and the measurement confirmed it).

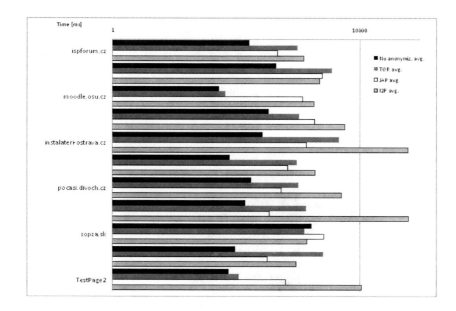

**Fig. 2.** Average latency (average from all results from the Testset 1) measured from webpage download – comparison of operation without anonymization (uppermost) and with anonymization tools (TOR, JAP, i2P from above) in logarithmic scale

**Table 3.** Summarized latency data as shown in the Fig. 1. All values except. the var. coeff. are in msec. The Item No. refers to the Table 1. Each column corresponds to the set of 15 measurements. Norm. case means no anonymization tool is applied.

| Item | Norm. | | | TOR | | | JAP | | | I2P | | |
|------|------|------|------|------|------|------|------|------|------|------|------|------|
| No. | avg. | std. dev. | var. coeff. | avg. | std. dev. | var. coeff. | avg. | std. dev. | var. coeff. | avg. | std. dev. | var. coeff. |
| 1 | 165 | 67 | 40.6% | 1008 | 242 | 24.0% | 471 | 109 | 23.2% | 1252 | 452 | 36.1% |
| 2 | 463 | 133 | 28.7% | 3655 | 1347 | 36.9% | 2485 | 1624 | 65.3% | 2269 | 1318 | 58.1% |
| 4 | 54 | 10 | 18.9% | 68 | 28 | 41.1% | 1205 | 1372 | 113.9% | 1846 | 1156 | 62.7% |
| 5 | 346 | 214 | 62.0% | 1074 | 480 | 44.7% | 1879 | 2182 | 116.2% | 5723 | 5096 | 89.0% |
| 6 | 273 | 34 | 12.4% | 4673 | 1990 | 42.6% | 1415 | 1578 | 111.5% | 61103 | 64 | 0.1% |
| 7 | 81 | 14 | 17.4% | 997 | 470 | 47.2% | 699 | 398 | 56.9% | 1940 | 760 | 39.1% |
| 8 | 182 | 42 | 23.1% | 1052 | 169 | 16.0% | 550 | 386 | 70.2% | 5076 | 2631 | 51.8% |
| 9 | 143 | 41 | 28.5% | 1383 | 417 | 30.2% | 354 | 143 | 40.2% | 62044 | 2233 | 3.6% |
| 10 | 1689 | 358 | 21.2% | 1307 | 301 | 23.0% | 2667 | 3222 | 120.8% | 1435 | 252 | 17.6% |
| 11 | 98 | 29 | 29.6% | 2591 | 1183 | 45.7% | 327 | 73 | 22.4% | 939 | 459 | 48.9% |
| 12 | 78 | 16 | 21.2% | 114 | 7 | 6.1% | 652 | 144 | 22.1% | 10849 | 10812 | 99.7% |

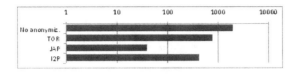

**Fig. 3.** Transmission speed distribution (in kbps) along the Testset 1 (in logarithmic scale)

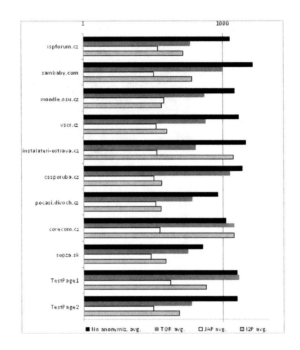

**Fig. 4.** Transmission speed – average from all webpages – comparison between traffic without anonymization (uppermost) and with anonymization tools (TOR, JAP, i2P from above) in logarithmic scale

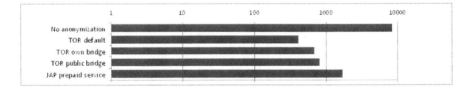

**Fig. 5.** Transmission speed of file download using various TOR configurations (upper bar – default, middle bar – own bridge, lower bar – public bridge)

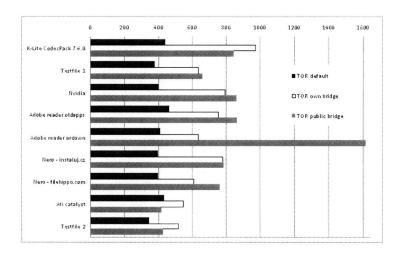

**Fig. 6.** Transmission speed of file download – comparison between normal traffic (uppermost) and anonymization tools (the sequence according to the list in section 2.2.B), in logarithmic scale

**Table 4.** Summarized data for transmission speed of webpage download (Testset 1) as shown in the Fig. 3. All values except. the var. coeff. are in msec. The Item No. refers to the Table 1. Each column corresponds to the set of 10 measurements. Norm. case means no anonymization tool is applied.

| Item No. | Norm. avg. | std. dev. | var. coeff. | TOR avg. | std. dev. | var. coeff. | JAP avg. | std. dev. | var. coeff. | I2P avg. | std. dev. | var. coeff. |
|---|---|---|---|---|---|---|---|---|---|---|---|---|
| 1 | 1484 | 300 | 20.2% | 199 | 29 | 14.4% | 39 | 1 | 2.1% | 137 | 59 | 43.1% |
| 2 | 4536 | 876 | 19.3% | 1001 | 238 | 23.7% | 32 | 2 | 7.2% | 214 | 250 | 116.6% |
| 4 | 1879 | 140 | 7.5% | 404 | 215 | 53.1% | 53 | 12 | 22.9% | 47 | 30 | 64.5% |
| 5 | 2267 | 253 | 11.2% | 429 | 152 | 35.3% | 36 | 4 | 10.4% | 60 | 44 | 74.4% |
| 6 | 3254 | 473 | 14.5% | 260 | 101 | 38.9% | 37 | 3 | 9.0% | 1688 | 172 | 10.2% |
| 7 | 2687 | 766 | 28.5% | 1469 | 109 | 7.4% | 32 | 0 | 0.8% | 47 | 45 | 96.4% |
| 8 | 802 | 112 | 14.0% | 221 | 0 | 0.0% | 35 | 1 | 1.9% | 46 | 37 | 80.9% |
| 9 | 1190 | 127 | 10.6% | 1846 | 733 | 39.7% | 43 | 3 | 6.0% | 1767 | 122 | 6.9% |
| 10 | 375 | 56 | 14.8% | 179 | 50 | 28.2% | 28 | 8 | 292% | 58 | 45 | 76.6% |
| 11 | 2066 | 498 | 24.1% | 2267 | 719 | 31.7% | 72 | 4 | 5.7% | 431 | 107 | 24.9% |
| 12 | 2078 | 161 | 7.8% | 214 | 48 | 22.6% | 31 | 5 | 15.1% | 112 | 80 | 72.0% |

# 4   Conclusions

As expected according to the results in [1], the measurement described in the article confirmed that each of the studied anonymization tools causes significant slowing down comparing normal WWW use. The slowing down includes both the increase of the RTT latency and the decrease of the transmission speed on the application layer. Some methods how do reduce the transmission speed decrease were studied too in case of TOR use. The success of this part of the study was limited but still considerable. Changing the TOR mode to bridge and proper selecting of the bridge it was shown that the transmission speed decrease ratio was approx. 2 times better (approx. 11 instead of almost 20 in default TOR configuration). The main conclusion of the study is that TOR as well as JAP and I2P offer feasible way how to anonymize the WWW traffic but their use should be limited to very special cases because the fact that the costs (expressed not in finances but in significantly longer time required for the same communication comparing normal use) are quite high. TOR is still the best available tool especially from the point of view of transmission speed, versatility and configurability despite the fact that it has been slightly overcame by JAP in latency as described in the article.

# References

1. Liška, T., Sochor, T., Sochorová, H.: Comparison between normal and TOR-anonymized web client traffic. Procedia – Social and Behavioral Sciences 9, 542–546 (2010)
2. Dingledine, R., Mathewson, N., Syverson, P.: Tor: The Second-Generation Onion Router,
   https://svn.torproject.org/svn/projects/
   design-paper/tor-design.pdf (quot. August 10, 2010)
3. I2P Anonymous Network – I2P, http://www.i2p2.de/index.html (quot. February 06, 2011)
4. JAP – Anonymity & privacy. Project AN.ON – Anonymity Online,
   http://anon.inf.tu-dresden.de/index_en.html (quot. February 06, 2011)
5. Ipoque. Internet Study 2008/2009,
   http://www.ipoque.com/sites/default/files/mediafiles/
   documents/internet-study-2008-2009.pdf (quot. February 06, 2011)
6. Empty Cache Button:: Add-ons for Firefox,
   https://addons.mozilla.org/en-US/firefox/addon/empty-cache-button/
7. Fasterfox:: Add-ons for Firefox,
   https://addons.mozilla.org/en-US/firefox/addon/RsccmanFasterfox/
8. Lori (Life-of-request-info):: Add-ons for Firefox,
   https://addons.mozilla.org/en-US/firefox/
   addon/lori-life-of-request-info/ (quot. February 06, 2011)
9. Wireshark – Go deep, http://www.wireshark.org/ (quot. February 06, 2011)
10. Download Statusbar:: Add-ons for Firefox,
    https://addons.mozilla.org/en-US/firefox/addon/download-statusbar/
    (quot. February 06, 2011)

# Real-Time Communication Network Concept Based on Frequency Division Multiplexing

Jacek Stój

Silesian University of Technology, Institute of Informatics
Akademicka 16, 41-100 Gliwice, Poland
jacek.stoj@polsl.pl
http://www.iinf.polsl.pl

**Abstract.** In all real-time systems the temporal determinism is a crucial aspect. It especially refers to communication networks where real-time communication protocols have to be used. In this area three main methods for achieving temporal determinism may be distinguished: token passing, master-slave and producer-distributor-consumer. All are based on Time Division Multiplexing TDM as a method for gaining the communication bus access. The paper presents a concept of a new communication network where the communication is based on Frequency Division Multiplexing FDM possibly making the data exchange more efficient than in most present solutions.

**Keywords:** communication network, frequency division multiplexing, FDM, protocol, real-time

## 1 Introduction

The most important characteristic feature of real-time systems is the temporal determinism. It means that it is possible to describe the operation of all the system elements in a function of time [1]. It especially refers to communication networks used in distributed systems where real-time communication protocols have to be used. In those networks the data flow may be precisely described and the maximum time needed for a given data exchange may be determined. That is the condition for the system response time calculation and is crucial for any real-time application (see: [2–4]).

Three main methods for achieving the temporal determinism used nowadays may be distinguished: token passing, master-slave and producer-distributor-consumer (PDC) [5–7]. All the existing known to the author real-time protocols use some variation of one or more of the above mechanisms. For example, Token-Ring is strictly token passing network (see: [8, 9]), Modbus is the master-slave network (see: [10, 11]) whereas Profibus it the combination of the two above as both the token passing and the master-slave mechanisms are used (Profibus is commonly described as multi-master communication network with token exchange between master subscribers [12, 13]). The Factory Instrumentation Protocol (FIP, known also as WorldFIP) is strictly the PDC network – the communication between producers and consumers of data is coordinated by a distributor

A. Kwiecień, P. Gaj, and P. Stera (Eds.): CN 2012, CCIS 291, pp. 247–260, 2012.

(see: [7, 14]). Similarly in the Can network, the data exchange is performed between producers and consumers of data, however there is no distributor and the
network access is based on priority bus arbitration [15–17].

Different real-time communication mechanisms are nothing else but sets of
rules which have to be obeyed while accessing the communication bus. In most
cases an exchange of the right for transmission between the network subscribers
is needed. Only subscribers that have got the transmission right may transmit
data in the network. It is passed over from one network subscriber to another
depending on the real-time mechanisms used:

- together with the token or
- according to the list of master-slave exchanges or
- according to the list of variables that should be sent between producers and
  consumers.

What is a characteristic feature of real-time networks is the existence of communication bus cycle. The cycle may be defined as the time which passes from one
to another realization of all kinds of cyclic data exchanges defined in a given communication network. This time includes also some period dedicated for acyclic
data exchanges realization. In token networks it is also the time of passing the
token among every network subscriber. The communication bus cycle duration
is one of the most important temporal parameters of real-time systems. The time
is also known as bus sweep time.

Generally speaking, all the industrial real-time protocols commonly used nowadays are Time Division Multiple Access approaches. The communication bus is
shared by all the network subscribers and only one subscriber may transmit at a
time. Moreover, significant part of the communication time (bus sweep time) is
consumed by realization of the passing of the transmission right. That increases
the bus sweep time and degrades the temporal parameters of the real-time system such as the system response time [18, 19]. To minimize this effect, the
communication process must be well designed. It can be done by for example
division of the user data onto more important data, which are cyclically sent
(e.g. in every bus sweep) and less important data sent less often or periodically
(on demand) [20]. Other mechanisms like shortening the automata cycle of programmable logic controller in order to make earlier or temporarily more frequent
data transfer or data reception may also be used [21].

Data exchange is in general more efficient in communication networks where
the passing of the transmission right takes little time. For example, in Profinet
IRT during the communication network startup, a schedule is prepared and
then distributed among the network subscribers. The subscribers may send their
messages in the network only according to the schedule [22–24].

Optimization of transmission right passing mechanisms improves the communication network throughput. Nevertheless, the network is still TDMA network,
so every subscriber has to wait for the transmission right at most one bus sweep
time.

An interesting solution would be to completely eliminate the need of passing
over the transmission right in order to allow data transmission. That would be

possible only when every system node have got its own communication channel on which it may transmit data at will. That would result in seemingly instant data delivery. The author claims it is possible using the Frequency Division Multiplexing FDM data exchange mechanism.

## 2    FDM Data Exchange Mechanisms Concept

Using Frequency Division Multiplexing it is possible to transfer multiple information signals simultaneously with separation into nonoverlapping frequency bands [25, 26]. However, the packing of multiple signals into one electrical signal is usually done by one transmitter. Whereas the concept which is presented here, is based on the idea that a different band is assigned to every single network subscriber. A given subscriber may transmit data on the band that it is assigned to in seemingly continuous manner with no drawback caused by passing of the transmission right.

In Figure 1 an example of a four node communication system with the FDM communication bus is presented. All four subscribers are connected with one communication bus with four frequency bands – one for every subscriber. Each frequency band is a separate logical communication channel. The subscribers transmit data in data blocks $DB$ using their frequency band $FB$ (pointed out by solid arrows in the figure). The transmission is done continuously without brakes, apart from small delays between consecutive data blocks. A given subscriber may also receive data from other subscribers by listening on other frequency bands (dashed arrows in the figure). For example, the subscriber A transmits data in data blocks $DB_A$ on frequency band $FB_A$. When the subscriber A needs some data that are being sent by the subscriber B, it listens on the $FB_B$ frequency band to receive the $DB_B$ data blocks.

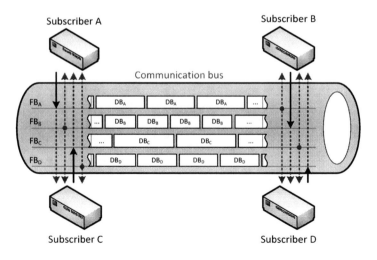

**Fig. 1.** FDM system with four frequency bands

One frequency band may be used for data transmission by only one network subscriber even if the subscriber transmits small amounts of data. The frequency band that a given subscriber should use for transmitting data may be understood as the network address of the subscriber.

To keep the hardware of the network subscribers' communication modules cost effective and practical Orthogonal Frequency Division Multiplexing OFDM could be used [27, 28]. The number of frequency bands determines the maximum number of the network subscribers that may be connected to the network. For that reason it should be high enough for keeping the communication system scalable. On the other hand the more frequency bands there are, the more expensive the communication modules hardware would be. In the author's opinion the number of 64 frequency bands should be considered as the most optimal, as 16 is probably too little for real life applications and 256 would be probably too expensive. Multi-band filters could also be used as described in [6] with (in this example) 22 different type of band-pass filters available for communication.

## 3   Real-Time Features

In the presented FDM communication network concept the implementation of real-time features should be a feasible task. Moreover, real-time parameters, such as the system response time, would be easy to calculate. That may save much of the design time.

In the FDM network every subscriber has its own logical communication channel. As a result no transmission right is needed to be passed between the network subscribers and they transmit data continuously without breaks. When a given subscriber wants to receive some data from the subscriber $x$, it listens to the frequency band $FB_x$ and gets all the data transferred in the $DB_{xi}$ data blocks. The transmission time $T_{DBx}$ of all the data blocks of the subscriber $x$ defines the frequency band sweep time $T_{FBS}$ (similarly to the communication bus sweep time). The maximum delay to receive all the data from subscriber $x$ by any subscriber on the network would be as follows:

$$T_{DBx} = T_{FBSx} + \max_{i=1}^{dbn}(T_{DBxi}) \ , \tag{1}$$

where:

$$T_{FBSx} = \sum_{i=1}^{dbn} T_{DBxi} \ , \tag{2}$$

and where:

$T_{DBx}$ – time needed for the reception of all data block transmitted by the subscriber $x$,
$T_{FBSx}$ – frequency band sweep time of the subscriber $x$,
$T_{DBxi}$ – time needed for the transmission of $DB_{xi}$ data block of the subscriber $x$ where $i$ is an index of the block,
$dbn$ – the number of data blocks transmitted by the given subscriber.

In the Equation (1), apart from the frequency band sweep time $T_{\text{FBS}x}$, the time of transmission of the longest data block of subscriber $x$ has to be included, as the receiving subscriber may start to listen to the frequency band just after the transmission of one of the data block was started. Then the listening subscriber has to wait for the block being transmitted to end in order to start data acquisition. In worst case analysis the longest possible block has to be considered. The $T_{\text{BD}xi}$ includes a small delay between consecutive data blocks which is the time of transmission of one character.

The greatest advantage of the FDM network is visible when the subscribers are capable of listening to all the frequency bands of the communication bus simultaneously. Then the time needed for the reception of all the data blocks being sent in the network by all network subscribers would be equal to the maximum $T_{\text{DB}x}$ time:

$$T_{\text{DR}} = \max_{x=1}^{sn}(T_{\text{DB}x}) = \max_{x=1}^{sn}\left[T_{\text{FBS}x} + \max_{i=1}^{dbn_x}(T_{\text{DB}xi})\right] , \tag{3}$$

where:

$T_{\text{DR}}$ – time needed for the reception of all data from all network subscribers,
$T_{\text{DB}x}$ – time needed for the reception of all data blocks from the subscriber $x$,
$T_{\text{FBS}x}$ – subscriber's $x$ frequency band sweep time,
$T_{\text{DB}xi}$ – the reception time of the data block $i$ from the subscriber $x$,
$sn$ – the number of subscribers in the network,
$dbn_x$ – the number of data blocks transmitted by the $x$ subscriber.

In other words, the time needed for the reception of all data blocks from the network depends only on one frequency band (one subscriber) where the sum of the data blocks is the longest. For more complicated cases of real-time analysis some formal methods has to be used together with simulation studies [29].

When the subscriber cannot simultaneously listen on all frequency bands, the data reception time would be as follows:

$$T_{\text{DR}ns} = \sum_{x=1}^{sn} T_{\text{DB}x} = \sum_{x=1}^{sn}\left[T_{\text{FBS}x} + \max_{i=1}^{dbn_x}(T_{\text{DB}xi})\right] . \tag{4}$$

In the Equation (4) the time $T_{\text{DR}ns}$ needed to receive all the data blocks from all the network subscribers without simultaneous reception on all frequency band is much greater than the time $T_{\text{DR}}$ shown in (3). The $T_{\text{DR}ns}$ time corresponds to networks with time division multiplexing TDM communication bus access mechanism. The subtraction result in the following:

$$T_{\text{FDM}} = T_{\text{DR}ns} - T_{\text{DR}} , \tag{5}$$

shows roughly the FDM network advantage over TDM communication networks. More precise calculation is presented in the Sect. 6.

# 4    Control and Information Frequency Bands

In the subscribers' data blocks it should be possible to transmit cyclic data as well as aperiodic data. So there have to be some means of requesting latter. It can be done in two ways. Firstly, by using the data block and the subscriber frequency band. For example, the subscriber A can transmit a request addressed to subscriber B to send some aperiodic data. The subscriber B receives the request by reading the subscriber A data block $DB_A$ on the frequency band $FB_A$ and then sends on its frequency band $FB_B$ a data block $DB_B$ with the requested data.

The second solution is to use a dedicated Control Frequency Band CFB. This band is a communication channel used by all the subscribers to transfer mainly control data. As there are more than one transmitter on the control frequency channel, a transmission right has to be passed over among the network subscribers in order to maintain real-time features. It is proposed to use the simplest possible solution such as token passing with implicit token (like in the Genius network [30, 31]). The CFB could be used for other purposes too: changing the subscriber configuration on the fly, realization of additional functions like communication bus switchover in systems with redundancy, etc.

Another communication channel should be useful for obtaining non-time-critical information data. It is defined as Information Frequency Band IFB. It would also be used by all network subscribers with communication bus access mechanism just like in the Control Frequency Band CFB. Using that channel it would be possible to transmit non-real-time data and for example obtain some detailed information about data being sent in the subscriber's data block like: identifiers of variables sent in different data fields of the subscriber's data block, type of the variables (e.g. integer, real), their size (applies to arrays etc.), range, precision etc.

With the Control Frequency Band CFB and the Information Frequency Band IFB, a sample communication bus will be as shown in Fig. 2. What is important is that the communication on the subscribers' frequency bands $FB_1$ to $FB_N$ does not depend on the CFB and IFB bands, as they only give some additional control and informational functionality.

# 5    Data Blocks

In the data block it is possible to transmit both cyclic and aperiodic data. The concept of the data block is shown in Fig. 3. The data block starts with a start character SC which may be used for receiver synchronization. Then the subscriber identifier SID is sent. It is followed by a field with the number of Cyclic Data Block. Then starts the user data, which are transmitted in the Cyclic Data Blocks CDB. Then follows the optional Aperiodic Data Blocks ADB in the same fashion as the CDBs. The data block ends with an end character EC. One subscriber may send many data block with different sets of cyclic CDB and aperiodic ADB data subblocks.

Communication bus

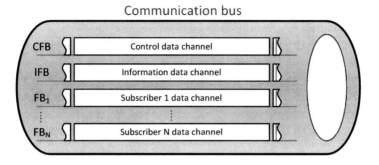

**Fig. 2.** FDM communication bus with the Control Frequency Band CFB and the Information Frequency Band IFB

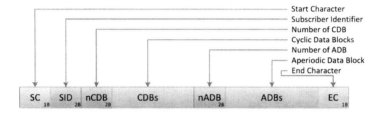

**Fig. 3.** Data block format

Both CDBs and ADBs fields have the same format as shown in Fig. 4. Each block includes the block identifier, the block length, a field with the user data and a CRC checksum. The requests for aperiodic data are also send in aperiodic block with proper identifier (the same as the requested data block), and with no data bytes.

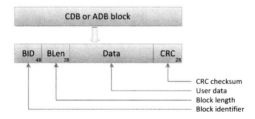

**Fig. 4.** Cyclic and aperiodic data block fields

The CRC checksum is calculated for every CDB and ADB separately. That makes it possible to apply the received subblock of data before the whole data block is received. Immediately after checking the CRC, the user data may be copied to the computer unit memory for later processing. With all the bytes sent in the network a parity bit is transmitted.

As every CDB and ADB subblocks have got the block identifier, it is possible to transmit more kinds of data blocks in one or several different massages (data blocks). It is important to mention however, that in the latter case, the frequency band sweep time $T_{\mathrm{FBS}}$ is the time of transfer of the whole list of data blocks which is illustrated in Fig. 5.

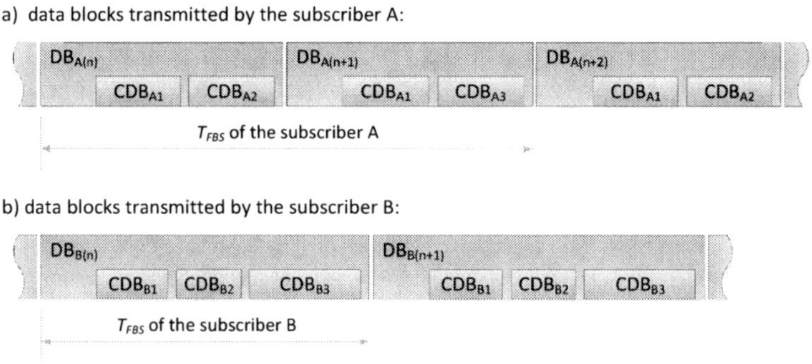

a) data blocks transmitted by the subscriber A:

| DB$_{A(n)}$ | | DB$_{A(n+1)}$ | | DB$_{A(n+2)}$ | |
|---|---|---|---|---|---|
| CDB$_{A1}$ | CDB$_{A2}$ | CDB$_{A1}$ | CDB$_{A3}$ | CDB$_{A1}$ | CDB$_{A2}$ |

$T_{FBS}$ of the subscriber A

b) data blocks transmitted by the subscriber B:

| DB$_{B(n)}$ | | | DB$_{B(n+1)}$ | | |
|---|---|---|---|---|---|
| CDB$_{B1}$ | CDB$_{B2}$ | CDB$_{B3}$ | CDB$_{B1}$ | CDB$_{B2}$ | CDB$_{B3}$ |

$T_{FBS}$ of the subscriber B

**Fig. 5.** Frequency band sweep time $T_{\mathrm{FBS}}$

In the example shown in Fig. 5, the subscriber A sends three kinds of cyclic data blocks: CDB$_A$1–CDB$_A$3. However, the block CDB$_{A1}$ is sent in every transmission, but the other blocks are sent in every other transmission. As a result the frequency band sweep time $T_{\mathrm{FBS}}$ includes two data blocks transmission (here marked as the $DB_{A(n)}$ and the $DB_{A(n+1)}$ transmission). On the other hand, the subscriber B sends all its cyclic data blocks in every transmission. The $T_{\mathrm{FBS}}$ time is the transmission of one $DB_B$ block.

The user data sent in the CDB and the ADB subblocks are by default broadcasted. However, the block identifier BID may be used to point out the destination subscriber that should process the data being sent. It may also include some information about the priority of the user data. It gives the subscriber recommendation about the processing order of different data blocks.

## 6    Protocol Efficiency

Based on the block format presented in the previous section, the protocol efficiency and the protocol overhead may be calculated. The number of control bytes in a given data block depends on the number of cyclic CDB and aperiodic ADB data subblocks being sent in that data block. The number of control bytes in the data block excluding control bytes of CDB and ADB data subblocks, according to the Fig. 3, is equal:

$$n_{\mathrm{DB}} = l_{\mathrm{SC}} + l_{\mathrm{SID}} + l_{n\mathrm{CDB}} + l_{n\mathrm{ADB}} + l_{\mathrm{EC}} = 8\,\mathrm{B}\ , \tag{6}$$

where:

$n_{DB}$ – Number of control bytes in the data block excluding control bytes of the subblocks,

$l_{\langle fn \rangle}$ – Length of the $\langle fn \rangle$ control field Fig. 3.

The number of control bytes in the cyclic CDB and aperiodic ADB data subblocks according to the Fig. 4 is:

$$n_{DBF} = l_{BID} + l_{BLen} + l_{CRC} = 8\,B \ . \tag{7}$$

The total number of control bytes in a data block is equal:

$$n_{CTRL} = n_{DB} + (v_{CDB} + v_{ADB}) \cdot n_{DBF} = 8 + c_{DB} \cdot 8 \ , \tag{8}$$

where:

$v_{CDB}$ – Number of cyclic data subblocks CDB sent in the given data block,
$v_{ADB}$ – Number of aperiodic data subblocks ADB sent in the given data block,
$c_{DB}$ – The total count of the data subblocks, sum of $v_{CDB}$ and $v_{ADB}$.

The total number of bytes in a data block is equal:

$$n_B = n_{CTRL} + n_{DATA} \ , \tag{9}$$

where the number of user data bytes $n_{DATA}$ may be defined as follows:

$$n_{DATA} = \sum_{i=1}^{v_{CDB}} v_{BLen(i)} + \sum_{i=1}^{v_{ADB}} v_{BLen(i)} \ , \tag{10}$$

where $v_{BLen(i)}$ is the value sent in the BLen fields of the data subblock, i.e. the length of the $i$ data subblock.

Taking the above into consideration, the protocol overhead defined a quotient of the number of control bytes to the number of user bytes sent in a data block is as follows:

$$p = \frac{n_{CTRL}}{n_{DATA}} \ . \tag{11}$$

The protocol efficiency defined as a quotient of the number of user data bytes to the total number of bytes is as follows [32]:

$$\eta = \frac{n_{DATA}}{n_B} = \frac{n_{DATA}}{n_{CTRL} + n_{DATA}} \ . \tag{12}$$

The efficiency for one frequency band calculated according to the Equation (12) is presented in the Fig. 6. Additional assumption was done for the needs of the presentation, that the maximum number of bytes being sent in one data subblock may not be more than 256. In the figure there is an additional Y axis which shows the number of subblocks used to transfer the useful (user) data.

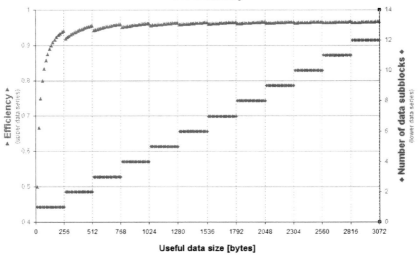

**Fig. 6.** FDM network efficiency

For a network subscriber that sends for example 3 cyclic data subblocks CDB with 64 bytes of user data each subblock, the protocol overhead and efficiency would be as follows:

$$p_{eg} = \frac{n_{CTRL}}{n_{DATA}} = \frac{8 + c_{DB} \cdot 8}{c_{DB} \cdot 64} = \frac{8 + 3 \cdot 8}{3 \cdot 64} = 16.7\% \ , \tag{13}$$

$$\eta_{eg} = \frac{n_{DATA}}{n_B} = \frac{c_{DB} \cdot 64}{8 + c_{DB} \cdot 8 + c_{DB} \cdot 64} = 85.7\% \ . \tag{14}$$

Time needed for the transmission of one data block is shown in the following Equation (15). For the needs of that example, the use of 38.4 kb/s baud rate network was assumed:

$$T_{DB} = \frac{n_B}{38.4} = \frac{224}{38.4} = 5.83 \, \text{ms} \ . \tag{15}$$

The time needed for reception of the mentioned data block according to the Equation (1) is:

$$T_{DBx} = 2 \cdot T_{DB} = 11.66 \, \text{ms} \ , \tag{16}$$

as there is only one block being transmitted.

What is important for the FDM network, is that according to the Equation (3), if there were for example 10 devices each sending one data block as described above, the total time needed for the reception of all the data from all the subscribers would be:

$$T_{DR} = \max_{x=1}^{sn} (T_{DBx}) = 2 \cdot T_{DB} = 11.66 \, \text{ms} \ . \tag{17}$$

This is true of course with the assumption that the receiver is capable of listening to all the frequency bands simultaneously. If that is not so, then the Equation (4) should be considered:

$$T_{\mathrm{DR}ns} = \sum_{x=1}^{10} (2 \cdot T_{\mathrm{DB}x}) = 116.60\,\mathrm{ms} \ . \tag{18}$$

The subscriber which cannot listen do more than one frequency band at a time uses the FDM network in Time Division Multiple Access manner. Then, the following subtraction results (according to the Equation (5)):

$$T_{\mathrm{FDM}} = T_{\mathrm{DR}ns} - T_{\mathrm{DR}} = 104.94\,\mathrm{ms} \ , \tag{19}$$

shows roughly the advantage of FDM network over TDM network – in the TDM network the reception time would be by about 105 ms greater than in the FDM network.

# 7    Main Points of Consideration

One of the most important points of consideration associated with the proposed FDM based communication network is electromagnetic compatibility EMC. Real-time communication is usually implemented in industrial environments where (among others) strong electromagnetic interferences are a common aspect. It should be analyzed whether implementation of such a communication network is possible in those environments. If it turns out that the network is vulnerable to those interferences, bit coding, which is an open manner yet, may be of help. However, possibly the whole concept should be dropped.

Secondly, the production costs of the FDM communication modules hardware have to be estimated. This is also a decisive factor on the possibility of real applications. There are many devices present on the market with the FDM implemented (like DSL routers), but the use of frequency division is quite different there, as they are used for point to point communication links. Besides, they are usually for personal, not industrial applications and the EMC requirements are less significant. The question is what will the final hardware cost be in case of industrial devices. Possibly FPGA would be of some help in this field [33].

Finally, present real-time systems have to be reliable. To achieve high reliability redundancy is commonly used. The great advantage of the proposed conceptual communication network is that its operation does not need initialization. It should be possible to just plug in the communication bus and have connection with all the network subscribers, every in *logically* different communication channel. That may be useful in systems with communication bus redundancy. Author's experience shows that the greatest temporal cost in case of failure in a system with communication bus redundancy, is the delay between losing communication on the primary communication bus and renewing it on the secondary bus. During some experiments concerned with Genius communication bus redundancy, the bus switchover delay was 10 to 20 times greater than the analytically calculated one [34]. Surprisingly, the communication process on the redundant communication bus during the switchover was not determined in any way – sometimes the connection was established sooner, sometimes later, with no apparent reason. In case of the proposed communication bus, this could never happen and the renewing of communication process will be seemingly *instantaneous*. However, to confirm that conjecture, further research is necessary.

## 8    Final Notes

In the paper the concept of real-time communication mechanism based on frequency division multiplexing is presented. It is meant as an alternative for presently known and used mechanisms like: token passing, master-slave and producer-distributor-consumer which are based on Time Division Multiple Access mechanisms.

The idea is based on maintaining communication between network subscribers on many *logical* communication channels using separate frequency bands for each of them. Every subscriber in the network has its own channel which eliminates the temporal cost of passing over the transmission right between subscribers. Two additional channels are also defined for control and information data exchange between all network nodes. However, they only introduce extra functionality and the main communication process does not depend on those channels.

The proposed FDM communication network should be far more effective in the means on transferring user data between the nodes of the communication system, as the user data from different subscribers are always available on the network on separate frequency bands with continuous transmission. The reception of the data block being sent on the network may be performed with little delay dependent only on the length of the data blocks being sent.

The presented communication network is only a concept planned to be implemented. The format of data block may still be changed during further research and development.

## References

1. IEEE Standard Glossary of Software Engineering Termonilogy. IEEE Std. 610.12-1990. IEEE Computer Society (September 1990)
2. Halang, W.A., Sacha, K.M.: Real-Time Systems. Implementation of Industrial Computerised Process Automation. World Scientific Publishing (1993)
3. Kopetz, H.: Real-Time Systems Design Principles for Distributed Embedded Applications, 2nd edn. Springer, Wien (2011)
4. Liu, J.W.: Real-Time Systems. Prentice Hall, New Jersey (2000)
5. Conti, M., Donatiello, L., Furini, M.: Design and Analysis of RT-Ring: a protocol for supporting real-time communications. In: IEEE International Workshop on Factory Communication Systems, pp. 91–88 (September 2000)
6. Miorandi, D., Vitturi, S.: Analysis of master-slave protocols for real-time-industrial communications over IEEE802.11 WLANs. In: 2nd IEEE International Conference on Industrial Informatics 2004, INDIN 2004, pp. 143–148 (June 2004)

7. Raja, P., Ruiz, L., Decotignie, J.D.: On the necessary real-time conditions for the producer-distributor-consumer model. In: IEEE International Workshop on Factory Communication Systems 1995, WFCS 1995, pp. 125–133 (October 1995)
8. Cao, J., Steele, R., Yao, L.J., Jia, W.: Design and simulation of a reliable token-ring protocol for realtime communications. In: Proceedings of IEEE International Computer Performance and Dependability Symposium, pp. 130–138 (September 1996)
9. Zhou, Z., Tang, B., Xu, C.: Design of Distributed Industrial Monitoring System Based on Virtual Token Ring. In: 2nd IEEE Conference on Industrial Electronics and Applications, ICIEA 2007, pp. 598–603 (May 2007)
10. Xuehua, S., Min, L., Hesheng, W., Hong, W., Fei, L.: The solution of hybrid electric vehicle information system by modbus protocol. In: International Conference on Electric Information and Control Engineering, ICEICE, pp. 891–894 (May 2011)
11. Sideng, H., Zhengming, Z., Yingchao, Z., Shuping, W.: A novel Modbus RTU-based communication system for adjustable speed drives. In: Vehicle Power and Propulsion Conference 2008, VPPC 2008, pp. 1–5. IEEE (September 2008)
12. Tovar, E., Vasques, F.: Guaranteeing real-time message deadlines in PROFIBUS networks. In: 10th Euromicro Workshop on Real-Time Systems 1998, pp. 79–86 (1998)
13. Bao, W., Zhang, H., Li, H., Huang, W., Peng, D.: Analysis and Research on the Real-Time Performance of Profibus Fieldbus. In: WRI World Congress on Software Engineering 2009, WCSE 2009, pp. 136–140 (May 2009)
14. Pedro, P., Burns, A.: Worst case response time analysis of hard real-time sporadic traffic in FIP networks. In: Ninth Euromicro Workshop on Real-Time Systems 1997, pp. 3–10 (June 1997)
15. Xian, G., Lin-shen, L.: Analysis and Research of Real Time Ability of Message Transmission in CAN Bus. In: International Conference on Control, Automation and Systems Engineering, CASE, pp. 1–3 (July 2011)
16. Zeng, R., Xiao, Y., Dai, W., Zhou, B., Zhang, L.: Design of technical condition monitoring system of vehicle based on CAN Bus. In: 3rd IEEE International Conference on Computer Science and Information Technology, ICCSIT, pp. 457–460 (July 2010)
17. Guohuan, L., Hao, Z., Wei, Z.: Research on designing method of CAN bus and Modbus protocol conversion interface. In: International Conference on Future BioMedical Information Engineering, FBIE 2009, pp. 180–182 (December 2009)
18. Griese, B., Brinkmann, A., Porrmann, M.: SelfS – A real-time protocol for virtual ring topologies. In: IEEE International Symposium on Parallel and Distributed Processing, IPDPS 2008, pp. 1–8 (April 2008)
19. Mahajan, A., Teneketzis, D.: Optimal Design of Sequential Real-Time Communication Systems. IEEE Transactions on Information Theory, 5317–5338 (November 2009)
20. Siegfried, C., Constantin, R., Stancescu, S.: Evaluation of protocol for industrial informatics systems. In: 8th International Conference on Communications, COMM, pp. 293–296 (June 2010)
21. Kwiecień, A., Sidzina, M.: The Method of Reducing the Cycle of Programmable Logic Controller (PLC) Vulnerable "to Avalanche of Events". In: Kwiecień, A., Gaj, P., Stera, P. (eds.) CN 2011. CCIS, vol. 160, pp. 379–385. Springer, Heidelberg (2011)
22. Hanzálek, Z., Burget, P., Šůcha, P.: Profinet IO IRT Message Scheduling With Temporal Constraints. IEEE Transactions on Industrial Informatics, 369–380 (August 2010)

23. Gunzinger, D., Kuenzle, C., Schwarz, A., Doran, H.D., Weber, K.: Optimising PROFINET IRT for fast cycle times: A proof of concept. In: 8th IEEE Int. Workshop on Factory Communication Systems, WFCS, pp. 35–42 (May 2010)

24. Prytz, G.: A performance analysis of EtherCAT and PROFINET IRT. In: IEEE International Conference on Emerging Technologies and Factory Automation, ETFA 2008, pp. 408–415 (September 2008)

25. Thompson, R.A., Tipper, D., Krishnamurthy, P., Kabara, J.: The physical Layer of Communications Systems. Artech House, Norwood, pp. 727–754 (2006)

26. Wylie-Green, M.P.: A power efficient continuous phase modulation – single carrier FDMA transmission scheme. In: Wireless Telecommunications Symposium, WTS 2008, pp. 267–272 (April 2008)

27. Smith, D.R.: Digital Transmission Systems, 3rd edn., pp. 412–414. Kluwer Academic Publishers

28. Gupta, A., Ghosh, D., Mohapatra, P.: Scheduling Prioritized Services in Multihop OFDMA Networks. IEEE/ACM Transactions on Networking 18(6), 1780–1792 (2010)

29. Sinha, P., Suri, N.: On the use of formal techniques for analyzing dependable real-time protocols. In: The 20th IEEE Real-Time Systems Symposium 1999, pp. 126–135 (December 1999)

30. Stój, J., Kwiecień, B.: Real-time system node model – some research and analysis. In: Contemporary Aspects of Computer Networks, vol. 3, pp. 231–240. WKŁ, Warszawa (2008)

31. Genius® I/O System and Communications. GE Fanuc Automation, document no: GEK-90486f1 (November 1994)

32. Gaj, P.: Pessimistic Useful Efficiency of EPL Network Cycle. In: Kwiecień, A., Gaj, P., Stera, P. (eds.) CN 2010. CCIS, vol. 79, pp. 297–305. Springer, Heidelberg (2010)

33. Monmasson, E., Idkhajine, L., Cirstea, M.N., Bahri, I., Tisan, A., Naouar, M.W.: FPGAs in Industrial Control Applications. IEEE Transactions on Industrial Informatics 7(2), 224–243 (2011)

34. Stój, J.: Temporal aspects of redundancy in distributed real-time computer systems (original title: Wpływ redundancji na zależności czasowe w rozproszonych informatycznych systemach czasu rzeczywistego). PhD Thesis, Silesian University of Technology, Gliwice, Poland (2009)

# Introduction to OPC UA Performance

Marcin Fojcik[1] and Kamil Folkert[2]

[1] Sogn og Fjordane University College
marcin.fojcik@hisf.no
[2] Silesian University of Technology, Institute of Informatics
kamil.folkert@polsl.pl

**Abstract.** Applications operating in the control layer of the industrial computer systems are designed not only to perform real-time data exchange between each other, but also to transfer some information to higher levels (SCADA, MES, ERP). OPC Unified Architecture (OPC UA) is an example of a standard that handles this type of communication. It is not a real-time protocol, but is designed rather to gather information about the transferred data with the occurrence time stamp and distribute that information on demand. Nevertheless, it is crucial to be able to estimate the performance of data transfer in this client-server model based protocol. For now, there is no documented research on the impact of number of clients and amount of transferred data on server's performance. The goal of this article is to find and describe the parameters of the OPC Unified Architecture protocol that are the most important for the system's performance.

**Keywords:** OPC UA, performance, network traffic, vertical communication.

## 1 Introduction

Contemporary industrial installations strongly depend on computer automation systems. Those systems are supposed to be easy to design and implement, reliable, versatile and user friendly. Moreover, those systems should be built from pre-designed modules, which can be freely combined for better scalability, while each of these parts should be able to cooperate with others [1].

One of the systems that fulfill the requirements specified above is OPC Unified Architecture (OPC UA) [2,3]. This standard defines both object-oriented data model and communication protocol. OPC UA originates from the previous versions of the OPC standard (Data Access, Alarms&Events, Historical Data Access), which were designed as communication protocol for Windows-based environments. Due to lack of versatility, security procedures and problems with handling various data types exchanges the new standard (free of those limitations) was created. OPC UA is a client-server communication model, as well as object data model and mechanisms for accessing those data. Services providing desired functionality can be accessed using optionally encrypted SOAP or binary

A. Kwiecień, P. Gaj, and P. Stera (Eds.): CN 2012, CCIS 291, pp. 261–270, 2012.
© Springer-Verlag Berlin Heidelberg 2012

encoded communication channel [4]. For that reason data can be exchanged independently from the underlying data source type, host operating system and network configuration [5,6].

OPC UA is based on Service Oriented Architecture. Clients and servers implement set of services that are used both for handling the communication and exchanging the actual data [2].

Services that handle the communication allow to locate and discover the OPC UA servers, browsing server's address space and maintaining the communication channels between clients and servers. Usually, client locates the server, connects to it and creates a session on the server (encrypted or not), and finally reads or modifies the values of servers variables or objects [7].

The data between client and server may be exchanged in two manners:

- single: the data is exchanged (read or written) on demand,
- repetitive (subscription-based): client is notified if the observed data has changed.

For single data exchange client calls proper service on the server and receives the response. Server should handle this kind of operation immediately, simply by accessing the data information model. This method is uncomplicated because the transmission handling operations are not being optimized.

Subscription-based data exchange is more complex operation. To access the data client creates a subscription for specific attributes of the OPC tags on the server. For the subscribed values *MonitoredItem* objects are constructed on the server, one for each tag. They are used for parametrization of the subscription mechanisms, e.g. sampling interval, update thresholds, queue size, notification delay and more. Once *MonitoredItems* are added to the subscription client starts the *Publish* service, which sends requests from client to server periodically. Server collects the requests in a separate queue for each session. When server acquires new value of any tag from the data source, it must update the values of all *MonitoredItems* associated with this tag. When *Publish* request is received and there are some *MonitoredItems* that were updated after the previous request, the notification to the subscribed clients is sent. The size of the requests queue is associated with *KeepAlive* parameter of the subscription. If it is bigger than 1 then server is not obliged to send a response for each request. According to that parameter server may wait for next request or sent response containing the updated values. The subscription mechanism is presented in Fig. 1.

If the values acquired from the data source are not changing for some time, there is no need of sending notification to the client. However, this situation could be interpreted by the client as server fault. To avoid this, server sends some *KeepAlive* information if the data is not changed.

If client has established a subscription on the server, the *Publish* request is an information for the server that client is connected and alive. If no requests are received server may close the session when a timeout expires.

The operations performed by the OPC UA client-server system may be divided into several stages:

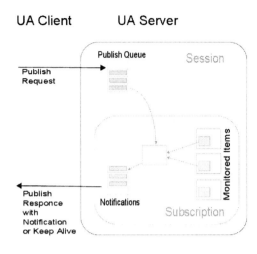

**Fig. 1.** Subscription handling

1. server start and creation of data model on the server,
2. preparation of data source handling,
3. client's connection start:
   (a) client connects,
   (b) client acquires the data model,
   (c) client creates the session,
4. creation of the subscription,
5. activation of the subscription,
6. *communication operations* (autonomous, according to the scenario, based on the implementation),
7. end of the subscription,
8. end of the session.

Point 6 (communication operations) consists of 3 main parts illustrated in Fig. 2. The communication model is also presented in the Fig. 3. OPC UA standard does not specify how to perform operations 1 (data source handling) and 2 (data queue for data source handling), it is server-specific behavior.

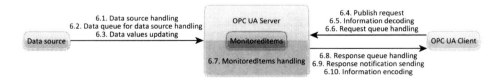

**Fig. 2.** Communication operations

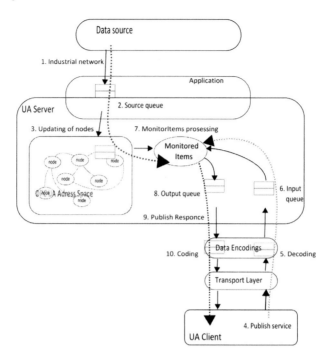

**Fig. 3.** OPC UA subscription communication model

## 2    Performance of OPC UA

OPC UA defines both client-server communication model based on TCP/IP and the server's data model [2,8]. It is commonly used in industrial applications. The specification suggests that this standard should be implemented and used in real industrial installations. However, the standard does not define how the server should acquire data from data source and how client may use the data received from the server. Considering subscription mechanism, the server is expected to deliver the subscribed data in a configured time period after data change occurs. Nevertheless, in practice OPC UA is not a real-time protocol and there are possible cases in which data will not be delivered on time. Fortunately OPC UA can handle this kind of situations by providing timestamp and quality markers, which may be used to identify the data that is not up-to-date.

The OPC UA specification does not determine the parameters of the communication, e.g. what is the maximum delay in subscribed data transfer from server to the client (depending on the amount and size of the data), how the number of the clients impacts on the communication (concerning the various amount of data and utilization of resources of server machine). Situation may be even more complicated when we assume that servers may exchange data between each other, which is possible in OPC UA.

For industrial applications it is crucial to define the parameters (e.g. communication delay, how many values may be lost in the communication, what is the impact of number of clients) not only for the OPC part of the system, but for the entire communication system.

The fact that specification does not refer to the communication with the rest of the system is correct from the versatility point of view. On the contrary, from the angle of final user it is not so convenient, because in real case OPC UA is never used as stand-alone system, it was designed as a vertical communication standard in bigger installations [9]. Unfortunately, there is lack of literature considering the performance issue of OPC UA. Only several aspects, like security (e.g. impact of data encryption [10]), are tested and described with details. Even in the articles [11,12] authors conduct only some simulations (without empirical test on real data) and they assume that "analysis of these (. . . ) is not enough to reach to some conclusions".

Undoubtedly there is a strong need for testing and/or simulating the performance of the OPC UA communication [13]. First step should be an attempt of determining the critical parts of the system. After analysis of the communication aspects described in Sect. 1 it is possible to define several aspects:

- configuration of communication with the data source (e.g. configuration of the field bus),
- updating of the data values in the data model on the server (in theory, server's CPU load should depend on the number of clients and amount of data to process),
- subscription interval (the data must be processes through the communication channel, input queue, some service handlers and the output queue),
- amount of the transferred data,
- number of the clients (each client has separate queue and communication channel).

## 3   Measurement Environment

The test environment was consisting of three main parts: the data source (the data was generated on a separate machine), the OPC UA server, OPC UA clients. All of the computers were connected together in the same segment of LAN network. The communication between the source and the server was provided by low-level socket-based binary protocol designed by the authors for this test case to avoid additional heavy traffic between the server and the data source.

The machines used in the test were PC-class computers with Pentium 8400 quad core 3.2 GHz and 4 GB RAM. The network was 100 Mbit Ethernet.

During the tests client were located on various number of computers (multiple clients can operate on one machine). However, clients location had no impact on server's updates times and network load.

OPC UA is not a real-time protocol, but it is designed for vertical communication handling. For that reason in the test authors used sampling intervals from

100 ms to 1000 ms. The amount of data was from 1 to 10,000 integer numbers. The maximum number of parallel clients was limited to 100.

The test applications were written in Java 7 using Prosys Client-Server Java Software Development Kit.

## 4    Tests of Performance

### 4.1    Impact of Communication with Underlying Data Source

First test was conducted to verify if updating server's tags may lead to increasing CPU load. Two versions of server applications were prepared. In first case the server was updating tag's values directly after acquiring new data from data source, both operations were performed in one thread. In second case there were two separate threads: first for gathering data from source in dedicated queue and second only for updating tags and associated *MonitoredItems*. As presented in Fig. 4 it is very important to use this proper model. In first case, where tags were updated without buffering, the CPU load for 20 clients was on the level of 60% and the CPU was not able to handle all of the processes. It had direct impact on communication quality, because threads responsible for communication maintaining were also blocked. In second case the CPU load for 50 clients was on the level of 15%. In this test case the CPU load was linearly dependent from number of clients (in both cases the same amount of data was exchanged).

In the second test we verified how server's update time and CPU load are depending on time parameters of source and clients. The results are presented in Fig. 5 and 6. There is only slight difference in performance (both in CPU load and network traffic) while changing client's subscription interval. However, the same change in source's data generating interval introduces slight delay in updating of the server, but the loss of performance is very significant, because

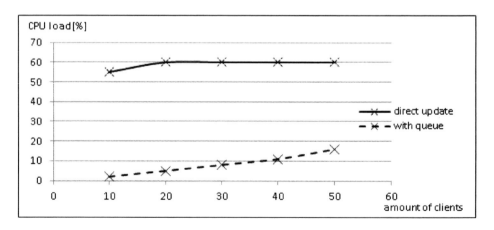

**Fig. 4.** Direct and indirect updating of Address Space

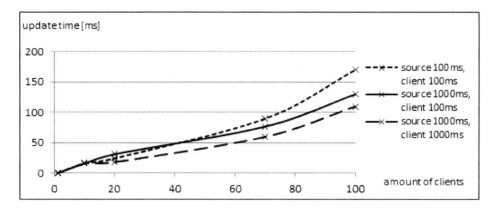

**Fig. 5.** Impact of source and client time parameters on server update time

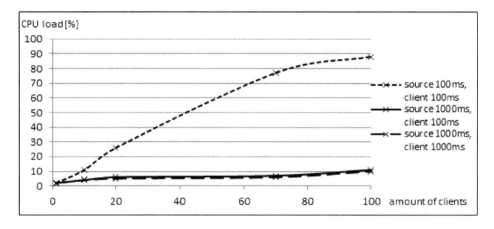

**Fig. 6.** Impact of source and client time parameters on servers CPU load

CPU load increases dramatically (the difference is 70% for 100 clients). The conclusion is that process of updating causes groundlessly high CPU load while the update time of server's data is almost unchanged. In both figures an inflexion point is visible (in neighborhood of values for 80 clients). In this point server is not able to perform correctly – for 100 ms client's subscription interval server is updating internal data model longer than 100 ms.

### 4.2    Impact of Number of Monitored Items

Third test were conducted to verify how server reacts for increasing number of data to be handled – what is the average network traffic generated. In this case both number of clients and amount of subscribed data were parameters. The results are presented in Fig. 7. For relatively small amount of data the dependency is linear. For bigger sets of subscribed data it is linear only to some

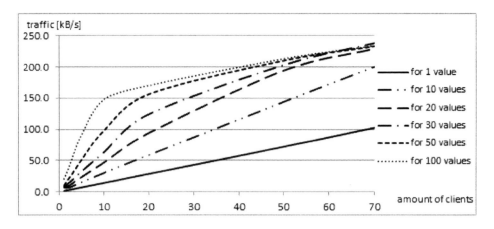

**Fig. 7.** Network traffic from server to clients

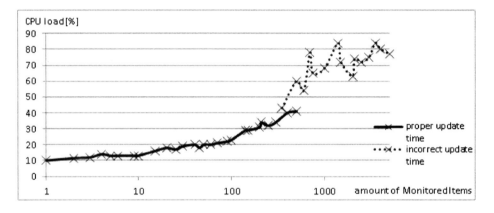

**Fig. 8.** Impact of amount of transferred values on servers CPU load

moment. Network is utilized less than 2.5%, so it seems server is not able to use available network bandwidth. We tested also if server's CPU is not used by other processes. Even artificially loaded, server has the same characteristic. Again, the conclusion is that in case of LAN network the server's internal update processes and resources utilization are a bottleneck. According to Fig. 8 it is important to keep CPU load below 30%–40%, because for those values server is able to handle updates in a reasonable time. In the figure, the dotted line indicates values, for which server's performance was unacceptable.

## 5    Summary and Conclusions

Based on the conducted research can be concluded that there are four main features of OPC UA communication based on Prosys Java SDK:

1. For standard LAN network based on 100 Mbit Ethernet network load reached about 2 Mb/s, which is only 2.5% of possible bandwidth utilization and may be considered as negligible in our test case. For that reason we can assume that bottleneck is not the network, but internal architecture of the server and utilization of hardware resources.
2. The CPU load of the server machine is almost linearly dependent from the number of *MonitoredItems* created by the clients.
3. If the communication with data source is not separated from updating server's tags (e.g. using buffering) the server's CPU load increases dramatically (for one client subscribing 2000 tags sampled every 100 ms CPU load is over 55%).
4. There are two possible main factors that have significant impact on the server's CPU load: 1) updating tags values and handling *MonitoredItems* in all subscriptions and 2) internal data encoding on protocol's communication stack (in our case that part was done by SDK). However, on this stage of research it is not possible to verify which factor is more important for the CPU's load.

Despite conducted research presents wide perspective on OPC UA performance issue, further research is indispensable. Specific tasks may be 1) determining the maximum number of *MonitoredItems* that server is able to handle without significant delays, 2) buffering of communication with the data source. The OPC UA standard does not specify how the change of subscription parameters should be realized concerning communication with data source. In most cases it is impossible to change the time parameters of industrial networks, while OPC UA specifies the sampling interval parameter, which is supposed to define interval of data exchange between server and data source. Modifying time parameters of communication between data source and server should not be possible in those cases and the data acquired from data source must be buffered in server for further usage.

There are multiple methods that may be used for further research on OPC UA performance. All of them need to be precised with more details. However, it may be matter for new research and articles because of diversity of issues related to that research area. Several general opportunities which authors would like to follow in the short term are:

- simulation of the communication system using queues modeling,
- detailed research on impact of amount and distribution of *MonitoredItems*,
- detailed research on methods of updating OPC UA variables with infomation acquired from data source,
- case study with different implementations of OPC UA communication (.NET, ANSI C),
- enhancing server's architecture to an enterprise-class solution.

**Acknowledgement.** This work was supported by the European Union from the European Social Fund (grant agreement number: UDA-POKL.04.01.01-00-106/09).

# References

1. de Souza, L.M.S., Spiess, P., Guinard, D., Köhler, M., Karnouskos, S., Savio, D.: SOCRADES: A Web Service Based Shop Floor Integration Infrastructure. In: Floerkemeier, C., Langheinrich, M., Fleisch, E., Mattern, F., Sarma, S.E. (eds.) IOT 2008. LNCS, vol. 4952, pp. 50–67. Springer, Heidelberg (2008)
2. OPC Foundation: OPC UA Specification: Parts 1–13 (2009)
3. Lange, J., Iwanitz, F., Burke, T.: OPC – From Data Access to Unified Architecture. VDE Verlag GmbH (2010)
4. Intel, AscoLab, Unified Automation: Reducing Product Development Effort for OPC Unified Architecture. White paper (2009)
5. Hannelius, T., Salmenpera, M., Kuikka, S.: Roadmap To Adopting OPC UA. In: 6th IEEE International Conference on Industrial Informatics, pp. 756–761 (2008)
6. Sauter, T., Lobashov, M.: How to Access Factory Floor Information Using Internet Technologies and Gateways. IEEE Transactions on Industrial Informatics (2011)
7. Gîrbea, A., Nechifor, S., Sisak, F., Perniu, L.: Design and implementation of an OLE for process control unified architecture aggregating server for a group of flexible manufacturing systems. IET Software (2011)
8. Virta, J., Seilonen, I., Tuomi, A., Koskinen, K.: SOA-Based Integration for Batch Process Management with OPC UA and ISA-88/95. In: 15th IEEE International Conference on Emerging Technologies and Factory Automation, ETFA, Bilbao, Spain (2010)
9. Mahnke, W., Leitner, S.H., Damm, M.: OPC Unified Architecture. Springer (2009)
10. Post, O., Seppälä, J.: The Performance of OPC-UA Security Model at Field Device Level. Department of Automation Science. Tampere University of Technology, Tampere, Finland
11. Cavalieri, S., Cutuli, G.: Performance Evaluation of OPC UA. In: 2010 IEEE Conference on Emerging Technologies and Factory Automation, ETFA (2010)
12. Cavalieri, S., Cutuli, G., Monteleone, S.: Evaluating impact of security on OPC UA performance. In: 3rd Conference on Human System Interactions, HSI, pp. 687–694 (2010)
13. Folkert, K., Fojcik, M., Cupek, R.: Efficiency of OPC UA Communication in Java-Based Implementations. In: Kwiecień, A., Gaj, P., Stera, P. (eds.) CN 2011. CCIS, vol. 160, pp. 348–357. Springer, Heidelberg (2011)

# Analysis of Challenge-Response Authentication in a Networked Control System

Wojciech Rząsa, Dariusz Rzońca, Andrzej Stec, and Bartosz Trybus

Rzeszow University of Technology
al. Powstańców Warszawy 12, 35-959 Rzeszów, Poland
{wrzasa,drzonca,astec,btrybus}@kia.prz.edu.pl
http://www.prz.edu.pl

**Abstract.** Analysis of authentication time in *challenge-response* protocol has been presented here. The protocol has been used in a prototype Networked Control System (NCS) with multiple Plant Control Units (PCU) and Supervisory Control Unit (SCU). The units are programmed in IEC 61131-3 compliant CPDev (Control Program Developer) environment. Global variable manager is used to share common data across the network. The paper presents theoretical estimation of the handshake time during authentication between SCU and PCU. Experiments have been carried out to compare the estimations with a real application.

**Keywords:** NCS, challenge-response, CPDev, IEC 61131-3.

## 1 Introduction

Modern control systems increasingly use the infrastructure of computer networks. They are no longer simple sensors and actuators connected to the computer, but groups of intelligent sensors or controllers, which exchange data (feedback control loop) via a computer network. They are called Networked Control Systems (NCS) [1,2] and are now more and more popular [3,4]. The components of such systems can be distributed on a large area, away from the central computer, and can perform different tasks depending on current needs. Therefore they are programmable, and their programming and configuration is carried out remotely.

Due to the use of public infrastructure or wireless connections, in places to which unauthorized persons have access, it becomes necessary to use data encryption [5]. Unfortunately the controllers used in such systems have often highly limited resources and their computing performance does not allow to implement solutions typically used in PCs. Furthermore, even minor disturbances in the network or a significant delay in data transmission can be exceptionally harmful to the operation of such systems. Therefore it is important to analyse the data exchange mechanisms in order to provide a solution that meets timing requirements.

A. Kwiecień, P. Gaj, and P. Stera (Eds.): CN 2012, CCIS 291, pp. 271–279, 2012.

## 2    System Architecture

As a result of the research, a prototype of small NCS system has been developed at Computer and Control Engineering Department of Rzeszów University of Technology. It consists of Supervisory Control Unit (SCU), responsible for general functioning of the whole system, and a number of Plant Control Units (PCU) which read sensors, process data and control actuators. The system evolved from the previous project (PACQ, [6,7]) focused primarily on collecting data from control modules and presenting them on a website. In the new solution, encryption of data transferred between the system components is introduced together with programming enhancements. The user can develop control software via one large project which incorporates control tasks executed by PCU and SCU modules. The task software for PCUs are distributed via SCU. The SCU module functions as a supervisory controller, process data hub (warehouse), or short-term database for external systems. The topology of the system is shown in Fig. 1.

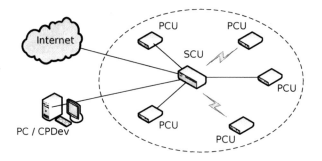

**Fig. 1.** Topology of the system

The SCU, being a device with the highest computing power, is an industrial PC running Linux and hosts software tools for programming the PCUs. The PCU modules are small, ARM7-based devices with low power consumption, but have strongly limited hardware and software resources. FreeRTOS operating system with uIP stack for network communication has been implemented in the modules. In the lab installation of the system the star topology has been applied. In case of many PCUs connected to the SCU additional switches that support wired or wireless communication can be used.

An industrial implementation of similar architecture is Mini-Guard Ship Control and Positioning System from Praxis A.T., Leiden, NL. [8]. Mini-Guard consists of seven types of dedicated controllers (PCUs) involving NXP ARM7 16/32-bit microcontrollers. The controllers communicate over Ethernet, external devices are connected through serial port or CAN bus. MarinePC equipped with Intel Atom and Windows Embedded operating system is the SCU of Mini-Guard.

# 3   Software Structure

Control software for the system is developed in CPDev (Control Program Developer) engineering environment according to IEC 61131-3 standard [9]. CPDev, created at Rzeszów University of Technology, integrates tools for programming, simulation, hardware configuration, on-line testing (commissioning) and running control applications. It is especially tailored for small- and medium-scale control systems. CPDev compiles control programs into intermediate, universal code executed by runtime interpreter at the controller side. The interpreter is called virtual machine (VM), and its intermediate language is Virtual Machine Assembler (VMASM) [10]. The VM code was written in C, thus it may run on different hardware, from 8-bit microcontrollers up to 32/64-bit general purpose processors. The prototype NCS system discussed in the paper, as well as Mini-Guard use ARM and x86 versions of the machine to execute CPDev programs.

CPDev has been equipped recently with the capability to incorporate many control tasks into a single project. This was first implemented and tested with FPGA implementation of the virtual machine [11]. The same approach, after some modifications, has now been applied to the NCS prototype. The CPDev programmer develops a single project for all the NCS system. The development process is done according to the IEC standard. First, hardware configuration is set (i.e. number and type of PCUs, communication channels etc.). Then Program Organization Units (POUs) are created (functions, function blocks or programs). POUs can be written in ST and IL textual languages (Structured Text, Instruction List) and in FBD, LD, SFC graphical ones (Function Block Diagram, Ladder Diagram, Sequential Function Chart) [12,13].

An important step of the development is to create control tasks. Each task will execute its own control algorithm compiled into VMASM and will run separately from the other tasks. This means that CPDev project with multiple tasks is executed by a group of *execution units*. In case of NCS system discussed here, the execution units are PCUs and SCU, each equipped with the CPDev virtual machine capable to execute a single task.

The user creates a task by selecting appropriate POUs and assigning them to the task. The POUs assigned to the task can be written in different IEC languages and are executed sequentially in the order defined by the user. Task execution mode can be cyclic (most common choice), endless loop or single execution. The user also specifies which unit will execute the particular task. Generally speaking, PCU tasks are plant-oriented, while SCU task provides supervisory control and monitoring.

As said, the control tasks are run independently by their own execution units. However, being parts of the same project, they usually communicate and exchange variables. Data exchange between the tasks is performed by means of global variables. The exchange mechanism allows sharing the globals among all units in the NCS network and utilizes CPDev *global variable manager* (GVM) to maintain the global storage hosted on the SCU.

Upon start of the project, GVM executes special initialization code generated by the CPDev compiler to set initial values of the global variables. This is done before any task is invoked by execution units. To avoid conflicts related to sharing global variables between tasks, the virtual machines in PCUs operate on so-called *process images*. At the start of the cycle, the task is provided with current copy of the global variables (local shadows). When the task is executed, only the shadows are used, so change of global values caused by other tasks does not affect calculations. When the cycle is about to end, the calculated shadows are stored in the global storage of GVM. Synchronization is done only for the variables that have been modified within the cycle.

As it can be seen, PCUs and SCU constantly communicate over the network to keep local shadows and global storage updated. The communication characteristics is crucial for the system overall performance, which can be additionally affected if encryption is introduced.

## 4    Analysis of Authentication Time in *Challenge-Response* Protocol

Authentication process between SCU and PCUs is based on *challenge-response* protocol (Fig. 2), as in [7], to avoid transmitting the password as plain text. Similar approach is used by Point to Point Protocol (PPP) with Challenge-Handshake Authentication (CHAP), specified in RFC 1994. The authentication scheme between SCU and PCUs is shown in Fig. 3.

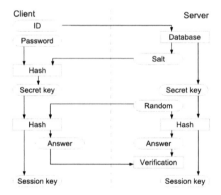

**Fig. 2.** *Challenge-response* protocol

The client (PCU) initiates the authentication by sending to the server (SCU) its *client ID* in *hello message* as plain text. The server replies with *authentication request* which contains a *random number* and a *salt* corresponding to the ID. The password is hashed by the client using the salt to make so-called secret key. Another secret key is read by the server from a database. Both sides independently hash their secret key with the random number. One part of the result calculated by the client is sent to the server as *authentication answer* and

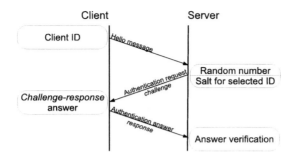

**Fig. 3.** Authentication scheme

then compared (*answer verification*). If the values are equal, the authentication is successful. The remaining part of the result becomes a *session key*, generated on both sides, and used later to encrypt the transmission.

Authentication process in the challenge-response protocol requires three consecutive messages to be successfully transmitted through the network. If one of the packets is lost the authentication fails, and must be performed again. To estimate handshake time in such case we will consider the network where every packet can be lost with a given probability $P_u$, so probability of successful transmission of single packet is $1 - P_u$. Thus probability of establishing the connection in single attempt $P_{n1}$ can be calculated as

$$P_{n1} = (1 - P_u)^3 . \tag{1}$$

Event of unsuccessful handshaking the connection in single attempt is the complement of establishing it in this attempt, thus its probability is $1 - P_{n1}$. Repeated attempts of authentication can be considered as independent Bernoulli trials, with probability of the success in every trial $P_{n1}$, and failure $1 - P_{n1}$. Denote as $N$ the random variable describing number of the trial where the connection has been established. Such variable has geometric distribution, and probability of establishing the connection exactly in $k$ trial is

$$P\{N = k\} = (1 - P_{n1})^{k-1} P_{n1} \text{ for } k = 1, 2, \ldots \wedge P_{n1} \in (0; 1) . \tag{2}$$

The expected value of number of trials needed to establish a connection is

$$E(N) = \frac{1}{P_{n1}} . \tag{3}$$

Basing on this expected value, average time of negotiating the connection $t_{\text{CONN}}$ can be calculated. Trials are repeated periodically after $t_{\text{TOUT}}$ time, so every unsuccessful attempt lasts $t_{\text{TOUT}}$. Duration of final (successful) attempt depends mainly on time of single packet transmission $t_{\text{NET}}$ (for simplification we assume here that time needed for processing the message and generating the reply is negligible in comparison with $t_{\text{NET}}$). Establishing the connection requires transmission of three consecutive messages, so the final trial lasts $3t_{\text{NET}}$. Therefore

$$t_{\text{CONN}} = (E(N) - 1)t_{\text{TOUT}} + 3t_{\text{NET}} \ . \tag{4}$$

Finally

$$t_{\text{CONN}} = \frac{1 - (1 - P_{\text{u}})^3}{(1 - P_{\text{u}})^3} t_{\text{TOUT}} + 3t_{\text{NET}} \ . \tag{5}$$

The above calculations assume that time delay associated with a single data transmission between the PCU and SCU is equal for all modules, regardless of additional intermediate switches.

## 5    Experiments

Experiments were performed in order to assess accuracy of the theoretical estimation of the challenge-response handshake time. The aim was to compare the theoretical estimations with connection establishment time for a real application.

The experiments were based on a test application designed to perform the required handshake and to allow time measurements. The application consisted of the *client* attempting to establish connection to the *server* according to the challenge-response protocol. The communication was performed using UDP over IP, thus packet losses caused by the computer network were not resolved by the underlying protocol suite (e.g. by retransmissions) and had required impact on the process of connection establishment.

In order to facilitate control of the experiment parameters and to take advantage of diverse methods of registering and processing the results, the experiments were performed using PCs acting as the server (SCU) and the client (PCU). This more convenient choice did not have negative impact on the credibility of the research, since not the hardware running the application, but parameters of the network were the crucial factor affecting the results.

The client and the server processes were located on different computers connected with a single 100 Mbps Ethernet switch. The client was attempting to establish subsequent connections every 100 ms (sending subsequent *hello messages*). Both processes were responding to the received packets according to the challenge-response handshake as soon as the packets were received, without any additional delay. The size of all packets transmitted in the experiments equalled 16 B.

The application was run under the control of the Linux 2.6 operating system. The parameters of the network required by the experiments were determined by the use of NetEm[1] being a part of Linux Traffic Control infrastructure[2] and described e.g. in [14]. The NetEm is designed to enable testing of network-related characteristics of applications, enabling emulation of different network parameters including delay, packet loss, packet corruption and packet reordering. Different probabilities and different statistical distributions can be used for

---

[1] http://www.linuxfoundation.org/collaborate/workgroups/networking/netem
[2] http://lartc.org/

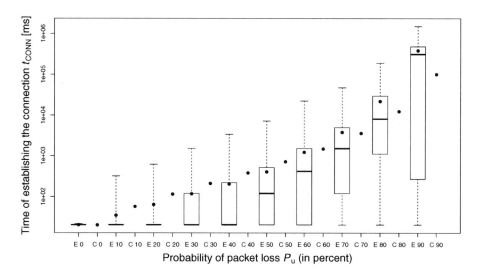

**Fig. 4.** Comparison of experiment results (E) and calculated values (C)

the network parameters. The module is frequently used in research concerning network parameters [15,16,17].

In the experiments two parameters of the network were configured using NetEm: delay and packet loss. Delay was set to 10 ms for each network interface and average round trip time for *ping* between the client and the server machines, including transmission time over the network, was equal to 20.197 ms with not significant standard deviation of 0.135 ms. Packet loss probability was set according to the requirements of subsequent experiments. All transmitted packets (regardless of their direction) had equal probability of being dropped, without any correlation with the preceding ones.

As seen in Fig. 3 time of negotiating the connection $t_{CONN}$ should be measured as the time between sending the first *hello message* by the client, and receiving final *response* by the server. Such measurement would not be trivial to perform, because client and server processes were run on different computers with different clocks. We decided to measure the time on the client side between sending the first *hello message*, and the final *response* (*final* means the one received by the server). Such time would be shorter from $t_{CONN}$ by $t_{NET}$, which could be added later. To measure it, the client had to determine if the packet with the response was received by the server, or lost. It would be easy with additional communication channel (without packet loosing) providing appropriate feedback from the server, but we prefer to avoid complicating the application by additional channels unless it is absolutely necessary. To resolve it, every response packet was numbered. Number of the last received response was stored on the server and included in the *challenges* it sent. Therefore, the client, after receiving the *challenge*, was able to determine if the *previous* trial of establishing the connection had ended successfully (i.e. the server had received the previous response), and calculate its time.

Experiments results were compared with $t_{CONN}$ calculated from the equations derived in the previous section for the following data $t_{TOUT} = 100\,\text{ms}$, $t_{NET} = 10\,\text{ms}$, and varying $P_u$. The comparison is shown in Fig. 4. The following symbols for statistical boxplot have been assumed – dots indicate mean values, upper edges (hinges) of the boxes indicate the upper quartile of the data set, and the lower hinges indicate the lower quartile, the lines in the boxes indicate the median value and the ends of the vertical lines (whiskers) indicate the minimum and maximum data values. For better legibility logarithmic scale on the ordinate axis has been assumed.

## 6    Conclusions

The prototype network control system presented in the paper implements continuous data exchange between execution units via secure channels. Performance of the system is limited therefore special care should be taken to avoid data loss during the communication. Theoretical calculations and experiments have been carried out to estimate the influence of probability of packet loss during *challenge-response* authentication process between the client and the server on time parameters.

Significant spread of the results obtained during the experiment can be observed (Fig. 4). It is not caused by much smaller measurement incidental errors, dependent mostly on precision of computer clock, but rather by the nature of the experiment. Connection establishing time resulting from packet round trip was about 30 ms, but a single packet loss in the least crucial stage increased the time by 100 ms. Consequently it was frequent to obtain value of standard deviation comparable to mean value of the measurements. This situation does not preclude usefulness of the experiment results, but it should be regarded while interpretation.

The research confirms correctness of the theoretical estimations. Figure 4 shows satisfactory convergence between the experiment mean values and theoretically estimated times. The differences may be caused by imprecision of pseudorandom generator responsible for dropping packets. The impact of this imprecision grows for higher data loss probabilities. The results for the highest probabilities have little practical significance anyway, since network with so high data loss is in fact unusable. For smaller probabilities the estimation proved its precision and may be exploited in practical applications.

The analysis focuses on the authentication only. Issues of data exchange between NCS devices after establishing the connection were not discussed. However, negotiating the connection plays an important role here. Each time a single message is lost during the communication, the connection is terminated automatically and its re-establishing is needed. Since the authentication procedure involves exchange of several messages between the two devices, the time it requires may be significant.

# References

1. Tipsuwan, Y., Chow, M.-Y.: Control methodologies in networked control systems. Control Engineering Practice 11, 1099–1111 (2003)
2. Lian, F.-L., Moyne, J., Tilbury, D.: Network Design Consideration for Distributed Control Systems. IEEE Transactions on Control Systems Technology 10(2), 297–307 (2002)
3. Hespanha, J.P., Naghshtabrizi, P., Xu, Y.: A Survey of Recent Results in Networked Control Systems. Proceedings of the IEEE 95(1), 138–162 (2007)
4. Baillieul, J., Antsaklis, P.J.: Control and Communication Challenges in Networked Real-Time Systems. Proceedings of the IEEE 95(1), 9–28 (2007)
5. Cheminod, M., Pironti, A., Sisto, R.: Formal Vulnerability Analysis of a Security System for Remote Fieldbus Access. IEEE Transactions on Industrial Informatics 7(1), 30–40 (2011)
6. Rzońca, D., Stec, A., Trybus, B.: Data Acquisition Server for Mini Distributed Control System. In: Kwiecień, A., Gaj, P., Stera, P. (eds.) CN 2011. CCIS, vol. 160, pp. 398–406. Springer, Heidelberg (2011)
7. Rzońca, D., Stec, A.: Small Prototype Acquisition System with Secure Remote Data Access. Annales UMCS Informatica AI 11(3), 87–100 (2011)
8. Mini-Guard Ship Control & Positioning System. Praxis Automation Technology B.V. (2010), http://www.praxis-automation.com
9. IEC. IEC 61131-3 standard: Programmable Controllers – Part 3. Programming Languages (2003)
10. Trybus, B.: Development and Implementation of IEC 61131-3 Virtual Machine. Theoretical and Applied Informatics 23(1), 21–35 (2011)
11. Hajduk, Z., Sadolewski, J., Trybus, B.: Multiple tasks in FPGA-based programmable controller. e-Informatica Software Engineering Journal 5(1) (2011)
12. Rzońca, D., Sadolewski, J., Stec, A., Świder, Z., Trybus, B., Trybus, L.: Mini–DCS System Programming in IEC 61131-3 Structured Text. Journal of Automation, Mobile Robotics & Intelligent Systems 2(3), 48–54 (2008)
13. Rzońca, D., Sadolewski, J., Stec, A., Świder, Z., Trybus, B., Trybus, L.: Open environment for programming small controllers according to IEC 61131-3 standard. Scalable Computing: Practice and Experience 10(3), 325–336 (2009)
14. Hemminger, S.: Network Emulation with NetEm. In: Proc. of the 2005 Linux Conference Australia, LCA 2005 (April 2005)
15. Burger, M., Kielmann, T.: MOB: zero-configuration high-throughput multicasting for grid applications. In: Proc. of the 16th International Symposium on High Performance Distributed Computing, Monterey, California, USA, June 25-29 (2007)
16. Sangtae, H., Injong, R., Lisong, X.: CUBIC: a new TCP-friendly high-speed TCP variant. ACM SIGOPS Operating Systems Review 42(5), 64–74 (2008)
17. Choe, Y.R., Schuff, D.L., Dyaberi, J.M., Pai, V.S.: Improving VoD server efficiency with bittorrent. In: Proc. of the 15th International Conference on Multimedia, Augsburg, Germany, September 25-29 (2007)

# Management of Industrial Networks Based on the FCAPS Guidelines

Andrzej Kwiecień and Karol Opielka

Silesian University of Technology, Institute of Informatics
Akademicka 16, 41-100 Gliwice, Poland
{andrzej.kwiecien,karol.opielka}@polsl.pl
http://www.polsl.pl

**Abstract.** The increasing complexity of distributed computer systems generates the need for continuous monitoring and analysis of their performance. This paper presents some aspects of implementation of a distributed network management system according to the network management requirements called FCAPS (fault, configuration, accounting, performance and security) outlined in the ISO and ITU-T documentations. The idea of utilizing of certain part of FCAPS management tasks by adding diagnostic data to network communications scenarios and providing this data via SNMP protocol is presented.

**Keywords:** distributed system, real-time, FCAPS, SNMP, management, industrial network.

## 1 Introduction

One of the fundamental aspects of industrial computer systems is to control network safety and reliability. Management and maintenance of such network is an essential aspect for the whole industrial system. Characteristic industrial working conditions bring many restrictions which are associated with the computer control systems that form a part of industrial systems. Increasing requirements for the industrial systems cause rapid development of industrial networks, this is connected also with the increasing requirements for the mechanisms to support their design and maintenance. Real-time requirements for control networks require a number of unique factors that must be taken into account in creating systems and tools used in industry. Those aspects and the distinctive architecture of the industrial network make it difficult to apply a standard network solution. Therefore, there is a need to do the analysis of a mechanisms supporting those systems.

The authors by analyzing industrial systems concluded that there is a need for defining the structure model that supports the maintenance of such installations. This model should provide foundations for constructing the industrial network systems with management functionality. The aim of this model is to help to develop ways of expanding existing solutions and to propose new solutions to ensure reliability and safety of real-time industrial network. Analysis of

A. Kwiecień, P. Gaj, and P. Stera (Eds.): CN 2012, CCIS 291, pp. 280–288, 2012.
© Springer-Verlag Berlin Heidelberg 2012

**Table 1.** FCAPS model [4]

| Fault Management | maintain accurate information on the software and hardware configuration of the network | Alarm handling, Alarm correlation, Alarm forwarding, Filter management, Log management, Threshold based reporting |
|---|---|---|
| Configuration Management | detect, archive and fix problems that occur in network | Auto Discovery, Network provisioning, Auto back up and recovery, Service Activation, Software upgrade to devices, Inventory management |
| Accounting Management | measure utilization and activities on the network | Service Usage, Service level agreements, Billing |
| Performance Management | maximize network performance monitoring and optimization | Performance monitoring, Report generation, Data collection and correlation |
| Security Management | control access to network and safety policy | Prevention, Authentication, System Access Control, Detection, Intrusion recovery, Containment and Recovery |

issues related to the management of computer networks allowed to select one of the available models, that is a general network management functional model FCAPS. FCAPS is a model defined by ISO [1] and ITU-T [2,3]. It describes management areas of communication networks and defines five of them: fault, configuration, accounting, performance and security. Attributes of them are presented in Table 1.

Fault management describes the rules for detecting faults in network and keeping a system with minimal breaking system's functioning. Effective fault management is based on locating of potential faults and taking steps to isolate them and to initiate recovery action. Other aspect connected with fault management is the capture of events, alarms and reporting of those occurrences. Configuration management identifies and records the configuration of network devices and monitors modifications of configuration and distribution of software to all network devices. Accounting management allows to examine the distribution of resources across the network. It defines the system efficiency by measuring utilization and activities on the network. Aim of performance management is to provide real-time access to the monitoring of network performance and resource allocation data. Its functionality should give opportunity to determine working condition of a network. The network performance encompasses its capacity, utilization, error rate, reaction time, etc. Monitoring of that information allows to define quality of network performance. In such way some potential problems can be identified before the occur. Security management is based on providing access to and control of network resources. Security indicates the safety policy by using authentication mechanisms, encryption and access control principles. Each of these five issues should be considered in terms of their practical functionality in industrial real-time systems and on that basis solutions and mechanisms for industrial network could be created. The scope of such studies should encompass the analysis of known applications of hardware and software systems used for their management, and also ways of determining of critical points in such installations. It should assist in development of mechanisms for support of all management areas.

## 2    FCAPS Guidelines and Industrial Networks

List of the FCAPS characteristics model and industrial computer network requirements rises a question of how is it possible to utilize this model in industrial network. Implementation of the FCAPS guidelines must be divided into several stages, where the final objective is to obtain a coherent management system based on the FCAPS model. The first stage is to generate information necessary to carry out network management, the second step is to obtain the data for the management system, next phase is to process data and react to decoding events, while at the same time it is essential to remember about such aspects like security and accounting. The listed stages should be analyzed in terms of implementing FCAPS model.

### 2.1    Data Generating

Fundamental element of network management is to generate information about its performance. It results from the necessity of analysis of the communication protocols in terms of generating of diagnostic data, the method of their acquisition and their scalability. Systems analysis is designed to specify the diagnostic information that should be included in the communication protocols, possibilities of obtaining system's data for later analysis, possibilities of using standard protocol and methodology of extending protocols and to add any other necessary information. Selection of a protocol extension method with diagnostic data depends on the chosen protocol. Industrial protocols described in IEC 61158 allow to various ways of realization the transmission of diagnostic data. Creating the data exchange scenario is possible by using built-in protocols mechanisms or open protocols functionality.

### 2.2    Data Obtaining

Generated data must be sent to the system responsible for managing of the network. This is connected with a need for modifications in the information exchange scenario, this aspect is particularly important in networks that are used to control industrial facilities. In many cases, the behavior of a required temporal determinism is necessary. For that reason it is essential to check the influence of additional data transmission on data flow in the control network in relation to hard real-time system requirements. It is also necessary to determine how to transfer and store information for its processing, analysis and presentation. These operations should identify ways in which to safely interact with the system in order to collect data. They should also show how to prioritize those tasks and incorporate them into a real-time system.

## 2.3   Analysis Data and Reaction

Effective data analysis is an essential element for the appropriate maintenance of the industrial network. It is a mechanism able to show the ways to determine the status and to vivificate the workflow in industrial networks. One of such ways is a data mining. This is the science discipline dealing with the problem of processing and analyzing data stored in the computer memory. It provides a mechanism for discovering of new and useful information for operators from existing data resource. Appropriate analysis of such information allows to derivate valuable information about the network and to control process whilst preserving a high level of security. It is important to prepare reaction mechanisms for an occurring event. The reaction may include either simply informing the user about the events or system's self-test mechanisms.

# 3   Example of Realization

Solutions in industrial networks typically consist of two basic parts of a system: hard real-time system directly responsible for control process and interactions with the industrial components, and a soft real-time part of a network [5]. Mechanisms of user interfaces like visualization and parameterization are usually based on later one. Hard real-time requires a detailed analysis of the amount and the range of diagnostic data relative to the time restrictions. Analysis and proposals of the solutions to systems of a hard real-time character were presented in [6]. Conception described in this paper is based on the extension of the protocol exchanges scenario by a diagnostic data in a hard real-time system. Methods of this extension, their impact on the system and ways of providing that information for system's analysis constitute the primary considerations of the authors. Here, base on that, the idea of generating diagnostic data by the individual system nodes, transferring information to the master unit and finally making the data available for the computer system in which there is acquisition and analysis of data is presented. One of the important factors is the proportion of cycle times [7], for the cycle with extended scenario of diagnostic data from the nodes and the cycle time excluding the transmission of additional data – the formula (1).

$$p = \frac{T_{\text{Control}} + T_{\text{Diag}}}{T_{\text{Control}}} \qquad (1)$$

where: $p$ – factor of network cycle time; $T_{\text{Control}}$ – time of control data transmission; $T_{\text{Diag}}$ – time of diagnostic data transmission.

It is important that each hard real-time system has a predetermined maximum value of this factor. This ratio depends on the scenario of information exchange in the network. Depending on the needs, data can be transmitted periodically, on demand, etc. In paper [6] conditions of choosing the most favorable scenarios for the exchange of data were determined, taking into account the cycle time of the network, the nodes load and the importance of information exchange. Solutions described in this paper allow to obtain the diagnostic information for system management from hard real-time part of system, which lay foundations for solutions presented in the current work. In the case of soft real-time part of system, currently, it is standard to use a bus based on Ethernet, so that further analysis is based on Ethernet connection. Typically, the visualization and other Human Machine Interface (HMI) devices have Ethernet interfaces and through those interfaces they allow for the diagnosis and monitoring of the individual components. The master Programmable Logic Controller (PLC) unit is equipped with an Ethernet connection that allows to connect and make it accessible for other nodes in the soft real-time network. In this way, the information generated in the PLC installation is provided by the master PLC controller. that in turn is a source of information about the hard real-time part of a system. Supervisory Control And Data Acquisition (SCADA) nodes, the report server or other network components such as switches are additional Ethernet network's elements. The existing mechanisms that can be used in managing of the presented structure network should be examined. Solution proposed by the authors is to employ for this purpose the Simple Network Management Protocol (SNMP) [8]. SNMP was designed to transport the control data. This protocol enables the exchange of management information such as status and diagnostic data between network components and Network Management System (NMS) [9]. SNMP is standard for the management of devices in a TCP/IP network. In SNMP management workstation there is so called the 'SNMP agent' in order to obtain information about the devices. Elements of device, which the SNMP agent can either access or modify, are described as SNMP objects. Collections of SNMP objects are contained in a logical database, the Management Information Base (MIB), and those objects are frequently referred to as MIB objects. In the Ethernet controller, SNMP embraces the common MIB as described in [10]. On the market, there are many network devices with SNMP protocol applicable in industrial networks. Turning on SNMP service enables agent which role is to provide the data collected about network nodes to NMS server where service called SNMP manager is operating. Collected data, stored in dedicated structures allow historical and current analysis of the network operating conditions. The structure of such network is presented in a Fig. 1. Standard exchange of messages is based on the inquiring of agents by the manager of the entries in the MIB (get, set commands), but it is also possible to send information without a request by the agents (traps command). The elementary functions provided by the SNMP protocol and their descriptions are presented in Table 2.

Using of SNMP protocol enables to reconstruct the network structure, diagnose: the status of each equipment's communications port, the speed of their operation, the bandwidth of the device, its load, the number of reported errors, the level of supply voltage, etc. SNMP allows not only to access the diagnostic data, but its functionality also includes the parameterization mechanisms such as remotely enable or disable ports, or set the speed. In short, SNMP provides a set of features to manage network from the designated stations. Range of options depends on the implementation stage of the protocol on each node of the system. The proposed solution is based on employing the SNMP protocol to transmit

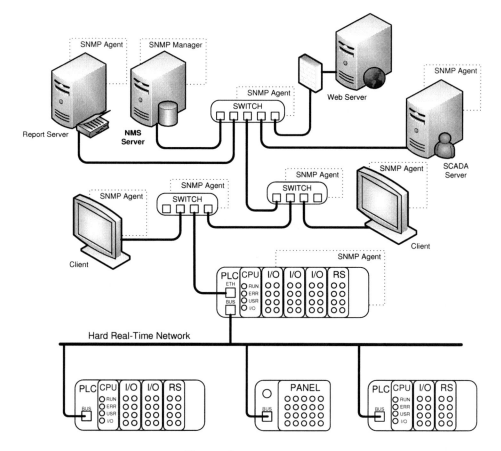

**Fig. 1.** System structure

**Table 2.** SNMP functions [11]

| Operation | Description | Sender |
|---|---|---|
| GetRequest | Accesses and retrieves the value of one or more instances of management information | Manager |
| GetNextRequest | Accesses and retrieves the value of the next instance of management information in lexicographical order | Manager |
| GetBulk | Accesses multiple values at one time | Manager |
| GetResponse | Reply to a GetRequest, GetNextRequest, and SetRequest operation | Agent |
| SetRequest | Stores and sets a value in a variable | Manager |
| Trap | An unsolicited message that is sent by an SNMP agent to an SNMP manager and indicates that some event has occurred | Agent |

information about the performance of the individual components operating in Ethernet network and retrieving information about PLC network operating condition through the master PLC with an Ethernet interface on which the SNMP functionality is implemented. The scope of information collected from the PLC consists in this case of the data about a PLC operating as an Ethernet node, but more importantly in this application it is enabling access to the data describing the performance of hard real-time network. Figure 2 illustrates this data flow. Information about a hard real-time part of network is stored in the MIB database, periodically updated by a dedicated PLC program task. Exchange of data using SNMP protocol in case of communication with the PLC master is presented in the Fig. 3. MIB located in the PLC includes data groups describing the work of this controller and PLC Data Group, which is used to provide information about other PLC nodes. PLC program cyclically records information like CPU status, I/O status, interfaces status and program status. NMS server by sending commands to SNMP agents receives in response the requested data packets, additionally the agent may under certain conditions, send the

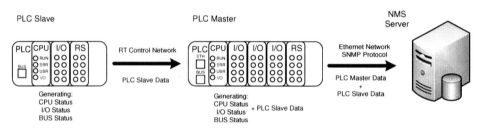

**Fig. 2.** Data flow

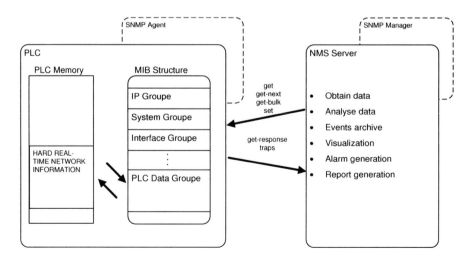

**Fig. 3.** SNMP communication

traps packets, indicating certain events. Taking into account the data contained in master PLC, this mechanism allows to monitor the whole network installation of an industrial plant. In addition to the SNMP protocol it is necessary to employ certain software tools in order to manage and provide the interface to the user. There are several commercial and free software applications that use SNMP for network management [11], but not all of them are suitable for industrial use. Industrial aspect of their use requires individual, detailed analysis. In this application a critical point is SNMP manager's configuration, in terms of exchange scenario and reaction to potential events. The proposed solution in addition to properly configured SNMP manager allows to carry out diagnostic data acquisition, analysis and logging of events, generate reports, notification of the occurrence of users' operations, detection of potential inconsistencies in the operating of the network, and even partial correction. Such system model fulfills certain requirements of FCAPS. There are some alternatives to mechanism based on SNMP protocol [12]. Lightweight Directory Access Protocol (LDAP) or OLE for Process Control (OPC) are examples of such mechanisms. Each of these solutions brings advantages and disadvantages. The advantage of SNMP is fact that it gives the simple ability to integrate the management of control network nodes with other components. Many network elements such as routers, switches have a built-in SNMP client service. Additionally, as noted above, there are many software solutions based on SNMP that can be used as a network management station, the OpenNMS tool for example [13].

## 4   Summary

The proposed solution presents one of the opportunities to realize part of the guidelines described in the FCAPS model. The advantage of using the SNMP protocol for transmission of diagnostic data about the operating condition of industrial networks is that it uses a well-known standard. This is based on the availability of a variety of applications, which are based on this protocol and provide functionality associated with the acquisition of data, and user interface. Implementing an SNMP agent on a node that is part of hard real-time control network provides the overview of the whole network installation in a NMS server. It is essential to outline compatibility of NMS with SCADA, reporting and other components. It generates a need for further analysis. It is also requires further verification of management aspects described in the FCAPS model. However, based on the carried out analysis, it is possible to claim that presented principle is a suitable approach to improving the efficiency of the industrial network management.

## References

1. ISO 7498-4, 1989 Information processing systems, Open Systems Interconnection, Basic Reference Model, Part 4: Management framework (1989)
2. ITU-T, 1996, M.3010 Principles for a telecommunications management network (1996)

3. ITU-T, 1997, M.3400 TMN management functions (1997)
4. Subramanian, M., Gonsalves, T., Usha Rani, N.: Network Management Principles and Practice. Pearson Education, India (2010)
5. Kwiecień, A.: Analiza przepływu informacji w komputerowych sieciach przemysłowych. Wydawnictwo Politechniki Śląskiej, Gliwice (2002) (in Polish)
6. Kwiecień, A., Opielka, K.: Industrial Networks in Explosive Atmospheres. In: Kwiecień, A., Gaj, P., Stera, P. (eds.) CN 2011. CCIS, vol. 160, pp. 367–378. Springer, Heidelberg (2011)
7. Gaj, P.: Pessimistic Useful Efficiency of EPL Network Cycle. In: Kwiecień, A., Gaj, P., Stera, P. (eds.) CN 2010. CCIS, vol. 79, pp. 297–305. Springer, Heidelberg (2010)
8. RFC 3411 An Architecture for Describing Simple Network Management Protocol (SNMP) Management Frameworks (2002)
9. Morris, S.: Network Management, MIBs and MPLS: Principles, Design and Implementation. Prentice Hall (2003)
10. RFC 1213 Version 2 of management information base (MIB-2) for network management of TCP/IP-based internets
11. Mauro, D., Schmidt, K.: Essential SNMP. O'Reilly Media (2005)
12. Sauter, T., Lobashov, M.: How to Access Factory Floor Information Using Internet Technologies and Gateways. IEEE Transaction on Industrial Informatics 7(4) (2011)
13. Chiang, C.-Y.J., Levin, G., Shihwei, L., Serban, C., Wolberg, M., Chadha, R., Hadynski, G., LaBarre, L.: Enabling Distributed Management for Dynamic Airborne Networks. In: IEEE International Symposium on Policies for Distributed Systems and Networks (2009)

# The Algorithms of Transmission Failure Detection in Master-Slave Networks

Marcin Sidzina[1] and Błażej Kwiecień[2]

[1] University of Bielsko-Biala, Department of Mechanical Engineering Fundamentals
msidzina@ath.bielsko.pl
[2] Silesian University of Technology, Institute of Informatics
blazej.kwiecen@gmail.com

**Abstract.** Authors present considerations on data interchange acceleration method in distributed control systems using dual data transmission line. The method is focused on the construction of a system based on two buses. However bus failure detection is very important in this method, because in such a case transmission on two buses must be stopped. From this point of view, failure detection is a fundamental goal. In article, authors intend to describe in general the problem and to propose some detection algorithms.

**Keywords:** PLC, distributed real time system, industrial computer network, time cycle of exchange data, PLC programming, avalanche of events.

## 1 Introduction

The purpose of introducing a redundancy is to protect communication system from data loss in case of failure of one or more of its components. Therefore, while designing such a system programmer-designer pays special attention to the maximum time of execution of all data exchange functions for the main communication bus. During analysis of such system functionality, the time for system respond is also checked. In case of occurring irregularities in its work switching the mode is suggested [1,2,3].

A failure, which appears in the system activates the mechanism of data transmission on stand-by bus or switching the receiver stations on stand-by bus, where the execution of the same exchange messages occurs, as on the main bus. In many considerations there are suggestions that such redundant system is uneconomic, and increases the cost connected with maintaining of the second inefficient bus [4]. However, these protections guarantee that, in case of a partial system failure the data will be read. Redundancy can be also used to integrate distributed systems of different levels [5,6].

The primary objective of using redundant communication systems is obviously to increase the reliability of transmission. The solution is to ensure continuity of

A. Kwiecień, P. Gaj, and P. Stera (Eds.): CN 2012, CCIS 291, pp. 289–298, 2012.

service in case of failure of one of the buses. Nevertheless, it increases the cost of entire system, because it has to be equipped with duplicated communication interfaces (communication coprocessors) and double transmission lines. Because the failures are the minor part of the system work, thus to reduce the relative cost of its construction one may attempt to use simultaneously the both buses for transmission during the normal work. Such idea was presented in [7], and this paper consists in its continuation. The aim of the authors is to improve two basic parameters, which determine time properties of real time distributed system. More precisely, it concerns about *useful throughput* $P_U$ and *useful efficiency* $\eta_U$ of computer networks. Considering the efficiency of networks, it is commonly known that this parameter has a large impact on the other ones, such as network cycle duration $T_{CS}$ [2], and data exchange duration $T_{WD}$ in distributed communication or measurement system [8,9]. Quantities $T_{CS}$ and $T_{WD}$ are industrial computer network parameters of great importance in ensuring that the *limit time* ($T_{GR}$) (connected with the transaction exchange) should be impassable. With an improvement of these parameters, it is possible to reduce significantly the relative cost of the system.

## 2    Idea of the Dual Bus Application

Redundant buses are an interesting alternative, which enables to reduce the time for data exchange in the communication systems. This can be achieved e.g. by divining the task and assigning them to the realization on individual buses working in the system. It results in the realization of at least two separable scenarios of exchange. These tasks result in shortening the time of data exchange, which in effect increases the throughput of the system, but not of the communication bus. Using stand-by bus, as an additional transmission medium in efficient communication system, reverses to some extent the role of a redundant system. The both buses are loaded by transaction exchange, so the argument of maintain a free bus become obsolete. In case of a failure, the task of the system is to switch all the data exchange transactions on the efficient network bus. In this case – regardless of which particular bus is damaged – the efficient one will take over the task of messages exchange in the discussed communication system.

From this point of view, the "idea" of using the stand-by bus as an additional communication link is interesting, because it enables to reduce the message exchange in distributed system. Moreover, transmission of various messages is executed on both buses, and a failure results in switching the communication on the efficient elements of the system (Fig. 1).

*Example 1.* Network consisted of four nodes was configured. Its work is based on the Modbus/RTU protocol. Each station consists of two network coprocessors. Node number 1 has two coprocessors, which work as Master stations. The remaining stations work as Slave stations. Scenario of all exchanges encompasses:

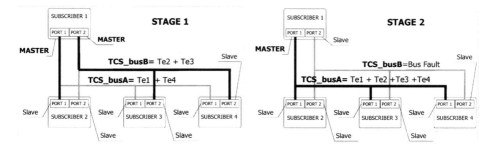

**Fig. 1.** Switching the work mode in case of one bus failure

| No | Type of periodical transaction of data exchange | Sample duration of transaction exchange in ms |
|---|---|---|
| $T_{e1}$ | Reading of five words from Slave station 4 | 32.5 |
| $T_{e2}$ | Reading of six words from Slave station 2 | 33.9 |
| $T_{e3}$ | Saving of 8 bits to Slave station 1 | 29.2 |
| $T_{e4}$ | Saving of 1 bit to Slave station 3 | 28.7 |

Total duration of all exchanges on one bus is 124.3 ms. In order to shorten the duration of all the exchanges, the scenario was divided into two scenarios, and was carried out by two coprocessors.

| No | Sample duration of transactions exchange in ms on the bus A | No | Sample duration of transactions exchange in ms on the bus B |
|---|---|---|---|
| $T_{e1}$ | 32.5 | $T_{e2}$ | 32.5 |
| $T_{e4}$ | 28.7 | $T_{e3}$ | 28.7 |
| $\sum$ | 61.2 | $\sum$ | 63.1 |

The situation where the stand-by bus is used continuously for messages exchanges, can results in improving the parameters of distributed system in a significant way. Assuming that there can be the failure of only one of the buses, then the deterioration of time parameters, in case of a failure, is a result of taking over the whole exchange scenario by one bus. Therefore, one of the important problems occurring in the configuration described above, is the way of switching (sticking together) into exchange scenario.

## 3    Definition of the Problem

Undoubtedly, usege of two buses for exchange of the messages benefits in shortening of the time of messages exchange. It is out of the question, that two parallelized data streams are able to convey more information than a single stream of data. However, in such system configuration, a problem with detection of a link failure appears. In most communication systems, the lack of response to a question sent is considered as a system failure. A specific parameter of above situation is *Timeout* parameter, which in case of industrial networks is configurable. Therefore, in most communication systems as a failure is considered the situation where a station will not response in a determined time (*Timeout*). As the research present in [10], *Timeout* parameter introduces the largest delay for switching of the communication among redundant system buses, and the remaining delays values are negligible small compared to the fixed *Timeout*.

*Example 2.* Example based on data [10]

| No | Description | ms |
|----|-------------|----|
| 1  | The measured duration of the message (writing of 16 digital outputs in Modbus/RTU) | 28 |
| 2  | Fixed Timeout | 1000 |
| 3  | The measured time of writing in case of damage of the bus | 1032 |

The measured duration of the message (writing of 16 digital outputs) in case of failure in the main bus, was 1032 ms. In such configuration of redundant system, *Timeout* parameter introduces the largest overhead of delay, in case of switching of the bus.

One of the many ways to shorten the system reaction can be continuous monitoring of the duration of transaction exchange. Having the knowledge of analytical dependences, times of exchange transmission and the maximum time response for each transaction, it is possible to begin sending in advance (before *Timeout* occurs) a control frame in order to determine the causes of irregularities. This action will reduce the reaction time of the system in case of bus or node failure.

Detection of a failure is the foundation of effectiveness of the presented method of synchronous transmitting with the use of redundant buses. When a failure is detected the following steps should be taken:

- hold immediately transmission in both buses;
- make a register of data, which were sent successfully;
- connect data streams into one, and bind appropriate label;
- start transmission via the efficient bus.

The authors propose several models of failure detection and some algorithms of procedure, which will be investigated in the future.

## 4    Algorithms of Transmission Failure Detection

Authors assume MODBUS network as a model. During no failure transmission in the network, data are transmitted and exchanged between subscribers through two buses (Fig. 2).

Authors indicate that bus redundancy (Sect. 2) allows for increasing network capacity and its efficiency. It is very important to study how communication

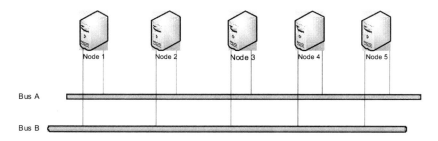

**Fig. 2.** Bus redundancy

system will work when failure occurs. When one of buses is damage it is necessary to stop data transmission through two buses, detect which part of data is already transmitted, start transmission through no failure bus. The time connected with failure service can be counted by assume "the worst accident". It is very important to estimate the value of delay because if this value will be too big it could disturb whole system which is a real-time communication system. The result of this estimation depends on kind of failure (bus or node failure). Of course network capacity and efficiency will decrease during failure but it doesn't matter from this point of view. In this article, we distinguish three work states:

– no failure work – both buses are working (Fig. 2);
– bus failure;
– node failure.

Regardless of which kind of failure occurs it is necessary to perform some procedures, which:

– enable shortest detection and kind of failure;
– enable to communication system stabilization;
– run required transmission procedures, depends on type of failure;
– enable communication without damaged part of system.

The most important element is very fast failure detection and to determine the kind of failure which implicates algorithm implementation and constant, which corrects system working. System failure could be caused be I/O device damage, communication bus failure, CPU device damage or errors in application program. In this article, two situations are considered which cause wrong system working:

– node failure;
– communication bus failure.

### 4.1   Communication Bus Failure

During "no failure" mode of communication system work, data are exchanged between nodes through two buses. However to ensure data security and consistency – it is necessary to prepare procedures which can secure communication system during one of buses failure (Fig. 3). The simplest way to detect this kind of failure is to send control frame through the same communication bus to the

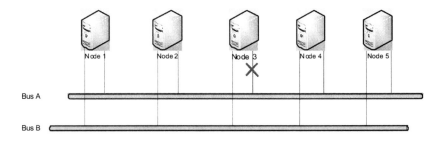

**Fig. 3.** Communication bus failure

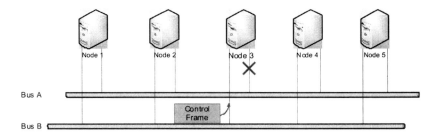

**Fig. 4.** Control frame sending

next node (Node 4). If these node responses to the control frame it means that the Node 3 is damage. More complicated situation is when the next nodes also don't respond. This situation may be caused by:

– communication bus failure;
– next nodes failure.

Of course, it is possible to ping all nodes to be sure that the communication bus is damaged. A question is, how long it will take and if the communication system is still real-time system?

The best way during a failure is action which enables the quickest kind of failure detection. If node (Node 3) doesn't exchange information through one of buses (Bus A) it is necessary to send control frame to this node (Node 3) through the second bus (Bus B) (Fig. 4). Assuming that in the same time all buses failure probability is near to zero, it is possible to diagnose very quickly if the bus or one of nodes are subjected to a failure (For example Node 3).

When the node responses to the control frame and sends acknowledge to the master station it is possible to confirm that the communication bus is damage. Simultaneously it is possible to configure control frame so as it contains data which have to be provided during broken transaction. This simple modification enables time failure reduction.

## 4.2   Node Failure

Situation is very similar, when one of the nodes fails (Fig. 5). If during normal work, suddenly one of nodes (Node 3) stops to exchange information in a computer system, it can be diagnosed and answered by:

- bus communication failure;
- node failure.

Failure detection, the same as for communication bus, must be done as quickly as possible. The easiest way to detect such a failure is to send a control frame to another node (Node 4) using the same communication bus. If the node responds, it means that Node 3 is damaged. If another node (Node 4) is not responding as well, similar to bus communication failure, we can face two issues:

- bus communication failure;
- another node failure.

The quickest way to diagnose such a failure seems to be, similar to bus communication failure, a control frame which is sent to affected node (Node 3) using second communication bus – i.e. Bus B (Fig. 6). If the node does not send a confirmation of receiving this control frame, it means that the node got damage and the communication bus is working fine (assuming that in the same time all buses failure probability is near to zero).

   The time frame of failure detection should be very long. But it is very important how the computer system behaves when a failure occurs. On one hand, situation with communication bus failure does not give a huge elbow room due to the fact that we cannot parallel information transmission process. On the other hand, situation with node failure seems to be more interesting. In such a case, the computer system which is tested, stops to fulfill its tasks (lack of node). If the node failure becomes uncritical for the whole system, no matter what the reason is, can continue its work using redundancy communication bus, which leads to parallel information exchange in the system.

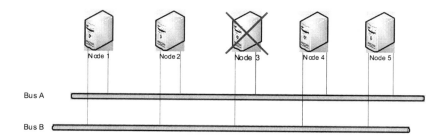

**Fig. 5.** Node failure

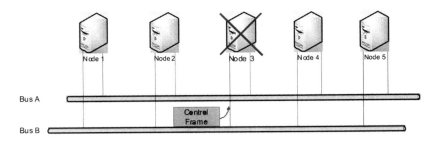

**Fig. 6.** Control frame sending

# 5   Proposal of Test Environment

In this article, which is continuation of [7] and [10], the concept of programmed
failure detection is included. This is a fundament of successful usage of second,
parallel, communication bus. This concept requires a lot of research which will
be done in the test environment, presented in figures below (Fig. 7, Fig. 8).
Necessary software is already implemented and it is now under introductory
tests. Received results enable us to perform discussion about limit of redundant
system usage to parallel data transmission in real-time systems.

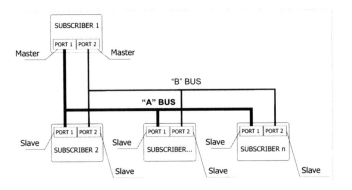

**Fig. 7.** Test environment [7]

Software testing procedure were performed in environment included 4 devices
which are working in the redundant Modbus/RTU communication system. Each
of devices has two communication ports RS-232 and RS-485. In case of RS-
232 port it is necessary to apply converter RS-232 to RS-485 to make possible
few devices connections. Double Master device is PAC model RX3i GE Intelligent Platforms. All devices have a possibility to implement any transmission
protocols.

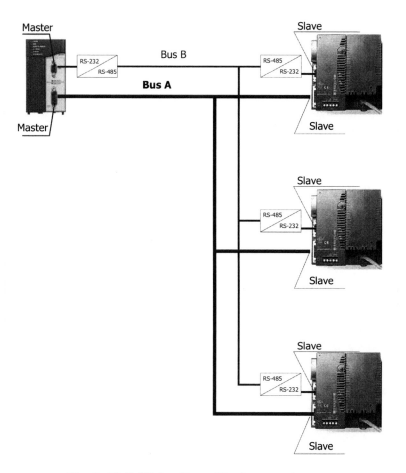

**Fig. 8.** PLC GE Intelligent Platforms environment

# 6   Conclusions

Presented method requires further research, which are planned by the authors in Master-Slave network environment. In case of positive test results, redundancy system model will be migrated to Client-Server communication system. We can assume, which has been also proven by single experiments, that a proposal of the method of system failure detection will be effective and helpful in getting high performance of data transmission in communication bus redundancy systems. In new system configuration, there is now place for definition of the connected with "stand-by" type of bus. Every bus, which is working in the system is an active bus. Only in case of failure situation, its status is being treated as an error situation.

# References

1. Stój, J., Kwiecień, B.: Real-time System Node Model – Some Research and Analysis. In: Contemporary Aspects of Computer Networks, vol. 2, pp. 231–240. WKŁ, Warszawa (2008)
2. Stój, J., Kwiecień, A.: The Response Time of a Control System with Communication Link Redundancy. In: Contemporary Aspects of Computer Networks, vol. 2, pp. 195–202. WKŁ, Warszawa (2008)
3. Gaj, P.: The Concept of a Multi-network Approach for a Dynamic Distribution of Application Relationships. In: Kwiecień, A., Gaj, P., Stera, P. (eds.) CN 2011. CCIS, vol. 160, pp. 328–337. Springer, Heidelberg (2011)
4. Kwiecień, A., Stój, J.: The Cost of Redundancy in Distributed Real-Time Systems in Steady State. In: Kwiecień, A., Gaj, P., Stera, P. (eds.) CN 2010. CCIS, vol. 79, pp. 106–120. Springer, Heidelberg (2010)
5. Jestratjew, A., Kwiecień, A.: Performance of HTTP Protocol in Networked Control Systems. IEEE Transactions on Industrial Informatics 99, 1 (2011), doi:10.1109/TII.2012.2183138
6. Kwiecień, B., Stój, J.: Network Integration on the Control Level. In: Kwiecień, A., Gaj, P., Stera, P. (eds.) CN 2011. CCIS, vol. 160, pp. 322–327. Springer, Heidelberg (2011)
7. Kwiecień, A., Sidzina, M.: Dual Bus as a Method for Data Interchange Transaction Acceleration in Distributed Real Time Systems. In: Kwiecień, A., Gaj, P., Stera, P. (eds.) CN 2009. CCIS, vol. 39, pp. 252–263. Springer, Heidelberg (2009)
8. Kwiecień, A.: Analiza przepływu informacji w komputerowych sieciach przemyslowych. Studia Informatica (2002) (in Polish)
9. Sidzina, M.: Dynamiczne modyfikacje programu aplikacji sterownika swobodnie programowalnego celem zwiekszenia czestosci wymian komunikatw w przemyslowych systemach rozproszonych czasu rzeczywistego. PhD Thesis. Silesian University of Technology, Institute of Informatics, Gliwice (2008) (in Polish)
10. Kwiecień, A., Stój, J., Sidzina, M.: Analiza wybranych architektur redundantnych z zastosowaniem sieci MODBUS/RTU. In: Kwiecień, A., et al. (eds.) Sieci Komputerowe. Aplikacje i zastosowania, vol. 2, pp. 359–367. WKiŁ, Warszawa (2007)
11. Kwiecień, A.: The improvement of working parameters of the industrial computer networks with cyclic transactions of data exchange by simulation in the physical model. In: The 29th Annual Conference of the IEEE Industrial Electronics Society, Roanoke, Virginia, USA, vol. 2, pp. 1282–1289 (November 2003) (Curr. ver. April 05, 2004)

# Model of the Threshold Mechanism with Double Hysteresis for Multi-service Networks

Maciej Sobieraj, Maciej Stasiak, and Piotr Zwierzykowski

Poznan University of Technology
Chair of Communications and Computer Networks
ul. Polanka 3, Poznań 60965, Poland
piotr.zwierzykowski@put.poznan.pl

**Abstract.** This paper presents a new generalized threshold model that can be used in cellular networks. In the proposed model, called *Double Hysteresis Model* (DHM), it is assumed that the amount of resources accessible for a new call of a given class can depend on three load areas of the system. The switching between areas is modulated by a three-state Markov chain that determines the average time the system spends in a particular load area. The results obtained for the discussed analytical model are compared with the results of the simulation of an exemplary WCDMA radio interface carrying a mixture of different multi-rate traffic streams. The research study confirms high accuracy of the proposed model.

**Keywords:** mobile networks, analytical model, threshold mechanism, hysteresis.

## 1 Introduction

Present-day communications and cellular networks offer a number of mechanisms that make expansion of traffic capacity of the system possible through a change in parameters of incoming or serviced calls. Such mechanisms include, for example, resource reservation, traffic compression and threshold mechanisms [1,2,3]. The aforementioned mechanisms can be utilized in the radio part of a cellular network because relatively low capacities of radio interfaces oftentimes pose a serious constraint on the effective optimization of 2G/3G/4G cellular networks [4].

In order to analyse and model traffic phenomena occurring in network systems in which different traffic management mechanisms are used, a great number of analytical models have been proposed. Models of systems with reservation are discussed in, for example [5,3], while models of systems with compression are addressed in [2,6,4]. The models in which an instance of exceeding a predefined load level (threshold) can be followed by a change, both in the volume of resources demanded by new calls and in the service time, are considered in [3,7,8,9]. These models are usually used for modeling systems with adaptive and elastic traffic [8]. The system in which one different threshold is introduced for some, or for all, traffic classes is called *Single-Threshold System* (STS). These

A. Kwiecień, P. Gaj, and P. Stera (Eds.): CN 2012, CCIS 291, pp. 299–313, 2012.
© Springer-Verlag Berlin Heidelberg 2012

systems are described by analytic models called *Single-Threshold Model* (STM). In many technological solutions, one common threshold is introduced for calls of all traffic classes [3]. If we introduce an appropriate set of thresholds for selected (or for all) classes of calls, then such a system can be labelled as the MTS system (*Multi-Threshold System*) and the appropriate analytical model – *Multi-Threshold Model* (MTM), respectively e.g. [7,9]. In the analysis of threshold mechanisms used in mobile networks we can also find threshold mechanisms in which a change in the way incoming calls are serviced is determined not only by the event of exceeding a predefined threshold, but also by the direction of the passage (transition). In the case of a single-threshold system, this means a replacement of a single-threshold with a pair of thresholds: one threshold for the direction of changes: low load – high load, and the other threshold or the direction of changes: high load – low load. These systems are called *Single-Hysteresis System* (SHS) [10].

To the best of the authors' knowledge, none of the analytical models of threshold systems presented in the literature of the subject take into consideration the phenomenon of double hysteresis. The paper proposes a new analytical double threshold model with hysteresis, further on in the paper called *Double Hysteresis Model* (DHM), that can be applied and implemented for modeling the *Call Admission Control* (CAC) function in communications and cellular networks.

The paper is divided into four sections. Section 2 presents a new analytical model of the threshold system with double hysteresis. In Section 3, a possible application of the proposed method for modeling of the radio interface with double hysteresis in the UMTS network is presented. The final section sums up the discussion.

## 2     Analytical Model of the System

Sections 2.1 to 2.3 discuss fundamental models that form the basis for the proposed DHM. The technical details of the DHM are then described in Sections 2.4 and 2.5.

### 2.1     Model of Full-Availability Group (FAG)

Let us assume that a system with complete sharing policy, called the full-availability group with multi-rate traffic, has the total capacity equal to $V$ Basic Bandwidth Units (BBUs). The value of BBU, i.e. $L_{BBU}$, is calculated as the greatest common divisor (GCD) of all resources demanded by traffic classes offered to the system [11]:

$$L_{BBU} = GCD(L_1, \ldots, L_M) \ . \tag{1}$$

The parameter $L_i$ is the amount of resources demanded by class $i$ call. The group is offered $M$ independent classes of Poisson traffic streams having the intensities:

$\lambda_1, \lambda_2, \ldots, \lambda_M$. The class $i$ call requires – an integer number of BBUs – equal to $t_i$ BBUs to set up a connection, where:

$$t_i = L_i/L_{\text{BBU}} \ . \tag{2}$$

The holding time for calls of particular classes has an exponential distribution with the parameters: $\mu_1, \mu_2, \ldots, \mu_M$. Thus, the average intensity of traffic offered to the system by the class $i$ traffic stream is equal to:

$$A_i = \lambda_i/\mu_i \ . \tag{3}$$

The occupancy distribution in the FAG can be described by the so-called Kaufman-Roberts recursion [12,13]:

$$n\left[P_n\right]_V = \begin{cases} 0, & \text{for} \quad n < 0 \ , \\ \sum_{i=1}^{M} A_i t_i \left[P_{n-t_i}\right]_V & \text{for} \quad 0 \leq n \leq V \ , \end{cases} \tag{4}$$

where $[P_n]_V$ is the occupancy probability of $n$ busy BBUs ($0 \leq n \leq V$). On the basis of Equation (4), the blocking probability $E_i$ for class $i$ stream can be expressed as follows:

$$E_i = \sum_{n=V-t_i+1}^{V} [P_n]_V \ . \tag{5}$$

The parameter $V$ is the total capacity of the group and is expressed in BBUs:

$$V = \lfloor V_{\text{phy}}/L_{\text{BBU}} \rfloor \ , \tag{6}$$

where $V_{\text{phy}}$ is the physical capacity of the group, which is expressed in the same unit as the resources $L$ demanded by calls of given classes.

On the basis of distribution $[P_n]_V$, the reverse transition rate $y_i(n)$ for a service stream of class $i$ outgoing from state $n$ can be calculated in the following way [12,14]:

$$y_i(n) = A_i \left[P_{n-t_i}\right]_V / \left[P_n\right]_V \ . \tag{7}$$

## 2.2   Multi-Threshold Model (MTM)

Let us consider a multi-threshold model [9,7] (MTM). This model is a generalization of the single-threshold model (STM) [3]. In the multi-threshold model we assume that for each call class a set of $q$ thresholds is introduced individually. For example, for class $j$ the set of adopted thresholds will be written in the following way $\{Q_{j,1}, Q_{j,2}, \ldots, Q_{j,q}\}$, where the first subscript indicates the class of a call, while the second subscript the number of threshold, with the initial assumption that: $\{Q_{j,1} \leq Q_{j,2} \leq \ldots \leq Q_{j,q}\}$, where $j \in \{1, \ldots, M\}$.

All states $n$, such that: $0 \leq n \leq Q_{i,1}$ are in the so-called *pre-threshold area*, the traffic stream of class $j$ ($j \in \{1, \ldots, M\}$) is defined by the parameters: $\{\lambda_j, t_{j,0}, \mu_{j,0}\}$. The next states $n$ ($Q_{i,k} < n \leq Q_{i,k+1}$) are in the so-called *post-threshold area* $k$, whereas the last states $n$, such that: $Q_{i,q} < n \leq V$ belong to the post-threshold area $q$ with the parameters $\{\lambda_j, t_{j,q}, \mu_{j,q}\}$.

Generally, in the models provided by the literature it is adopted that $t_{j,0} > t_{j,k}$ and $\mu_{j,0}^{-1} < \mu_{j,k}^{-1}$, where $0 < k \leq q$. This situation occurs in the case of calls belonging to the so-called elastic traffic classes [8]. In the case of adaptive traffic, the parameter $\mu_j$ does not undergo any changes, i.e. $\mu_{j,0} = \mu_{j,k} = \mu_j$ [8].

The occupancy distribution in threshold systems $[P_n]_{V,Q}$ can be approximated by the so-called generalized Kaufman-Roberts recursion [1,3,9] which, in the case of MTM, takes on the following form [9]:

$$n\,[P_n]_{V,Q} = \sum_{j=1}^{M} \sum_{k=0}^{q} A_{j,k} t_{j,k} \sigma_{j,k}(n - t_{j,k}) \left[P_{n-t_{j,k}}\right]_{V,Q} , \qquad (8)$$

where $k$ is equal to zero for the pre-threshold area. The following notation is adopted in Equation (8): $A_{j,k} = \lambda_j/\mu_{j,k}$ denotes average traffic offered to the system by the call stream of class $j$ and $\sigma_{j,k}(n)$ are the transition coefficients:

$$\sigma_{j,k}(n - t_{j,k}) = \begin{cases} 0 & \text{for } n - t_{j,k} \leq Q_{j,k} , \\ 1 & \text{for } n - t_{j,k} > Q_{j,k} . \end{cases} \qquad (9)$$

The parameter $\sigma_{j,k}(n)$ determines in which areas traffic streams are described by the parameters $\lambda_j, t_{j,k}, \mu_{j,k}$.

If we assume that the system services two sets of classes: $M_1$ classes for which the threshold has not been introduced, and $M_2$ classes that undergo the threshold mechanism, then the occupancy distribution (Equation (8)) for this case can be rewritten as follows:

$$n\,[P_n]_{V,Q} = \sum_{i=1}^{M_1} A_{i,0} t_{i,0} \left[P_{n-t_{i,0}}\right]_{V,Q} +$$
$$+ \sum_{j=1}^{M_2} \sum_{k=0}^{q} A_{j,k} t_{j,k} \sigma_{j,k}(n - t_{j,k}) \left[P_{n-t_{j,k}}\right]_{V,Q} , \qquad (10)$$

where $t_{i,0} = t_i$ and $A_{i,0} = A_i$ for $1 \leq i \leq M_1$.

The blocking probability in MTM for calls of classes that do not undergo threshold mechanism ($1 \leq i \leq M_1$) can be calculated on the basis of Equation (5), whereas the blocking probability for calls of classes that undergo threshold mechanism ($1 \leq j \leq M_2$) depends on the location of the threshold [9], and can be expressed as follows:

$$E_j = \sum_{k=0}^{q} E_{j,k} , \qquad (11)$$

where $E_{j,k}$ is the blocking probability of calls of class $j$ in the threshold area $k$. This probability is determined in the following way [4]:

$$
E_{j,k} = \begin{cases} 0 & \text{for} \quad \begin{cases} V - t_{j,k} \geq Q_{j,k+1} \ , \\ V - t_{j,k} > Q_{j,k} \ , \end{cases} \\[2em] \sum_{n=V-t_{j,k}+1}^{Q_{j,k+1}} [P_n]_{V,Q} & \text{for} \quad \begin{cases} V - t_{j,k} < Q_{j,k+1} \ , \\ V - t_{j,k} > Q_{j,k} \ , \end{cases} \\[2em] \sum_{n=Q_k+1}^{Q_{j,k+1}} [P_n]_{V,Q} & \text{for} \quad \begin{cases} V - t_{j,k} < Q_{j,k+1} \ , \\ V - t_{j,k} \leq Q_{j,k} \ , \end{cases} \end{cases} \tag{12}
$$

where $Q_{j,q+1} = V$.

The reverse transition rate in MTM is expressed by Equation (7) for calls not undergoing threshold mechanism $(1 \leq i \leq M_1)$, and for calls with threshold $(1 \leq j \leq M_2)$, we obtain [4,9]:

$$
y_{j,k}(n) = A_{j,k}\, \sigma_{j,k}(n - t_{j,k}) \left[P_{n-t_{j,k}}\right]_{V,Q} \Big/ [P_n]_{V,Q}
$$
$$
\text{for} \quad 0 \leq k \leq q \quad \text{and} \quad Q_{j,k} + t_{j,k} < n \leq Q_{j,k+1} + t_{j,k} \ . \tag{13}
$$

### 2.3 Modified Multi-Threshold Model (MMTM)

Consider a threshold system in which, for a given class $j$, the following set of thresholds $\{Q_{j,1}, Q_{j,2}, \ldots, Q_{j,q}\}$, has been introduced. Let us further assume that the thresholds correspond to the last state of the system from which a transition to next states is still possible. The adoption of such an assumption makes it possible to make the definition of the threshold system conditional on newly arriving calls (call stream) and on currently serviced calls (service stream). An example of a threshold system determined relative to the call stream is the MTS system whose MTM model is described in Sect. 2.2. Let us consider now a case of a threshold system determined relative to the service stream [10].

In line with the adopted definition, state $Q_{j,k}$ is the last state in which servicing of a call of class $j$ demanding $t_{j,k}$ BBUs can be terminated. This means that in state $Q_{j,k}$ there is still a transition to state $Q_{j,k} - t_{j,k}$ possible. Further on in the paper, the threshold system defined relative to the service stream will be called *Modified Multi-Threshold System* (MMTS), and its analytical model *Modified Multi-Threshold Model* (MSTM). Analysing the relation between MTM and MMTM models is easily noticeable [10]:

$$
Q'_{j,k} = Q_{j,k} - t_{j,k} - 1 \ , \tag{14}
$$

where $Q'_{j,k}$ is a threshold for the MTM model corresponding to the threshold $Q_{j,k}$ for the MMTM model. Equation (14) makes it possible to determine the characteristics of the MMTS system on the basis of the characteristics of the

MTS system (with threshold $Q'_{j,k}$), whose analytical model (MTM) is discussed in Sect 2.2. Further on in the paper the occupancy distribution for the MMTM model will be denoted by the symbol $[P_n]_{V,Q'}$.

## 2.4  Double Hysteresis Model (DHM)

Consider the proposed DHS system (Double Hysteresis System). Figure 1 presents the operation of such system. Thresholds $Q_1$ and $Q_3$ for the direction of changes in the load from low to high and thresholds $Q_2$ and $Q_4$ for the direction of changes in the load from high to low are introduced to the system. As long as the system is in the so-called first load area (below threshold $Q_1$ ), each call of class $j$ is serviced with the demanded number of $t_{j,0}$ BBUs. After an increase in the load and the passage of threshold $Q_1$, the system is in the second load area in which the CAC function assigns each call of class $j$ a lower number of BBUs, equal to $t_{j,1}$ ($t_{j,1} < t_{j,0}$). Whereas when the load of the system exceeds the threshold $Q_3$, the system will be in the so-called third load area, then the admission control function assigns to each call of class $j$ the lowest number of BBUs, equal to $t_{j,2}$ ($t_{j,2} < t_{j,1}$). A decrease in the load of the system and the instance of exceeding the threshold value $Q_4$ will cause the system to return to the second load area, as indicated by the leftward "1" and upward "2" arrows in Fig. 1. A new call of class $j$ will be serviced with an increased demanded number of $t_{j,1}$ BBUs. If the load of the system decreases even more and exceeds the threshold $Q_2$ (leftward "3" and upward "4" arrows), then the system will be in the first load area. A new call of class $j$ will be serviced with the initial number of demanded $t_{j,0}$ BBUs. With such an approach, the changes in the load of the system causing two hysteresis of the allocated resources of particular calls depend on the direction of changes in the load.

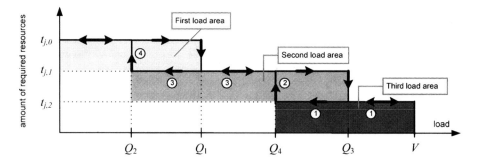

**Fig. 1.** Threshold system with double hysteresis – working idea

Let us assume theoretically that the class $j$ is serviced exclusively in the exemplary system and that calls receive the number of BBUs equal to $t_{j,0} = 4$ in the load area $<0, Q_1>$, $t_{j,1} = 2$ in the area $<Q_2, Q_3>$ and $t_{j,2} = 1$ in the area $<Q_4, V>$ to set up a connection. Figure 2 shows diagrams of the Markovian processes for a given call class that determine the decomposition of the service process in DHM into three service processes approximating the changes in the load of the system in the direction: from first to the second and then to the third load area (Fig. 2(a)), from the second to the first and then to the third load area (Fig. 2(b)), and in the direction: from the third to the second and to the first load area (Fig. 2(c)).

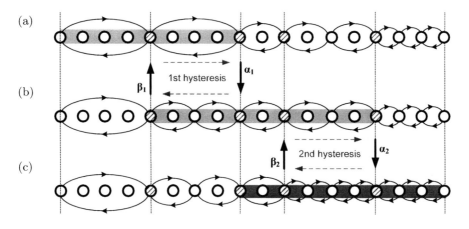

**Fig. 2.** Approximation of a two-threshold system with double hysteresis into three two-threshold systems for the direction: (a) from the first to the second and then to the third load area (i.e. low load – high load); (b) from the second to the first and then to the third load area; (c) from the third to the second and to the first load area (i.e. high load – low load)

Let us consider then two-threshold systems that correspond to the directions in load changes in the system. Let us assume that each of the systems services $M_1$ traffic classes that do not undergo the threshold mechanism, and $M_2$ traffic classes that undergo threshold mechanism. The occupancy distribution in the multi-threshold system $[P_n]_{V,Q}$ for the direction of changes: low load – high load (MTM, Fig. 2(a), can be thus described by the distribution (10) for each class $j$ ($1 \leq j \leq M_2$) the threshold $Q_{j,k}$ is equal to $Q_k$, where $k \in \{1,3\}$. Similarly, the occupancy distribution for the direction of the changes: high load – low load (MMTM, Fig. 2(c) can be expressed by the distribution (10) for each call of class $j$ ($1 \leq j \leq M_2$), the threshold $Q_{j,k}$ is equal to $Q'_k = Q_k - t_{j,k} - 1$, where $k \in \{2,4\}$. This distribution is designated by $[P_n]_{V,Q'}$. The occupancy distribution for the direction of the changes: form the second load area to the first load area and then to the third load area, is calculated on the basis of (10), where the thresholds are defined in the following way:

$$\forall_{j \in \{1,...,M_2\}} Q_{j,k} = \begin{cases} Q_k & \text{for } k = 3 , \\ Q'_k = Q_k - t_{j,k} - 1 & \text{for } k = 2 . \end{cases} \quad (15)$$

This distribution is designated by $[P_n]_{V,QQ'}$.

## 2.5  Switched Markov Process in DHM

Let us note that the distributions $[P_n]_{V,Q}$ and $[P_n]_{V,Q'}$ can be considered as certain boundary distributions for the occupancy distribution in a system with hysteresis. The real occupancy distribution in the DHS system is a complex matter. A passage of a threshold value, for instance $Q_1$ (from the first load area to the second load area), is followed by a transfer of serviced calls of class $j$ demanding $t_{j,0}$ BBUs to the first load area. The distribution $[P_n]_{V,Q'}$ does not take into consideration these "residual" calls in the second load area (Fig. 2). In a similar way, we can analyse the reverse situation, i.e. a transition of the threshold $Q_2$ and the transfer of calls to the low load area. In the distribution $[P_n]_{V,Q}$, the "residual" calls of class $j$ that occupy $t_{j,1}$ BBUs are not taken into consideration (Fig. 2).

Let us assume that the distributions $[P_n]_{V,Q}$, $[P_n]_{V,QQ'}$ and $[P_n]_{V,Q'}$ are approximate distributions that represent the execution of the service process in the DHS system with the preponderance of calls of class $j$ demanding $t_{j,0}$ BBUs, $t_{j,1}$ and $t_{j,2}$ BBUs to set up a connection. Thus, we can approximate the occupancy distributions $[H_n]_V$ in the DHM by a three-state Markovian switching process between the distributions $[P_n]_{V,Q}$, $[P_n]_{V,QQ'}$ and $[P_n]_{V,Q'}$:

$$[H_n]_V = P(1)\,[P_n]_{V,Q} + P(2)\,[P_n]_{V,QQ'} + P(3)\,[P_n]_{V,Q'} \quad , \tag{16}$$

where the probabilities $P(1)$, $P(2)$ and $P(3)$ are the probabilities of the execution of the service process in the DHM by respective service processes (see Fig. 2). These probabilities can be interpreted as the average time which the service process spends in a particular load areas of DHM.

The proposed model assumes that the parameters $P(1)$, $P(2)$ and $P(3)$ will be determined on the basis of the three-state Markovian process that switches service processes. This modulating process is presented in Fig. 3. The figure shows the parameters $\alpha_1$ , $\beta_1$ and $\alpha_2$ and $\beta_2$ which are the intensities of the switching between the considered processes. On the basis of the diagram shown

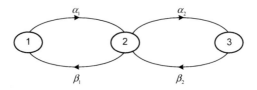

**Fig. 3.** Modulating Markov process in threshold system with double hysteresis

in Fig. 3, it is easy to determine the probabilities $P(1)$, $P(2)$ and $P(3)$:

$$P(1) = \frac{1}{1 + \frac{\alpha_1}{\beta_1} + \frac{\alpha_1 \alpha_2}{\beta_1 \beta_2}} \quad , \qquad P(2) = \frac{\alpha_1}{\beta_1} P(1) \ , \qquad P(3) = \frac{\alpha_1 \alpha_2}{\beta_1 \beta_2} P(1) \ . \tag{17}$$

The parameters $\alpha_1$ and $\beta_1$ in the proposed method are approximated by the streams that move the system from area $<0,Q_1>$ to area $<Q_2,Q_3>$ and by the streams that move the system from area $<Q_2,Q_3>$ to area $<0,Q_1>$. The parameters $\alpha_2$ and $\beta_2$ in the proposed method are respectively approximated by the streams that move the system from area $<Q_1,Q_3>$ to area $<Q_4,V>$ and by the streams that move the system from area $<Q_4,V>$ to area $<Q_2,Q_3>$. In the adopted approach, switching between appropriate service processes in the DHM is equated with switching between working areas of the DHS system, i.e. between the first $<0,Q_1>$ and the second $<Q_2,Q_3>$ load areas, and between the second $<Q_2,Q_3>$ and the third $<Q_4,V>$ load areas, respectively. Such an approach has been verified by numerous simulation experiments carried out by the authors for systems with different parameters [10].

The streams between load areas can be determined by the analysis of the transitions in MTM and MMTM models related to thresholds $Q_1$ and $Q_3$, respectively. A transition of MTM from state $n \leq Q_1$ to state $Q_3 \geq n > Q_1$ simultaneously determines a transition from the load area $<0,Q_1>$ to $<Q_2,Q_3>$. In turn, a transition from the occupancy area $<Q_2,Q_3>$ to the area $<0,Q_1>$ can be determined on the basis of the MMTM transitions from state $Q_3 \geq n \geq Q_2$ to state $n < Q_2$.

It can be noticed that the total stream that transfers the system from the area $<0,Q_1>$ to the area $<Q_2,Q_3>$, can be determined by the following equation[10]:

$$\alpha_1 = \sum_{n=Q_1-t_{\max}+1}^{Q_1} \left( \sum_{i=1}^{M_1} A_{i,0} t_{i,0} \varphi_{i,0}(n) + \sum_{j=1}^{M_2} A_{j,0} t_{j,0} \varphi_{j,0}(n) \right) . \tag{18}$$

The parameter $\varphi_{z,0}(n)$ is calculated in the following way:

$$\varphi_{z,0}(n) = \begin{cases} 1 & \text{for} \quad n > Q_1 - t_{z,0} , \\ 0 & \text{for} \quad n \leq Q_1 - t_{z,0} , \end{cases} \tag{19}$$

where $1 \leq z \leq M_1$ for calls that do not undergo hysteresis and $1 \leq z \leq M_2$ for calls that undergo hysteresis.

In Equations (18)–(19), only those streams that transfer the process to states $n$, such that $Q_3 > n > Q_2$, are taken into consideration. In state $Q_1$, all streams transfer the process to the area $<Q_2,Q_3>$, while in state $Q_1 - t_{\max} + 1$, where $t_{\max}$ is the number of BBUs demanded by a call that requires the highest number of BBUs.

Let us consider the transitions of the process from the occupancy area $<Q_2,Q_3>$ to the area $<0,Q_1>$. The appropriate outgoing stream from area $<Q_2,Q_3>$ can be expressed as follows [10]:

$$\beta_1 = \sum_{n=Q_2}^{Q_2+t_{\max}-1} \sum_{i=1}^{M_1} y_{i,0}(n) t_{i,0} \varphi_{i,0}(n) + \sum_{n=Q_2}^{Q_2+t_{\max}-1} \sum_{j=1}^{M_2} \sum_{k=0}^{1} y_{j,k}(n) t_{j,k} \varphi_{j,k}(n) , \tag{20}$$

where for calls of class $1 \leq i \leq M_1$:

$$\varphi_{i,0}(n) = \begin{cases} 1 & \text{for} \quad n < Q_2 + t_{i,0} , \\ 0 & \text{for} \quad n \geq Q_2 + t_{i,0} , \end{cases} \tag{21}$$

and for calls of class $1 \leq j \leq M_2$:

$$\varphi_{j,k}(n) = \begin{cases} 1 & \text{for} \quad n < Q_2 + t_{j,k} \ , \\ 0 & \text{for} \quad n \geq Q_2 + t_{j,k} \ . \end{cases} \tag{22}$$

The values of the service streams $y_{i,0}(n)$ and $y_{j,k}(n)$ in Equation (20) can be determined on the basis of the dependencies (7) and (13). The parameters determined by Equations (21)–(22) represent the service streams that move the process to such states $n$ for which $n < Q_2$, whereby in state $Q_2$ all streams transfer the process to the area $<0,Q_1>$, while in state $Q_2 + t_{\max} - 1$, where $t_{\max}$ is the number of BBUs demanded by a call that requires the highest number of BBUs.

The values of the parameters $\alpha_2$ and $\beta_2$ can be determined on the basis of Equations (18)–(22), where the thresholds $Q_1$ and $Q_2$ will be replaced by the threshold $Q_4$ and $Q_3$, respectively. Additionally in Equation (20), the parameter $k$ should be modified: $1 \leq k \leq 2$.

Having determined the probabilities $P(1)$, $P(2)$ and $P(3)$, we can finally determine the occupancy distribution $[H_n]_V$ in the threshold system with hysteresis on the basis of Equation (16).

## 3   The Application of DHM for the Radio Interface

Let us discuss a possibility of introducing DHS to the radio interface in the UMTS network. The Wideband Code Division Multiple Access (WCDMA) radio interface is an important element of the UMTS Terrestrial Radio Access Network (UTRAN) presented in Fig. 4. The following notation has been adopted in Fig. 4: RNC is the Radio Network Controller, WCDMA is a radio interface, Iub is the interface connecting Node B and RNC [15,16].

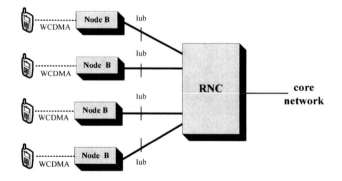

**Fig. 4.** Elements of the UMTS Terrestrial Radio Access Network [17]

In the planning process of the UMTS network, an appropriate dimensioning of the the WCDMA radio interface between the user and Node B, and the Iub connections between Node B and the RNC, has a particular significance. In the paper we will concentrate on the radio interface. In this section we attempt to estimate a possible influence of DHS on the operation of the WCDMA radio interface in the UMTS network.

## 3.1    Radio Interface in the UMTS Network

The capacity of the WCDMA radio interface in cellular systems is seriously limited due to the occurrence of different types of interference [16]. Therefore, the WCDMA interface is called the noise limited interface and this property is also described as the so-called *soft capacity* of the radio interface.

Accurate signal reception is possible only when the ratio of energy per bit $E_b$ to noise spectral density $N_0$ is appropriate. A too low value of $E_b/N_0$ will cause the receiver to be unable to decode the received signal, while a too high value of the energy per bit relative to the noise spectral density will be perceived as interference for other users of the same radio channel. The relation $E_b/N_0$ for a user of class $s$ call can be calculated as follows [15]:

$$\left(\frac{E_b}{N_0}\right)_s = \frac{W}{\nu_s R_s} \frac{P_s}{I_{\text{total}} - P_s} \ ,$$

(23)

where: $P_s$ – signal power received from a user of the class $s$ connection, $W$ – chip rate of spreading signal, $\nu_s$ – activity factor of a user of class $s$, $R_s$ – bit rate of a user of class $s$, $I_{\text{total}}$ – total received wideband power, including thermal noise power.

The mean power of a user of the class $s$ connection can be expressed by the following equation:

$$P_s = L_s I_{\text{total}} \ ,$$

(24)

where $L_s$ is the so-called load factor for a user of the class $i$ connection:

$$L_s = \left(1 + \frac{W}{\left(\frac{E_b}{N_0}\right)_s R_s \nu_s}\right)^{-1} \ .$$

(25)

The parameter $L_s$ is non-dimensional and is expressed in the ratio of the occupancy of the radio interface.

Sample values $E_b/N_0$ for different traffic classes and corresponding values of the load factor $L_s$, with the Equation (25) taken into account, are shown in Table 1.

Knowing the load factors for a single traffic source, it is possible to determine the load $\eta_{\text{UL}}$ for the uplink direction of the radio interface in UMTS systems [15,16]:

$$\eta_{\text{UL}} = (1 + \delta) \sum_{s=1}^{M} N_s L_s \ ,$$

(26)

where $N_s$ is the number of serviced traffic sources of class $s$, and $\delta$ is the average value of the ratio of the interference from other cells to the own interference for a given cell.

**Table 1.** Examples of $E_b/N_0$, $\nu_s$ and $L_s$ for different classes of calls [18]

| Class of traffic $(s)$ | Speech (class 1) | Video (class 2) | Data 1 (class 3) | Data 2 (class 4) |
|---|---|---|---|---|
| $W$ [Mcps] | 3.84 | | | |
| $R_s$ [kbit/s] | 12.2 | 64 | 144 | 384 |
| $\nu_s$ | 0.67 | 1 | 1 | 1 |
| $(E_b/N_0)_s$ [dB] | 4 | 2 | 1.5 | 1 |
| $L_s$ | 0.0053 | 0.0257 | 0.0503 | 0.1118 |

The total load for the downlink direction of the WCDMA radio interface can be written in the following way [15,16]:

$$\eta_{\text{DL}} = \sum_{s=1}^{M} N_s L_s \left(1 - \xi + \delta\right) , \tag{27}$$

where $\xi$ is the orthogonality factor. It indicates the degree of interference reduction between the users of the same cell through the application of channel codes based on the OVSF (*Orthogonal Variable Spreading Factor*) [16].

### 3.2 Numerical Study

In order to confirm the proposed method of blocking probability calculation in a cell with the WCDMA radio interface in the uplink direction, the results of the analytical calculations were compared with the results of the simulation experiments. In the simulation, 95% confidence intervals were calculated after the $t$-Student distribution that are almost included within the marks plotted in the figures.

The study was carried out for users demanding a set of classes described in Table 1. In the exemplary system it was assumed that the threshold mechanism with double hysteresis was applied to the second traffic class. In all presented example it was also assumed that:

- $L_{\text{BBU}}$ is equal to $10^{-4}$,
- the number of BBUs required by calls of particular classes (Equation (2)):
  $t_{1,0} = \frac{L_{1,0}}{L_{\text{BBU}}} = 53$ BBUs,
  $t_{2,0} = \frac{L_{2,0}}{L_{\text{BBU}}} = 257$ BBUs, $t_{2,1} = 129$ BBUs, $t_{2,2} = 64$ BBUs,
  $t_{3,0} = \frac{L_{3,0}}{L_{\text{BBU}}} = 503$ BBUs, $t_{4,0} = \frac{L_{4,0}}{L_{\text{BBU}}} = 1118$ BBUs.
- traffic of particular classes is offered to the system in the following exemplary proportions: $A_{1,0}t_{1,0} : A_{2,0}t_{2,0} : A_{3,0}t_{3,0} : A_{4,0}t_{4,0} = 1 : 1 : 1 : 1$,
- the maximum theoretical load capacity in uplink direction $V_{\text{phy}}^{\text{UL}}$ is non-dimensional and equal to 1,
- $\delta$ is assumed to be equal to 0.25 and the maximum uplink capacity of the WCDMA radio interface is equal: $V = \lfloor 1/1.25 \cdot 10^{-4} \rfloor = 8000$ BBUs,
- the hysteresis thresholds are assumed to be equal to, respectively: $Q_1 = 50\%$ (4000 BBUs), $Q_2 = 35\%$ (2800 BBUs), $Q_3 = 85\%$ (6800 BBUs) and $Q_4 = 70\%$ (5600 BBUs) of the radio interface capacity.

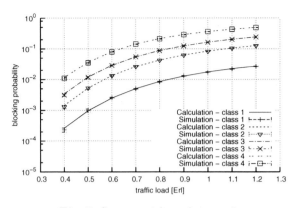

**Fig. 5.** System without hysteresis

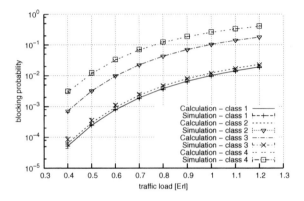

**Fig. 6.** System with double hysteresis for the second class of traffic

Figure 5 presents the calculation and the simulation results for the radio interface without hysteresis. Figure 6 shows the exemplary system in which double hysteresis was introduced for the second class. All figures present results in relation to traffic offered per BBU in a particular system.

Comparing the results in Figs. 5 and 6 it is observable that the introduction of the threshold mechanism with double hysteresis is followed by a decrease in the blocking probability for all traffic classes. This phenomenon is related to a decrease in the number of BBUs allocated to calls that undergo hysteresis (after exceeding the thresholds $Q_1$ and $Q_3$). Such a diminished number of BBUs is allocated to calls until the thresholds $Q_2$ and $Q_4$ are reached, respectively.

All the presented results show the robustness of the proposed method of blocking probability calculation. In each case, regardless of the offered traffic load, the results are characterized by fair accuracy. It is worthwhile to emphasize that the accuracy of the proposed calculation method is independent of the load of the system, the total number of traffic classes serviced in the system and, additionally, does not depend on the number of classes which undergo the threshold mechanism with double hysteresis.

For the purpose of a more complex evaluation of the WCDMA system, a special purpose-made simulator was worked out. The simulator was implemented in C++ language and it used the process interaction simulation method [19]. The simulation model, however, does not take into consideration many technological properties of the UMTS system, such as the propagation model of the radio channel or the mobility of users. The devised simulator is intended for modelling the traffic capacity of the system, therefore for modelling of a system at the so-called call level [11], and at this level technological parameters have no significant influence upon the precision of mapping of a modelled system [20].

## 4    Conclusion

This paper proposes a new analytical model that makes it possible to determine traffic characteristics (such as the occupancy distribution and the blocking probability) of a system in which the so-called threshold mechanism with double hysteresis is used for traffic management. The paper presents an analytical model that corresponds to a threshold system in which two pairs of thresholds have been introduced – called the double hysteresis system (DHS). The threshold mechanism in the DHS system allows the blocking probability to be decreased for particular traffic classes and leads to a reduction in fluctuations in the load of the system during transitions between different load areas.

The paper presents a possibility of the application of the mechanism with double hysteresis for traffic control in the UMTS network. For this purpose, a special purpose-made simulator for call service simulation in the WCDMA radio interface was constructed. All the presented simulation experiments for the WCDMA radio interface in the UMTS network confirm good accuracy of the proposed analytical model. Summing up, the double hysteresis mechanism can be used in the call admission control function in communications and cellular networks.

The proposed DHM was validated under the assumption that all call streams had Poisson distribution (Erlang traffic), but this can be easily generalized for BPP (Binomial-Poisson-Pascal) traffic using the methods described in [4].

## References

1. Roberts, J.: Teletraffic models for the Telcom 1 integrated services network. In: Proceedings of 10th International Teletraffic Congress, Montreal (1983)
2. Rácz, S., Gerö, B.P., Fodor, G.: Flow level performance analysis of a multi-service system supporting elastic and adaptive services. Journal of Performance Evaluation 49(1-4), 451–469 (2002)
3. Kaufman, J.: Blocking with retrials in a completly shared recource environment. Journal of Performance Evaluation 15, 99–113 (1992)
4. Stasiak, M., Głąbowski, M., Wiśniewski, A., Zwierzykowski, P.: Modeling and dimensioning of mobile networks: from GSM to LTE. John Wiley and Sons, Ltd., Chichester (2011)

5. Stasiak, M., Zwierzykowski, P., Parniewicz, D.: Modelling of the WCDMA interface in the UMTS network with soft handoff mechanism. In: Proceedings of IEEE Global Communications Conference, GLOBECOM, Honolulu, Hawaii, USA (2009)
6. Stasiak, M., Wiewióra, J., Zwierzykowski, P., Parniewicz, D.: Analytical Model of Traffic Compression in the UMTS Network. In: Bradley, J.T. (ed.) EPEW 2009. LNCS, vol. 5652, pp. 79–93. Springer, Heidelberg (2009)
7. Moscholios, I., Logothetis, M., Kokkinakis, G.: Connection-dependent threshold model: a generalization of the Erlang multiple rate loss model. Journal of Performance Evaluation 48, 177–200 (2002)
8. Vassilakis, V.G., Moscholios, I.D., Logothetis, M.D.: The extended connection-dependent threshold model for elastic and adaptive traffic. In: 5th International Symposium on Communication Systems, Networks and Digital Signal Processing, Patras, Greece, pp. 42–45 (2006)
9. Głąbowski, M., Kaliszan, A., Stasiak, M.: Modeling product-form state-dependent systems with bpp traffic. Journal of Performance Evaluation 67(3), 174–197 (2010)
10. Sobieraj, M., Stasiak, M., Weissenberg, J., Zwierzykowski, P.: Analytical model of the single threshold mechanism with hysteresis for multi-service networks. IEICE Transactions 95-B(1), 120–132 (2012)
11. Roberts, J. (ed.): Performance Evaluation and Design of Multiservice Networks, Final Report COST 224. Commission of the European Communities, Brussels (1992)
12. Kaufman, J.: Blocking in a shared resource environment. IEEE Transactions on Communications 29(10), 1474–1481 (1981)
13. Roberts, J.: A service system with heterogeneous user requirements – application to multi-service telecommunications systems. In: Pujolle, G. (ed.) Proceedings of Performance of Data Communications Systems and their Applications, pp. 423–431. North Holland, Amsterdam (1981)
14. Głąbowski, M.: Modelling of state-dependent multi-rate systems carrying BPP traffic. Annals of Telecommunications 63(7-8), 393–407 (2008), doi:10.1007/s12243-008-0034-5
15. Holma, H., Toskala, A.: WCDMA for UMTS. Radio Access For Third Generation Mobile Communications. John Wiley and Sons, Ltd. (2000)
16. Laiho, J., Wacker, A., Novosad, T.: Radio Network Planning and Optimization for UMTS. John Wiley and Sons, Ltd., Chichester (2006)
17. Parniewicz, D., Stasiak, M., Zwierzykowski, P.: Multicast Connections in Mobile Networks with Embedded Threshold Mechanism. In: Kwiecień, A., Gaj, P., Stera, P. (eds.) CN 2011. CCIS, vol. 160, pp. 407–416. Springer, Heidelberg (2011)
18. Stasiak, M., Wiśniewski, A., Zwierzykowski, P., Głąbowski, M.: Blocking probability calculation for cellular systems with WCDMA radio interface servicing PCT1 and PCT2 multirate traffic. IEICE Transactions on Communications E92-B(4), 1156–1165 (2009)
19. Tyszer, J.: Object-Oriented Computer Simulation Of Discrete-Event Systems. Kluwer Academic Publishers Group (1999)
20. Wiśniewski, A.: Modelling of the mobile systems with WCDMA radio interface. PhD thesis, Poznan University of Technology, Faculty of Electronics and Telecommunications, Poznan, Poland (2009)

# Modeling of Energy Consumption for Mobile Wireless Ad Hoc and Sensor Networks

Jerzy Martyna

Institute of Computer Science, Faculty of Mathematics and Computer Science
Jagiellonian University, ul. Prof. S. Łojasiewicza 6, 30-348 Cracow, Poland

**Abstract.** In this paper, we study energy consumption of mobile two-tier wireless ad hoc and sensor networks in comparison with flat mobile wireless ad hoc and sensor networks. Based on idealized wireless network model, we study the total energy expenditure of both these architectures. We provide general formulas for the trajectory of the movement of the mobile nodes in two-tier mobile wireless ad hoc and sensor networks. Finally, we investigate the case with multiple mobile nodes both with their cooperation and without. Simulation results are presented for commonly used network architectures.

**Keywords:** ad hoc networks, wireless sensor networks, energy consumption.

## 1 Introduction

Wireless ad hoc networks [1,2] are decentralized, self-organized networks capable of forming a communication network without any fixed infrastructure. In these networks all nodes are equipped with a radio transmitter and receiver which allow it to communicate with other nodes over wireless channels. Ad hoc networks allow for a multi-hop transmission of data between nodes outside the direct radio links.

Wireless sensor networks [3] are a particular type of ad hoc networks [4], in which the nodes are equipped with advanced sensing functionalities (acoustic, pressure, thermal, etc.), a small processor, and a short-range wireless transceiver. Wireless sensor networks are used for building a global view of the monitored region, which is accessible to the external user(s) from outside through one or more gateway nodes.

Node mobility is a prominent feature of ad hoc and sensor networks. Mobile ad hoc networks (MANETs) [5] are driven mainly by their ability to provide instant wireless networking solutions in situations where infrastructures do not exist. It is obvious that that these networks are more robust than their wired or cellular infrastructures. While the deployment of MANETs is yet to come, extensive research efforts are being currently undertaken to provide a high throughput and low-energy wireless access to resources of such networks.

The sensor network with mobile agents was at first proposed by Tong et al. [6]. In the paper by W. Zhao and M.Ammar [7], the concept of using mobile

A. Kwiecień, P. Gaj, and P. Stera (Eds.): CN 2012, CCIS 291, pp. 314–323, 2012.
© Springer-Verlag Berlin Heidelberg 2012

nodes for message ferrying is considered, where the objective is to use mobiles to provide nonrandom proactive routers. Many different approaches and protocols have been presented by Basagni et al. [5], the authors of a valuable textbook. An analysis of energy consumption for ad hoc wireless networks using a bit-meter-per-oule metric was suggested by Gao [8]. Currently, an analytical result on the average number of hops between any two nodes if the transmission is successful and the average number of hops traversed by packets before being dropped if the transmission is unsuccessful was presented by Z. Zhang et al. [9]. The energy correlations between node core components, including processors, RF modules and sensors was given by Hai-Ying Zhou et al. [10].

Two layer networks, often called two-tier networks, belong to a typical form of wireless ad hoc and sensor networks. The first layer is formed by ordinary network nodes which are responsible for gathering the information from the monitored region. Some network nodes as belonging to a backbone network form a dominant set. To these nodes belong all clusterhead nodes which are treated as the second layer. They are natural places for an aggregated and compressed traffic converging from many ordinary nodes (sensors).

The main goal of this paper is to analyse energy of flat ad hoc and sensor networks in comparison with the two-tier ad hoc and sensor networks. On the other hand, we examine the dependencies between the total energy expenditure of these two architectures as well. Furthermore, we study the effect of the node density and mobility on energy consumption. It allows us to design the network parameters, such as the mobile nodes trajectory or radio coverage area.

The rest of the paper is organized as follows. Section 2 presents the background and the related work. Section 3 details the energy consumption analysis over flat mobile wireless ad hoc and sensor networks. In Section 4, we provide an analysis of energy consumption over two-tier mobile ad hoc and sensor networks. Section 5 presents some numerical results of energy consumption in mobile wireless ad hoc and sensor networks. In Section 6, we summarize the findings of this paper and present conclusions.

## 2    Background and Related Work

In our model, we assume that a certain non-zero minimum level of power will be radiated regardless of how short a radio link is. The total power required for communicating over distance $r$ is given by [11], namely

$$p(r) = \max\{p_{\min}, \beta r^{\alpha}\} + p_{\mathrm{rx}} \tag{1}$$

where $\alpha$ is the power index for the channel path loss in the field of the antenna, $p_{\min}$ is the minimum transmitter power, $\beta$ represents the minimum transmission power required to communicate at a reference near-field distance of 1 meter, $p_{\mathrm{rx}}$ is the fixed overhead for receiving data. We can follow the remark of Equation (1) as all the nodes are closer than $r_{\min} = (p_{\min}/\beta)^{\frac{1}{2}}$. Then, we obtain the power requirement as a constant at $p_{\min}$ watts.

We now apply the square region analysis to the node location and data generation. We assume that the sensor distribution is a two-dimensional Poisson Field with a constant mean and variance of $\lambda$ nodes/m$^2$. Therefore, the probability distribution for the number of nodes located within an area of $A$ m$^2$ is given by

$$P(N(A) = k) = \frac{(\lambda A)^k}{k!} e^{-\lambda A} . \tag{2}$$

Thus, within any given area the location of each node is a uniformly distributed two-dimensional random vector.

The assumed model allows us to formulate the dependence on the relative position between the phenomenon and the nodes itself. We model the volume of the node data generated by each node as $\mu(x, y)$ [bits], where $(x, y)$ are the location parameters of the node. Thus, the average traffic load is given by the expression $\rho = \lambda \cdot \mu(x, y)$ [bits/m$^2$].

We now determine the overall energy $E$ consumed in one time unit. We recall that this energy is composed of three parts, the first energy, $E_D$ – energy spent in data transmission, the second $E_{PHY}$ – energy required by the MAC physical layer for synchronization and channel acquisition, the third energy, $E_{TR}$ – associated with the transmission protocol activity. Thus, we can define [12] the three parameters of energy efficiency

$$\eta_D = \frac{E_D}{E} \qquad \eta_{PHY} = \frac{E_{PHY}}{E_D} \qquad \eta_{TR} = \frac{E_{TR}}{E_D} .$$

In this paper we will consider only the proactive networks. By proactive networks we mean that all the routes and links between the source and the destination are maintained regardless of the data traffic. The reactive networks, on the contrary, the route and the link are found when a message must be delivered to the destination. It is intuitive that the reactive networks are energy efficient when the traffic is light and inefficient when the traffic is heavy. Therefore, the proactive type of a network is more useful than the heavy load traffic. A comparison between the energy consumption in both types of networks was presented by Q. Zhao et al. in the paper [12].

## 3    Energy Consumption Analysis over Flat Mobile Wireless Ad Hoc and Sensor Networks

In this section, we will analyze energy consumption over flat wireless mobile ad hoc and sensor networks. We will compute the energy consumption measured by means of the power-per-meter metrics.

In the analysis, it is assumed that $N$ nodes are deployed in a random and uniform way on a disk of radius $R$. The node density $\rho$ is given by $N/\pi R^2$. In order to compute the consumed energy during the transmission of one packet originated at a randomly chosen node to a gateway node located at the center of the disk in a flat ad hoc/ sensor network, we assume that nodes are capable of adjusting the transmission power in order to cover the neighbourhood of radius $r$.

In order to simplify our model we assume that a minimum-energy route between the chosen node and the center of the disk has been established at no cost.
Every transmission is successful and all transmissions are without a significant
overhead in energy consumption. Moreover, we assume that nodes are in the
sleep state by turning off most of their transceiver circuits. They are woken up
by the signal from the mobile node or by the event, but not the transmission of
other nodes. In our analysis, we ignore the energy consumed by the node in the
sleeping state.

We consider the radio model following the methodology of W. Heinzelman
et al. [13], which analyzed energy-efficient communication protocols for wireless
sensor networks. When a node is receiving one packet, it consumes $E_{rx}$ Joule/bit.
The transmission of one packet to a neighbouring node in distance $r$ consumes
$E_{tx}(r)$ Joule/bit. From the paper by E. Shih et al. [11] the value of this transmission energy is given by

$$E_{tx}(r) = e_{tx} + \max\{e_{min}, e_{out} \cdot r^{\alpha}\} \tag{3}$$

where $e_{tx}$ is the energy consumed by the transmitter circuits, $e_{min}$ is the antenna
output energy to reach the destination node with an acceptable SNR, $e_{out}$ is the
minimum energy radiated regardless of the transmission range.

Let $X$ be the distance from the chosen node to the gateway node in the flat
mobile ad hoc/sensor network. Thus, the probability density function of $x$ is
given by

$$p_X(x) = \frac{2}{R^2}x, \quad 0 < x \leq R . \tag{4}$$

The total energy $E_{AdHoc/WSN}(r)$ consumed by moving one packet to the gateway
node with an optimal transmission range $r$ is given by

$$E_{AdHoc/WSN}(r) = \min_{r \geq r_{min}} \int_0^R p_X(x)dx = \min_{r \geq r_{min}} \frac{2}{R^2} \int_0^R E(r)h(x,r)x dx \tag{5}$$

where $r_{min}$ is the minimum transmission range for guaranteeing network connectivity, $h(x,r)$ is the number of hops for a packet in order to reach the gateway
node, $x$ is the meters away.

Further, we can obtain for the two-dimensional flat static ad hoc/sensor network

$$E_{AdHoc/WSN}^{stat} = \min_{r \geq r_{min}} \frac{2R}{3r} \left( E_{tx}(r) + (N-1)\frac{r^2}{R^2}E_{rx} \right)$$

$$= \begin{cases} O(\sqrt{N \log N}) & , \rho \uparrow, r_0 = 0 \\ O(N) & , \rho \uparrow, r_0 > 0 \\ O(\sqrt{N(\log N)^{\alpha-1}}) & , R \uparrow \end{cases} \tag{6}$$

In the two-dimensional network a node has on average $(N-1)\frac{r^2}{R^2}$ neighbours
who listen to that node transmission. Thus, the energy consumed in one hop for

a normally dense network is given by

$$E_{\text{AdHoc/WSN}}^{\text{stat}}(r) = E_{\text{tx}}(r) + (N-1)\frac{r^2}{R^2}E_{\text{rx}} \tag{7}$$

where $R$ is the radius of a disk, $E_{\text{tx}}$ is the energy that covers neighbourhood of radius $r$.

In case of mobile network, we assume the simplest model of mobility, namely the so-called Random Waypoint (RWP) model [14]. Thus, we have the minimum transmission range for guaranteeing network connectivity egual to

$$r_{\min} = \max\left\{r_0, \frac{1}{P_{\text{pause}}}\sqrt{\frac{\log N}{\pi N}}\right\} \tag{8}$$

where $P_{\text{pause}}$ is equal to $\frac{t_{\text{p}}}{t_{\text{p}}+\frac{0.521405}{v}}$, $t_{\text{p}}$ is the pause time and $v$ is the velocity of node.

Thus, the total energy $E_{\text{AdHoc/WSN}}(r)$ consumed by moving one packet packet to the gateway node for the two-dimensional flat mobile ad hoc/sensor network is given by

$$E_{\text{AdHoc/WSN}}^{\text{mob}} = E_{\text{tx}}(r) + (N-1)\frac{\left(t_{\text{p}} + \frac{0.521405}{v}\right)^2}{t_{\text{p}}^2 R^2} \cdot E_{\text{rx}} \ . \tag{9}$$

# 4    Energy Consumption Analysis over Two-Tier Multi-hop Mobile Wireless Ad Hoc and Sensor Networks

In this section, we investigate the energy consumption over a two-tier multihop mobile ad hoc and sensor networks.

We assume that our multi-hop wireless ad hoc/sensor network is two-tier. The first tier is formed by the ordinary nodes, and the second is built by the clusterhead nodes. We analyze the following scenario: the first one, in which the clusterhead nodes are static, and the second one with mobile clusterhead nodes.

## 4.1    Static Clusterhead Nodes without Overlapping the Clusters and an Immobile Gateway

The energy consumption in a packet transmission from an ordinary node to the clusterhead, for instance, with a perfect wake-up scheme, is given by

$$E_{\text{nc}}^{\text{stat}} = \min_{r_{\text{c}} \geq r_{\min}} h\left(E_{\text{tx}} + \max\left\{e_{\min}, e_{\text{out}}r^2\right\} + E_{\text{rx}}\right) \cdot \frac{E[x]}{r_{\text{c}}} \tag{10}$$

where $r_{\text{c}}$ is the radius of the cluster, $E_{\text{tx}}$ is the energy consumed by the radio transmitter, $e_{\min}$ the minimum energy radiated regardless of the transmission, $e_{\text{out}}$ is the antenna output energy in order to reach, $E_{\text{rx}}$ is the receiving energy,

$\frac{E(x)}{r_c}$ is a constant independent of the network size, $h$ is equal to 1 or 2 and means the number of hops in the cluster.

We assume the existence of a two-tier network in which all packets are provided to the gateway. Thus, we can compute the average energy needed for data receiving in this network. This energy is equal to

$$E^{2-\text{tier}}_{\text{AdHoc/WSN}} = M \cdot \bar{n}_c \cdot E^{\text{stat}}_{\text{nc}}(N_c - 1) + M \cdot \xi \cdot E^{\text{stat}}_{\text{nc}}(N_c - 1) \qquad (11)$$

where $M$ is the number of packets generated by the ordinary nodes in the cluster, $\bar{n}_c$ is the mean number of clusters used for transmission in the network, $N_c$ is the number of clusterhead nodes, $\xi$ is the loss coefficient of energy associated with data aggregation.

## 4.2   The Static Clusterhead Nodes and One Mobile Gateway

In this case we assume that the mobile gateway can collect the data only from the clusterhead nodes. Each of the clusterhead nodes consumes the energy with the perfect wake-up scheme, in such way as in the previous case. Thus, the consumed energy is equal to

$$E^{2-\text{tier}}_{\text{stat}} = E^{\text{stat}}_{\text{nc}} + E^{\text{stat}}_{\text{m}} \qquad (12)$$

where $E_{\text{m}}$ is the energy needed for transmission all the packets from the clusterhead nodes to the mobile gateway. This energy is given by

$$E^{\text{stat}}_{\text{m}} = \frac{r^2}{R^2} \cdot N_c \cdot E_{\text{rx}} + E_{\text{tx}}(d) \qquad (13)$$

where the first concerns the energy consumed by all the clusterhead nodes in the coverage area, $N_c$ is the number of clusterhead nodes, $r$ is the radius of coverage of the mobile node, $R$ is the radius of the disk, $E_{\text{rx}}$ is the energy consumed in the reception of packet, $E_{\text{tx}}$ is the energy needed for packet transmission, $d$ is the distance between the clusterhead node and the gateway node.

## 4.3   The Two-Tier Network and Mobile Gateways without Cooperation

In this case two or more gateways without overlapping are considered. Similarly as in the previous cases, we assumed the perfect wake-up scheme for all the nodes in the network. We will first analyze the energy consumption of the batteries in a two-tier network mode. Let one of the nodes located in the area of the cluster transmit its data to the cluster. This transmission needs one or two hops. It means that the total energy consumed by the data transfering from the cluster area to the clusterhead node demands the minimal value of energy, namely

$$E_{\text{m}} = \min_{y \geq y_0} \frac{y^2}{R^2} N_c E_{\text{rx}} + E_{\text{tx}}(H)$$

$$= \begin{cases} O(1), & \rho^\uparrow, \ y_0 = 0 \\ O(N), & \rho^\uparrow, \ y_0 > 0 \\ O(1), & R^\uparrow, \end{cases} . \qquad (14)$$

Further, we added the energy consumed by the mobile nodes and the clusterhead nodes. This energy is given by

$$E_{\text{stat,noncoop}}^{2-\text{tier}} = n_{\text{c}} \cdot E_{\text{clust}} + n_{\text{m}} \cdot E_{\text{m}} \tag{15}$$

where $n_{\text{m}}$ is the number of the mobile nodes, $n_{\text{c}}$ is the number of cluster in the network.

### 4.4   The Two-Tier Network and Mobile Gateways with Cooperation

In this case, we examine the situation where the mobile nodes have access to each other's signals. It means that the mobile nodes cooperate in the data reception. Two cooperative mobile nodes with an overlap can be described by two parameters $c_1$ and $c_2$, where $c_1$ and $c_2$ are the numbers of clusterhead nodes which are visible by the first and second mobile nodes, respectively. Thus, the number of the overlapped clusterhead nodes is equal $M - (c_1 + c_2)$, where $M$ is the total number of all clusterheads in the network. In general, this number is equal to $M - \sum_{i=1}^{N} c_i$, where $N$ is the number of the cooperating mobile nodes. The consumed energy in this case is given by

$$E_{\text{stat,coop}}^{2-\text{tier}} = \left( M - \sum_{i=1}^{N} c_i \right) E_{\text{clust}} + \sum_{i=1}^{N} c_i \cdot E_m^{\text{coop}} \tag{16}$$

where $E_m^{\text{coop}}$ is the energy consumed by the cooperating mobile nodes.

## 5   Trajectory Planning of Mobile Nodes

One of the most important metrics in the data collection in the wireless mobile ad hoc and sensor networks is the amount of the gathered data. The trajectories and coverage areas of the mobile nodes decide about the value of this metric.

We propose two trajectories for one mobile node. The first of them (see Fig. 1(a)) is devoted to the flat ad hoc immobile and mobile network. The coverage of the mobile node must be adjusted to the distance between two neighbouring branches in this trajectory. The mobile node has to scan area $L/2r$, where $L$ is the width of the network field, $r$ is the radius of the mobile node. The total travel time in this network is equal to $L^2/2r$, assuming that the travel time along the $y$ axis is about $L/v$, where $v$ is the speed of the mobile node.

In the second case we propose the trajectory given in Fig. 1(b). This trajectory guarantees that the mobile nodes scan the data from all of the clusterhead nodes. The total travel time for this trajectory is given by $\sqrt{2} \cdot L/v$, where $v$ is the speed of the mobile node.

## 6   Numerical Results

We simulate an ad hoc and sensor network in a square area of $1000\,\text{m} \times 1000\,\text{m}$. In the flat multihop case, the gateway node is located at the center, (500, 500).

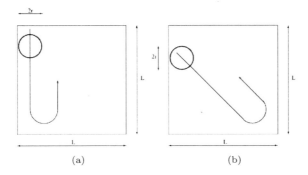

**Fig. 1.** Trajectories of mobile node (a) in flat multi-hop and (b) two-tier ad hoc/wireless sensor networks, respectively

Thus, the data are collected over the shortest multihop routes to the gateway node. In the mobility case we assumed that one mobile nodes are traveling across the deployment region. For the flat multihop wireless networks we assumed the trajectory given in Fig. 2(a). In the experiment we measured the number of packets transmitted by all the nodes in the network in order to achieve this goal. The simulation was repeated for 200 random topologies and averaged. Fig. 2(b) shows the total number of packets transmitted with the increasing density of nodes and the fixed transmission range of 50 meters and 30 meters for ad hoc and the sensor network, respectively. The unit of node density is the number of nodes within the radio coverage area of a node, i.e., number of nodes in area $\pi r^2$.

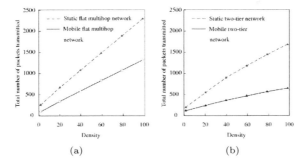

**Fig. 2.** Total number of packets transmissions required to collect one packet from each node for (a) flat multihop and (b) two-tier ad hoc/wireless sensor networks, respectively

Figures 3(a) and 3(b) present the average energy consumption per packet as the transmission radius increases for a constant number of nodes which are fixed as 50 for the ad hoc network and 100 for the static (and mobile) flat and static (and mobile) two-tier network, respectively. For both architectures we computed the total energy consumed in network in order to collect one packet

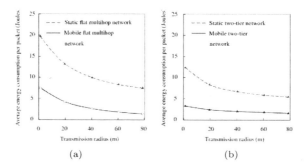

**Fig. 3.** Average energy consumption per packet for (a) flat multihop and (b) two-tier ad hoc/wireless sensor networks, respectively

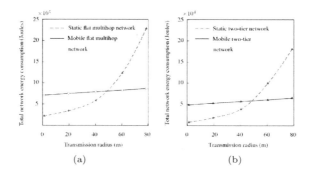

**Fig. 4.** Total network energy consumption for (a) flat multihop and (b) two-tier ad hoc/wireless sensor networks, respectively

from each node in the same circumstances as previous - see Fig. 4(a) and 4(b). We can seen that with a smaller transmission radius, the number of packets transmitted by the network decreases when the mobile nodes are used. We note also that when the two-tier network is used, all nodes are within one or two hops from the clusterhead nodes. It causes smaller energy consumption in comparison with the flat architecture of the ad hoc or the sensor network. Also the energy consumption in all the mobile networks is smaller than in the immobile networks.

## 7    Conclusion

We studied the architectures for mobile wireless ad hoc and sensor networks. We first investigated their energy consumption in the flat structure, and compared it with that of the two-tier architecture. We proved that the two-tier architecture of these networks achieves significant energy savings by means of shifting the responsibility of routing from the ordinary nodes to the clusterhead and mobile nodes. It allows us to find the parameters of the trajectory of the mobile nodes. Additionally, we also analyzed the case with multiple mobile nodes which are cooperative or noncooperative in data reception.

It raises a question: whether it is realistic to consider the transmission between the mobile nodes which are ground-based or the low-altitude aerial and immobile nodes. New technologies used in these networks proved that this concept can be employed in many situations and environments. The key to the implementation of these networks lies in the adjustment the power efficiency of the nodes to the distance and speed of the nodes which would normally be reachable.

# References

1. Murthy, C.S.R., Manoj, B.: Ad Hoc Wireless Networks: Architectures and Protocols. Prentice Hall PTR (May 2004)
2. Perkins, C.E.: Ad Hoc Networking. Addison-Wesley Professional (December 2000)
3. Akyildiz, I.F., Su, W., Sankarasubramaniam, Y., Cayirci, E.: Wireless Sensor Networks: A Survey. Computer Networks 38, 393–422 (2002)
4. Santi, P.: Topology Control in Wireless Ad Hoc and Sensor Networks. John Wiley and Sons, Hoboken (2006)
5. Basagni, S., Conti, M., Giordano, S., Stojmenovic, I.: Mobile Ad Hoc Networking. IEEE Press/Wiley Interscience (2004)
6. Tong, L., Zhao, Q., Adireddy, S.: Sensor Networks with Mobile Agents. In: Proc. Int. Symp. Mil. Commun., Boston, MA, vol. 1, pp. 688–693 (October 2003)
7. Zhao, Q., Tong, L.: Distributed Opportunistic Transmission for Wireless Sensor Network. In: Proc. ICASSP, vol. 3, pp. 833–836 (May 2004)
8. Gao, J.: Analysis of Energy Consumption for Ad Hoc Wireless Sensor Networks Using a Bit-meter-per-Joule Metric IPN. Prog. Rep. 42–150 (August 2002), http://ipnpr.jpl.nasa.gov/progress_report/42-150/150L.pdf
9. Zhang, Z., Mao, G., Anderson, B.D.O.: On the Effective Energy Consumption in Wireless Sensor Networks. In: Proc. of the Wireless Communication and Networking Conference, WCNC (2010)
10. Zhou, H.-Y., Lun, D.-Y., Gao, Y., Zun, D.-C.: Modeling of Node Energy Consumption for Wireless Sensor networks. Wireless Sensor Networks 3, 10–23 (2011)
11. Shih, E., Cho, S., Ickes, N., Min, R., Sinha, A., Wang, A., Chandrakasan, A.: Physical Layer Driven Protocol and Algorithm Design For Energy-Efficient Wireless Sensor Networks. In: Proc. ACM MobiCom, Rome, Italy, pp. 271–286 (July 2001)
12. Zhao, Q., Tong, L.: Energy Efficiency of Large-Scale Wireless Networks: Proactive Versus Reactive Networking. IEEE Journal on Selected Areas in Communications 23(5), 1100–1112 (2005)
13. Heinzelman, W.B., Chandrakasan, A., Balakrishnan, H.: Energy-Efficient Communication Protocols for Wireless Microsensor Networks. In: Proc. 33rd Hawaii Int. Conf. Syst. Sci., Maui, HI, pp. 1–10 (January 2000)
14. Johnson, D., Maltz, D.: Dynamic Source Routing in Ad Hoc Wireless Networks, Mobile Computing, pp. 153–181. Kluwer Academic Publishers

# Simulation Study of the Mobility Models for the Wireless Mobile Ad Hoc and Sensor Networks

Jerzy Martyna

Institute of Computer Science, Faculty of Mathematics and Computer Science
Jagiellonian University, ul. Prof. S. Lojasiewicza 6, 30-348 Cracow, Poland

**Abstract.** One of the most important methods for evaluating the performance of ad hoc and wireless sensor networks is through the use of simulation. Simulation provides a number of significant parameters and metrics of wireless networks. Additionally, simulation allows the effects of some parameters, such as mobility, data traffic or transmission range, to be studied in detail, while other metrics are held constant. This paper provides a set of well understood mobility models that can be used to simulate mobile ad hoc and sensor networks with independent node movements. Simulation results show that the use of pre-defined pathways has a significant impact on the performance of mobile ad hoc and sensor networks.

**Keywords:** mobility models for wireless networks, ad hoc networks, wireless sensor networks.

## 1 Introduction

Wireless self-organizing networks consisted of mobile nodes and requiring no fixed infrastructure will be to as *mobile ad hoc networks* (MANETs). These networks are characterized by dynamic topology. In other words, nodes can joint or leave the network as well as being able to change the range of their transmission. Furthermore, because of their distributed nature, MANETs are more robust than their cellular counterparts against single-point failure. While the notions of distributed nature can be formalized, the notion of robustness to mobility is still vague. Consequently, existensive research efforts are currently underway to enhance the operation and interconnection of wireless devices of such networks [1,2].

Mobile sensor networks are composed of a distributed collection of nodes, each of which in addition to sensing, computation and communication also has locomotion capabilities. The use of mobile sensor networks includes applications ranging from urban combat scenarios and emergency environment monitoring. Data from the nodes are transmitted to a base station located safety outside the building, where they are assembled to form a live map showing the connection of hazardous compounds within the building. Mobile wireless sensor networks can also be an integral part of military command, control, communications, computing, reconnaissance and targeting systems. They are used for monitoring friendly

A. Kwiecień, P. Gaj, and P. Stera (Eds.): CN 2012, CCIS 291, pp. 324–333, 2012.
© Springer-Verlag Berlin Heidelberg 2012

forces, battlefield survillance, etc. Sensor networks can also be incorporated into the guidance systems of intelligent ammunition. More details on mobile sensor networks can be found in the papers [3,4,5,6].

Nodes mobility is the main added value of wireless ad hoc and sensor network. In general, mobility has a central role in these networks. The effect of mobility on the system management and protocols is relevant at many layers: protocols, applications, used data units, etc. Therefore, mobility plays a central role in the model design and is a relevant background effect to be modeled in every simulation analysis of wireless ad hoc and sensor networks. Moreover, the effect of mobility introduces adaptive behaviours of users, protocols, and applications.

Two types of mobility models can be adopted in the simulation of wireless ad hoc and sensor networks: *motion traces* and *synthetic traces* [7].

*Motion traces* are significant descriptions about the steady-state mobility of a node only if the motion samples are collected for significant time intervals. If the sample frequency is low, approximated solutions (e.g. interpolation) can be obtained. Moreover, motion traces can be collected only for existing systems. In general, this method provides accurate and realistic information about mobility patterns, but requires large log files, depending on the number of observed nodes and the time granularity of samples. The motion traces technique in the mobile ad hoc was first introduced by S. Shah et al. [8].

*Synthetic models* represent the mobility of nodes in a realistic way, without using motion traces. The main requirement for such models was mathematical tractability instead of realism. Such model are used in many simulation studies, mainly due to meet the requirements of mobile networks' modeling and simulation [9,10]. Three degrees of 'randomness' in the classification of these models have been distinguished [9], namely:

- models that allow nodes to move anywhere in the simulation area,
- models that bound the movement of nodes (like street, etc.) but still alow for pseudorandom selection of direction and speed,
- *Manhattan model* based on predefined paths (deterministic paths).

We survey a number of synthetic mobility models used in mobile ad hoc and sensor network simulations in the Section 2.

The main objective of this paper is to investigate, using a simulator, a network of basic dependencies occurring in the various traffic models for sensor and ad hoc networks. Furthermore, the aim of this work is experimental comparison of the basic parameters of these models for the same data, which, among others, allows determination of the impact of speed of the nodes on the value of the number of hops in these models. Knowledge of these parameters renders them easier to use, and enables the selection of values for each parameter.

The next section presents the most important tools for the simulation of wireless ad hoc and sensor networks. Section 3 describes the basic mobility models at issue in this study. Section 4 presents the most significant results of simulation tests. Section 5 gives conclusions and suggestions for further research.

## 2    Network Simulators for Mobile Ad Hoc and Sensor Networks

In this section, we provide the basic review several network simulators for mobile ad hoc and sensor networks.

OPNET (Optimized Network Engineering Tool) [11] is an environment for the specification, simulation and performance analysis of wired and wireless networks. It provides tools to assist user to build the simulation model and for output data analysis. Nevertheless, the scalability is a major problem in this tool. Additionally, it takes too long to run the simulation. The results provided by OPNET are significantly different. It is definitely unclear which is the 'right' behaviour. Furthermore, OPNET is a commercial product, and one has to pay a large amount of money to buy it.

OMNeT++ is a discrete-event simulator written in C++ [12]. It provides NEP, a network descriptor language to assist the modeler in the model definition. It is a freely distributed. It can be used for traffic analysis, queueing network modeling, hardware system architectures, and, more in general, modeling any other system that can be mapped to active components that communicate by passing messages. This fact determines that OMNeT++ comes with far less pre-built protocols than the other network simulators. Moreover, some network models and protocols are definitely missing. This fact dramatically limits the use of OMNeT++.

The *ns-2* network simulator [13] is a object-oriented discrete event simulator target at networking research and available as *public domain*. Its first version (*ns-1*) began in 1989 as a variant of REAL network simulator [14] and was developed by the Network Research Group at the Lawrence Berkeley National Laboratory (LBNL), USA. *ns-2* uses OTcl, an object-oriented version of Tcl, as a command and configuration interface. The interface code to the OTcl interpreter is separate from the main simulator. The results provided by *ns-2* are widely used in the networking research community and has found large acceptance as a tool to experiment new ideas, protocols, etc. At the time being, *ns-2* is well-suited for packet switched networks and wireless networks (ad hoc, sensor networks), and is also used mostly for small scale simulations of queueing and routing algorithms, congestion control, traffic analysis, etc. Moreover, *ns-2* is the most used simulator for studies on mobile networks. Although *ns-2* is quite easy to use, it is not a fine-tuned and finished product. It requires a large amount of time to study the inside of *ns-2* before a simulation modeler develops new model.

## 3    Mobility Models for Ad Hoc and Sensor Networks

In this section, we present a number of synthetic mobility models that have been used in the simulation of mobile ad hoc and sensor networks.

The *Random Walk Mobility Model* (RWMM) is based on random directions and speeds. In this mobility model, an mobile node moves from current location to a new location by randomly choosing a direction and speed in which to travel. Each movement in the Random Walk Mobility Model occurs in either a constant time interval $t$ or a constant distance traveled $d$, at the end of which a new direction and speed are calculated. If the mobile node reaches a simulation boundary, it 'bounces' off the simulation border with an angle determined by the incoming direction. The mobile node moves along this new path. This mobility model is well known: its properties are studied in the papers by C. Bettstetter [15,7].

*Random Waypoint Model* (RWP) has been introduced by D. Johnson and D. Maltz (1996) [16]. In this model, each node chooses uniformly at random a destination point (the waypoint) within the deployment region, and moves toward it along a straight line. Node speed is chosen uniformly at random in the interval $[v_{min}, v_{max}]$, where $v_{min}$ and $v_{max}$ are the miniumum and maximum node speeds. When the node arrives at destination, it remains stationary for predefined pause time, and then starts moving again according to the same patters, RWP mobility model has been deeply studied by C. Bettstetter et al. [17,18], J. Yoon et al. [19].

The *Random Walk Model* and the *Random Direction Model* are variants of the Random Waypoint Model. The algorithm of the *Random Walk Model* (random walk is often substituted for *Brown motion*) can be divided into the following steps: (i) select a random speed uniformly distributed in $[v_{min}, v_{max}]$, (ii) select a random direction uniformly distributed, (iii) move into that direction for a predefined amount of time or a certain distance. If the border of the simulated area is reached, select a new direction. Next, (iv) wait some time and go back to step one. The *Random Direction Model* is a small modification of Random Waypoint Model, defined in order to avoid the mobile node concentration in the center of the simulated area. To obtain a uniform number of neighbours for each mobile node, the modeler should be careful about the model parameters. The model is similar to Random Waypoint: before a movement period, a speed and direction is uniformly selected, to be maintained up to the area boundaries will be reached [20]. Another variation is the Random Direction Model in the Boundless Simulation Area Mobility Model, in which the mobile nodes that reach one side of the simulation are continue moving and wreappear on the opposite side of the simulation area. This technique creates a *torus-shaped simulation* and allows to travel unobstructed.

## 4   Simulation Results

This section presents the most significant results of simulation tests for different traffic models used in ad hoc and sensor networks.

With the use of *ns-2* simulator, simulations were conducted of the Random Waypoint Model, Random Walk Model, Random Walk with torus boundary (with wrapping). For all models, identical parameters were assumed, namely: simulation area, whose dimensions are 1000 m × 1000 m, number of nodes 50. Simulation time was specified at 300 seconds, transmission range of each node 250 m, pause length (except for Random Walk Model) 5 s, maximum speed of nodes equal to 20 m/s. It was assumed that average speed of nodes is 10 m/s. Of course, in the case of study on the impact of speed on network parameters, the tests were conducted for various speeds, similarly as for pause length and number of nodes. For individual models, two routing protocols were used: DSDV (Dynamic destination Sequenced Distance Vector) and AODV (Ad hoc On-demand Distance Vector) [21].

The main simulation script is written in OTcl. Files with the established network traffic (TCP and CBR e.g. *constant bit rate*) and studied traffic models were attached to it. As a result of the simulation files were obtained with the extension *.tr*, containing all the events which occurred during the simulation, the files with selected data needed to draw charts and files with the extension *.nam* containing data needed for visualization of the simulation. An example of visualization of the simulation, referring to the first simulation scenario, is shown in Fig. 1.

The simulation tests do not take into account the Random Direction Model, as it was regarded here as unrealistic. The model assumes the object stopping within the boundaries of the simulation area, which typically does not correspond to actual displacement of mobile nodes in mobile network, in which stopping within the limits of observation does not occur. Therefore, the model was omitted in simulation tests.

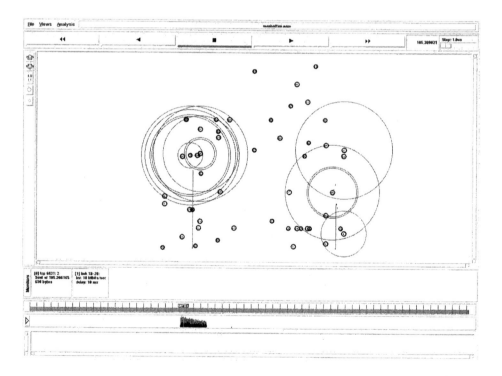

**Fig. 1.** A toolkit of our simulation program

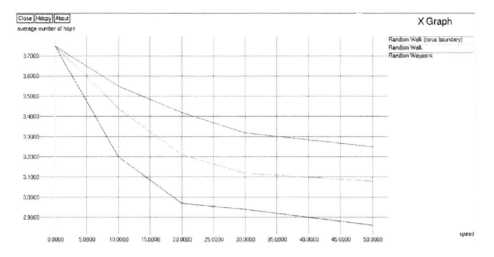

**Fig. 2.** Average numberof hops in the Random Walk Model, Random Waypoint Model, Random Walk Model with torus boundary versus the speed of mobile nodes in m/s

In the first simulation scenario, the Random Waypoint Model, Random Walk Model, Random Walk Model with torus boundary were used. For speed equal to zero, average number of hops is the same for all models. This follows from the same initial data for these models. With increasing speeds the number of hops for each of the models began to increase, see Fig. 2, where it was found that the Random Walk Model takes lowest values. This is due, among others, to the fact that the nodes in the Random Walk Model model tend to stay closer to the center of the simulation area than the nodes in other models. For example, in the Random Walk Model with wrapping, the used wrapping method in the case of the boundaries of the simulation field causes the nodes to be more evenly distributed in the simulation area than in the case of rebound or movement.

In the second scenario, the average number of hops was compared, depending on the number of nodes for Random Walk Model and Random Waypoint Model. It was noted that with the growing number of mobile network nodes the average number of hops also increases (see Fig. 3). It was found that it grows slightly faster in the case of the Random Walk Model.

In the third simulation scenario, the Random Waypoint Model was examined more closely. Among other things, the average number of link changes was determined experimentally, depending on the speed of nodes (Fig. 4). The obtained results show that doubling the speed causes an almost double increase in the number of link changes. For this mobility model, the average number of hops was also examined, depending on the length of the pause time (Fig. 5). It was noted that when the pause time is 0, the Random Waypoint Model behaves like the Random Walk Model. The reason is that the lack of a pause time causes constant motion of the node, thus continually creating new connections and the number of hops reaches its maximum value for the given motion parameters.

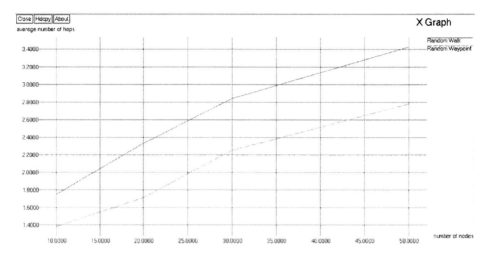

**Fig. 3.** Average number of hops in the Random Walk Model and Random Waypoint Model versus the number of nodes

In the fourth simulation scenario, the impact of node speed on the duration of the connection was examined. The Random Waypoint Model with two different radio transmitting ranges 250 m and 150 m was taken into account. Since nodes are mobile, the connections are formed and broken as soon as nodes enter and leave the coverage area of transmission. Throughout the simulation, new connections are made. Due to their use, routes for routing are created. With the increase in speed of nodes, the average duration of created connections decreases, which is presented in a graph in Fig. 6.

The obtained results allow us to provide the following observations:

1. The Random Walk Model at very small values of the parameters is similar to the known Brownian motion model. On the other hand, it may also be stated that it becomes similar to the Random Waypoint model with pause equal to 0. The main difference between these models is the tendency of mobile nodes in the RWM model to reside in the center of the simulation area.

2. The Random Waypoint Model, which is most often used for the analysis of multiple network protocols, can also be used to create a realistic motion model, such as a moving person. The disadvantage of this model is the movement of mobile node to its destination in a straight line.

3. The average number of hops decreases most quickly depending on the average speed of nodes for the Random Waypoint model, then for the Random Walk Model, and then for the Random Walk with torus boundary model. This means that the Random Waypoint Model is most dependent on average speed of nodes as compared to other models.

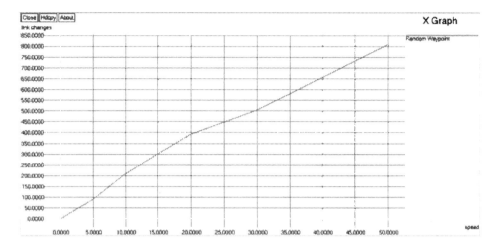

**Fig. 4.** Average number of link changes in the Random Waypoint Model versus the speed of mobile nodes in m/s

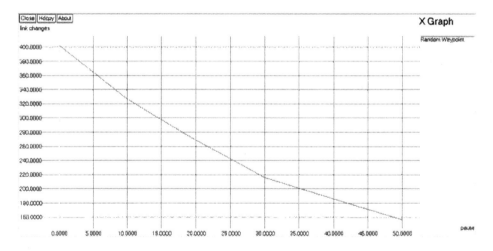

**Fig. 5.** Average number of link changes in the Random Waypoint Model versus the lenght of pause time

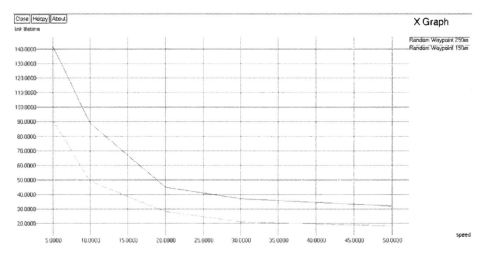

**Fig. 6.** Average duration of created connections in the Random Waypoint Model versus the duration of the connection for the various transmitting radio ranges

## 5 Conclusions

The obtained results allow to choose the appropriate traffic model to determine the parameters of a given routing protocol in mobile ad hoc network or sensor network. Comparative results for different traffic models allow to estimate how the simulation results for a given routing protocol change when using other traffic model. This is important when examining the effectiveness parameters of various routing protocols for mobile ad hoc networks and sensor networks.

It is expected that future studies will be extended to other traffic models, such as Boundless Simulation Area traffic model, in which the mobile node is allowed to travel in the simulation area without the effect of boundaries. On the other hand, instead of this model the RW model with torus boundary might be used, in which one node always moves in one direction, and every now and then passes the same neighboring nodes. However, such a case has not been the subject of this simulation study and will be subject to further research.

## References

1. Ramanathan, R., Redi, J.: A Brief Overview of Ad Hoc Networks. Challenges and Directions. IEEE Communication Magazine, 20–22 (May 2002)
2. Goldsmith, A.J., Wicker, S.B.: Design Challenges for Energy-Constrained Ad Hoc Wireless Networks. IEEE Wireless Communication 9, 8–27 (2002)
3. Pottie, G.J., Kaiser, W.J.: Wireless Integrated Network Sensors. CACM 43(5), 551–558 (2000)
4. Poduri, S., Sukhatme, G.S.: Constrained Coverage in Mobile Sensor Networks. In: Proc. IEEE Int. Conf. Robotics and Automation, ICRA 2004, New Orleans, LA, pp. 40–50 (April-May 2004)

5. Howard, A., Matari, M.J., Sukhatme, G.S.: An Incremental Self-Deployment Algorithm for Mobile Sensor Networks. Autonomous Robots. Special Issue on Intelligent Embedded Systems 13(2), 113–126 (2002)

6. Kahn, J.M., Katz, R.H., Pister, K.S.J.: Next Century Challenges: Mobile Networking for Smart Dust. In: Proc. 5th Annual ACM/IEEE Int. Conf. Mobile Computing and Networking, MOBICOM 1999, Seattle, WA (August 1999)

7. Camp, T., Boleng, J., Davies, V.: A Survey of Mobility Models for Ad Hoc Network Research. Wireless Communication and Mobile Computing. Special Issue on Mobile Ad Hoc Networking: Research, Trends and Applications (2002)

8. Shah, S., Hernandez, E., Helal, A.: CAD-HOC: A CAD Like Tool for Generating Mobility Benchmarks in Ad Hoc Networks. In: Proc. of SAINT 2002, Nara, Japan (February 2002)

9. Bettstetter, C.: Mobility Modeling in Wireless Networks: Categorization, Smooth Movement, and Border Effects. Mobile Computing and Communication Review 5(3) (July 2001)

10. Bettstetter, C., Hartenstein, H., Pérez-Costa, X.: Stochastic Properties of the Random Waypoint Mobility Model: Epoch Length, Direction Distribution and Cell-Change Rate. In: Proc. of the 5th ACM Int. Workshop, MSWiM 2002 (September 2002)

11. OPNET Simulation Tool, http://www.opnet.com

12. Varga, A.: OMNET++. IEEE Network Interactive 16(4) (2002)

13. The Network Simulator – ns-2, http://www.isi.edu/nsnam/ns/

14. REAL Network Simulator (1997),
    http://www.cs.cornell.edu/skeshav/real/overview.html

15. Bettstetter, C.: Smooth is Better than Sharp: a Random Mobility Model for Simulation of Wireless Networks. In: Proc. of ACM Int. Workshop on Modeling, Analysis and Simulation of Wireless and Mobile Systems, MSWiM 2001, Rome (July 2001)

16. Johnson, D., Maltz, D.: Dynamic Source Routing in Ad Hoc Wireless Networks, Mobile Computing, pp. 153–181. Kluwer Academic Publishers (1996)

17. Bettstetter, C., Krauze, O.: On Border Effects in Modeling and Simulation of Wireless Ad Hoc Networks. In: Proc. IEEE Int. Conf. on Mobile and Wireless Communication Networks, MWCN, Recife (2001)

18. Bettstetter, C., Resta, G., Santi, P.: The Node Distribution of the Random Waypoint Mobility Model for Wireless Ad Hoc Networks. IEEE Trans. on Mobile Computing 2(3), 257–269 (2003)

19. Yoon, J., Liu, M., Noble, B.: Random Waypoint Considered Harmful. In: Proc. IEEE INFOCOM, pp. 1312–1321 (2003)

20. Royer, E.M., Melliar-Smith, P.M., Moser, L.E.: An Analysis of the Optimum Node Density for Ad Hoc Mobile Networks. In: Proc. of IEEE Int. Conf. on Communications, ICC, Helsinki (June 2001)

21. Santi, P.: Topology Control in Wireless Ad Hoc and Sensor Networks. John Wiley and Sons Inc., Hoboken (2005)

# Realistic Model of Radio Communication in Wireless Sensor Networks

Mariusz Słabicki, Bartosz Wojciechowski, and Tomasz Surmacz

Wrocław University of Technology
Institute of Computer Engineering, Control and Robotics
Wybrzeże Wyspiańskiego 27, 50-370 Wrocław
{mariusz.slabicki,bartosz.wojciechowski,tomasz.surmacz}@pwr.wroc.pl

**Abstract.** Many theoretical works analysing Wireless Sensor Networks make ad hoc assumptions regarding their physical properties. In our work we measure physical properties of WSNs, such as node power consumption and effective error rates in many realistic scenarios, e.g. in open field, forest and urban environments. We use popular MicaZ and IRIS motes compliant to ZigBee standard to provide a good framework for further research on organization, routing and management algorithms for energy efficient and robust environment monitoring. Using our measurements we find realistic parameters for abstract models to be used in WSN simulations and provide radio transmission model for near-ground node placement.

**Keywords:** Wireless Sensor Networks, power-efficiency, modelling.

## 1 Introduction

Wireless Sensor Networks (WSNs) are not a new concept, but in spite of much research, their widespread adoption is yet to come. Among the possible applications of WSNs are crop monitoring, animal tracking, battle-field monitoring, earthquake detection and warning. Each application is characterized by a set of requirements, e.g. the expected frequency of measurements, communication range or a set of sensors. Also, in many applications, if the deployment is difficult or costly, there are additional requirements for energy efficiency, reliability of the nodes and resilience from adverse ambient conditions e.g. moisture, heat or high concentration of reactive chemicals. Individual nodes often have very limited energy resources and communication range. For example, popular radio transceivers that comply to IEEE 802.15.4 standard and operate in the 2.4 GHz ISM band have range limited to about 100 m. It is far worse when there is no line-of-sight between the nodes, or when they are placed on the ground and the effective range drops quickly to single meters. Consequently, to cover large area and provide the required level of reliability, a requirement of large number of nodes communicating in a multi-hop manner exists.

Massive number of nodes used in one network means that meeting the design requirements is hard. This is not only due to the limited capabilities of a single

A. Kwiecień, P. Gaj, and P. Stera (Eds.): CN 2012, CCIS 291, pp. 334–343, 2012.

node but also a vast number of design decisions that need to be made (e.g. choosing the right number of nodes to be deployed, algorithms for self-organization, management, routing, synchronisation etc.). Decisions taken on different stages of the design process are usually linked and influence all properties of the resulting network. Hence, optimisation of any parameter cannot focus on one aspect of the design but needs to analyse it entirely and propose comprehensive solutions. The advantage of having large number of nodes is an improved reliability of the network, coming from redundant nodes, and therefore – extra available routing paths. In most cases failure of one or more nodes does not prevent the network from performing its tasks.

Limited energy resources is an intrinsic characteristic of all but few WSN applications. Low price and device miniaturization put hard restrictions on the available energy sources. Therefore, energy conservation is a first-order constraint, both for node designers and network architects. Even if nodes are as energy-efficient as possible, wrong network parameters (i.e. too sparse, or too dense network, or nodes operating without duty cycling) will lead to short lifespan of such an application.

Due to the massively parallel nature of WSNs, much research in this subject is done by means of simulation. Therefore, it is of highest importance that widely used tools, simulators and models do not lead to wrong conclusions if they substantially diverge from reality. As always, there is a trade-off between flexibility, simplicity and simulation speed on one side and accuracy on the other. Simplified models should be used carefully.

In this work we are considering properties of WSN nodes from the point of view of environment monitoring applications. The main requirement in such applications is long operation of the network, so most of this paper is devoted to measurements of communication properties and power consumption of real-life devices with application to realistic modelling.

## 2   Related Work

Both power usage and communication models of WSNs are widely described in literature [1,2]. However, the knowledge about the real power consumption and communication parameters of popular nodes does not proliferate to some works on algorithms for network organization and management. The most dangerous to the relevance of the results is the assumption about effective communication range being only dependent on transmission power $P_{TX}$. Many authors use the oversimplified model of signal power at the receiver, depending only on the distance $d$ and the environmental coefficients $\alpha$ and $\varepsilon$ [3,4]:

$$P_{RX} = \varepsilon \frac{P_{TX}}{d^\alpha} \sim \frac{P_{TX}}{d^\alpha} \ . \tag{1}$$

As nodes can receive a packet if $P_{RX}$ is above their sensitivity threshold and noise level, the above model suggests that transmission power should increase with the power of distance $d$. However, in real-world WSNs a lot of factors influence the basic node-to-node communication and cause the above model to be too general, inaccurate and difficult to use. Imperfections of antenna coverage (antennas are not perfectly omnidirectional), random background noise, interference from neighbouring nodes, or obstacles in line-of-sight [5] cause precision loss. Power at the receiver is also subject to gaussian noise [6] and multi-path propagation that is often ignored (e.g. in [2]). Applicability requires knowledge of the distance between communicating parties but this is usually unknown and has to be determined. Consequently, the right transmission power can not be calculated based on a single demand-response packet exchange, leading to additional overhead.

Although power used by wireless sensor nodes was measured earlier [7,8], and modelling energy consumption is a topic of ongoing work [9], there is a number of papers that inaccurately model power consumption related to transmission and reception of data. For example [3] assumes that overall energy consumed by a node when transmitting and receiving can be approximated as

$$E_{TX}(d) = E_{elec} \cdot k + \varepsilon_{amp} \cdot k \cdot d^2$$
$$E_{RX} = E_{elec} \cdot k$$

$$(2)$$

where $k$ is the number of bits, $E_{elec} \cdot k$ represent energy consumed by transmitter/receiver electronics and $\varepsilon_{amp} \cdot k \cdot d^2$ is related to power amplifier cost for transmitting over the distance $d$. This model was later extended to distinguish between short and long range communication. For short distances signal fades with $d^2$ (as in (2)), while for $d > d_0$ path loss exponent equals 4. In both models packet reception costs are always smaller than cost of transmission, while measurements of real life WSN nodes (e.g. [8]) reveal that for a number of settings this is not true (cf. Table 1 and Table 2). Still, the above two models are used in a number of papers that evaluate energy efficiency of Wireless Sensor Networks.

**Table 1.** Power used by MicaZ node in transmission, reception and sleep modes

| Tx power [dBm] | Power used [mW] | Sleep mode | Power used [mW] |
|---|---|---|---|
| -25 | 48.0 | Idle | 11.07 |
| -15 | 51.6 | ADC Noise Reduction | 21.75 |
| -10 | 56.1 | Power-down | 0.87 |
| -7 | 60.3 | Power-save | 1.20 |
| -5 | 64.2 | Reserved | – |
| -3 | 67.8 | Reserved | – |
| -1 | 72.0 | Standby | 0.90 |
| 0 | 75.9 | Extended Standby | 1.23 |
| Receive | 69.0 | – | – |

**Table 2.** Power used by Iris node in transmission and receiving modes

| Tx power [dBm] | Power used [mW] | Tx power [dBm] | Power used [mW] |
|---|---|---|---|
| -17 | 39.0 | -1 | 50.4 |
| -12 | 40.5 | 0 | 51.9 |
| -9 | 42.0 | 0.7 | 53.1 |
| -7 | 43.5 | 1.3 | 54.3 |
| -5 | 44.7 | 1.8 | 55.5 |
| -4 | 46.2 | 2.3 | 57.0 |
| -3 | 47.7 | 2.8 | 57.9 |
| -2 | 49.2 | 3.2 | 58.8 |
| Receive | 54.0 | Sleep | 13.0 |

The case when receiving cost exceeds transmission is presented in [2]. It takes into account efficiency of controlled power amplifier (PA) in downlink as well as low noise amplifier (LNA) and gain control (GC) in uplink. This facilitates situations when for small transmission powers the overall power consumed by a node is smaller than the power consumed when receiving. This difference results from the fact that power consumed by PA depends on transmission power used, while LNA and GC in uplink consume constant amounts of energy. However, as mentioned earlier, paper by Wang et al. does not take into account multi-path propagation of the signal nor energy consumption of real nodes.

# 3   Measurement Setup

## 3.1   Communication Measurements

All our measurements have been conducted with two MicaZ nodes at a time, one of which was programmed as a *master* node, mostly sending packets. Another one was a *slave* node, responsible for receiving packets sent by the master node and sending replies. Both nodes were placed on tripods, with slave node running on batteries. Before each test the distance between the nodes as well as the height above the ground was verified with a measuring tape.

Each test consisted of sending many groups of 20 packets and measuring how many responses from the slave the master node received in each group. Such tests were repeated until steady results could be obtained, i.e. the response ratio of the last group did not differ by more than 10% from both the total and the running average. Each test could also be abandoned manually if the responses were unsatisfactory, e.g. the distance was too big for a given transmission power and no transmissions came through. The tests were repeated with transmission power settings ranging from −25 dBm to 0 dBm.

In order to determine how much it affected practical radio transmissions, communication tests were performed in various environments. These included: corridor in a building (free of obstacles, but "polluted" with WiFi transmissions), open space (grass surface in a city park, asphalt surface along a low-traffic suburban road), and a forest where trees obscured the line of sight of nodes.

## 3.2   Power Measurements

Power consumption measurements have been conducted for MicaZ and Iris nodes in two differing setups. For current measurement in sleep mode we have used Agilent 34410A precision multimeter connected between the measured node and the stabilized voltage source. During the sleep mode residual current consumption is constant and such measurement is satisfactory. In transmission mode, however, momentary current consumption varies all the time, as shown in Fig. 1. To measure these fast-changing currents we have used MAX4373 voltage amplifier responding to the voltage drop at the shunt resistor. The measurement setup is shown in Fig. 2. Energy consumption of data transmissions depends on many

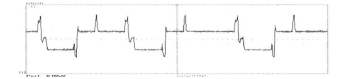

**Fig. 1.** Characteristic of energy consumption in transmission mode

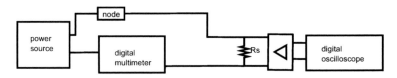

**Fig. 2.** Measurement set for energy consumption in fast changing modes

factors, but very often [3,10,11] it is linked only to the range of radio transmission, i.e. the antenna-dissipated power dependent in turn on the transmission level. This is reasoned by the low energy demands between consecutive transmissions. Our measurements show that the real behaviour of WSN nodes does not adhere to this simplified model.

MicaZ nodes operate on ATMega128 processor which has 8 different sleep modes. Power consumption measurements of these sleep modes are shown in Table 1. Modes named "Reserved" were not used. On Iris there is only one sleep mode, which was measured (Table 2).

Difference of powers used at different transmission power levels is noticeable. It is important in situations when successful transmission rate has big variance (cf. Table 3). Sometimes it is not worth sending packets with higher transmit power, as the overall cost for all the retransmissions will be higher than with the lower transmit power level. That is why it is necessary to set control power policy in WSN. Even bigger difference of power requirements can be noticed between sleep modes and transmission modes. Power which is necessary to send and receive messages is over 15 times higher than the power used by node in sleep mode, thus taking advantage of sleep modes whenever possible can have big impact on energy savings.

## 4   Communication Model

Figure 3 presents measurement results obtained from our ping tests performed at a wide flat surface of lawn in a city park. The figures show average number of returned transmissions as a function of transmission power and the height above the ground of both communicating nodes. As we can see, for a distance of 15 meters and height of 1 meter (Fig. 3(a)) there is substantial loss of packets. The same occurs for 30 meters distance and the height of 1.4 meters (Fig. 3(b)). This phenomenon is caused by reflections of electromagnetic waves when they

**Table 3.** Mean number and standard deviation of received packets in forest environment, $h = 1.0$ m for sets consisting of 20 packets

| d [m] | $P_{TX}$ [dBm] | | | | | | | | |
|---|---|---|---|---|---|---|---|---|---|
| | -25 | -15 | -10 | -7 | -5 | -3 | -1 | 0 | |
| 2.5 | 98.5 | 100 | 100 | 100 | 100 | 100 | 100 | 100 | Percentage of packets received [%] |
| 5 | 22.7 | 100 | 100 | 100 | 100 | 100 | 100 | 100 | |
| 10 | 0 | 0 | 0.6 | 90.3 | 100 | 100 | 100 | 100 | |
| 15 | 0 | 0 | 0 | 37.1 | 90.6 | 100 | 100 | 100 | |
| 20 | 0 | 0 | 0 | 8.8 | 89.5 | 100 | 100 | 100 | |
| 25 | 0 | 0 | 0 | 1.8 | 93.0 | 100 | 100 | 100 | |
| 30 | 0 | 0 | 0 | 70.0 | 100 | 100 | 100 | 100 | |
| 35 | 0 | 0 | 33.7 | 100 | 100 | 100 | 100 | 100 | |
| 40 | 0 | 0 | 0 | 55.0 | 85.6 | 99.4 | 100 | 100 | |
| 45 | 0 | 0 | 0 | 0 | 13.1 | 86.9 | 100 | 100 | |
| 50 | 0 | 0 | 0 | 0 | 0 | 9.5 | 90.0 | 98.3 | |
| 55 | 0 | 0 | 0 | 0 | 0 | 5.5 | 61.0 | 91.6 | |
| 58 | 0 | 0 | 0 | 0 | 0 | 0 | 0 | 1.1 | |
| 2.5 | 0.4697 | 0 | 0 | 0 | 0 | 0 | 0 | 0 | Variance $\sigma^2$ |
| 5 | 2.2926 | 0 | 0 | 0 | 0 | 0 | 0 | 0 | |
| 10 | 0 | 0 | 0.3536 | 1.9834 | 0 | 0 | 0 | 0 | |
| 15 | 0 | 0 | 0 | 6.2935 | 3.5814 | 0 | 0 | 0 | |
| 20 | 0 | 0 | 0 | 3.8000 | 2.1192 | 0 | 0 | 0 | |
| 25 | 0 | 0 | 0 | 0.8951 | 1.8522 | 0 | 0 | 0 | |
| 30 | 0 | 0 | 0 | 1.8516 | 0.3536 | 0 | 0 | 0 | |
| 35 | 0 | 0 | 3.1074 | 0 | 0 | 0 | 0 | 0 | |
| 40 | 0 | 0 | 0 | 8.8741 | 2.5877 | 0.3536 | 0 | 0 | |
| 45 | 0 | 0 | 0 | 0 | 3.7517 | 3.1193 | 0 | 0 | |
| 50 | 0 | 0 | 0 | 0 | 0 | 3.7253 | 0.8165 | 0.9129 | |
| 55 | 0 | 0 | 0 | 0 | 0 | 2.1191 | 4.6234 | 1.7017 | |
| 58 | 0 | 0 | 0 | 0 | 0 | 0 | 0 | 0.5789 | |

propagate near the surface, leading to out-of-phase signals to reach the receiver and have partial cancelling effect.

Double-ray propagation model over ideal smooth earth surface states that the received power is a sum of two rays. The first ray is directly propagating between the transmitter and the receiver, the second one is a ray reflected from the ground. If the distance $d$ between antennas is much bigger than heights $h_1$ and $h_2$ of antennas above the ground (such that $\sqrt{1 + (\frac{h_1+h_2}{d})^2}$ can be approximated as $1 + \frac{1}{2} \cdot (\frac{h_1+h_2}{d})^2$), the difference of rays' paths is

$$\Delta R = \frac{2h_1 h_2}{d} \qquad (3)$$

leading to a phase shift of:

$$\Delta \phi = \frac{4\pi h_1 h_2}{\lambda d} \quad . \qquad (4)$$

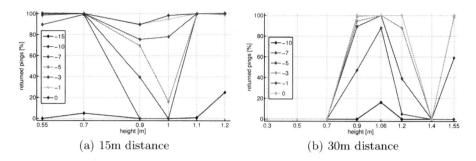

(a) 15m distance    (b) 30m distance

**Fig. 3.** Average number of pings returned at the distance of (a) 15m and (b) 30m as a function of transmission powers and transmitter and receiver height

If the angle between the ground and the reflected ray is smaller than Brewster angle (which is the case if $d >> 5h_1h_2$), phase in the reflected ray is shifted by $\pi$. In our experiment this condition is true. In worst case these two rays will cancel each other and the received power will be zero. That happens if phase shift is $n\pi$ for odd values of $n$. Setting $\Delta\phi = n\pi$ and solving (4) for $d$ gives

$$d = \frac{4h_1h_2}{n\lambda} \quad . \tag{5}$$

For odd values of $n$ in (5) rays are in opposite phases and the power at the receiver drops to 0. This effect can be observed in Fig. 3(b), but is most clearly seen in Fig. 4. Height of transceiver and receiver was 1.4 m, and at the distance of 30 m the mean number of returned pings is almost zero. For distances bigger or smaller than 30 meters, the ratio of returned pings was substantially higher.

Double ray model is an improvement of the basic open space model. In a simple open space model the number of received packets decreases with distance between antennas, while in reality, the number of received packets may sometimes increase with increasing distance (Fig. 4).

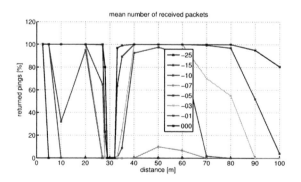

**Fig. 4.** Average number of correct transmissions as a function of distance and transmission power (in dBm)

Double ray model can be used for modelling transmission in open area with flat ground (like open street, airfield etc.). If environment isn't similar to open area, double ray model isn't right. In forest environment, transmission with high power has characteristic similar to free space, but transmission with lower power behaves accordingly to double ray model. Probably it is because rays with higher power are reflected from trees at least once before they reach the receiver, while low power reflected rays fade out.

This is consistent with observations regarding Fresnel zones [12,13]. These state, that even though the radio waves travel in a straight line from the transmitter to the receiver, the obstacles near the path cause reflected waves to arrive to the receiver out of phase. If the phase difference is between $\frac{\pi}{2}$ and $\frac{3}{4}\pi$, this will have the cancelling effect. Consecutive Fresnel zones are ellipsoid-shaped and the radius of the $n$-th zone at $d_1$ distance from one of the communicating points (where $d_2 = d - d_1$ is a distance to the other endpoint) is given by [13]:

$$F_n = \sqrt{\frac{n\lambda d_1 d_2}{d_1 + d_2}} \ . \tag{6}$$

Waves travelling directly through odd-numbered Fresnel zones arrive with a phase shift between $-\frac{\pi}{2}$ and $\frac{\pi}{2}$ ($0 \ldots \frac{\pi}{2}$ in $F_1$), thus adding to the received signal, while waves received through even-numbered zones arrive phase-shifted between $\frac{\pi}{2}$ and $\frac{3}{4}\pi$, cancelling the main signal (thus, the obstacles in odd-numbered zones weaken the signal, while obstacles in even-numbered zones actually improve its strength). If we assume $h_1 = h_2$, and phase change at signal reflection, the strength of such a signal at the receiver can be expressed as

$$P_{\mathrm{RX}} = \varepsilon P_{\mathrm{TX}} \cdot \left( d^{-\alpha} - (d^2 + 4h^2)^{\frac{-\alpha}{2}} \cos \frac{4\pi h^2}{\lambda d} \right) \ . \tag{7}$$

Attenuation function att $= 10 \cdot \log(P_{\mathrm{RX}}/P_{\mathrm{TX}})$ based on (7) is shown in Fig. 5. Substantial energy loss that can be observed at $(h = 0.9, d = 15)$ and $(h = 1.4, d = 30)$ is consistent with measurements shown in Fig. 3. As packet loss ratio depends on signal attenuation, the data shown in Fig. 3 can be seen as a cross-section of Fig. 5 at $d = 15\,\mathrm{m}$ and $d = 30\,\mathrm{m}$ respectively. For experiments with low $P_{\mathrm{TX}}$ power, local optima for $(h, d) \in \{(0.7, 15), (1.2, 15), (1, 30), (1.6, 30)\}$ observed in measured data can be also explained by the attenuation function.

Existence of Fresnel zones has another important impact on communication model. As it is shown in Fig. 3(b), when antennas' height over ground is small, communication range shortens significantly. It is because a lot of energy transmitted from antenna is absorbed by the ground.

The model described in (7) does not consider random noise. It can be added to the signal attenuation as a stochastic variable with its probability distribution. As it is shown in Fig. 6, standard deviation is noticeable, especially on slopes of attenuation function (cf. cross-section in Fig. 5).

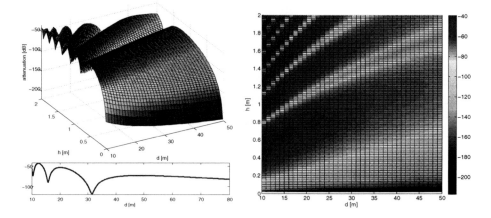

**Fig. 5.** Attenuation of the signal as function of $(d, h)$ for $\lambda = 12.5\,\text{cm}$ and $\alpha = 2$ and its cross-section at $h = 1.4\,\text{m}$

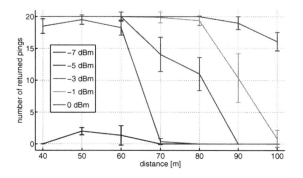

**Fig. 6.** Transmission success rates and standard deviation on a flat surface for $h = 1.4\,\text{m}$

## 5    Conclusions

Results of our experiments show that when considering transmission ranges for Wireless Sensor Networks one must consider the attenuation explained by Fresnel Effect and 2-ray propagation model to allow for anomalies that are present at some communication distances. This is extremely important in environment monitoring scenarios where nodes are randomly spread over a large area and lay on the ground. The near-zero height above the ground vastly reduces each node's transmission range and significantly reduces the network lifespan. Appropriate physical design of nodes that allows elevating their antennas above the ground level may help solving this problem. The most commonly used models for data communication based on power dissipation only are not enough. Our model provides more accurate estimation of received signal power. Inside a node an adaptive power selection algorithm will be preferred over the model, as the distance between nodes is not known, but in a simulated environment, when placement of nodes is known, the attenuation model can be used to predict more accurate transmission success rates and appropriate overall transmission costs can be calculated, based on measurements shown in Tables 1 and 2.

**Acknowledgment.** This work was partially supported by grant no. N 516 483740 from the National Science Centre.

# References

1. Aslam, S., Farooq, F., Sarwar, S.: Power consumption in wireless sensor networks. In: Proceedings of the 7th International Conference on Frontiers of Information Technology, FIT 2009, pp. 14:1–14:9. ACM, New York (2009)
2. Wang, Q., Hempstead, M., Yang, W.: A realistic power consumption model for wireless sensor network devices. In: 3rd Annual IEEE Communications Society on Sensor and Ad Hoc Communications and Networks, SECON 2006, vol. 1, pp. 286–295 (September 2006)
3. Heinzelman, W.R., Chandrakasan, A., Balakrishnan, H.: Energy-efficient communication protocol for wireless microsensor networks. In: Proceedings of the 33rd Annual Hawaii International Conference on System Sciences, vol. 8, pp. 3005–3014. IEEE Computer Society, Washington, DC (2000)
4. Heinzelman, W., Chandrakasan, A., Balakrishnan, H.: An application-specific protocol architecture for wireless microsensor networks. IEEE Transactions on Wireless Communications 1(4), 660–670 (2002)
5. Xu, L., Yang, F., Jiang, Y., Zhang, L., Feng, C., Bao, N.: Variation of received signal strength in wireless sensor network. In: 2011 3rd International Conference on Advanced Computer Control (ICACC), pp. 151–154 (January 2011)
6. Mao, G., Fidan, B., Anderson, B.D.: Wireless sensor network localization techniques. Computer Networks 51(10), 2529–2553 (2007)
7. Krämer, M., Geraldy, A.: Energy measurements for MicaZ node. In: 5. GI/ITG KuVS Fachgespräch Drahtlose Sensornetze. University of Kaiserslautern (2006)
8. Shnayder, V., Hempstead, M., Chen, B.-R., Allen, G.W., Welsh, M.: Simulating the power consumption of large-scale sensor network applications. In: Proceedings of the 2nd International Conference on Embedded Networked Sensor Systems, SenSys 2004, pp. 188–200. ACM, New York (2004)
9. Zogović, N., Dimić, G., Bajić, D.: Channel loss based energy consumption model for low-power wireless communications. In: 2011 4th IFIP International Conference on New Technologies, Mobility and Security (NTMS), pp. 1–5 (February 2011)
10. Manish, B., Garnett, T., Chandrakasan, A.P.: Upper bounds on the lifetime of wireless sensor networks. In: IEEE International Conference on Communications, ICC 2001, vol. 3, pp. 785–790 (2001)
11. Younis, O., Fahmy, S.: HEED: a hybrid, energy-efficient, distributed clustering approach for ad hoc sensor networks. IEEE Transactions on Mobile Computing, 366–379 (2004)
12. Parsons, J.D.: The Mobile Radio Propagation Channel, 2nd edn. Wiley (November 2000)
13. Haslett, C.: Essentials of Radio Wave Propagation. Cambridge University Press, New York (2008)

# The Evaluation of Unconstrained Multicast Routing Algorithms in Ad-Hoc Networks

Maciej Piechowiak[1] and Piotr Zwierzykowski[2]

[1] Kazimierz Wielki University, Bydgoszcz, Poland
[2] Poznan University of Technology, Poznan, Poland
mpiech@ukw.edu.pl, pzwierz@et.put.poznan.pl

**Abstract.** Wireless ad-hoc networks are playing an important role in extending the implementation of traditional wireless infrastructure (cellular networks, wireless LAN, etc). Routing design in ad-hoc networks is a challenge because of limited node resources. Thus efficient data transmission techniques like multicasting are under scrutiny. The article analyzes and explores the performance of multicast heuristic algorithms without constraints and quality of multicast trees in ad-hoc networks and proves the thesis that well-known multicast heuristic algorithms designed for packet networks have a good performance in ad-hoc networks with grid structure.

**Keywords:** ad-hoc networks, multicast, routing algorithms, network topology.

## 1 Introduction

Ad-hoc networks are sets of nodes that form temporary networks without any additional infrastructure and no centralized control. The nodes in an ad-hoc network can represent an end-user devices such as smartphones or laptops in traditional networks. In some measurement systems nodes can represent an autonomous sensors or indicators. These nodes generate traffic to be forwarded to some other nodes (unicast) or a group of nodes (multicast). Due to a dynamic nature of ad-hoc networks, traditional network routing protocols are not viable.

In some measurement systems nodes can represent an autonomous sensors or indicators. Mesh networks can be used to collect of sensor data for data processing for a wide range of applications such as tensor systems, air pollution monitoring, and the like. Nodes in these networks generate traffic to be forwarded to some other nodes (unicast) or a group of nodes (multicast).

Mobile ad-hoc networks (MANET) and mesh networks are closely related, but MANET also have to deal with the problems introduced by the mobility of the nodes (nodes may represent mobile devices). Similarly to the mesh networks, nodes act both as an end system (transmitting and receiving data) and as a router (allowing traffic to pass through) resulting in multihop routing. Networks are *in motion* – nodes are mobile and may go out of range of other nodes in the network.

A. Kwiecień, P. Gaj, and P. Stera (Eds.): CN 2012, CCIS 291, pp. 344–351, 2012.

The implementation of multicasting requires solutions of many combinatorial problems accompanying the building of optimal transmission trees [1]. In the optimization process it can be distinguished: MST (*Minimum Steiner Tree*), and SPT (*Shortest Path Tree*) – tree with the shortest paths between the source node and each of the destination nodes. Finding the MST, which is a $\mathcal{NP}$-complete problem, results in a structure with a minimum total cost. The relevant literature provides a wide range of heuristics solving this problem in polynomial time and dedicated mostly for paket networks [2,3,4,5,6]. In case of MANET multicast protocols, two basics architectures are used: tree-based protocols, where MAODV (*Multicast Ad-hoc On-demand Distance Vector routing*) [7] is the most discussed tree-based protocol and mesh based protocol: ODMRP (*On-Demand Multicast Routing Protocol*) [8]. Multicast routing with QoS was examined in [9,10].

The main thesis of this article is that well-known multicast heuristic algorithms designed for packet networks and analyzed by authors in earlier works [11,12] have also a good performance in ad-hoc networks with grid structure.

The article discusses the effectiveness of the most commonly used constrained multicast heuristic algorithms as well as their comparative usefulness in ad-hoc networks. The article structure is as follow: Section 2 formulates an optimization problem and defines multicast routing algorithms. Section 3 describes network topology generator proposed by authors. Section 4 presents the methodology of research. Section 5 includes the results of the simulation of the implemented algorithms along with their interpretation.

## 2   Problem Formulation

The simplest heuristic approach, solves the Minimum Steiner Tree problem called the Shortest Path Tree (SPT), and relies on computation the shortest paths between the source and receivers. Individual paths have the minimum length, but a multicast distribution tree created in this way is not optimal.

Let us assume that physical network topology is represented using an undirected, connected graph $G = (V, E)$, with $V$ representing the set of nodes, and $E$ being the set of communication links. Let us assume that, for a given graph $G$, there is the function $c\colon E \to \mathcal{R}^+$ that ascribes weights (costs) to all edges and the cost of a connection represents the usage of the link resources. The cost of a given path $p = (v_0, v_1, \ldots, v_k)$ is defined as the sum of the component edges: $c(p) = \sum_{i=0}^{k-1} c(v_i, v_{i+1})$. The distance from the initial vertex $s$ to the final vertex $t$ is defined as:

$$\delta(s,t) = \begin{cases} \min\{c(p) : p \text{ path from } s \text{ to } t\} & \text{path exists}, \\ \infty & \text{else}. \end{cases} \tag{1}$$

The shortest path from vertex $s$ to vertex $t$ is a path $p$ whose cost $c(p)$ is equal to distance $\delta(s,t)$ from $s$ to $t$.

The optimization problem related to the construction of the path between the sending (source) node $s$ and a member of the multicast group $m_x \in M$, known as *Shortest Path Tree* problem (SPT), can be formulated in Table 1.

**Table 1.** Shortest Path Tree problem formulation

---

**notation**

| | |
|---|---|
| $e \in E$ | edge of the graph representing links, |
| $v \in V$ | vertex of the graph representing the node, |
| $m_x \in M$ | vertex that belongs to the multicast group, |
| $p(s, m_x)$ | path between vertex – source $s$, |
| | and the receiving node $m_x \in M$, |

**constant**

| | |
|---|---|
| $e_{ij} \in \{0, 1\}$ | edge of the graph between vertices $i$ and $j$, |
| | ($0$ – edge does not exist, $1$ – edge exists), |

**variable**

| | |
|---|---|
| $c_{ij} \in \mathbb{R}^+$ | cost of edge (link), |
| $c_T \in \mathbb{R}^+$ | cost of Steiner tree, |
| $c_p \in \mathbb{R}^+$ | cost of path $p(s, m_x)$, |

**goal**

**minimalize** $c_p = \sum_{(i,j) \in p(s, m_x)} c_{ij} e_{ij}$ .

---

KMB [13] algorithm presents a more complex but also more efficient approach. It computes complete graph connecting source $s$ with all group members $M$ for a given network $G$. Then, the minimum spanning tree is computed for this graph and, finally, mapped into the original network $G$. This is a somehow simplified description of KMB logic, nevertheless it gives good understanding of this algorithm. Time complexity of this algorithm is $\mathcal{O}(\Delta |V|^2)$.

SPT and KMB algorithms were selected by the authors as representative algorithms for multicast trees efficiency evaluation.

## 3   Network Topology Generators

Ad-hoc networks were analyzed in many works, including [14,15]. Mesh networks are under scrutiny of WING Project (*Wireless Mesh Network for Next-Generation Internet*) [16,17,18]. These publications provide detailed analysis on modeling topologies for ad-hoc networks as well as sensor networks, methods for controlling topologies, models of mobility of nodes in networks and routing protocols in wireless ad-hoc networks. Wireless ad-hoc networks are formed by devices that have mobile energy source with limited capacity. It is essential then for the energy consumption to be maintained at a possibly low level in order to prolong the time duration of autonomous operation of the device. The adopted model of the costs of links between the devices takes into consideration an Euclidean distance between them as an energy used by the antenna system of a device. The proposed implementation assumes that network devices have isotropic radiators and are stationary.

The proposed generator provides grid regular topologies. It starts with construction of a square grid $N = \lceil \sqrt{n} \rceil^2$ with evenly distributed nodes, where $n$ is the target number of nodes. In the next step, the superfluous nodes are removed in a pseudo-random way. Then, the nodes are joined by edges depending on the Euclidean distance between them and the ranges of the nodes $r \in \langle 1, 2 \rangle$, where $r \in \mathbb{N}$.

The *average node degree* for ad-hoc grid-like networks ($D_{av}$) is 3.6 (Figs. 1–4). Average node degree is defined as:

$$D_{av} = \frac{2k}{n} \qquad (2)$$

where $n$ – number of nodes, $k$ – number of links.

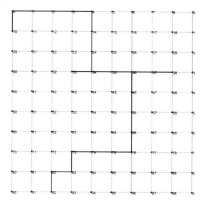

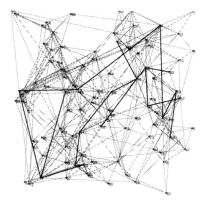

**Fig. 1.** Visualization of multicast tree and network topology obtained using the proposed ad-hoc generator ($n = 100$, $k = 360$)

**Fig. 2.** Visualization of multicast tree and network topology obtained using the Waxman model ($n = 100$, $k = 360$, $\alpha = 0.15$, $\beta = 0.05$)

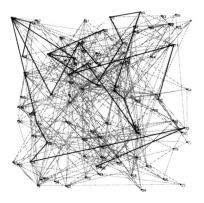

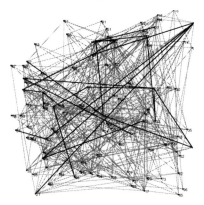

**Fig. 3.** Visualization of multicast tree and network topology obtained using the Waxman model ($n = 100$, $k = 360$, $\alpha = 0.15$, $\beta = 0.15$)

**Fig. 4.** Visualization of multicast tree and network topology obtained using the Waxman model ($n = 100$, $k = 360$, $\alpha = 0.15$, $\beta = 0.95$)

In order to reliable comparison of multicast algorithms into network topologies with different properties, a flat random graph constructed graphs according to the Waxman method was used [19]. This method was also adopted in author's simulation application.

Figures 1–4 are exemplary visualizations of multicast trees constructed by KMB algorithm. Two types of network topologies have been used to implement KMB algorithm: grid network obtained with an application of generator proposed by authors (Fig. 1) and Waxman model also implemented in authors application (Figs. 2–4). All networks have the same parameters: 100 nodes and 360 links. Increase of the parameter $\beta$ increases the ratio of the long links against the short ones thus multicast tree has the greater delay (Fig. 4). It should be also noticed that multicast tree obtained in network with grid topology has lowest cost. The results presented in Sect. 5 show effectiveness of multicast trees in these topologies in details.

## 4   Simulation Methodology

The works related to simulation studies of multicast routing algorithms define the receiving node group $M$ randomly through a choice $m$ of network nodes from all available nodes $n$ of the network ($n = |N|$) [20,21]. The source node $s$ is also chosen randomly from the $n$ number of nodes in the network.

The research work was conducted with the application of the networks generated by the above-mentioned methods that were appropriately adopted and unified [11,12]. In both models (grid and Waxman) nodes were arranged on a square grid with the size of $1000 \times 1000$. It was an important element during the simulation process to maintain a steady average node degree of the graph (for each of the generated networks). The fixed value $\alpha = 0.15$ gives the desirable number of network links $k$ to achieve average node degree $D_{av} = 3.6$ in networks generated with an application of Waxman model comparable with mesh topologies (Waxman parameters: $\alpha = 0.15$, $\beta = 0.05 - 0.95$).

Onto the existing network of connections (represented by adjacency matrix), the cost matrix $C(u, v)$ was applied (resulting from the Euclidean distance between the nodes).

## 5   Simulation Results

Due to a wide range of solutions presented in the literature of the subject, the following representative algorithms were chosen: KMB [3] and SPT [2] algorithms. Their popularity, in applications and the number of citations in literature were decisive in their selection. Such a set of algorithms includes solutions potentially most and least effective in terms of costs of constructed trees. This, however, will make the results of the comparison more distinct, even with comparisons with the applications of different methods of generating network topologies (random graphs and regular ad-hoc structures).

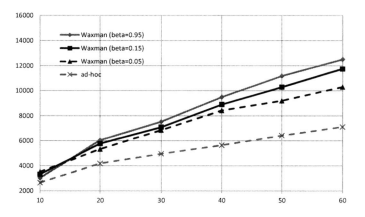

**Fig. 5.** Average cost of multicast tree obtained with an application of KMB algorithm in relation to the number of multicast nodes $m$ ($n = 100$, $k = 360$)

In the first phase of the experiment (Fig. 5) the dependency between the average cost of the constructed trees with an application of KMB algorithm and the number of multicast nodes $m$ was examined. The influence of the $\beta$ parameter in Waxman model on the costs of the obtained trees is not so significant as in the case when the ad-hoc network is applied.

The KMB algorithm constructs multicast trees with the total cost of 43% lower in ad-hoc networks, on average, as opposed to the same algorithm implemented in Waxman networks ($\beta = 0.95$), and 30% in Waxman networks ($\beta = 0.05$) respectively.

In the second phase of the experiment (Fig. 6) the dependency between the average cost of the constructed trees with an application of SPT algorithm and the number of multicast nodes $m$ was examined. The SPT algorithm constructs multicast trees with the total cost of 41% lower in ad-hoc networks, on average, as opposed to the same algorithm implemented in Waxman networks ($\beta = 0.95$) and 28% in Waxman networks ($\beta = 0.05$) respectively.

For each point of the plot shown in Figs. 5–6 the average cost of the tree was obtained for 5000 networks generated in 5 series for each routing algorithm. The results of the simulations are shown in the charts in the form of marks with 95% confidence intervals that were calculated after the *t-Student* distribution. 95% confidence intervals of the simulation are almost included within the marks plotted in the figures.

## 6   Conclusions

The survey of literature shows proposals of many routing protocols designed for ad-hoc networks. Unicast protocols are dominating set of whole routing solutions while multicast routing algorithms and protocols are in minority and they are still an open topic. First authors' approach evaluates multicast routing

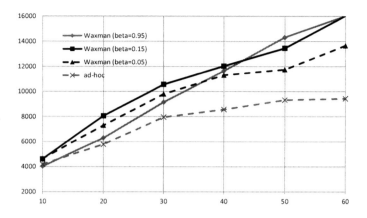

**Fig. 6.** Average cost of multicast tree obtained with an application of SPT algorithm in relation to the number of multicast nodes $m$ ($n = 100$, $k = 360$)

algorithms (designed especially for packet networks with Internet-like topologies) implemented in mesh networks with regular structure (grid topology). The results of algorithms obtained in the mesh networks were compared with the results obtained in random graphs. Conducted studies confirmed the effectiveness of examined heuristic algorithms in ad-hoc topologies. Thus the paper proves the thesis statement set in Introduction.

The simulation research methodology proposed earlier [6,11] permit to model networks with wide range of nodes and many network topology parameters. This will constitute the next stage in the authors' research work aiming to define a methodology for testing multicast heuristic algorithms in mesh networks and compare their efectiveness with dedicated algorithms and protocols.

Authors believe that the inclusion of the methods of mesh topology generation as well as the basic parameters of the test network are necessary conditions to have the existing and new multicast routing algorithms compared in a reliable way.

# References

1. Hakimi, S.L.: Steiner's Problem in Graphs and Its Implications. Networks 1, 113–133 (1971)
2. Crawford, J.S., Waters, A.G.: Heuristics for ATM Multicast Routing. In: Proceedings of 6th IFIP Workshop on Performance Modeling and Evaluation of ATM Networks, pp. 5/1–5/18 (July 1998)
3. Vachaspathi, P., Kompella, J.P., George, C.P.: Multicasting for Multimedia Applications. In: INFOCOM, pp. 2078–2085 (1992)
4. Mokbel, M.F., El-Haweet, W.A., El-Derini, M.N.: A Delay Constrained Shortest Path Algorithm for Multicast Routing in Multimedia Applications. In: Proceedings of IEEE Middle East Workshop on Networking. IEEE Computer Society (1999)
5. Piechowiak, M., Stasiak, M., Zwierzykowski, P.: The Application of K–Shortest Path Algorithm in Multicast Routing. Theoretical and Applied Informatics 21(2), 69–82 (2009)

6. Piechowiak, M., Stasiak, M., Zwierzykowski, P.: Analysis of the Influence of Group Members Arrangement on the Multicast Tree Cost. In: Proceedings of The Fifth Advanced International Conference on Telecommunications, AICT (2009)
7. Royer, E.M., Perkins, C.: Multicast Ad hoc On-Demand Distance Vector (MAODV) Routing. Network Working Group, RFC: draft (July 2000)
8. Lee, S., Su, W., Gerla, M.: On-Demand Multicast Routing Protocol (ODMRP) for Ad Hoc Networks. Network Working Group, RFC: draft (July 2000)
9. Bur, K., Ersoy, C.: Ad Hoc Quality of Service Multicast Routing. Computer Communications 29(1), 136–148 (2005)
10. Bur, K., Ersoy, C.: Performance Evaluation of a Mesh-Evolving Quality-of-Service-Aware Multicast Routing Protocol for Mobile Ad Hoc Networks. Performance Evaluation 66(12), 701–721 (2009)
11. Zwierzykowski, P., Piechowiak, M.: Performance of Fast Multicast Algorithms in Real Networks. In: Proceedings of EUROCON 2007 the International Conference on: Computer as a Tool, Warsaw, Poland, pp. 956–961 (September 2007)
12. Zwierzykowski, P., Piechowiak, M.: Efficiency Analysis of Multicast Routing Algorithms in Large Networks. In: Proceedings of The Third International Conference on Networking and Services, ICNS 2007, Athens, Greece, pp. 101–106 (June 2007)
13. Kou, L., Markowsky, G., Berman, L.: A Fast Algorithm for Steiner Trees. Acta Informatica (15), 141–145 (1981)
14. Santi, P.: Topology Control in Wireless Ad Hoc and Sensor Networks. ACM Computing Surveys 37, 164–194 (2005)
15. Rajaraman, R.: Topology Control and Routing in Ad Hoc Networks: A Survey. ACM SIGACT News 33, 60–73 (2002)
16. WING – Wireless Mesh Network for Next-Generation Internet (2012), http://www.wing-project.org
17. Riggio, R., Rasheed, T., Testi, S., Granelli, F., Chlamtac, I.: Interference and Traffic Aware Channel Assignment in WiFi-based Wireless Mesh Networks. In: Elsevier Ad Hoc Networks (2010)
18. Gomez, K., Riggio, R., Rasheed, T., Miorandi, D., Chlamtac, I., Granelli, F.: Analysing the Energy Consumption Behaviour of WiFi Networks. In: Proceedings of IEEE Greencom 2011 (2011)
19. Waxmann, B.: Routing of Multipoint Connections. IEEE Journal on Selected Area in Communications 6, 1617–1622 (1988)
20. Rouskas, G., Baldine, I.: Multicast Routing with End-to-End Delay and Delay Variation Constraints. IEEE Journal on Selected Areas in Communications 15, 346–356 (1997)
21. Wei, L., Estrin, D.: The Trade-Offs of Multicast Trees and Algorithms. In: Proceedings of ICCCN 1994, pp. 55–64. IEEE (1994)

# Performance Evaluation of Cellular Communication Systems for M2M Communication in Smart Grid Applications

Ganesh Man Shrestha[1] and Jürgen Jasperneite[1,2]

[1] inIT-Institut Industrial IT, Liebigstraße 87, D-32657 Lemgo
{ganesh.shrestha,juergen.jasperneite}@hs-owl.de
[2] Fraunhofer IOSB-INA, Langenbruch 6, D-32657 Lemgo
juergen.jasperneite@iosb-ina.fraunhofer.de

**Abstract.** The increasing power demands and growing awareness for sustainable and green energy has led to distributed generation of power from different sources. This transition from centralized to a distributed power generation has increased the necessity to upgrade the traditional grid. The future grid, i.e. *Smart Grid*, should offer two way flow of power and information. Smart grid needs to intelligently manage the power generation, transmission, and distribution to generate optimal power resources and adapt consumers to those power resources. In addition, it should support smart metering and monitoring to reduce energy consumption and cost. This intelligent management demand near real time communication between the power generators, consumer utilities and the control center. Thus machine-to-machine (M2M) communication is the necessity of future smart grid applications.

Smart grid is a huge infrastructure and its components are located at far-off locations. Hence, wired and short range wireless communication solutions would not be ideal for smart grid applications. This paper presents the performance evaluation of different cellular communication systems as a solution for M2M communication in smart grid applications.

**Keywords:** smart grid, machine-to-machine (M2M) communication, cellular communication.

## 1 Introduction

The traditional power grids are some of the biggest and most reliable systems built in the 20th century and has served its purpose for almost a century. The hierarchical structure of the traditional grids, where power is centrally distributed from a few number of large scale generators to a large number of consumers, is not suitable for present power demand. The limited interconnection between the grids and one way flow of power without the balance between the generation and consumption has led to the underutilization of power. So, the power generation and distribution should be based on the real-time consumption.

Moreover, the growing awareness for sustainable and green energy has resulted in distributed generation of power from different sources like sun, wind,

A. Kwiecień, P. Gaj, and P. Stera (Eds.): CN 2012, CCIS 291, pp. 352–359, 2012.

household waste, power-heat coupling etc. This transition from centralized to a distributed power generation has resulted in the need for intelligent transmission, distribution and management of the consumer demands. This requires an infrastructure where all the distributed generators, transmission line, distribution substations, and the consumer utilities can be interconnected as a single network or "smart grid". The focal concept of a smart grid is to automate the power generation, distribution, and consumption with automated metering, monitoring, and management [1,2]. Smart grid is a vast infrastructure which consists of mechanical, electrical, electronic and communication systems. In this paper, we focus only on the communication perspective of the smart grid.

A general architecture of a smart grid is shown in Fig. 1. As seen in the figure the smart grid is the extension of the existing power grid with added communication infrastructure and the distributed power generator empowering the two-way flow of power and information. The electricity consumers like industrial factories, buildings, homes, and electric cars can also be an electricity producer. The produced electricity can also be put back into the grid. For example, a smart home may be able to generate electricity using solar panels and put it back into the grid, or electric vehicles may also put power back into the grid to help balance loads when demand is high.

The smart metering and monitoring features are available in the smart grid. The smart metering enables the consumer utilities to optimize power consumption. The smart monitoring enables the distributor to monitor the real-time power demands for a reliable power transmission and also monitor grid status to detect mechanical failures. Thus, smart metering and monitoring demand secure, uninterrupted and near real-time communication between the electric utilities

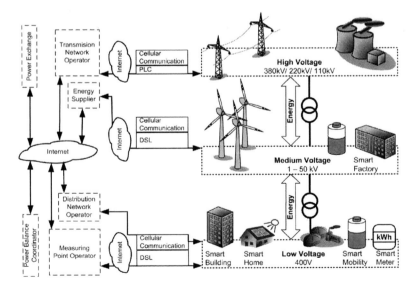

**Fig. 1.** Smart Gird Architecture

**Table 1.** Smart grid communication technologies [6,7]

| Communication Technology | | | Data Rate | Coverage Range | Remarks |
|---|---|---|---|---|---|
| Power Line Communication (PLC) | | | 2–3 Mbps | 1–3 km | Harsh and noisy channel environment |
| Digital Subscribers Line (DSL) | Ethernet | | Up to 1 Gbps | 100 m | High installation cost and less flexible |
| | Optical Fiber | | Up to 14 Tbps | 160 km | Extremely costly to realize in a distributed system like smart grid |
| Wireless | WLAN[a] | | Up to 54 Mbps | 200–400 m | Easy installation but high interference, short range |
| | Zigbee | | 250 kbps | 30–50 m | Easy installation but low data rate and short range |
| | Cellular | 2G[b] | Up to 170 kbps | 1–10 km | Easy installation but low data rates, licensed band |
| | | 3G[c] | 384 kbps–14.4 Mbps | 1–10 km | Easy installation but licensed band |
| | | 4G[d] | Up to 42 Mbps | 1–10 km | Easy installation but licensed band |

[a] Wireless Local Area Network.
[b] Global Packet Radio System (GPRS).
[c] High Speed Downlink Packet Access (HSDPA).
[d] Long Term Evolution (LTE).

without any human assistance. Thus machine-to-machine (M2M) communication is the necessity of future smart grid applications. In M2M communication, the communicating utilities are able to exchange information and make decision based on the exchanged information without human assistance. The M2M communication using public networks becomes more important and some of potential areas are smart grids, vehicular telematics, health care, industrial plants etc. [3,4,5].

The dedicated connection between the consumer utilities, measuring points, distribution network, energy supplier and the transmission network would be very complex, expensive and difficult to maintain. Internet-based communication would be much easier to implement and maintain, especially to interconnect geographically distributed substations. Thus, the internet based communication would be the ideal communication solution in smart grid applications. Table 1 present an overview of smart grid communication technologies.

The power line communication (PLC) suffers from high attenuation due to unpredictable voltage transients and harmonics. The need for physical connections in DSL communication reduces flexibility and increases the installation cost in the field of substations. The wireless communication is the suitable solution for the smart grid applications because of its low cost, easy installation, and flexibility. There is no need for physical communication between the communicating utilities but the short communication range of wireless standards like WLAN and Zigbee make it unsuitable to use it as a general solution for smart grid communication. The WLAN and Zigbee can only be used for communication between the utilities within a home or building environment. Hence, cellular communication is the ideal solution for the smart grid communication because of its large coverage range, easy installation and flexibility. In this paper we want to share our results of evaluating cellular communication systems for this purpose.

The rest of the paper is organized as follows: Section 2 presents an overview of recent works on the smart grid communications. Section 3 presents cellular communication setup for M2M communication in a smart grid. The results of the communication setup of Sect. 3 are presented in Sect. 4. Section 5 finally concludes the paper.

## 2   Recent Works

The high cost of installing wired communication setup in smart grid applications has inclined more research towards the wireless solution. An overview of applicability of wireless M2M communication in a smart grid is presented in [3] and it further presents a network architecture of home energy management system (HEMS) in smart grid. Different areas of M2M communication such as smart grid, E-health, vehicular telematics are presented in [4] and the paper further presents the motivation for the transition from current cellular based M2M communication solution to the embedded internet based M2M communication. The paper presented in [8] presents the communication challenges, security and privacy issues for different smart grid communication networks. The paper also presents the smart metering standardization activities in Europe.

The distribution feeder level communication requirements in smart grid applications are presented in [9] and the paper further proposes three-layer wireless communication architecture to increase the reliability and reduce the latency. A comparative study of different communication technologies applicable for smart grid and an overview of the smart grid standards are presented in [6]. The theoretical study on different wireless communication technologies, such as WLAN, WiMAX, Zigbee, Cellular etc., and the challenges for their deployment in smart grid application is presented in [10].

A general overview of wireless network architecture for smart grid applications is presented in [11] and the paper further present two network planning approaches for NAN communication using 3G cellular technology. In [12], the wireless technologies – Wi-Fi, ROF (Radio-over-fiber), and 4G cellular network – are presented as a possible solution for smart grid communication.

The above works illustrate the considerable advantages of wireless communication over wired communication in smart grid applications but most of the paper presents only the conceptual and theoretical research. Some of the paper presented the simulation results but the physical realization can be much more challenging. In this paper, we present the performance evaluation of M2M communication in smart grid application using cellular communication technology.

## 3   M2M Communication Setup

The idea behind the M2M communication setup used in our experiment is to use the Internet and public cellular network as a communication system between the smart grid utilities. The stationary utilities in the home/building will be connected to the Internet while the utilities in the far-off location and mobile utilities (for e.g. electric car) will be connected to the cellular network. This idea is realized in M2M communication setup shown in Fig. 2. The sender resembles the stationary utilities connected to internet while the responder resembles the utilities in the far-off location. Cisco 1800 series routers were used as sender and responder machines because of its built-in QoS measurement capabilities. The sender was installed in the test lab inside the university premises and connected to the internet via the university network. The responder was installed in a resident home in Lemgo, Germany and connected via a cellular modem to a German ISP (Internet Service Provider). The experiments were performed over a period of five days for different cellular technologies (2G, 3G, and 4G). The GPRS, HSDPA, and LTE services offered by a German telecommunication provider were used for 2G, 3G, and 4G technology respectively. The download and upload data rate of the different technologies are shown in Table 2.

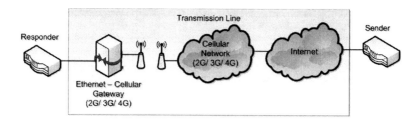

**Fig. 2.** M2M Communication Setup(2G, 3G, and 4G)

**Table 2.** Data rate of different cellular technology

| Cellular Technology | Data Rate | |
|---|---|---|
| | Download | Upload |
| 2G (GPRS) | 110 kbps | 26.8 kbps |
| 3G (HSDPA) | 7.2 Mbps | 5.76 Mbps |
| 4G (LTE) | Not specified | Not specified |

## 4   Measurement Results

The round trip time (RTT) and jitter were used as the performance metrics of our M2M communication setup. The cisco built-in IP SLA (service level agreement) support was used to measure the RTT and the jitter.

## 4.1    Round Trip Time (RTT)

RTT is the time taken by a signal from the sender to reach the responder plus the time taken by the acknowledgment of that signal to be received back by the sender at the IP level (OSI layer 3). To measure the RTT, the sender periodically sends an ICMP echo request with a payload size of 60 byte at a transmission interval of 10 second and measures the time after successfully receiving the ICMP echo response from the responder. The measured values stored in text files on a daily basis.

The RTT measured for duration of five days with 2G, 3G and 4G cellular technology is plotted in Fig. 3. The highest RTT was measured for 2G (GPRS) and the lowest RTT was measured for LTE. The maximum fluctuation in RTT in case of GPRS is due to high number of active users. The mean RTT was measured as 300 ms, 126.7 ms, and 69.86 ms for GPRS, HSDPA, and LTE respectively. The low RTT shows that new generation cellular technology will offer good performance for near real-time M2M communication.

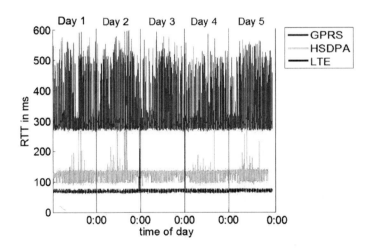

**Fig. 3.** RTT for GPRS (2G), HSDPA (3G), and LTE (4G) Cellular Communication Setup

## 4.2    Jitter

Jitter is the time difference between the maximum and minimum packet transmission time from sender to responder. The sender periodically sends an UDP data burst of 20 packets at a transmission interval of 30 seconds to responder (OSI layer 4). The responder measures the time and stores the measured time and the time stamp received from the sender. The measured values are stored in text files on a daily basis. The jitter was calculated for GPRS and HSDPA for duration of five days and is shown in Fig. 4. The low jitter was measured during early and late hours of the day while significant amount of jitter was measured

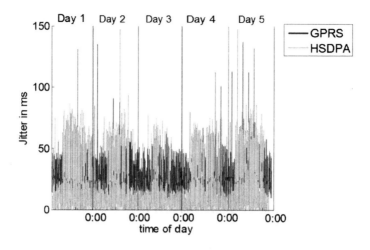

**Fig. 4.** Jitter for GPRS (2G), and HSDPA (3G) Cellular Communication Setup

in the other hours of the day. The mean jitter was measured 25.09 ms and 8.4 ms for GPRS and HSDPA respectively. For LTE no jitter was measured because the measurements were performed at a time when the LTE-technology was just introduced in the region and very few users existed of this technology.

## 5 Conclusion

The paper presented the idea of using the internet and the public cellular networks for M2M communication in smart grid applications. We used 2G (GPRS), 3G (HSDPA) and 4G (LTE) technology provided by a German telecommunication provider for a performance evaluation. The RTT and jitter was measured to demonstrate the performance of our M2M communication setup. The lowest RTT was measured for LTE and the highest RTT was measured for GPRS. The results demonstrate the possibility of realization of near real-time M2M communication using the internet and the public cellular network.

## References

1. Fang, X., Misra, S., Yang, D.: Smart Grid – The New and Improved Power Grid: A Survey. IEEE Communications Surveys & Tutorials 99 (2011)
2. Lo, C., Ansari, N.: The Progressive Smart Grid System from Both Power and Communications Aspects. IEEE Communications Surveys & Tutorials 99 (2011)
3. Niyato, D., Xiao, L., Wang, P.: Machine-to-Machine Communications for Home Energy Management System in Smart Grid. IEEE Communications Magazine 49(4), 53–59 (2011)
4. Wu, G., Talwar, S., Johnsson, K., Himayat, N., Johnson, K.D.: M2M:From Mobile to Embedded Internet. IEEE Communications Magazine 49(4), 36–43 (2011)

5. M2M Journal (2010, 2011), http://www.m2m-alliance.com (visited February 2, 2012)
6. Güngör, V.C., et al.: Smart Grid Technologies: Communication Technologies and Standards. IEEE Transactions on Industrial Informatics 7(4), 529–539 (2011)
7. http://www.ntt.co.jp/news/news06e/0609/060929a.html (visited February 3, 2012)
8. Fan, Z., et al.: Smart Grid Communications: Overview of Research Challenges, Solutions, and Standardization. IEEE Communications Surveys & Tutorials 99 (2012)
9. Aravinthan, V., Karimi, B., Namboodiri, V., Jewell, W.: Wireless Communication for Smart Grid Applications at Distribution Level – Feasibility and Requirements. In: 2011 IEEE Power and Energy Society General Meeting, pp. 1–8 (2011)
10. Parikh, P.P., Kanabar, M.G., Sidhu, T.S.:Opportunities and Challenges of Wireless Communication Technologies for Smart Grid Applications. In: IEEE Power and Energy Society General Meeting, pp. 1–7 (2010)
11. Wietfeld, C., Georg, H., Groening, S.,Lewandowski, C., Mueller, C., Schmutzler, J.: Wireless M2M Communication Networks for Smart Grid Applications. In: 11th European Wireless Conference 2011 – Sustainable Wireless Technologies (European Wireless), pp. 1–7 (2011)
12. Feng, Z., Yuexia, Z.: Study on Smart Grid Communications System based on New Generation Wireless Technology. In: International Conference on Electronics, Communications and Control (ICECC), pp. 1673–1678 (2011)

# The Weather Impact on Speech Quality in GSM Networks

Jan Rozhon, Petr Blaha, Miroslav Voznak, and Jan Skapa

VSB-Technical University of Ostrava
17. Listopadu 15, 70833 Ostrava-Poruba, Czech Republic
{jan.rozhon,petr.blaha,miroslav.voznak,jan.skapa}@vsb.cz

**Abstract.** Since the cellular GSM networks are one of the most commonly used communication technologies today, the quality of speech in these networks becomes a topic of great significance. Many advances and approaches have been introduced in the field of speech quality during the last decade, most of them focusing on the IP networks, where the speech quality is influenced by every single network node through which the communication passes. This paper focuses on the cellular network environment, through perceptual analysis of the speech sample using the Perceptual Evaluation of Speech Quality algorithm we found a bond between the speech quality in GSM networks and current weather conditions. We have obtained both meteorological data and Mean Opinion Score value specifying the current speech quality in the GSM networks and by utilization of highly advanced techniques in data mining and data analysis, our team found out the correlation between current rain density and the speech quality.

**Keywords:** GSM, IP networks, PESQ, Speech Quality Assessment.

## 1 Introduction

Through years of successful expansion GSM and UMTS technologies have become a commonplace and paved their way to every aspect of human activity. We can now see the cell phones being used by almost all people to communicate, exchange messages or even access the Internet. Although the variety of services the cell phones now offer has been vastly extended, the voice communication still occupies the position of a most important kind of communication.

Since both technologies are needed to provide constantly high quality of service in all areas, the voice communication cannot be excluded because of its leading position. To successfully measure and control the quality of service, or quality of speech to be more specific, in GSM/UMTS environment we can take advantage of multiple algorithms that have been invented for the IP based networks, which would allow us to create a flexible and low cost platform for GSM/UMTS speech quality measurement. By embracing this approach and creation of such a platform we would be allowed to measure influence of almost any

A. Kwiecień, P. Gaj, and P. Stera (Eds.): CN 2012, CCIS 291, pp. 360–369, 2012.
© Springer-Verlag Berlin Heidelberg 2012

signal interference from attenuation to high number of subscribers. The influence of the actual weather conditions on the speech quality in the GSM/UMTS networks is also measureable and it is a main topic of this paper.

In further sections we will try to describe the measuring mechanism and the whole platform as well as the algorithm used for evaluation of the collected results and will also try to find whether there is a correlation between the speech quality and the current weather conditions.

## 2   State of the Art

There are two main categories of speech quality assessment techniques – subjective and objective, the output of which is the Mean Opinion Score (MOS), which is a five degree scale for speech quality evaluation developed by ITU-T. The main goal of objective methods is as precise as possible estimation of the MOS value as it would be obtained by the subjective methods with the number of participants high enough to perform reasonable statistical analysis. We distinguish two separate sub-groups in the objective methods – Intrusive and Nonitrusive [1,2]. The intrusive methods use the original voice sample as it has entered the communication chain and compare it with the degraded one as it has been outputted by this communication chain, the following list contains the most important intrusive algorithms.

- PSQM (Perceptual Speech Quality Measurement),
- PAMS (Perceptual Analysis Measurement System),
- PESQ (Perceptual Evaluation of Speech Quality),
- P.OLQA (Perceptual Objective Listening Quality Assessment).

From the mentioned algorithms PESQ is currently the most common one [3]. It combines the advantages of PAMS (robust temporal alignment techniques) and PSQM (exact sensual perception model) and is described in ITU-Ts recommendation P.862 [4]. Since this algorithm is most widely used these days and since it generates results with reasonable accuracy and efficiency we have decided to use this algorithm as the keystone of our testing platform. The algorithms basic philosophy is depicted on the Fig. 1. The last mentioned algorithm, P.OLQA, is intended to be a successor of the PESQ and it tries to avoid the weaknesses of the PESQs model and to incorporate the possibility of high bandwidth signal analysis.

Contrary to intrusive methods which need both the output (degraded) sample and the original sample, non-intrusive methods do not require the original sample. This is why they are more suitable to be applied in real time. Yet, since the original sample is not included, these methods frequently contain far more complex computation models. Examples of these types of measurements frequently use INMD (in-service non-intrusive measurement device) that has access to transmission channels and can collate objective information about calls in progress without disrupting them. These data are further processed using a particular method, with a MOS value as the output. The method defined by

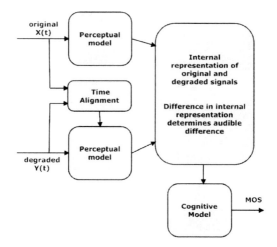

**Fig. 1.** Basic diagram of PESQ algorithm

ITU-T recommendation P.563 or a more recent computation method E-model defined by ITU-T recommendation G.107 are examples of such measurements.

Today we can see various implementations of the speech quality testing mechanisms and algorithms mainly in business solutions. These solutions are conformant with the specifications of the International Telecommunication Union only from some part therefore the measurements cannot be compared without the thorough knowledge of the used algorithms [5,6]. On the other hand the telecommunication union itself presents on its websites the simple implementation of one of the most advanced algorithms in the field of speech quality measurement. Several open source programs are also built upon these algorithms and the companies (Optikom, Psytechnics), which developed the source code also offer their own services based on this source code and its modifications. Regarding the speech quality in GSM and 3G networks mainly the first named company Optikom offers some services [7], but no one has performed the long term measurement with the focus on determination of weather influence on the speech quality in the GSM networks.

## 3   Testing Platform and Data Analysis

Our main goal was to create a testing platform that would be able to generate GSM calls automatically in regular intervals and together with that to log the actual weather conditions and analyze the voice sample in accordance to P.862. By using this platform we would be able to generate statistically significant amount of input data to perform the data mining analysis and find the possible correlation between one or multiple weather attributes and the obtained MOS value describing the quality of speech.

## 3.1   Platform

To successfully achieve this goal, we decided to use a Linux based system due to its natural effectiveness in the field of automation and process monitoring. Moreover this would allow the whole platform to be highly cost-effective.

On the Linux machine, we configured the most commonly used VoIP PBX Asterisk in the way that it was able to generate one call every five minutes over the SIP (Session Initiation Protocol) and RTP (Real-time Transport Protocol). These calls were routed to a SIP peer, which represented a SIP/GSM+UMTS gateway with two separate SIM cards. Each call was routed from the SIM card in module one, to the SIM in module two, thus allowing to use single BTS station in the building nearby and consequently minimize the interference caused by the BTS switching or long signal routes to best possible minimum since the measured is being performed on the closest BTS. The GSM gateway then routed the incoming calls back to Asterisk PBX, which recorded the voice data to a separate WAV file. This way the loop was created allowing the Linux system with Asterisk PBX to have access to both original WAV file and the degraded one, which is necessary for successful implementation of PESQ algorithm. The hardware requirements were low and are listed below:

- Low-end HW server with Ethernet interface,
- SIP/GSM+UMTS gateway with two separate modules working on 1800 MHz (DCM-1800).

As the speech files we used the samples of calibrated voices specifically designed for the use with PESQ algorithm with the sampling frequency of 8 000 Hz, encoded using PCM into 128 kbps stream of one channel audio. The cooperation between Asterisk PBX and PESQ algorithm was achieved via the System command in Asterisk dialplan [8,9] that fires the python script, which performs degraded file modification, fetching the current meteorological data, performing the PESQ evaluation and storing the data into the database.

The degraded file modification was made necessary due to about 2 seconds of leading period in the degraded file, where two ring tones from Asterisk were recorded. If this was left unhandled, the results from the PESQ analysis would be highly inaccurate and therefore not useful.

The meteorological data were obtained from the local meteorological station working in the university campus about 300 meters from the BTS station. The actual data transmission was implemented by HTTP communication, where python script asked meteorological station for current conditions in text format, which then was parsed and data were stored.

As a database the SQLite engine is used because of two reasons. Firstly, the configuration and manipulation of the database are quite simple in this database engine but still providing enough features to successfully complete the given task without any complication caused by the engine limitations. And secondly, the whole database can be backed up easily [10]. This is allowed by the fact, that the whole database is stored one user-defined file. Especially the latter is important due to the long period of measurement.

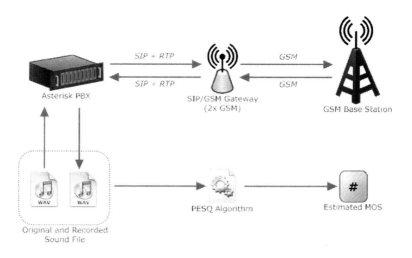

**Fig. 2.** Scheme of testing platform with the detail on MOS computation

Storing data into the database is useful also from the data analysis point of view, because the values of measured MOS can be displayed in almost real time with the use of the conditional database lookup for any month, day or even hour or minute thanks to the used timestamp. Moreover the aggregate functions allow quick data analysis even on the huge number of stored values.

The software equipment used to perform the measurement consists of:

- Ubutnu 10.04.1 x64,
- Asterisk PBX 1.6.2.16,
- SQLite3.

The testing platform scheme is depicted on the Fig. 2, where the basic communication and protocol are displayed in the clearer detail.

### 3.2   Platform Limits and Simplification

From the previously presented architecture it is clear that the GSM network and the parameters affecting it is not the only factor that contributes to the final speech quality. In addition the status of the network over which Asterisk PBX and SIP/GSM gateway communicate can influence the speech quality. Moreover the codec translation between used codecs (G.711A, GSM) can have deteriorating effect on the speech quality devaluing the final results. Because these inherent effects cannot be eradicated basic assumptions applied to measurement had to be designed.

Firstly the communication between Asterisk PBX and SIP/GSM gateway takes place on the 100 Mbps Ethernet line, where no other traffic is allowed. Therefore the network will never reach the congestion state and all the information on the network is exchanged as quickly as it is physically possible for this type of interconnection. Because of this the delay of packets and its variation is minimal and does not need to be counted with.

The codec translation is not a parameter, which can be easily dealt with. Since the SIP/GSM gateway supports only two codes for the VoIP communication (namely G.711 and G.729) and the GSM communication is built upon the GSM EFR codec the translation will take place every time the call is made.

Therefore the measured MOS value will always be affected by this process. To eliminate or at least diminish the influence to speech quality we do a calibration of the system. This process takes the ideal MOS value of speech quality using the GSM EFR codec as the basis, to which we compare the best result taken from several hundred measurements. The difference between the ideal and best case result is then identified as the distortion caused by the codec translation and this value is then added to all the measured values. This way the codec translation influence is limited and the measurement will provide reasonable results. Even if this countermeasure was not performed the trend in the MOS values would still be preserved and the correlation between weather condition and measured values could still be found.

Other approach could be built upon creating the reference codec translation result in the exclusively IP based network, but as it has been explained in the previous text, this step could be skipped and the reasonable data still can be obtained.

### 3.3   Evaluation of Results and Discovering the Correlation

As the convenient and efficient way of data analysis clustering algorithms allow for standardized and confirmed data segregation into the groups with similar attributes. The simplest definition is shared among all and includes one fundamental concept: the grouping together of similar data items into clusters. These obtained clusters should reflect some mechanism at work in the domain from which instances or data points are drawn, a mechanism that causes some instances to bear a stronger resemblance to one another than they do to the remaining instances.

Let $X \in R^{m \times n}$ a set of data items representing a set of $m$ points $x_i$ in $R_n$. The goal is to partition $X$ into $K$ groups $C_k$ such every data that belong to the same group are more alike than data in different groups. Each of the $K$ groups is called a cluster. The result of the algorithm is an injective mapping $X \rightarrow C$ of data items $X_i$ to clusters $C_k$ [11].

Since there are multiple clustering methods and algorithms, the efficient and most precise one for our purposes needed to be found empirically. Through several testing runs of several algorithms ($K$-means clustering, EM clustering, ... ) $K$-means served best in our case meaning that the clusters incorporated the logically most correct data and did not suffer from the algorithms tendency to create the clusters of equal or similar size. The definition of $K$-means clustering looks as follows.

Let $S = \{X_1, X_2, \ldots, X_N\}$ be a dataset with $n$ observations, each of which is $p$-dimensional. The objective in $K$-means clustering is to group these observations into categories $C_1, C_2, \ldots, C_K$ for given $K$, such that the objective function

$$O_K = \sum_{i=1}^{n} \sum_{k=1}^{K} I(X_i \in C_k)(X_i - \mu_k)'(X_i - \mu_k)$$

is minimized. Here $\mu_k$ represents the mean vector of observations from $C_k$, $\mu_k = 1/n_k \sum_{i \in C_k} X_i$ where $n_k = |C_k|$ is the number of observations in $C_k$ and $I(X \in C_k)$ is an indicator function specifying whether observation $X$ belongs to the $k$-th group. Further, note that the following $||x|| = \sqrt{x'x}$ denotes the Euclidian norm of $p$-dimensional vector $x$.

## 3.4   Results

As we put the pieces together we can finally analyze the results. Using the mentioned K-means clustering method we have performed analysis of all available possible influencer of MOS value in GSM/UMTS environment. To be more specific, we explored the influence of the Current Temperature, Humidity, Rain, Dew Point, Wind Speed and Atmospheric Pressure.

In addition we had even other parameters at our disposal, such as THW Index, Wind Direction and so on, but these are expected to have no impact on the measured MOS value because of their signal non-interfering nature.

Through series of data mining operation we came to conclusion, that the self-correcting mechanisms implemented in the GSM/UMTS technology prevent call from being interfered by weather condition. The fact that there is no statistically significant relation between MOS and Humidity, which is the most influencing factor from the signal strength point of view, discouraged us from the thought that there might be a relation of any kind. Because of this we didnt expect that the other parameters connected directly to humidity like current rain rate would be influencing the voice quality with a measurable significance. However, this correlation can be found in the statistical data.

The Table 1 shows the actual probability of MOS value depending on the current rain density. Second row in the table can be seen as the most important and tells us the following: if the current rain density is between 28.5 and 33.9 mm/h we can expect MOS to be lower than 2.1 with the probability of 64%. Other rows can be read similarly. These results in the following, with the increasing rain activity, the MOS value drops significantly. Especially when the high rain density is reached (greater than 5 mm/h) the MOS value drops to level, where the user can experience very bad speech quality and low comprehensibility. The last row is actually a conjunction of two separate clusters, which were identified as distinct areas by the algorithm. However they are mutually complementary and therefore they were combined into single row.

If we put that together with our knowledge of low humidity influence, we can state that this quality drop occurs in the beginning of the rain, while the air

**Table 1.** The influence of current rain density on MOS

| CurRainRate [mm/h] | Favors | Relative Impact |
|---|---|---|
| $\geq 33.899$ | $< 2.099$ | 100 |
| $28.502 - 33.899$ | $< 2.099$ | 64 |
| $15.133 - 28.502$ | $2.099 - 2.408$ | 100 |
| $4.883 - 15.133$ | $2.408 - 2.781$ | 100 |
| $< 4.883$ | $\geq 2.781$ | 100 |

humidity is still low and the effect diminishes while the humidity rises. This can be caused by the slow adaptation of the network or mobile station to the rain.

The data of high significance (with rain density) can be interpreted as the sequence resulting in the chart on the Fig. 3. Here we can see the data subset containing the samples with the significant rain density (60 samples with rain density higher than 2.2 mm per hour) ordered from the highest to the lowest rain density. As we can observe the humidity varies independently on the current rain density, however MOS factor is influenced measurably.

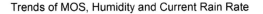

Trends of MOS, Humidity and Current Rain Rate

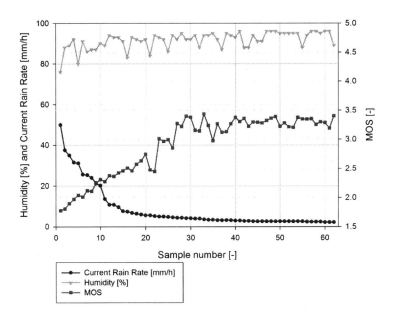

**Fig. 3.** Relation between MOS, Humidity and Current Rain Rate

Figure 3 clearly confirms the stated information about the rain influence on the quality of speech in GSM/UMTS environments. As already stated this is the only significant relation, therefore no further analysis makes sense at this point.

To complete the picture, it is necessary to state that the measured data were collected from August to November 2011 and the total data pool contains about

20 000 data rows. Since the BTS station and all the other parts of the network, through which the signal traverse (MSC, IMS Core), are not under our control and therefore the information about their load is not possible to obtain, the statistical methods were used to eliminate the effect of variable load of these network elements using huge amount of input data.

## 4   Conclusion

Through long term measurement we have obtained both meteorological data and MOS values specifying the current speech quality in the GSM networks. By utilization of advanced techniques in data mining and data analysis, our team found the correlation between current rain density and the speech quality. This shows us a great decrease of speech quality in the earliest phases of the rain, where the humidity is still low. Particularly, we can observe a 50 percent decrease in measured MOS parameter when comparing sample results obtained during the heavy rain and those obtained during a mild shower. This decrease is reflected in the worsened speech quality by glitches in the speech, low comprehensibility and other communication difficulties. The reason for this behavior can be found in the BTS transmitted power adjustment, which can modify the level in a range of up to 20 dB [12,13].

The slow response of the correction procedure in the BTS can cause problems with the same nature as we have witnessed during our measurement. Other possible explanation includes the influence of the number of subscribers logged in the particular BTS at a particular time, but this would result in quality deterioration mainly during the busy hour and this behavior was not observed. No other significant bond was found.

By performing this measurement, we have proved that the low cost measuring platform can be developed and used for the speech quality measurement in the GSM networks. We have successfully taken advantage of our teams knowledge in the IP telephony and transited this knowledge to cellular networks.

The greatest possible improvement of our method as we see it now is to perform the measurement during the whole year to have complete knowledge of the speech quality trends during all the possible weather conditions and to modify the testing platform by introducing OpenBTS solution, which would allow for gaining the full control over the transmission chain.

**Acknowledgments.** This work was supported by IT4Innovations Centre of Excellence project, reg. no. CZ.1.05/1.1.00/02.0070 within Operational Program Research and Development for Innovations conducted at VSB-Technical University of Ostrava, Czech Republic.

## References

1. Rozhon, J., Voznak, M.: Development of a speech quality monitoring tool based on ITU-T P.862. In: Proceedings 34th International Conference on Telecommunications and Signal Processing, Budapest, pp. 62–66 (2011)

2. ITU-T Recommendation G.107, The E-model: A computational model for use in transmission planning, Geneva (April 2009)
3. Wallace, K.: Implementing Cisco Unified Communications Voice over IP and QoS (CVOICE) Foundation Learning Guide: (CCNP Voice CVoice 642-437), 4th edn. Cisco Press, Indianapolis (2011)
4. ITU-T: P. 862: Perceptual evaluation of speech quality (PESQ): An objective method for end-to-end speech quality assessment of narrow-band telephone networks and speech codecs (2001)
5. Pravda, I., Vodrazka, J.: Voice quality planning for NGN including mobile networks. In: 12th International Conference on Personal Wireless Communications (PWC 2007), Prague (2007)
6. Voznak, M., Tomes, M., Vaclavikova, Z., Halas, M.: E-model Improvement for Speech Quality Evaluation Including Codecs Tandeming. In: Advances in Data Networks, Communications, Computers, Faro, pp. 119–124 (2010)
7. Opticom: Opticom quality testing products page,
   http://www.opticom.de/products/product-overview.php
8. Meggelen, J., Smith, J., Madsen, L.: Asterisk: The Future of Telephony, 2nd edn. O'Reilly, Sebastopol (2007)
9. Meggelen, J., Madsen, L., Bryant, R.: Asterisk: The Definite Guide, 3rd edn. O'Reilly, Sebastopol (2011)
10. Owens, M., Allen, G.: The Definite Guide to SQLite, 2nd edn. Apress, New York (2010)
11. Fung, G.: A Comprehensive Overview of Basic Clustering Algorithms (2001)
12. Mouly, M., Pautet, M.B.: The GSM System for Mobile Communications. Telecom Publishing (1992)
13. Saunders, S.R., Zavala, A.A.: Antennas and Propagation for Wireless Systems, 2nd edn. Wiley (2007)

# Tandem Retrial Queueing System with Correlated Arrival Flow and Operation of the Second Station Described by a Markov Chain

Chesoong Kim[1], Alexander Dudin[2,*], and Valentina Klimenok[2]

[1] Sangji University, Wonju, Kangwon, 220-702, Korea
dowoo@sangji.ac.kr
[2] Belarusian State University, 4, Nezavisimosti Ave., Minsk , 220030, Belarus
dudin@bsu.by, vklimenok@yandex.ru

**Abstract.** Tandem queues are good mathematical models for description of information transmission in various communication systems and networks. These queues play also an important role for the validation of different decomposition algorithms designed for investigating more general queueing networks. So, their investigation is interesting for theory and applications. In this paper, we consider tandem queue suitable for modeling the systems and networks where information flows are correlated and bursty what is typical for many modern telecommunication networks. Possible correlation of customers inter-arrival times and batch arrivals are taken into account via of consideration of the *Batch Markovian Arrival Process* (*BMAP*) as input stream to the system. The system consists of two stations. The service time at the station 1 is assumed to be generally distributed. There is no buffer at this station, and customers who meet the busy server repeat attempts to enter the system in random time intervals. The service process at the station 2 is assumed to be described by the continuous time Markov chain with a finite state space. This assumption holds good, e.g., if the station 2 has a finite buffer, consists of a finite number of identical or heterogeneous servers where the service time distribution is assumed to be of *PH* (PHase) type. Markov chain embedded at service completion epochs at the station 1 and the process of system states at arbitrary time are under study. Ergodicity condition and algorithms for computing the steady state probabilities are presented.

**Keywords:** queueing networks, batch Markovian arrival process, phase type service, tandem retrial queue.

## 1 Introduction

Tandem queues are widely used for capacity planning and performance evaluation of computer and communication systems, service centers, contact centers,

---

* Corresponding author.

A. Kwiecień, P. Gaj, and P. Stera (Eds.): CN 2012, CCIS 291, pp. 370–382, 2012.

security and authentication systems, manufacturing systems, etc. Some examples of their application to real systems can be found in [1].

The theory of tandem queues is quite well developed, for more references see, e.g. [1,2,3,4]. The overwhelming majority of the papers is devoted to the queues with the stationary Poisson arrival process and exponential service time distribution. But it is well known that the arrival flows in many real world systems and in telecommunication networks in particular have more general nature, e.g., the successive inter-arrival times may be dependent, see, e.g., [5] and [6]. The $BMAP$ can be used to model such flows.

The $BMAP$ was introduced as versatile Markovian point process ($VMPP$) by M. Neuts in [7]. Further, his notations were simplified greatly by [8] and ever since these processes bear the name batch Markovian arrival process ($BMAP$). The class of $BMAP$s includes many input flows considered in literature previously, such as stationary Poisson ($M$), Erlangian ($E_k$), Hyper-Markovian ($HM$), Phase-Type ($PH$), Interrupted Poisson Process ($IPP$), Markov Modulated Poisson Process ($MMPP$), etc.

Tandem queues with the $BMAP$ input with an infinite or a finite buffer where considered in [9,10,11,12,13,14,5,15,16,17,18,19]. Much less attention was paid to queues with retrials (repeated calls). Single server retrial queues with the $BMAP$ input and arbitrary service time distribution were first investigated in [20,21]. In recent papers [22,23,24], tandem retrial queues with the $BMAP$ input are analyzed. In this paper, we consider tandem retrial queue with more abstract description of operation of the station 2 of the system. The presented results can be used as a background for investigation of retrial tandems with complicated Markovian behavior of station 2 of a tandem. E.g., this station can consist of a finite number of inhomogeneous unreliable servers with phase type or Markov Service Process with a finite capacity of a buffer. Cross-traffic arriving directly at station 2, group occupation of the servers, negative customers, breakdowns, interruptions, vacations and disasters can be easily incorporated to the model. Results of [22,23,24] can be derived as particular simple cases from the corresponding results presented in this our paper.

The rest of the paper is organized as follows. In Section 2, the model under consideration is described. In Section 3, the embedded Markov chain at service completion epochs at station 1 is introduced and its transition probabilities are specified. Section 4 contains the results concerning the stationary distribution of the Markov chain. The stationary state distribution of the tandem queue at arbitrary times is calculated in Sect. 5. Finally, Section 6 concludes the paper.

## 2   The System Model

We consider tandem queues consisting of two stations. Operation of station 1 is described by the $BMAP/G/1$ retrial queueing system. This means the following. The system has a single server. The input flow is described by a $BMAP$. The $BMAP$ is defined by means of the underlying process $\nu_t, t \geq 0$, which is an irreducible continuous time Markov chain with state space $\{0, \ldots, W\}$ where

$W$ is some finite integer. Arrivals of customers may occur in batches at epochs when the process $\nu_t,\ t \geq 0$ makes the jumps. The intensities of jumps from one state into another, which are accompanied by an arrival of a batch consisting of $k$ customers, are combined into the matrices $D_k,\ k \geq 0$ of size $(W+1) \times (W+1)$. The matrix generating function of these matrices is $D(z) = \sum_{k=0}^{\infty} D_k z^k, |z| \leq 1$. The matrix $D(1)$ is the infinitesimal generator of the process $\nu_t, t \geq 0$. The stationary distribution vector $\boldsymbol{\theta}$ of this process satisfies the equations $\boldsymbol{\theta} D(1) = \mathbf{0}, \boldsymbol{\theta}\mathbf{e} = 1$. Here and in the sequel $\mathbf{0}$ is the zero row vector and $\mathbf{e}$ is the column vector of appropriate size consisting of ones. In case the dimensionality of the vector is not clear from context, it is indicated as a suffix, e.g. $\mathbf{e}_{\overline{W}}$ denotes the unit column vector of dimensionality $\overline{W} = W + 1$.

The average intensity $\lambda$ (fundamental rate) of the $BMAP$ is defined as

$$\lambda = \boldsymbol{\theta} D'(z)|_{z=1} \mathbf{e} \ ,$$

and the intensity $\lambda_g$ of group arrivals is defined as $\lambda_g = \boldsymbol{\theta}(-D_0)\mathbf{e}$. The variance $v$ of intervals between group arrivals is calculated as $v = 2\lambda_g^{-1}\boldsymbol{\theta}(-D_0)^{-1}\mathbf{e} - \lambda_g^{-2}$, while the coefficient of correlation $c_{\text{cor}}$ of intervals between successive group arrivals is given by

$$c_{\text{cor}} = (\lambda_g^{-1}\boldsymbol{\theta}(-D_0)^{-1}(D(1) - D_0)(-D_0)^{-1}\mathbf{e} - \lambda_g^{-2})/v \ .$$

For more information about the $BMAP$, its history, properties, special cases and related research see [8] and the survey paper [25].

If the arriving batch meets a free server upon arrival, one customer immediately starts a service. The service time distribution function is assumed to be $B^{(1)}(t)$ with the finite first moment $b_1^{(1)} = \int_0^{\infty} t\mathrm{d}B^{(1)}(t)$. If the server is busy or the batch consists of more than one customer, the customers, which did not start a service, go to the so called orbit.

Orbit is a virtual pool from which customers try their luck later on after a random amount of time. The inter-retrial times of an individual customer do not depend on the behavior of other customers in the orbit as well as on the arrival and service processes. The times between the retrials made by all customers are exponentially distributed with some parameter $\alpha_i,\ i > 0$, when $i$ customers stay in the orbit, $\alpha_0 = 0$. The most popular dependencies of $\alpha_i$ on $i$, which are considered in literature, have the following forms: $\alpha_i = i\alpha,\ \alpha > 0$, (classic retrial strategy), $\alpha_i = \gamma,\ \gamma > 0$, (constant retrial rate), $\alpha_i = i\alpha + \gamma,\ \alpha > 0, \gamma > 0$, (linear strategy of retrials). Because we intend to develop an algorithmic analysis of the model, we do not need to fix a concrete dependence of $\alpha_i$ on $i$. We only assume that $\lim_{i \to \infty} \alpha_i = \infty$.

The customers from the orbit are assumed be absolutely persistent. They repeat attempts until they succeed to meet a free server. Service time distribution function of a customer from the orbit is assumed to be $B^{(2)}(t)$ with the finite first moment $b_1^{(2)} = \int_0^{\infty} t\mathrm{d}B^{(2)}(t)$.

After the service at station 1, the customer has to get the service at station 2. The operation of station 2 is defined by irreducible Markov chain $\eta_t = (m_t, \psi_t)$, $t \geq 0$, with a finite state space

$$\mathcal{M} = \{m_t \in \{0, \ldots, M\}, \psi_t \in \{1, \ldots, Z_{m_t}\}\} \ .$$

We may interpret the component $m_t$ of the Markov chain $\eta_t$, $t \geq 0$, as the number of customers at station 2 at the moment prior the moment $t$ and the component $\psi_t$ can be interpreted a vector process describing behavior of different underlying Markov chains affecting the value of the component $m_t$. E.g., the component $\psi_t$ can be interpreted as the set of the states of the underlying processes of a service or repair at the servers of station 2, the states of the underlying cross-traffic process at this station, etc.

Behavior on the Markov chain $\eta_t$ is described as follows. If the state of the component $m_t$ of the Markov chain $\eta_t = (m_t, \psi_t)$ prior the service completion moment at station 1 is not equal to $M$, the chain $\eta_t$ makes at this moment a transition according to a fixed one-step transition probability matrix $\mathcal{S}$ of order $\tilde{Z} = \sum\limits_{m=0}^{M} Z_m$. The customer, which has completed the service at station 1, starts the service at station 2.

Till the next service completion moment at station 1, transitions of the Markov chain $\eta_t$ are defined by the intensities

$$Q_{(m,\psi),(m',\psi')}, \ m, m' \in \{0, \ldots, M\}, \ \psi \in \{1, \ldots, Z_m\}, \ \psi' \in \{1, \ldots, Z_{m'}\}$$

combined into the matrices $Q_{m,m'}$, $m, m' \in \{0, \ldots, M\}$ of order $Z_m \times Z_{m'}$. We denote by $Q$ the matrix (generator) consisting of blocks $Q_{m,m'}$, $m, m' \in \{0, \ldots, M\}$.

If the state of the component $m_t$ of the Markov chain $\eta_t = (m_t, \psi_t)$ prior the service completion moment at station 1 is equal to $M$, with probability $q$, $0 \leq q \leq 1$, the customer completed the service at station 1 leaves the system permanently (is lost) and the Markov chain $\eta_t$ does not make any transition. With supplementary probability $1 - q$, operation of station 1 is blocked. Server of station 1 does not take any customer for service. Arriving primary customers, if any, move to the orbit. Customers from the orbit can not access the server. During the blocking time, transitions of the Markov chain $\eta_t$ are defined by the generator $Q^{(b)}$ having the same dimension and structure as the generator $Q$, but probably another blocks $Q^{(b)}_{m,m'}$, $m, m' \in \{0, \ldots, M\}$. Station blocking time lasts until the component $m_t$ will take the value that is not equal to $M$. At that moment, the customer, which completed the service at station 1 and caused the blocking, starts the service at station 2 and the Markov chain $\eta_t$ makes a transition according to a transition probability matrix $\mathcal{S}^{(b)}$.

In the following, we analyze the proposed system.

# 3    The Process of the System State and the Embedded Markov Chain

Consider the process

$$\zeta_t = \{i_t, r_t, \eta_t, \nu_t\}, t \geq 0 \ ,$$

where $i_t$, $i_t \geq 0$, is the number of customers in the orbit, $\eta_t$, $\eta_t \in \mathcal{M}$, is the state of the process describing operation of station 2, $\nu_t \in \{0, \ldots, W\}$ is the state of the $BMAP$ underlying process, the component $r_t$ is a random which takes values 0, 1, 2, 3 depending on whether the server of station 1 is idle, serves a primary customer, a customer from the orbit or is blocked at time $t$.

The process $\zeta_t$, $t \geq 0$ is non-Markovian. Thus, to investigate this process, we first consider the embedded Markov chain at the service completion epochs $t_n, n \geq 1$, at the first phase of service, namely

$$\xi_n = \{i_n, \eta_n, \nu_n\}, n \geq 1 \ ,$$

where

$$i_n = i_{t_n+0}, \ i_n \geq 0, \ \eta_n = \eta_{t_n-0} \in \mathcal{M}, \quad \nu_n = \nu_{t_n}, \ \nu_n \in \{0, \ldots, W\} \ .$$

Here $t_n+0$ stands for the time instant just after the epoch $t_n$ while $t_n-0$ denotes the time instant before the epoch $t_n$.

Define the one-step transition probabilities of the Markov chain $\xi_n, n \geq 1$, by

$$P\{(i, m, \psi, \nu) \rightarrow (i', m', \psi', \nu')\} =$$

$$P\{i_{n+1} = i', m_{n+1} = m', \psi_{n+1} = \psi', \nu_{n+1} = \nu' | i_n = i, m_n = m, \psi_n = \psi, \nu_n = \nu\}$$

and the matrices $P_{i,i'}$ formed by these probabilities listed in the lexicographic order.

**Lemma 1.** *The transition probability matrix of the chain $\xi_n$, $n \geq 1$, has the block structure $P = (P_{i,j})_{i \geq 0, \ j \geq \max\{0, i-1\}}$, where*

$$P_{i,j} = ((S\hat{I} + q\bar{I}) \otimes I_{W+1}) \mathcal{A}_i \left( \alpha_i \Omega_{j-i+1}^{(2)} + \sum_{k=1}^{j-i+1} \tilde{D}_k \Omega_{j-i-k+1}^{(1)} \right) +$$

$$+ (1-q) \sum_{n=0}^{j-i+1} diag \left\{ O_{(W+1)Z_0}, \ldots, O_{(W+1)Z_{M-1}}, \int_0^\infty e^{Q_{M,M}^{(b)} t} \bar{Q}^{(b)} \otimes P(n, t) dt \right\} \times$$

$$\times (S^{(b)} \otimes I_{W+1}) \mathcal{A}_{i+n} \left( \alpha_{i+n} \Omega_{j-i-n+1}^{(2)} + \sum_{k=1}^{j-i-n+1} \tilde{D}_k \Omega_{j-i-n-k+1}^{(1)} \right)$$

*where*

$$\mathcal{A}_i = \int_0^\infty e^{-\alpha_i t} e^{Q \oplus D_0 t} dt = (\alpha_i I - Q \oplus D_0)^{-1}, \ i \geq 0 \ ,$$

$$\Omega_n^{(r)} = \int\limits_0^\infty e^{Qt} \otimes P(n,t) \mathrm{d}B^{(r)}(t), \; r = 1, 2, \; n \geq 0 \; ,$$

$P(n,t)$, $n \geq 0$, $t \geq 0$, is the matrix function defined by the expansion:

$$\sum_{n=0}^\infty P(n,t) z^n = e^{D(z)t} \; ,$$

$\otimes$ and $\oplus$ denote the Kronecker product and sum for matrices, respectively,

*I. (O.) is the identity (zero) matrix of a dimension defined by the suffix. If the dimension is clear from context, the suffix can be omitted,*

$diag\{\dots\}$ *is the block diagonal matrix with diagonal entries listed in the brackets,*

$$\hat{I} = diag\{I_{Z_0}, \dots, I_{Z_{M-1}}, O_{Z_M}\}, \; \bar{I} = I - \hat{I}, \; \tilde{D}_k = I_{\bar{Z}} \otimes D_k, \; k \geq 0,$$

$\bar{Q}^{(b)}$, $Q_{M,M}^{(b)}$ *are the blocks of the generator $Q^{(b)}$ partitioned as*

$$Q^{(b)} = \begin{pmatrix} \hat{Q}^{(b)} & \tilde{Q}^{(b)} \\ \bar{Q}^{(b)} & Q_{M,M}^{(b)} \end{pmatrix}.$$

*Proof.* The proof of lemma 1 is implemented by means of the analysis of the chain transitions taking into account the probabilistic sense of involved matrices. The entry $(P(n,t))_{\nu,\nu'}$ of the matrix $P(n,t)$ defines probability that $n$ customers arrive in the $BMAP$ during time $t$ and the state of the underlying process of the $BMAP$ at the moment $t$ is $\nu'$ conditional that the state of this process at the moment $0$ was $\nu$. Correspondingly, the entries of the matrix $\Omega_n^{(r)} = \int\limits_0^\infty e^{Qt} \otimes P(n,t) \mathrm{d}B^{(r)}(t)$ define probability of $n$ customers arrival and corresponding transitions of the Markov chain $\eta_t$, which describes operation of station 2, and the underlying process $\nu_t$ of the $BMAP$ during a service time of a primary (for $r = 1$) or repeated (for $r = 2$) customer at station 1. The entries of the matrix $\mathcal{A}_i \alpha_i$ define probability that the idle period of the server at station 1 will be finished by arrival of a customer from the orbit (where $i$ customers currently have been staying) and the Markov chain $\eta_t$, which describes operation of station 2, and the underlying process $\nu_t$ of the $BMAP$ make the corresponding transitions during this period. The entries of the matrix $\mathcal{A}_i \tilde{D}_k$ define probability that the idle period of the server at station 1 will be finished by arrival of a primary customer in a batch consisting of $k$ customers and the Markov chain $\eta_t$, which describes operation of station 2, and the underlying process $\nu_t$ of the $BMAP$ make the corresponding transitions during this period.    □

**Corollary 1.** *The Markov chain $\xi_n$, $n \geq 1$, belongs to the class of asymptotically quasi-Toeplitz Markov chains (AQTMC), see [26].*

*Proof.* It can be verified that the following limits exist

$$Y_l = \lim_{i \to \infty} P_{i,i+l-1} = ((\mathcal{S}\hat{I} + q\bar{I}) \otimes I_{W+1}) \Omega_l^{(1)} +$$

$$+(1-q)\sum_{n=0}^{l} diag\left\{O_{(W+1)Z_0},\ldots,O_{(W+1)Z_{M-1}}, \int_0^{\infty} e^{Q_{M,M}^{(b)}t}\bar{Q}^{(b)} \otimes P(n,t)\mathrm{d}t\right\}\times$$

$$\times(S^{(b)} \otimes I_{W+1})\Omega_{l-n}^{(1)}, \quad l \geq 0 \ .$$

This means that Markov chain $\xi_n$, $n \geq 1$, belongs to the class of $AQTMC$. $\square$

It is easy to see that the matrix generating function $Y(z) = \sum_{l=0}^{\infty} Y_l z^l$, $|z| \leq 1$, has the following form:

$$Y(z) = \Big[((S\hat{I} + q\bar{I}) \otimes I_{W+1})+$$

$$+(1-q)diag\left\{O_{(W+1)Z_0},\ldots,O_{(W+1)Z_{M-1}}, \int_0^{\infty} e^{Q_{M,M}^{(b)}t}\bar{Q}^{(b)} \otimes e^{D(z)t}\mathrm{d}t\right\}\times$$

$$\times(S^{(b)} \otimes I_{W+1})\Big]\Omega^{(2)}(z) \ ,$$

where

$$\Omega^{(2)}(z) = \sum_{n=0}^{\infty} \Omega_n^{(2)} z^n = \int_0^{\infty} e^{Qt} \otimes e^{D(z)t}\mathrm{d}B^{(2)}(t) \ .$$

## 4    Stationary Distribution of the Embedded Markov Chain

According to assumptions about the underlying Markov chains, the matrix $Y(1)$ is an irreducible one. So, as follows from [26], the sufficient condition for ergodicity of the chain $\xi_n$, $n \geq 1$, is the fulfillment of the inequality

$$\boldsymbol{y}Y'(1)\boldsymbol{e} < 1 \ , \tag{1}$$

where $\boldsymbol{y}$ is the unique solution of the system

$$\boldsymbol{y}Y(1) = \boldsymbol{y}, \quad \boldsymbol{y}\boldsymbol{e} = 1 \ .$$

It is easy to verify that the vector $\boldsymbol{y}$ can be represented in a form

$$\boldsymbol{y} = \boldsymbol{\delta} \otimes \boldsymbol{\theta}$$

where $\boldsymbol{\theta}$ is the vector of stationary probabilities of the underlying process $\nu_t$ of customers arrival and $\boldsymbol{\delta}$ is the unique solution to the system

$$\boldsymbol{\delta}\boldsymbol{e} = 1, \quad \boldsymbol{\delta} =$$

$$\delta\left[(\mathcal{S}\hat{I}+q\bar{I})+(1-q)diag\left\{O_{Z_0},\ldots,O_{Z_{M-1}},(-Q^{(b)}_{M,M})^{-1}\bar{Q}^{(b)}\right\}S^{(b)}\right]\int\limits_0^\infty e^{Qt}\mathrm{d}B^{(2)}(t)$$

and inequality (1) has the following form:

$$\rho = \lambda\left(b_1^{(2)} + (1-q)\delta diag\left\{O_{Z_0},\ldots,O_{Z_{M-1}},(-Q^{(b)}_{M,M})^{-1}\right\}e\right) < 1 . \quad (2)$$

In what follows we assume that condition (2) holds good. Then the following stationary state probabilities of the embedded Markov chain $\xi_n = \{i_n, \eta_n, \nu_n\}$, $n \geq 1$, exist:

$$\pi(i,\eta,\nu) = \lim_{n\to\infty} P\{i_n = i, \eta_n = \eta, \nu_n = \nu\}, \quad i \geq 0, \ \eta \in \mathcal{M}, \ 0 \leq \nu \leq W .$$

Let us introduce the vectors of stationary probabilities $\boldsymbol{\pi}(i,\eta)$ listed in lexicographic order of components $\nu$, the vectors of stationary probabilities $\boldsymbol{\pi}(i,m)$ listed in lexicographic order of components $\psi$, and the vectors

$$\boldsymbol{\pi}_i = (\boldsymbol{\pi}(i,0),\ldots,\boldsymbol{\pi}(i,M)), \quad i \geq 0 .$$

To compute the vectors $\boldsymbol{\pi}_i$, $i \geq 0$, the numerically stable algorithm developed for the $AQTMC$ in [26] can be applied. The algorithm is described in the following

**Theorem 1.** *The probability vectors $\boldsymbol{\pi}_i$, $i \geq 0$, are calculated as follows:*

$$\boldsymbol{\pi}_l = \boldsymbol{\pi}_0\Phi_l, \ l \geq 1 ,$$

*where the matrices $\Phi_l$, $l \geq 1$ are calculated using the recurrent formulae:*

$$\Phi_l = (\bar{P}_{0,l} + \sum_{i=1}^{l-1}\Phi_i\bar{P}_{i,l})(I - \bar{P}_{l,l})^{-1}, \ l \geq 1 ,$$

*and the vector $\boldsymbol{\pi}_0$ is computed as the unique solution of the system*

$$\boldsymbol{\pi}_0(I - \bar{P}_{0,0}) = \mathbf{0}, \quad \boldsymbol{\pi}_0(I + \sum_{l=1}^{\infty}\Phi_l)e = 1 .$$

*Here the matrices $\bar{P}_{i,l}$ are calculated by*

$$\bar{P}_{i,l} = P_{i,l} + \sum_{n=l+1}^{\infty} P_{i,n}G_{n-1}G_{n-2}\ldots G_l, \ l \geq i, \ i \geq 0 ,$$

*and the matrices $G_i$ are computed using recursion:*

$$G_i = (I - \sum_{l=i+1}^{\infty} P_{i+1,l}G_{l-1}G_{l-2}\ldots G_{i+1})^{-1}P_{i+1,i}, \ i \geq 0 .$$

*Proof.* The proof of the theorem was implemented in [26] by means of derivation and solving the following alternative with respect to Chapman-Kolmogorov
system of equilibrium equations for the vectors $\boldsymbol{\pi}_i$, $i \geq 0$:

$$\boldsymbol{\pi}_l = \sum_{i=0}^{l} \boldsymbol{\pi}_i \bar{P}_{i,l}, \ l \geq 0 \ .$$

This system was derived by means of sequential constructing the Markov chains
which are censored with respect to the $AQTMC$ $\xi_n, n \geq 1$, (see, e.g., [27]) with
different level of censoring. The entry $g_i(\eta, \nu; \eta', \nu')$ of the matrix

$$G_i = (g_i(\eta, \nu; \eta', \nu'))_{\eta, \eta' \in \mathcal{M}, \nu, \nu' = 0, \ldots, W} \ ,$$

is the conditional probability that the chain $\xi_n$, $n \geq 1$ eventually hits the set
$\{(i, \tilde{\eta}, \tilde{\nu})\}$, $\tilde{\eta} \in \mathcal{M}, \tilde{\nu} = 0, \ldots, W$, in the state $(i, \eta', \nu')$ given that it starts from
the state $(i + 1, \eta, \nu)$.                                                      □

## 5   The Stationary Distribution at Arbitrary Times

We have solved the problem of calculating the stationary distribution of the
Markov chain $\xi_n$, $n \geq 1$. Now we are able to calculate the stationary distribution
of the non-Markovian continuous time process $\zeta_t = \{i_t, r_t, \eta_t, \nu_t\}$, $t \geq 0$, defined
in Sect. 3.

Let

$$p(i, r, \eta, \nu) = \lim_{t \to \infty} P\{i_t = i, r_t = r, \eta_t = \eta, \nu_t = \nu\} \ ,$$

$$i \geq 0, \ \eta \in \mathcal{M}, \ \nu \in \{0, \ldots, W\}, \ r = \overline{0, 3} \ ,$$

be the steady-state probabilities of the process $\zeta_t, t \geq 0$.

Define    the    vectors    of    these    probabilities    $\boldsymbol{p}(i, r, \eta)$    =
$(p(i, r, \eta, 0), \ldots, p(i, r, \eta, W))$,    vectors $\boldsymbol{p}(i, r, m)$ consisting of the vectors
$\boldsymbol{p}(i, r, \eta)$ listed in the lexicographic order of the components of the vectors $\psi$,
and vectors $\boldsymbol{p}(i, r)$ consisting of the vectors $\boldsymbol{p}(i, r, m)$ listed in the lexicographic
order of the component $m$, $m \in \{0, 1, \ldots, M\}$.

**Theorem 2.** *The non-zero stationary probability vectors $\boldsymbol{p}(i, r)$, $i \geq 0$, $r = \overline{0, 3}$,
of the process $\zeta_t$, $t \geq 0$, are related to the stationary probability vectors $\boldsymbol{\pi}_i$, $i \geq 0$,
of the embedded Markov chain $\xi_n$, $n \geq 1$, in the following way:*

$$\boldsymbol{p}(i, 0) = \tau^{-1} \left[ \boldsymbol{\pi}_i((S\hat{I} + q\bar{I}) \otimes I_{W+1}) + (1 - q) \sum_{l=0}^{i} \boldsymbol{\pi}_l \times \right.$$

$$\left. \times diag \left\{ O_{(W+1)Z_0}, \ldots, O_{(W+1)Z_{M-1}}, \int_0^\infty e^{Q_{M,M}^{(b)} t} \bar{Q}^{(b)} \otimes P(i-l, t) dt \right\} (S^{(b)} \otimes I_{W+1}) \right] \mathcal{A}_i,$$

$$\boldsymbol{p}(i,1) = \tau^{-1} \sum_{l=0}^{i} \boldsymbol{\pi}_l \left[ ((\mathcal{S}\hat{I} + q\bar{I}) \otimes I_{W+1})\mathcal{A}_l \sum_{k=1}^{i-l+1} \tilde{D}_k \tilde{\Omega}^{(1)}_{i-l-k+1} + \right.$$

$$+ (1-q) \sum_{n=0}^{i-l} diag\left\{ O_{(W+1)Z_0}, \ldots, O_{(W+1)Z_{M-1}}, \int_0^{\infty} e^{Q^{(b)}_{M,M}t} \bar{Q}^{(b)} \otimes P(n,t)dt \right\} \times$$

$$\left. \times (\mathcal{S}^{(b)} \otimes I_{W+1})\mathcal{A}_{l+n} \sum_{k=1}^{i-l-n+1} \tilde{D}_k \tilde{\Omega}^{(1)}_{i-l-n-k+1} \right] ,$$

$$\boldsymbol{p}(i,2) = \tau^{-1} \sum_{l=0}^{i+1} \boldsymbol{\pi}_l \left[ ((\mathcal{S}\hat{I} + q\bar{I}) \otimes I_{W+1})\mathcal{A}_l \alpha_l \tilde{\Omega}^{(2)}_{i-l+1} + \right.$$

$$+ (1-q) \sum_{n=0}^{i-l+1} diag\left\{ O_{(W+1)Z_0}, \ldots, O_{(W+1)Z_{M-1}}, \int_0^{\infty} e^{Q^{(b)}_{M,M}t} \bar{Q}^{(b)} \otimes P(n,t)dt \right\} \times$$

$$\left. \times (\mathcal{S}^{(b)} \otimes I_{W+1})\mathcal{A}_{l+n} \alpha_{l+n} \tilde{\Omega}^{(2)}_{i-l-n+1} \right) \right] ,$$

$$\boldsymbol{p}(i,3) = \tau^{-1}(1-q) \times$$

$$\times \sum_{l=0}^{i} \boldsymbol{\pi}_l diag\left\{ O_{(W+1)Z_0}, \ldots, O_{(W+1)Z_{M-1}}, \int_0^{\infty} e^{Q^{(b)}_{M,M}t} \otimes P(i-l,t)dt \right\} ,$$

*where $\tau$ is a mean inter-departure time from station 1 defined by*

$$\tau = \sum_{i=0}^{\infty} \boldsymbol{\pi}_i \left[ ((\mathcal{S}\hat{I} + q\bar{I}) \otimes I_{W+1})\mathcal{A}_i (I + \alpha_i I b_1^{(2)} - \tilde{D}_0 b_1^{(1)})\boldsymbol{e} + \right.$$

$$+ (1-q) \left[ diag\{O_{Z(0)}, \ldots, O_{Z(M-1)}, -(\hat{Q}_{M,M})^{-1}\} + \right.$$

$$+ \sum_{n=0}^{\infty} diag\left\{ O_{(W+1)Z_0}, \ldots, O_{(W+1)Z_{M-1}}, \int_0^{\infty} e^{Q^{(b)}_{M,M}t} \bar{Q}^{(b)} \otimes P(n,t)dt \right\} \times$$

$$\left. \left. \times (\mathcal{S}^{(b)} \otimes I_{W+1})\mathcal{A}_{i+n} (I + \alpha_{i+n} I b_1^{(2)} - \tilde{D}_0 b_1^{(1)}) \right] \boldsymbol{e} \right] ,$$

$$\tilde{\Omega}^{(r)}_i = \int_0^{\infty} e^{Qt} \otimes P(i,t)(1 - B^{(r)}(t))dt, \; i \geq 0 .$$

*Proof.* The process $\zeta_t$, $t \geq 0$, is a semi-regenerative one, see, e.g., [28], with embedded Markov renewal process $\{\xi_n, t_n\}$, $n \geq 1$. By the theory of semi-regenerative processes, the stationary state distribution $p(i,r,\eta,\nu)$, $i \geq 0$, $\eta \in \mathcal{M}$, $\nu \in \{0, \ldots, W\}$, $r = 0,1,2,3$, of the process $\zeta_t$ exist if the process $\{\xi_n, t_n\}$ is an irreducible aperiodic recurrent process and the value $\tau$ of the mean inter-departure time at station 1 is finite. It is easily verified that in our case all these conditions hold true if inequality (2) is satisfied.

Let us find the vectors $\mathbf{p}(i, r, \eta)$ of the stationary state probabilities $p(i, r, \eta, \nu)$. To calculate these vectors, we will use the ergodic theorem for semi-regenerative processes given in [6]. By this theorem, the stationary distribution of the process $\zeta_t$, $t \geq 0$, can be related to the stationary distribution of the embedded Markov chain $\xi_n$, $n \geq 1$.

Introduce the matrices

$$K_{l,r,\eta}(i, \eta', t), \ l, i \geq 0, \ \eta, \eta' \in \mathcal{M}, \ r = 0, 1, 2, 3 \ ,$$

whose $(\nu, \nu')$-th entry is defined as a conditional probability that, given time 0 is an instant of the service completion at station 1 and the embedded Markov chain $\xi_n$ is in the state $(l, \eta, \nu)$ at that time, the next service completion epoch at station 1 occurs later than $t$ and the process $\zeta_t$, $t \geq 0$, is in the state $(i, r, \eta', \nu')$ at time $t$.

The vectors $\mathbf{p}(i, r, \eta)$ are expressed in terms of the stationary distribution $\boldsymbol{\pi}(i, \eta), i \geq 0, \ \eta \in \mathcal{M}$, of the chain $\xi_n$ as follows:

$$\mathbf{p}(i, r, \eta') = \tau^{-1} \sum_{l=0}^{\infty} \sum_{\eta \in \mathcal{M}} \boldsymbol{\pi}(l, \eta) \int_{0}^{\infty} K_{l,r,\eta}(i, \eta', t)\mathrm{d}t, \quad i \geq 0, \ \eta' \in \mathcal{M}, \ r = 0, 1, 2, 3 \ .$$

The value $\tau$ of the mean inter-departure time interval at station 1 is defined by

$$\tau = \sum_{i=0}^{\infty} \sum_{\eta \in \mathcal{M}} \sum_{\nu=0}^{W} \boldsymbol{\pi}(i, \eta, \nu) \boldsymbol{\tau}_{i,\eta,\nu} \ ,$$

where $\boldsymbol{\tau}_{i,\eta,\nu}$ is a column vector of the mean values of an inter-departure time interval at station 1 conditional $\xi_n = (i, \eta, \nu)$ at the beginning of this interval. The rest of the proof consists of derivation of expressions for matrices $K_{l,r,\eta}(i, \eta', t)$, and is omitted here.

## 6   Conclusion

In this paper, the problem of calculation of the stationary distribution at arbitrary and embedded epochs for tandem queue with the first station modeled by the $BMAP/G/1$ retrial queueing system, operation of station 2 described by multidimensional continuous time Markov chain, customers loss or station 1 blocking is solved.

The results can be exploited for capacity planning, performance evaluation and optimization of real-life two-phase tandem queues in the case of correlated bursty arrival process and complicated description of the service mechanism at station 2. Ignorance of taking into account correlation in the arrival process and assumption that the arrival process is stationary Poisson can lead to huge errors in prediction of performance measures of the real world systems described by tandem queues, see e.g. [17,23,18] where impressive numerical results were presented for the partial cases of the system under study in this paper.

**Acknowledgments.** This research was supported by Basic Science Research Program through the National Research Foundation of Korea (NRF) funded by the Ministry of Education, Science and Technology (Grant No. 2010-0003269).

# References

1. Balsamo, S., Persone, V.D.N., Inverardi, P.: A review on queueing network models with finite capacity queues for software architectures performance prediction. Performance Evaluation 51, 269–288 (2003)
2. Gnedenko, B.W., Konig, D.: Handbuch der Bedienungstheorie. Akademie Verlag, Berlin (1983)
3. Hall, N.G., Sriskandarajah, C.: A survey of machine scheduling problems with blocking and no-wait in process. Operations Research 44, 510–525 (1996)
4. Perros, H.G.: A bibliography of papers on queueing networks with finite capacity queues. Performance Evaluation 10, 255–260 (1989)
5. Heindl, A.: Decomposition of general queue networks with $MMPP$ inputs and customer losses. Performance Evaluation 1, 117–136 (2003)
6. Klemm, A., Lindermann, C., Lohmann, M.: Modelling IP traffic using the batch Markovian arrival process. Performance Evaluation 54, 149–173 (2003)
7. Neuts, M.F.: A versatile Markovian point process. Journal of Applied Probability 16, 764–779 (1979)
8. Lucantoni, D.M.: New results on the single server queue with a batch Markovian arrival process. Communications in Statistics – Stochastic Models 7, 1–46 (1991)
9. Breuer, L., Dudin, A.N., Klimenok, V.I., Tsarenkov, G.V.: A two-phase $BMAP/G/1/N \to PH/1/M - 1$ system with blocking. Automation and Remote Control 65, 117–130 (2004)
10. Dudin, A.N., Kim, C.S., Klimenok, V.I., Taramin, O.S.: A dual tandem queueing system with a finite intermediate buffer and cross traffic. In: Proceedings of the 5th international Conference "Queueing Theory and Network Applications", pp. 93–100 (2010)
11. Gomez-Corral, A.: A tandem queue with blocking and Markovian arrival process. Queueing Systems 41, 343–370 (2002)
12. Gomez-Corral, A.: On a tandem G-network with blocking. Advances in Applied Probability 34(3), 626–661 (2002)
13. Gomez-Corral, A.: A matrix-geometric approximations for tandem queues with blocking and repeated attempts. Operations Research Letters 30, 360–374 (2002)
14. Heindl, A.: Decomposition of general tandem networks with $MMPP$ input. Performance Evaluation 44, 5–23 (2001)
15. Kim, C.S., Dudin, A.N., Klimenok, V.I., Taramin, O.S.: A tandem $BMAP/G/1 \to M/N/0$ queue with group occupation of servers at the second station. Mathematical Problems in Engineering ID 324604 (2012), doi:10.1155/2012/324604
16. Kim, C.S., Dudin, S.: Priority Tandem Queueing Model with Admission Control. Computers and Industrial Engineering 60, 131–140 (2011)
17. Kim, C.S., Klimenok, V.I., Tsarenkov, G.V., Breuer, L., Dudin, A.N.: The $BMAP/G/1 \to PH/1/M$ tandem queue with feedback and losses. Performance Evaluation 64, 802–818 (2007)
18. Klimenok, V.I., Breuer, L., Tsarenkov, G.V., Dudin, A.N.: The $BMAP/G/1/N \to PH/1/M - 1$ tandem queue with losses. Performance Evaluation 61, 17–60 (2005)

19. Lian, Z., Liu, L.: A tandem network with $MAP$ inputs. Operations Research Letters 36, 189–195 (2008)
20. Dudin, A.N., Klimenok, V.I.: Queueing system $BMAP/G/1$ with repeated calls. Mathematical and Computer Modelling 34, 115–128 (1999)
21. Dudin, A.N., Klimenok, V.I.: A retrial $BMAP/SM/1$ system with linear repeated requests. Queueing Systems 34, 47–66 (2000)
22. Kim, C.S., Klimenok, V.I., Taramin, O.S.: A tandem retrial queueing system with two Markovian flows and reservation of channels. Computers and Operations Research 37(7), 1238–1246 (2010)
23. Kim, C.S., Park, S.H., Dudin, A., Klimenok, V., Tsarenkov, G.: Investigaton of the $BMAP/G/1 \rightarrow /PH/1/M$ tandem queue with retrials and losses. Applied Mathematical Modelling 34, 2926–2940 (2010)
24. Klimenok, V.I., Taramin, O.S.: Tandem service system with batch Markov flow and repeated calls. Automation and Remote Control 71, 1–13 (2010)
25. Chakravarthy, S.R.: The batch Markovian arrival process: a review and future work. In: Krishnamoorthy, A., et al. (eds.) Advances in Probability Theory and Stochastic Processes, pp. 21–49. Notable Publications, NJ (2001)
26. Klimenok, V.I., Dudin, A.N.: Multi-dimensional asymptotically quasi-Toeplitz Markov chains and their application in queueing theory. Queueing Systems 54, 245–259 (2006)
27. Kemeni, J., Shell, J., Knapp, A.: Denumerable Markov Chains. Van Nostrand, New York (1966)
28. Cinlar, E.: Introduction to stochastic processes. Prentice-Hall, NJ (1975)

# On the Stationary Distribution of Tandem Queue Consisting of a Finite Number of Stations

Valentina Klimenok[1], Alexander Dudin[1], and Vladimir Vishnevsky[2]

[1] Department of Applied Mathematics and Computer Science
Belarusian State University
Minsk 220030, Belarus
{klimenok,dudin}@bsu.by
[2] ZAO Research & Development Company
Information and Networking Technologies (INSET), Moscow, Russia
vishn@inbox.ru

**Abstract.** In this paper we consider a tandem queueing system consisting of a finite number of multi-server queues (stations) without buffers. Customers arrive to the first station in the $MAP$ (Markovian Arrival Process). The service times at the servers of the tandem are exponentially distributed with different parameters for different stations. We describe the output flows from the stations of the tandem as $MAP$s and present the simple method for recursive constructing the infinitesimal generator of the multi-dimensional Markov chain describing the operation of the system, calculating the stationary distributions and loss probabilities associated with the tandem. It is shown that the marginal stationary distribution of any station can be calculated as stationary distribution of the system of $MAP/M/N/N$ type.

**Keywords:** tandem queueing system, multi-server stations, Markovian Arrival Process, stationary state distribution, decomposition, loss probabilities.

## 1 Introduction

Theory of tandem (multi-stage) queueing systems represent a bridge between the theory of queues, which considers the problems related to performance evaluation of an isolated queue, and the theory of queueing networks which considers an arbitrary set of connected queueing systems and is popular model for description of computer and communication networks. Tandem queues can be considered as a simplest case of queueing networks having a linear topology. So, tandem queues are a popular subject for research, for references see, e.g., [1,2,3].

Most of the papers devoted to tandem queues are restricted to the case of dual tandem queues, i.e., queues consisting of exactly two sequential servers with the stationary Poisson arrival process of customers. In this paper we consider tandem system consisting of any finite number of stations with the $MAP$. Assumption that an arrival process of customers is defined by the $MAP$ instead

A. Kwiecień, P. Gaj, and P. Stera (Eds.): CN 2012, CCIS 291, pp. 383–392, 2012.

of stationary Poisson arrival process allows to effectively take into account variance and correlation of inter-arrival times typical for information flows in the modern telecommunication networks, see, e.g., [4,5].

The customer arriving to the system under study has to get sequentially the service at all stations of the tandem. However, due to lack of waiting area at the stations the customer can be lost at each of them. So, the quality of a service in the system is essential defined by the probability of successful service of an arbitrary customer arriving at the tandem by all stations. Also, the probability of losses at different stations and subsystems of a tandem are important for performance evaluation of a tandem network and, discovering and avoiding so called bottlenecks in the network.

As the related works, the papers [6,7] can be referred to. In these papers, the $GI/M/1/1 \to (/M/1/1)^{n-1}$ type queue was considered. Arriving flow is assumed having there an arbitrary distribution of inter-arrival times. So, in some sense this flow is more general that the $MAP$ considered in our paper. However, in turn, the $MAP$ can be considered be more general than the arrival flow in [6,7] because in the $MAP$ the sequential inter-arrival times may be dependent. Also, in contrast to [6,7] where all stations are represented by the single server queues, we assume that the stations can have an arbitrary finite number of servers.

The rest of the paper is organized as follows. In Section 2, the considered model is exactly described. In Section 3, the problem of calculating stationary distributions in the tandem is formulated in terms of a multi-dimensional Markov chain. Output flows from the stations specified. Section 4 contains the results concerning the stationary distributions of the queue of the $MAP/M/N/N$ type which models operation of an arbitrary system of tandem. The problem of calculation of the stationary distribution of the tandem and its fragments is solved in Sect. 5. Formulas for computation of loss probability in the tandem, its fragments and separate systems are given in Sect. 6. Finally, Section 7 concludes the paper.

## 2   Model Description

We consider a tandem queueing system consisting of $R$, $R > 1$, stations in series. In terms of Kendall's notation this system can be coded as

$$MAP/M/N_1/N_1 \to \cdot/M/N_2/N_2 \to \ldots \to \cdot/M/N_R/N_R .$$

The $r$-th station $(r = 1, \ldots, R)$ is represented by the $N_r$ – server queue without a buffer. The servers belonging the same station are identical and independent. The service time of a customer at a server of the $r$-th station is exponentially distributed with parameter $\mu_r$.

Customers arrive at the first station in the $MAP$ (Markovian Arrival Process). Arrival of customers in the $MAP$ is directed by the underlying random process $\nu_t$, $t \geq 0$. The process $\nu_t$, $t \geq 0$, is an irreducible continuous time Markov chain with state space $\{0, 1, \ldots, W\}$. The sojourn time of this chain in the state $\nu$ is exponentially distributed with the positive finite parameter $\lambda_\nu$. When the sojourn time in the state $\nu$ expires, with probability $p_{\nu,\nu'}^{(0)}$, the process $\nu_t$ jumps into the state $\nu'$ without generation of customers, and with probability $p_{\nu,\nu'}^{(1)}$ it jumps into the state $\nu'$ with generation of a customer $\nu, \nu' \in \{0, \ldots, W\}$.

The behavior of the $MAP$ is completely described by the matrices $D_0$, $D_1$ defined by their entries $(D_1)_{\nu,\nu'} = \lambda_\nu p_{\nu,\nu'}^{(1)}$, $\nu, \nu' \in \{0, \ldots, W\}$, and

$$(D_0)_{\nu,\nu} = -\lambda_\nu, \nu \in \{0, \ldots, W\}, \ (D_0)_{\nu,\nu'} = \lambda_\nu p_{\nu,\nu'}^{(0)}, \ \nu, \nu' \in \{0, \ldots, W\}, \nu \neq \nu' \ .$$

The matrix $D = D_0 + D_1$ is an infinitesimal generator of the process $\nu_t, t \geq 0$. The fundamental arrival rate $\lambda$ is defined by

$$\lambda = \boldsymbol{\theta} D_1 \mathbf{e}$$

where $\boldsymbol{\theta}$ is the row vector of a stationary probabilities of the Markov chain $\nu_t, t \geq 0$. The vector $\boldsymbol{\theta}$ is the unique solution to the system

$$\boldsymbol{\theta} D = \mathbf{0}, \ \boldsymbol{\theta} \mathbf{e} = 1 \ .$$

Here and in the sequel $\mathbf{e}$ is the column-vector of appropriate size consisting of 1's and $\mathbf{0}$ is the row-vector of appropriate size consisting of zeros.

The coefficient of variation, $c_{\text{var}}$, of intervals between successive arrivals is defined by $c_{\text{var}}^2 = 2\lambda\boldsymbol{\theta}(-D_0)^{-1}\mathbf{e} - 1$. The coefficient of correlation, $c_{\text{cor}}$, of the successive intervals between arrivals is given by

$$c_{\text{cor}} = (\lambda\boldsymbol{\theta}(-D_0)^{-1}(D - D_0)(-D_0)^{-1}\mathbf{e} - 1)/c_{\text{var}}^2 \ .$$

For more information about the $MAP$ see, e.g., [8].

If a customer arriving at the first station or proceeding to the $r$-th, $r = 2, \ldots, R$, station after service at the $(r-1)$-th station meets all servers busy it leaves the tandem forever.

Our aim is to study the output flows from the stations of the tandem, calculate stationary distribution of its fragments and loss probabilities.

## 3 Problem of Calculating Stationary Distributions in the Tandem. Output Flows from the Stations

The process of the system states is described in terms of the irreducible multi-dimensional continuous-time Markov chain $\xi_t = \{n_t^{(1)}, n_t^{(2)}, \ldots, n_t^{(R)}, \nu_t\}, t \geq 0$, where $n_t^{(r)}$ is the number of busy servers at the $r$-th station, $n_t^{(r)} \in \{0, \ldots, n^{(r)}\}$; $\nu_t$ is the state of the $MAP$ underlying process at time $t$, $\nu_t \in \{0, \ldots, W\}$.

The state space, $S$, of the chain is defines as

$$S = \{\{0, 1, \ldots, N_1\} \times \{0, 1, \ldots, N_2\} \times \ldots \times \{0, 1, \ldots, N_R\} \times \{0, 1, \ldots, W\}\} \ .$$

The row vector $\mathbf{p}$ of the steady state probabilities of the chain has dimension

$$(W + 1) \prod_{r=1}^{R}(N_r + 1)$$

and is calculated as the unique solution of the system of linear algebraic equations

$$\mathbf{p}Q = \mathbf{0}, \ \mathbf{p}\mathbf{e} = 1 \ . \tag{1}$$

The matrix $Q$ is an infinitesimal generator of the chain $\xi_t$, $t \geq 0$. Construction of this matrix can be performed in routine way by means of the standard technique used in the theory of queueing networks and is not too difficult task. But this work in the case of more or less large value of $R$ is rather time-consuming. So, it is interesting to find a way for calculating the stationary distribution of the states of whole tandem or its parts (fragments) or marginal stationary distribution of the states of any station without writing down the explicit expression of the generator $Q$. It is worth to note that the operation of a fragment of the tandem consisting of any number of stations and located at the head of the tandem does not depend on the operation of the rest of stations. So, it is intuitively clear that some kind of decomposition can be exploited to calculate the stationary distribution of the tandem and its fragments without complete construction of a generator.

In this study, we develop the simple, exact and user-friendly method for calculating the collection of marginal stationary distributions of the parts of a tandem as well as of the whole tandem and the corresponding loss probabilities.

The method is essentially based on the results of investigation of output flows from the stations of the tandem. These flows are characterized as follows.

**Theorem 1.** *The input flow at the $(r+1)$-th station (output flow from the $r$-th station), $r \in \{1, 2, \ldots, R-1\}$, belongs to the class of Markovian arrival processes. This MAP is defined by the matrices $D_0^{(r+1)}$ and $D_1^{(r+1)}$ which are calculated using the following recurrent formulas:*

$$D_0^{(r+1)} = -\mu_r diag\{0, 1, \ldots, N_r\} \otimes I_{K_r} +$$

$$+ \begin{pmatrix} D_0^{(r)} & D_1^{(r)} & 0 & \cdots & 0 & 0 \\ 0 & D_0^{(r)} & D_1^{(r)} & \cdots & 0 & 0 \\ 0 & 0 & D_0^{(r)} & \cdots & 0 & 0 \\ \vdots & \vdots & \vdots & \ddots & \vdots & \vdots \\ 0 & 0 & 0 & \cdots & D_0^{(r)} & D_1^{(r)} \\ 0 & 0 & 0 & \cdots & 0 & D_0^{(r)} + D_1^{(r)} \end{pmatrix}, \tag{2}$$

$$D_1^{(r+1)} = \begin{pmatrix} 0 & 0 & \cdots & 0 & 0 & 0 \\ \mu_r & 0 & \cdots & 0 & 0 & 0 \\ 0 & 2\mu_r & \cdots & 0 & 0 & 0 \\ \vdots & \vdots & \ddots & \vdots & \vdots & \vdots \\ 0 & 0 & \cdots & (N_r - 1)\mu_r & 0 & 0 \\ 0 & 0 & \cdots & 0 & N_r\mu_r & 0 \end{pmatrix} \otimes I_{K_r}, \ r = 1, 2, \ldots, R-1 \ , \tag{3}$$

*with initial condition defined by*

$$D_0^{(1)} = D_0, \ D_1^{(1)} = D_1 \ .$$

*Here $\otimes$ is the symbol of Kronecker's product of matrices, see [9], $diag\{0, 1, ..., N_r\}$
denotes the diagonal matrix with the diagonal entries listed in the brackets,
the value $K_r$ is calculated by $K_r = (W + 1) \prod_{r'=1}^{r} (N_{r'} + 1), r = 1, 2, \ldots, R,$
and $I_{K_r}$ stands for identity matrix of size $K_r$.*

*Proof.* Let $r = 1$. It is evident that the process $\{n_t^{(1)}, \nu_t\}, t \geq 1$, describing the
operation of the first station is a Markov chain. Enumerate the states of this
chain in lexicographic order, i.e., as follows: $(0, 0), (0, 1), \ldots, (0, W), (1, 0), (1, 1),$
$\ldots, (1, W), \ldots, (N_1, 0), (N_1, 1), \ldots, (N_1, W)$. Then, it is easy to see that transitions of the chain, that do not lead to service completion at the first station
(and generation of arrival at the second station) are defined by the matrix $D_0^{(2)}$
calculated by formula (2). Transitions of the chain leading to service completion
at the first station (and generation of arrival at the second station) are defined
by the matrix $D_1^{(2)}$. According to definition of a $MAP$, this means that arrival
process at the second station is the $MAP$ defined by the matrices $D_0^{(2)}$ and $D_1^{(2)}$.
The rest of the proof is implemented by induction. □

*Remark 1.* In what follows, we will denote the $MAP$ describing input flow at
the $r$-th station as $MAP^{(r)}, r = 1, 2, \ldots, R$. Note that $MAP^{(1)}$ is the $MAP$ of
arrivals at the tandem. Output flow from the $R$-th station will be referred to
as $MAP^{(R+1)}$. The matrices $D_0^{(R+1)}$ and $D_1^{(R+1)}$, which define $MAP^{(R+1)}$, are
given by formulas (2) and (3) when $r = R + 1$.

Using the result of Theorem 1 we can calculate the marginal stationary distribution of the $r$-th station of the tandem as the stationary distribution of the
queueing system $MAP^{(r)}/M/N_r/N_r, r = 1, 2, \ldots, R$. For the sake of reader's
convenience, in the next section we describe in brief the procedures for calculating the steady state distribution for such a type of queue. For brevity, we
omit index $r$ in denotation of matrices describing the $MAP$ and intensity of the
service.

## 4   Stationary Distributions of the Queue $MAP/M/N/N$

Operation of the $MAP/M/N/N$ system is described by the Markov chain $\eta_t = \{n_t, \nu_t\}$ where $n_t$ is the number of busy servers and $\nu_t$ is the state of the $MAP$
underlying process at time $t$, $n_t \in \{0, \ldots, N\}$, $\nu_t \in \{0, \ldots, W\}$. Infinitesimal
generator, A, of the chain is defined as

$$A = \begin{pmatrix} D_0 & D_1 & 0 & \cdots & 0 & 0 \\ \mu I & D_0 - \mu I & D_1 & \cdots & 0 & 0 \\ 0 & 2\mu I & D_0 - 2\mu I & \cdots & 0 & 0 \\ \vdots & \vdots & \vdots & \ddots & \vdots & \vdots \\ 0 & 0 & 0 & \cdots & D_0 - (N-1)\mu I & D_1 \\ 0 & 0 & 0 & \cdots & N\mu I & D - N\mu I \end{pmatrix}.$$

Let us enumerate the states of the chain $\eta_t$ in lexicographic order and let $\mathbf{q}$ be the row vector of the steady state probabilities of the chain. This vector is calculated as the unique solution of the system of linear algebraic equations

$$\mathbf{q}A = \mathbf{0}, \quad \mathbf{q}\mathbf{e} = 1 \ .$$

In the case of large dimension of the system it is reasonable to use special algorithms for solving the system. The most relevant among them are described below.

Let us partition the vector $\mathbf{q}$ as $\mathbf{q} = (\mathbf{q}_0, \mathbf{q}_1, \ldots, \mathbf{q}_N)$ where the vectors $\mathbf{q}_i$, $i = 0, \ldots, N$, are of size $W + 1$.

**Algorithm 1.** *The vectors $\mathbf{q}_i$, $i = 0, \ldots, N$, can be calculated as*

$$\mathbf{q}_i = \mathbf{q}_0 F_i, \quad i = 0, \ldots, N \ ,$$

*where the matrices $F_i$ are computed recursively*

$$F_0 = I, \quad F_i = \frac{1}{i\mu}[F_{i-1}((i-1)\mu I - D_0) - (1 - \delta_{i,1})F_{i-2}D_1], \quad i = 1, \ldots, N \ ,$$

*and the vector $\mathbf{q}_0$ is the unique solution of the system*

$$\mathbf{q}_0(F_{N-1}D_1 + F_N(D - N\mu I)) = 0, \quad \mathbf{q}_0 \sum_{i=0}^{N} F_i \mathbf{e} = 1 \ .$$

*Here $\delta_{i,1}$ is Kronecker's delta.*

For more details about this algorithm see [10].

It is seen that in the recursion for $F_i$ involves the subtraction operation. This implies that, in the case when the value $N$ is large, the described algorithm can be the numerically unstable. In such a situation, an algorithm based on probabilistic meaning of the matrix $A$ can be applied (see [11]). This algorithm adopted to the form of the matrix $A$ is described as follows.

**Algorithm 2.** *The stationary probability vectors $\mathbf{q}_i$, $i = 0, \ldots, N$, are computed by*

$$\mathbf{q}_l = \mathbf{q}_0 \Phi_l, \quad l = 1, \ldots, N \ ,$$

*where the matrices $\Phi_l$ are computed recursively:*

$$\Phi_0 = I, \quad \Phi_i = \Phi_{i-1}D_1(i\mu I - D_0 - (1 - \delta_{i,N})D_1 G_i)^{-1}, \quad i = 1, \ldots, N \ ,$$

*the matrices $G_i, i = \overline{0, N-1}$, are computed from the backward recursion*

$$G_i = (i+1)\mu[(i+1)\mu I - D_0 - D_1 G_{i+1}]^{-1}, \quad i = N-2, N-3, \ldots, 0 \ ,$$

*with the terminal condition*

$$G_{N-1} = N\mu(N\mu I - D)^{-1} \ ,$$

*the vector* $\mathbf{q}_0$ *is calculated as the unique solution to the following system of linear algebraic equations:*

$$\mathbf{q}_0(D_0 + D_1 G_0) = \mathbf{0}, \quad \mathbf{q}_0 \sum_{i=0}^{N} \Phi_i \mathbf{e} = 1 \ .$$

Note, that the subtraction operation is avoided in this algorithm and all inverted matrices exist and are nonnegative. Thus, the algorithm is highly stable numerically.

## 5   Calculation of the Stationary Distribution of the Tandem and Its Fragments

In this section, we present the method of calculating the stationary distribution of the tandem and its fragments based on the results of investigation of output flows given by Theorem 1.

Let $\langle r, r+1, \ldots, r' \rangle$ denote a fragment of the tandem consisting of the $r$-th, $(r+1)$-th,..., $r'$-th stations, $1 \leq r \leq r' \leq R$.

**Theorem 2.** *The stationary distribution of the fragment* $\langle r, r+1, \ldots, r' \rangle$ *of the tandem can be calculated as the stationary distribution of the tandem queue* $MAP^{(r)}/M/N_r/N_r \rightarrow \cdot/M/N_{r+1}/N_{r+1} \rightarrow \ldots \rightarrow \cdot/M/N_{r'}/N_{r'}$ *where* $MAP^{(r)}$ *is defined by formulas (2)–(3).*

**Corollary 1.** *The marginal stationary distribution of the* $r$-th *station of the tandem is calculated as the stationary distribution of the queueing system* $MAP^{(r)}/M/N_r/N_r$, $r = 1, 2, \ldots, R$.

This distribution can be computed based on the algorithms presented in the previous section.

**Theorem 3.** *The joint stationary distribution* $\mathbf{p}^{(1,\ldots,r)}$ *of the first* $r$ *stations of the tandem can be calculated as the stationary distribution of the* $MAP^{(r+1)}$, *i.e.,*

$$\mathbf{p}^{(1,\ldots,r)} = \boldsymbol{\theta}^{(r+1)}$$

*where the vector* $\boldsymbol{\theta}^{(r+1)}$ *is the unique solution of the system*

$$\boldsymbol{\theta}^{(r+1)}(D_0^{(r+1)} + D_1^{(r+1)}) = \mathbf{0}, \ \boldsymbol{\theta}^{(r+1)}\mathbf{e} = 1, \ r = 1, 2, \ldots, R \ ,$$

*and the matrices* $D_0^{(r+1)}$, $D_1^{(r+1)}$ *are calculated using recurrent formulas (2)–(3).*

This system can be also effectively solved based on the numerically stable algorithm elaborated in [11].

It is clear that the vector $\mathbf{p}$ of the stationary distribution of the whole tandem coincides, in terms of notation of Theorem 3, with the vector $\mathbf{p}^{(1,\ldots,R)}$.

**Corollary 2.** *The vector* **p** *of the stationary distribution of the tandem is calculated as the unique solution of the system of linear algebraic equations*

$$\mathbf{p}(D_0^{(R+1)} + D_1^{(R+1)}) = \mathbf{0}, \quad \mathbf{pe} = 1 \ . \tag{4}$$

It follows from system (4) that the matrix $D_0^{(R+1)} + D_1^{(R+1)}$ coincides, up to a constant multiplier, with infinitesimal generator $Q$ of the Markov chain $\xi_t = \{n_t^{(1)}, n_t^{(2)}, \ldots, n_t^{(R)}, \nu_t\}, t \geq 0$, describing the operation of the tandem. Further, it is not difficult to verify that $Q = D_0^{(R+1)} + D_1^{(R+1)}$ exactly.

Thus, using the result of analysis of output flows from the stations of the tandem we have constructed the matrix $Q$ avoiding time-consuming work required in the direct approach to the construction of this matrix.

## 6   Loss Probabilities

Having the stationary distributions of the tandem and its fragments calculated, we can find a number of stationary performance measures of the system. The most important of them are various loss probabilities. According to ergodic theorems for Markov chains, see, e.g. [12], probabilities of customers losses in fragments of the tandem can be computed as the ratio of the rates of input flow to the fragment and output flow from this fragment. Thus, the following theorem is valid.

**Theorem 4.** *The loss probability in the fragment* $\langle r, r+1, \ldots, r' \rangle$ *of the tandem is calculated by*

$$P_{\text{loss}}^{(r,\ldots,r')} = \frac{\lambda_r - \lambda_{r'+1}}{\lambda_r} \tag{5}$$

*where* $\lambda_r$ *is the fundamental rate of customers arriving to the $r$-th station according to the* $MAP^{(r)}$, $r = 1, \ldots, R$.

*The value of* $\lambda_r$ *is calculated as*

$$\lambda_r = \boldsymbol{\theta}^{(r)} D_1^{(r)} \mathbf{e}$$

*where the vector* $\boldsymbol{\theta}^{(r)}$ *is the unique solution of the system*

$$\boldsymbol{\theta}^{(r)}(D_0^{(r)} + D_1^{(r)}) = \mathbf{0}, \quad \boldsymbol{\theta}^{(r)}\mathbf{e} = 1, \ r = 1, 2, \ldots, R+1 \ .$$

**Corollary 3.** *The loss probability,* $P_{\text{loss}}^{(r)}$, *of an arbitrary customer at the $r$-th station of the tandem is calculated by formula*

$$P_{\text{loss}}^{(r)} = \frac{\lambda_r - \lambda_{r+1}}{\lambda_r} \ .$$

**Corollary 4.** *The loss probability,* $P_{\text{loss}}$, *of an arbitrary customer in the whole tandem is calculated by formula*

$$P_{\text{loss}} = \frac{\lambda_1 - \lambda_{R+1}}{\lambda_1} \ .$$

The obtained results give the simple elegant way for calculating the stationary distributions related to the tandem without intermediate buffers. However, this way does not reduce the dimension of the system of linear algebraic equations (system (1) has the same rank $K_R$ as the system (4)) that should be solved to compute the stationary distribution of the states of a whole tandem.

If even the algorithms like the numerically stable algorithm 2, which effectively exploits the sparse structure of the generator and is described above, can not solve this system, some kind of heuristics can be used. E.g., when we recursively compute the matrices $D_0^{(r)}$ and $D_1^{(r)}$ and the size of the matrices $D_0^{(r')}$ and $D_1^{(r')}$ for some $r' \in \{1, \ldots, R-1\}$ becomes too large, we can try to approximate this $MAP$ by the $MAP$ having the same intensity, several initial moments of distribution (at least, the same coefficient of variation) and the same coefficient of correlation, but much smaller dimension, see, e.g., [13,14,15]. After that, we can continue procedure for calculation of the matrices $D_0^{(r)}$ and $D_1^{(r)}$ starting from the corresponding matrices that define this $MAP$ having smaller dimension of underlying process.

## 7    Conclusion

We have analyzed the tandem queue with many stations in series defined by the multi-server queues without buffers. Arrival flow to the tandem is modeled by the Markovian Arrival Process what allows to take into account possible dependence of customers inter-arrival times. Algorithmic way for recursive construction of the generator of the multi-dimensional Markov chain that describes behavior of tandem is offered. This way is based on analysis of departure processes from the stations of a tandem. Algorithms for computation of marginal stationary distribution of stations of tandem are presented. Simple expressions for computing variety of loss probabilities are presented.

Presented results theoretically can be more or less easily extended to the case of more general so called $PH$ – phase type distribution of the service times, see, e.g., [16] and systems with finite buffers. However computer realization in this case becomes much more time consuming due to the essential increase of the state space of the Markov chain under study that will additionally include components indicating the current phases of the service at each busy server or the number of servers providing currently different phases of the service at each station.

## References

1. Balsamo, S., Persone, V.D.N., Inverardi, P.: A review on queueing network models with finite capacity queues for software architectures performance prediction. Performance Evaluation 51, 269–288 (2003)
2. Gnedenko, B.W., Konig, D.: Handbuch der Bedienungstheorie. Akademie Verlag, Berlin (1983)

3. Perros, H.G.: A bibliography of papers on queueing networks with finite capacity queues. Performance Evaluation 10, 255–260 (1989)
4. Heyman, D.P., Lucantoni, D.: Modelling multiple IP traffic streams with rate limits. IEEE/ACM Transactions on Networking 11, 948–958 (2003)
5. Klemm, A., Lindermann, C., Lohmann, M.: Modelling IP traffic using the batch Markovian arrival process. Performance Evaluation 54, 149–173 (2003)
6. Bromberg, M.A.: Multi-phase systems with losses with exponential servicing. Automation and Remote Control 40, 27–31 (1979)
7. Bromberg, M.A., Kokotushkin, V.A., Naumov, V.A.: Service by a cascaded network of instruments. Automation and Remote Control 38, 60–64 (1977)
8. Lucantoni, D.M.: New results on the single server queue with a batch Markovian arrival process. Communications in Statistics-Stochastic Models 7, 1–46 (1991)
9. Graham, A.: Kronecker Products and Matrix Calculus with Applications. Ellis Horwood, Chichester (1981)
10. Klimenok, V.I.: Calculation of characteristics of a multiserver queue with rejections and burst-like traffic. Automatic Control and Computer Sciences 33(6), 35–43 (1999)
11. Klimenok, V., Kim, C.S., Orlovsky, D., Dudin, A.: Lack of invariant property of Erlang loss model in case of the MAP input. Queueing Systems 49, 187–213 (2005)
12. Skorokhod, A.: Probability theory and random processes. High School, Kiev (1980)
13. Alfa, A.S., Diamond, J.E.: On approximating higher order MAPs with MAPs of order two. Queueing Systems 34, 269–288 (2000)
14. Heindl, A., Mitchell, K., van de Liefvoort, A.: Correlation bounds for second order MAPs with application to queueing network decomposition. Performance Evaluation 63, 553–577 (2006)
15. Heindl, A., Telek, M.: Output models of $MAP/PH/1(/K)$ queues for an efficient network decomposition. Performance Evaluation 49, 321–339 (2002)
16. Neuts, M.: Matrix-geometric Solutions in Stochastic Models – An Algorithmic Approach. Johns Hopkins University Press, USA (1981)

# Busy Period Characteristics
# for Single Server Queue
# with Random Capacity Demands

Oleg Tikhonenko[1] and Magdalena Kawecka[2]

[1] Institute of Theoretical and Applied Informatics, Polish Academy of Sciences
Baltycka 5, 44-100 Gliwice, Poland
oleg.tikhonenko@gmail.com
[2] Czestochowa University of Technology, Institute of Mathematics
Ul. Dabrowskiego 69, 42-201 Czestochowa, Poland
kawecka_magdalena@wp.pl

**Abstract.** We investigate single server queueing systems with batch
Poisson arrivals and without demands losses under assumption that each
demand has some random capacity (generally, each demand is characterized by $l$-dimensional indication vector). Service time of the demand
arbitrary depends on its capacity (indications). The Laplace-Stieltjes
transform of total capacities (random vector of sum of indications) of
demands that were served during a busy period of the system is determined.

**Keywords:** queueing system, busy period, demand capacity, demand
indication vector, total demands capacity, vector of total sums of demands indications.

## 1   Introduction

We consider single server queue $BM/G/1/\infty$, where the notation $BM$ defines
the stationary Poisson process with batch arrivals. Let the rate of the Poisson
process be $a$ and the probability that the batch size equals $k$ be $g_k = P\{\nu = k\}$,
where $\nu$ is the number of demands in arbitrary batch, $k \geq 1$. Denote as $G(z) =$
$= \sum\limits_{k=1}^{\infty} g_k z^k$ the generation function of the batch size. Suppose that 1) each demand is characterized by some non-negative random capacity $\zeta$ which is independent of other demands capacities; 2) the demand service time $\xi$ depends on
its capacity only.

Such systems have been used to model and solve the various practical problems
occurring in the design of computer and communicating systems [1,2].

We assume that the common distribution of $\zeta$ and $\xi$ random variables is
determined by the following joint distribution function:

$$F(x,t) = P\{\zeta < x, \xi < t\} .$$

A. Kwiecień, P. Gaj, and P. Stera (Eds.): CN 2012, CCIS 291, pp. 393–400, 2012.
© Springer-Verlag Berlin Heidelberg 2012

Let $L(x) = F(x, \infty)$ be the distribution function of the demand capacity and $B(t) = F(\infty, t)$ be the distribution function of its service time. Denote as

$$\alpha(s, q) = \int_0^\infty \int_0^\infty e^{-sx - qt} \mathrm{d}F(x, t)$$

the two dimensional (with respect to $x$ and $t$) Laplace-Stieltjes transform (LST) of the function $F(x, t)$, where $\mathrm{Re}(s) > 0$, $\mathrm{Re}(q) > 0$. Now we can determine mixed moments (if they exist) of the random vector $(\zeta, \xi)$:

$$\alpha_{ij} = E(\zeta^i \xi^j) = (-1)^{i+j} \frac{\partial^{i+j}}{\partial s^i \partial q^j} \alpha(s, q) \Big|_{s=0, q=0}.$$

Denote as $\Gamma$ the total sum of capacities of demands which were served during the busy period of the system. Let $\Gamma(x) = P\{\Gamma < x\}$ be the distribution function of the random variable $\Gamma$ and $\gamma(s)$ be the LST of the function $\Gamma(x)$, i.e.

$$\gamma(s) = \int_0^\infty e^{-sx} \mathrm{d}\Gamma(x).$$

The main purpose of the article is to determine the function $\gamma(s)$.

## 2    The Equation for the Function $\gamma(s)$

Further we shall use some modification of the additional event method [3]. Suppose that $s > 0$ and any demand having capacity $x$ is red with probability $e^{-sx}$ or blue with probability $1 - e^{-sx}$ independently of other demands. Then the LST $\varphi(s) = \alpha(s, 0)$ of the function $L(x)$ have the sense of probability that an arbitrary demand is red. Suppose that independently of the system behaviour some events take place. These events named catastrophes make up the stationary Poisson process with parameter $q > 0$. Then $\alpha(s, q)$ is the probability that an arbitrary demand is red and there will be no catastrophes during its service time. In this case $\gamma(s)$ is the probability that only red demands will be served during the busy period.

It's obvious that characterstics of the random variable $\Gamma$ don't depend on the order of demands service inside the busy period. Assume that 1) the order of demands service is arbitrary inside any batch; 2) batches entering to the system during service of demands from the batch opening the busy period are served in the opposite to their entering order and have an advantage over demands from the opening batch; 3) if service of some demand from the batch entering inside the busy periof begins, then all demands of this batch will be served in the first place.

Now we can connect with each batch coming inside the busy period the time interval from the epoch of beginning service of the first demand from this batch to the epoch of finishing service of all demands from this batch and batches entering later. This interval we shall name the busy period corresponding to the batch. We shall name a batch (coming inside of the busy period) good, if only red demands were served during its corresponding busy period. It's obvious that an arbitrary "inside" batch is good with probability $\gamma(s)$ for the discipline under consideration.

Let $P(x)$ be the conditional probability that only good batches come to the system during the service of the demand from the "opening" batch under condition that its capacity is equal to $x$. Let us introduce the notation $B(t|\zeta = x) = P\{\xi < t|\zeta = x\}$. Then we have

$$P(x) = \int_0^\infty \sum_{k=0}^\infty \frac{(at\gamma(s))^k}{k!} e^{-at} dB(t|\zeta = x) = \int_0^\infty e^{-(a-a\gamma(s))t} dB(t|\zeta = x) .$$

So, we obtain for probability $\kappa(s)$ that only good batches come to the systems during servicing of a demand from the "opening" batch:

$$\kappa(s) = \int_0^\infty e^{-sx} P(x) dL(x) = \int_0^\infty e^{-sx-(a-a\gamma(s))t} dF(x,t) = \alpha(s, a - a\gamma(s)) .$$

Then for the probability $\gamma(s)$ we obtain the following functional equation:

$$\gamma(s) = \sum_{k=1}^\infty g_k(\kappa(s))^k = G(\alpha(s, a - a\gamma(s))) , \tag{1}$$

which is true for such $s$ that $\operatorname{Re}(s) > 0$, as it follows from the analytical continuation principle. In particular, if each batch consists of a single demand (i.e. for the system $M/G/1/\infty$) we have $G(z) = z$ and

$$\gamma(s) = \alpha(s, a - a\gamma(s)) .$$

## 3    Solution of the Equation

Denote as $\beta(q) = \alpha(0, q)$ the LST of the function $B(t)$ and as $\beta_1 = -\beta'(0)$ the first moment of service time.

**Theorem 1.** *a) The functional Equation (1) determines a single analytical in the domain $\operatorname{Re}(s) > 0$ function $\gamma(s)$, so that $|\gamma(s)| < 1$ in this domain;*

*b) the function $\gamma(s)$ can be represented in the following form:*

$$\gamma(s) = \int_0^\infty e^{-sx} d\Gamma(x) ,$$

*where $\Gamma(x)$ is the proper distribution function, if $\rho = a\beta_1 G'(1) \le 1$, and defective, if $\rho > 1$ (it means that $\Gamma(\infty) = 1$, if $\rho \le 1$, and $\Gamma(\infty) < 1$, if $\rho > 1$).*

*Proof.* Let us introduce the function $\psi(s, z) = G(\alpha(s, a - az)) - z$. This function is convex with respect to $z$, if $s$ is fixed, where $\operatorname{Re}(s) > 0$. Indeed, the function $G(\alpha(s, q))$ is completely monotone [4] (pp. 439–442) with respect to each of its arguments, because it is an LST of some distribution function $R(x, t) = $

$$= \sum_{k=1}^{\infty} g_k R_k(x,t),$$ where $R_k(x,t)$ is the distribution function with LST being equal to $(\alpha(s, a - a\gamma(s)))^k$. Then, for real $s \geq 0$, $q \geq 0$ the inequality

$$\frac{\partial^2 G(\alpha(s,q))}{\partial q^2} \geq 0$$

takes place and we have, obviously,

$$\frac{\partial^2 \psi(s,z)}{\partial z^2} = a^2 \frac{\partial^2 G(\alpha(s,q))}{\partial q^2}\bigg|_{q=a-az} \geq 0 \ .$$

Therefore, the function $\psi(s,z)$, being continuous with respect to $z$ (if $s \geq 0$ is fixed), takes in the point of $z = 0$ and $z = 1$ the positive and negative value consequently.

It follows from the convexity of the function $\psi(s,z)$ with respect to $z$ that a single solution $\gamma(s)$ of the functional equation $G(\alpha(s, a - az)) - z = 0$ exists, when $s \geq 0$, $0 \leq z \leq 1$.

Let us introduce the following sequence of functions:

$$\gamma_0(s) = 0, \quad \gamma_{k+1}(s) = G(\alpha(s, a - a\gamma_k(s))), \ k \geq 0 \ . \tag{2}$$

By mathematical induction we shall prove that

$$0 \leq \gamma_k(s) \leq \gamma_{k+1}(s) \leq 1 \tag{3}$$

and the functions $\gamma_k(s)$ are completely monotone for all $k \geq 0$.

So far as the function $G(\alpha(s,a))$ is completely monotone, the truth of the inequality (3) when $k = 0$ follows from (2) and from the evident fakt that function $\gamma_0(s) = 0$ is completely monotone. Suppose that the inequality (3) takes place for some $k$ and the function $\gamma_k(s)$ is completely monotone. Then, as it follows from the inequalities

$$0 \leq \gamma_k(s) = G(\alpha(s, a - a\gamma_{k-1}(s))) \leq G(\alpha(s, a - a\gamma_k(s))) = \gamma_{k+1}(s) \leq 1$$

and from the properties of completely monotone functions [4], the inequality (3) takes place for all $k \geq 0$, and for all such $k$ the functions $\gamma_k(s)$ are completely monotone.

Then, the sequence of completely monotone functions $\{\gamma_k(s)\}_{k=1}^{\infty}$ is monotonous and bonded, therefore the limit $\gamma(s) = \lim_{k \to \infty} \gamma_k(s)$ exists. It's obvious that $|\gamma(s)| < 1$ and the function $\gamma(s)$ is completely monotone. Hence, it can be represented in the form

$$\gamma(s) = \int_0^{\infty} e^{-sx} d\Gamma(x) \ ,$$

where $\Gamma(x)$ is some measure on $\mathbb{R}$.

It's clear that $z = \gamma(0)$ is the minimal solution of the equation $G(\alpha(s, a - az)) - z = 0$. But $z = 1$ is the solution of the equation too (the both solutions coincide in the case of $\gamma(0) = 1$). It follows from the convexity of the function $G(\alpha(0, a - az)) = G(\beta(a - az))$ with respect to $z$ that the second solution $z < 1$ exists, if and only if $-aG'(1)\beta'(0) = aG'(1)\beta_1 > 1$. Therefore, only in this case the distribution function corresponding to the LST $\gamma(s)$ will be defective (i.e. $\Gamma(\infty) < 1$). If $aG'(1)\beta_1 \leq 1$, then $\gamma(0) = 1$ and $\Gamma(\infty) = 1$, i.e. the corresponding distribution function is proper.

The theorem is proved.

Let $\varphi_i$ and $\beta_i$ be the $i$th moment of demand capacity and service time consequently. Then, as it follows from (1), when $\rho < 1$ we have the next formulae for the first ($\gamma_1$) and second ($\gamma_2$) moment of the random value $\Gamma$:

$$\gamma_1 = \frac{G'(1)\varphi_1}{1 - \rho}, \quad \gamma_2 = \frac{G'(1)\varphi_2}{1 - \rho} + \frac{2a(G'(1))^2\varphi_1\alpha_{11}}{(1 - \rho)^2} + \frac{\varphi_1^2[G''(1) + a^2\beta_2(G'(1))^3]}{(1 - \rho)^3},$$

or for the case of $G(z) = z$ we have

$$\gamma_1 = \frac{\varphi_1}{1 - \rho}, \quad \gamma_2 = \frac{\varphi_2}{1 - \rho} + \frac{2a\varphi_1\alpha_{11}}{(1 - \rho)^2} + \frac{a^2\beta_2\varphi_1^2}{(1 - \rho)^3}.$$

We can calculate moments of arbitrary order (if they exist) of the random variable $\Gamma$ by the same way.

## 4   Some Generalization of Theorem 1

Suppose that each demand is characterized by $l$-dimensional random non-negative indication vector $\boldsymbol{\zeta} = (\zeta_1, \ldots, \zeta_l)$.

We assume, that the common distribution of an indication vector $\boldsymbol{\zeta}$ and service time $\xi$ of the demand is done by the following joint distribution function:

$$F(\mathbf{x}, t) = P\{\zeta_1 < x_1, \ldots, \zeta_l < x_l, \xi < t\} = P\{\boldsymbol{\zeta} < \mathbf{x}, \xi < t\},$$

where $\mathbf{x} = (x_1, \ldots, x_l)$ and $x_i$, $i = \overline{1, l}$, are non-negative real numbers.

Let $L(\mathbf{x}) = F(\mathbf{x}, \infty)$ be the distribution function of demand $l$-dimensional indication vector, $B(t) = F(\infty, t)$ be the distribution function of its service time, where $\infty = (\infty, \ldots, \infty)$ is an $l$-dimensional vector. Denote as

$$\alpha(\mathbf{s}, q) = \int_0^\infty \cdots \int_0^\infty \int_0^\infty e^{-(\mathbf{s}, \mathbf{x}) - qt} dF(\mathbf{x}, t)$$

the $(l+1)$-dimensional Laplace-Stieltjes transform (LST) of the function $F(\mathbf{x}, t)$, where $\mathbf{s} = (s_1, \ldots, s_l)$, $(\mathbf{s}, \mathbf{x}) = s_1 x_1 + \ldots + s_l x_l$; $\mathrm{Re}(s_i) > 0$, $i = \overline{1, l}$; $\mathrm{Re}(q) > 0$. Now we can determine mixed moments (if they exist) of the random vector $(\boldsymbol{\zeta}, \xi)$ or its arbitrary partial vector:

$$E(\zeta_1^{i_1} \ldots \zeta_l^{i_l} \xi^j) = (-1)^{i_1 + \ldots + i_l + j} \frac{\partial^{i_1 + \ldots + i_l + j} \alpha(\mathbf{s}, q)}{\partial s_1^{i_1} \ldots \partial s_l^{i_l} \partial q^j}\bigg|_{\mathbf{s} = \mathbf{0}, q = 0},$$

where $i_k, j = 1, 2, \ldots ; k = \overline{1, l}$; $\mathbf{0} = (0, \ldots, 0)$ is an $l$-dimensional vector.

Denote as $\Gamma_i$ the sum of $i$th components of indication vectors of demands that were served during a busy period of the system under consideration, $i = \overline{1, l}$, $\boldsymbol{\Gamma} = (\Gamma_1, \ldots, \Gamma_l)$. Let $\Gamma(\mathbf{x}) = P\{\Gamma_1 < x_1, \ldots, \Gamma_l < x_l\}$ be the distribution function of the random vector $\boldsymbol{\Gamma}$, $\gamma(\mathbf{s})$ be the $l$-dimensional LST of the distribution function $\Gamma(\mathbf{x})$, i.e.

$$\gamma(\mathbf{s}) = \int_0^\infty \ldots \int_0^\infty e^{-(\mathbf{s}, \mathbf{x})} d\Gamma(\mathbf{x}) \ .$$

By the same way we can prove that for the function $\gamma(\mathbf{s})$ the following functional equation takes place:

$$\gamma(\mathbf{s}) = G(\alpha(\mathbf{s}, a - a\gamma(\mathbf{s}))) \ , \tag{4}$$

which is true for such $s_i$, $i = \overline{1, l}$, that $\mathrm{Re}(s_i) > 0$.

In this case we can easily obtain the generalization of Theorem 1:

**Theorem 2.** *a) The functional Equation (4) determines a single analytical in the domain* $\mathrm{Re}(s_i) > 0$, $i = \overline{1, l}$, *function* $\gamma(\mathbf{s})$, *so that* $|\gamma(\mathbf{s})| < 1$ *in this domain;*
*b) the function* $\gamma(\mathbf{s})$ *may be represented in the following form:*

$$\gamma(\mathbf{s}) = \int_0^\infty \ldots \int_0^\infty e^{-(\mathbf{s}, \mathbf{x})} d\Gamma(\mathbf{x}) \ ,$$

*where* $\Gamma(\mathbf{x})$ *is the proper distribution function, if* $\rho = a\beta_1 G'(1) \leq 1$, *and defective, if* $\rho > 1$ *(it means, that* $\Gamma(\infty) = 1$, *if* $\rho \leq 1$, *and* $\Gamma(\infty) < 1$, *if* $\rho > 1$).

Theorem 1 is obviously a special case of Theorem 2, when $l = 1$.

## 5   Some Special Cases

### 5.1   Case 1

Let $l = 1$. Suppose that the random variables $\zeta$ and $\xi$ are independent, i.e. $\alpha(s, q) = \varphi(s)\beta(q)$, where $\beta(q) = \alpha(0, q)$ is LST of the distribution function $B(t)$, $\mathrm{Re}(q) > 0$. Then from (1) we have $\gamma(s) = G(\varphi(s)\beta(a - a\gamma(s)))$. In particular, if we suppose that all demands capacities are equal to one, then the total capacity of demands being served during the busy period coincides with their number. In this case we have $\varphi(s) = e^{-s}$. Hence, denoting as $R(z)$ the generation function of number of demands being served during busy period, we have, obviously, that $\gamma(s) = R(e^{-s})$, and obtain the well known equation

$$R(z) = G(z\beta(a - aR(z)))$$

for the generation function $R(z)$.

## 5.2   Case 2

Let $l = 1$. Suppose that demand service time is proportional to its capacity, i.e. $\xi = c\zeta$, where $c = const > 0$. It's known [1], that in this case we have $a(s, q) = \varphi(s + cq)$. So, the Equation (1) looks as

$$\gamma(s) = G(\varphi(s + ca - ca\gamma(s))) \ . \tag{5}$$

In particular, if demand capacity equals numerically to its service time (in this case we have $c = 1$ and $\varphi(s)$ must be changed by $\beta(q)$ in the Equation (5)), then the total capacity of demands being served during the busy period coincides with the duration of the period (i.e. $\gamma(s)$ in (5) must be changed by LST $\pi(q)$ of the busy period). Then from (5) we obtain the well known equation for the function $\pi(q)$:

$$\pi(q) = G(\beta(q + a - a\pi(q))) \ .$$

## 5.3   Case 3

Let $l = 2$. Suppose that for each demand we have $\zeta_1 = 1$, and $\zeta_2$ is the demand capacity. Let service time be proportional to the demand capacity ($\xi = c\zeta_2$, $c > 0$). In this case we have

$$a(\mathbf{s}, q) = a(s_1, s_2, q) = e^{-s_1}\varphi(s_2 + cq)$$

and (4) equation looks as

$$\gamma(s_1, s_2) = G(e^{-s_1}\varphi(s_2 + ca - ca\gamma(s_1, s_2))) \ . \tag{6}$$

Suppose that $\zeta_2$ equals numerically to service time of the demand and $c = 1$. Then the components $\Gamma_1$, $\Gamma_2$ of the random vector $\boldsymbol{\Gamma}$ have the following sense: $\Gamma_1$ is the number of demands being served during the busy period, $\Gamma_2$ is the duration of the period. Then the Equation (6) has the following form:

$$\gamma(s_1, q) = G(e^{-s_1}\beta(q + a - a\gamma(s_1, q))) \ . \tag{7}$$

Let

$$g(z, q) = E(z^\eta e^{-q\Pi}) = \sum_{n=1}^\infty z^n P\{\eta = n\} \int_0^\infty e^{-qt} dP\{\Pi < t | \eta = n\} \ .$$

Then we obviously have $\gamma(s_1, q) = g(e^{-s_1}, q)$, and for the function $g(z, q)$ that describe the common distribution of the busy period duration and the number of demands being served during the period, we obtain from (7) the next well known equation:

$$g(z, q) = G(z\beta(q + a - ag(z, q))) \ .$$

So, the last equation is the consequence of the Equation (4) too.

# 6 Conclusion

Study of busy period characteristics is very important in queueing theory. For example, it has aplications in investigations of non-stationary queues and different service disciplines.

Our approch to this problem is based on the theory of queueing systems with non-homogeneous demands (especially, demands having some random capacity). In the paper we obtain total sums characteristics of indications of demands served during busy period of $M/G/1/\infty$ queue under assumption that service time depends on the demand indications. These characteristics reflect the amount of information being processed during processor continious action.

We show that all classical characteristics of $M/G/1/\infty$ busy period are the special cases of our results.

Note that results obtained in the paper are true for every conservative service discipline including processor sharing [5].

# References

1. Tikhonenko, O.M.: Queueing Models in Computer Systems. Universitetskoe, Minsk (1990) (in Russian)
2. Tikhonenko, O.: Computer Systems Probability Analysis. Akademicka Oficyna Wydawnicza EXIT, Warsaw (2006) (in Polish)
3. Klimov, G.P.: Stochastic Service Systems. Nauka, Moscow (1966) (in Russian)
4. Feller, W.: An Introduction to Probability Theory and Its Applications, vol. 2. Wiley, New York (1971)
5. Yashkov, S.F., Yashkova, A.S.: Processor Sharing. A Survay of the Mathematical Theory. Autom. Remote Control 68(9), 1662–1731 (2007)

# A CPU-GPU Hybrid Approach to the Uniformization Method for Solving Markovian Models – A Case Study of a Wireless Network

Beata Bylina, Marek Karwacki, and Jarosław Bylina

Institute of Mathematics, Marie Curie-Skłodowska University
Pl. M. Curie-Skłodowskiej 5, 20-031 Lublin, Poland
{beata.bylina,jaroslaw.bylina}@umcs.pl, marek.karwacki@gmail.com

**Abstract.** One of the most rapidly developing technologies are wireless means of communications with particular emphasis on wireless Internet. There are many methods to investigate the efficiency of wireless networks, some of them are based on Markov chains.

Solving Markovian models of wireless networks may pose some computational problems. They are connected with the size (usually a very huge one) and the time of computations. In this paper we deal with one of the methods for finding transient probabilities in Markovian models – namely the uniformization. We analyze the uniformization method from both algorithmic and implementational perspective to see how the CPU-GPU architecture can be used to accelerate the computations.

We present two parallel algorithms of the uniformization method – the first one utilizing only a multicore machine (CPU) and the second one, with the use of not only a multicore CPU, but also a graphical processor unit (GPU) for the most time-consuming computations.

**Keywords:** Markov chains, Markovian models, uniformization method, GPU, heterogeneous computations, parallel computing, wireless network models.

## 1 Introduction

There is a lot of interest in wireless networks [1], especially wireless Internet networks which are developing rapidly. This is due to high flexibility and growing demand for bandwidth. However, they still need a lot of improvement because of their innate unreliability – especially in comparison to cable networks. Thus, there is a need to construct and investigate models describing properties of such wireless networks. One of a tools for such modeling are Markov chains which return wireless networks characteristics well [2,3].

To investigate wireless networks' operations within some time interval we use homogeneous irreducible Markov chains with finite space of states. Here, the transition rate matrix $Q$ of the Markov chain describing the behavior of the wireless network is independent of time. Thus, we can compute transient probabilities of states at any moment of time by solving the system of first order

A. Kwiecień, P. Gaj, and P. Stera (Eds.): CN 2012, CCIS 291, pp. 401–410, 2012.

ordinary differential equations and then computing various characteristics of the model (like the utilization of the channel or probability of full queue) from the probabilities $x(t)$.

In this paper we investigate whether it is possible to accelerate the uniformization method used for solving Markovian models of wireless networks using GPU in conjunction with more traditional CPU.

From a set of various methods we chose the uniformization method [4,5] (developed also in [6,7]), because this method usually achieves better results than others; moreover, it is numerically stable and allows us to control truncation errors. However, it is quite difficult to use with huge matrices, because it takes a lot of matrix-vector multiplications.

As we said, the main and the most time-consuming operation of this method is the matrix-vector multiplication [8]. This operation can be accelerated by its parallelization – and we can do this by using a multicore CPU or GPU. GPUs contains many parallel processors which can simultaneously run thousands of threads executing the same code on different input data. GPU can not act alone, it requires host platform powered by the CPU, which enables the CPU and the GPU to cooperate and perform CPU-GPU heterogeneous computations.

In the paper we present two parallel algorithms realizing the uniformization method. One of them works on a multicore CPU and the other one works on a joint CPU-GPU architecture, utilizing both a multicore CPU and (for more time-consuming tasks) a manycore GPU. We also consider two formats adjusted to store sparse matrices (both on GPUs and CPUs) – namely the CSR and HYB formats. Some numerical tests of efficiency are conducted for these algorithms and formats with the use of data from wireless networks models – described in details in [9].

The article outline is as follows. Section 2 presents briefly the uniformization methods and the GPU architecture. In Section 3 we propose parallel algorithms (for a CPU only and for a CPU-GPU system) to accelerate the uniformization method. Section 4 shows some numerical results. Section 5 concludes the paper.

## 2    Background

### 2.1    Transient Probabilities Analysis

Analyzing transient probabilities we obtain a system of first order ordinary differential equations (ODE) which is to be solved to find the probabilities. The system can be written (as a matrix-vector ODE) as follows:

$$\frac{dx(t)}{dt} = Q^T x(t) \tag{1}$$

where $Q$ is a transition rate matrix of a Markov chain and $x(t)$ is an unknown column vector of states' probabilities in the moment $t$ (usually denoted by a row vector $\pi(t)$ – here: $x(t) = \pi(t)^T$).

For the equation (1) there exists an analytical solution of the form:

$$x(t) = e^{Q^T t} \cdot x(0) \tag{2}$$

where $x(0)$ is the initial value of the probabilities vector. The main difficulty is in computing the expression $e^{Q^T t}$ which (from Taylor formula for exponential function) we can write as an infinite series:

$$e^{Q^T t} = \sum_{k=0}^{\infty} \frac{(Q^T t)^k}{k!} . \tag{3}$$

The cause of the problem is appearance of high powers of the matrix $Q^T$ computing which is a numerically unstable task.

**Uniformization.** The uniformization method serves solving ODE systems (1) for continuous-time Markov chains (CTMC). The idea of the method is the discretization of the CTMC, that is replacing the CTMC by a DTMC (a discrete-time Markov chain) and a Poisson process. The discretization of the CTMC is based on replacing its transition rate matrix $Q$ by a stochastic matrix $P$ of transitions' probabilities between states in a given time interval.

The probabilities matrix $P$ can be obtained from $Q$ by:

$$P = I + \frac{1}{\alpha} Q \qquad \text{where} \qquad \alpha = \max_i |a_{ii}| .$$

Thus, the problematic expression from Equation (2) can be written as follows:

$$e^{Q^T t} = e^{\alpha P^T t - \alpha I t} = e^{-t\alpha} e^{(t\alpha) P^T} .$$

Multiplying the equation by $x(0)$ and using (3) we get:

$$x(t) = \left( \sum_{k=0}^{\infty} \frac{(\alpha t)^k}{k!} e^{-\alpha t} (P^T)^k \right) \cdot x(0) . \tag{4}$$

**Truncation Error.** Among merits of the uniformization method we can find truncation error control facility. For the numerical computations we need to replace the infinite series (4) with a finite one, what means fixing number of operations and introducing an truncation error determining the accuracy of the computations.

Let:

$$x^*(t) = \left( \sum_{k=0}^{L} \frac{(\alpha t)^k}{k!} e^{-\alpha t} (P^T)^k \right) \cdot x(0) . \tag{5}$$

The truncation error is expressed as $\delta(t) = x(t) - x^*(t)$. Hence, we can obtain a number $L$ defining the length of the finite series (5) needed to achieve a given accuracy $\varepsilon$:

$$||x(t) - x^*(t)||_\infty = 1 - \sum_{k=0}^{L} \frac{(\alpha t)^k}{k!} e^{-\alpha t} \leq \varepsilon . \tag{6}$$

Moreover, in practical implementations we will subdivide the integration domain $\langle 0, t \rangle$ to prevent overflow issues. We divide the interval $\langle 0, t \rangle$ in $l$ intervals of equal lengths $t/l$.

## 2.2 GPU Architecture and CUSP

Graphics Processing Units (GPUs) have recently been used for many applications beyond graphics, introducing the term *general-purpose computation on graphics processors* (GPGPU), owing to (among others) the CUDA (*Compute Unified Device Architecture*) [10] prepared by NVIDIA. An application prepared with the CUDA is composed of a parallel program performed by the CPU and some parts (known as 'kernels') performed in parallel by the GPU.

Graphic processing units are manycore computing systems, being able to deal with thousands of threads running in parallel, processing some data.

In the experiments we used Tesla M2050 based on Fermi architecture with 448 CUDA Cores with 1.15 GHz clock speed. This model contains 3072 MB of total amount memory with clock rate at 1546 MHz. Unlike the previous CUDA architectures, Fermi offers much better performance during double precision computations.

In contrast to CPUs, GPUs have a relatively higher effective memory clock in comparison to its computational core clock.

Starting of the kernel is done from the CPU. A GPU program consists of many threads grouped in blocks – and they are created and managed by the GPU. As in other parallel programs, on CUDA architecture the operations performed on each data needs to be mutually independent, than can be efficiently computed in parallel.

One of the major challenges in developing GPGPU algorithms is to create techniques which fully use pipelining, many cores and high memory bandwidth. It is not an easy task to build an efficient algorithm which uses all the GPU's features. Therefore, we propose using ready-to-use libraries wherever it is possible – what means that problems of pipelining, memory access and block size do not involve us directly. In this paper we choose CUSP [11] library for benchmarking, which supports multiple sparse matrices storage formats.

## 3     Algorithms and Implementation Details

Algorithm 1 computes the states' probabilities vector $x(t)$ in the moment $t$ from the Equation (5). In this algorithm inputs are: the matrix $Q^T$, the initial probabilities vector $x(0)$ and the truncation criterion $\varepsilon$. This algorithm is a multithreaded version using a multicore CPU with OpenMP directives and BLAS (Basic Linear Algebra Subprograms) functions, namely Sparse from the MKL (Math Kernel Library) [12] – extensively threaded mathematic routines for applications that require maximum performance.

Algorithm 2 presents a hybrid version of the uniformization method adjusted to a CPU-GPU ensemble. In this algorithm some operations are performed by

---

**Algorithm 1.** The uniformization method for a multicore CPU

---

**Require:** $Q^T, \varepsilon, t, pt = x(0)$

**Ensure:** $pt = \left( \sum\limits_{k=0}^{L} \frac{(\alpha t)^k}{k!} e^{-\alpha t} (P^T)^k \right) \cdot x(0)$

  1: $\alpha = \max\limits_{i} |q_{ii}|$, *parallelized by OpenMP*
  2: $P^T = \frac{1}{\alpha} Q^T$, *parallelized by MKL BLAS*
  3: $P^T = I + P^T$, *parallelized by OpenMP*
  4: $\Theta = 100; l = \alpha t/\Theta; t = t/l; at = \alpha t$, *single thread*
  5: Compute $L$ from formula 6 using $\varepsilon$, *single thread*
  6: **for** $i = 1$ to $l$ **do**
  7:     $m = pt$,
  8:     **for** $k = 1$ to $L$ **do**
  9:         $r = P^T \cdot m$, *parallelized by MKL Sparse BLAS*
 10:         $m = \frac{at}{k} \cdot r$, *parallelized by MKL BLAS*
 11:         $pt = pt + m$, *parallelized by MKL BLAS*
 12:     **end for**
 13:     $pt = pt \cdot e^{-at}$, *parallelized by MKL BLAS*
 14: **end for**

---

a multicore CPU and others are performed by a manycore GPU. Here, we marked elements for the CPU and for the GPU.

In addition to the previous algorithm we used self-written CUDA kernels and CUSP library on GPU, which performs the most time-consuming operation, namely the matrix-vector multiplication. CUSP library requires to store sparse matrices in one of five standardized formats: Coordinate (COO), Compressed Sparse Row (CSR), Diagonal (DIA), ELLPACK (ELL) or Hybrid (HYB).

Coordinate format is the simplest and most flexible format for general sparse matrices. However, compared with other formats, it is very memory inefficient and computationally intensive.

Compressed Sparse Row format is a general-purpose sparse matrix format (just like COO). It is, however, more efficient, though somewhat less elastic.

ELLPACK [13] stores a sparse matrix on two arrays (one for values, other for column indices) of dimension $M \times Z$, where $M$ is the number of rows and $Z$ is the maximum number of non-zero elements per row. Rows which contain less than $Z$ elements are padded with zeros. This format is well suited for sparse matrices with similar density in each row.

Hybrid format combines efficient memory bandwidth of ELL and flexibility of COO. It is very often the most efficient format for general sparse matrices. HYB stores the most common number of non-zeros per row in ELL and the rest part in COO. The number of optimal non-zeros per row can be computed using a histogram of the row sizes.

In this paper we test two formats for storing the matrix $Q$ – CSR (for both versions) and HYB (for the CPU-GPU hybrid version). The format CSR was chosen because it is well implemented both for CPU and GPU, so both algorithms can

---

**Algorithm 2.** A CPU-GPU hybrid implementation of the uniformization method

---

**Require:** $Q^T, \varepsilon, t, pt = x(0)$

**Ensure:** $pt = \left( \sum_{k=0}^{L} \frac{(\alpha t)^k}{k!} e^{-\alpha t} (P^T)^k \right) \cdot x(0)$

1: $\alpha = \max_i |q_{ii}|$, *parallelized by OpenMP*
2: $P^T = \frac{1}{\alpha} Q^T$, *parallelized by MKL BLAS*
3: $P^T = I + P^T$, *parallelized by OpenMP*
4: $\Theta = 100; l = \alpha t / \Theta; t = t/l; at = \alpha t$, *single thread*
5: Compute $L$ from formula 6 using $\varepsilon$, *single thread*
6: $CPU \rightarrow GPU$ copy of $Q^T$ matrix
7: $CPU \rightarrow GPU$ copy of $pt$ vector
8: **for** $i = 1$ to $l$ **do**
9:     $m = pt$, $GPU \rightarrow GPU$ copy
10:    **for** $k = 1$ to $L$ **do**
11:        $r = P^T \cdot m$;, *parallelized by CUSP on GPU*
12:        $m = \frac{at}{k} \cdot r$, *parallelized by CUDA kernel on GPU*
13:        $pt = pt + m$, *parallelized by CUDA kernel on GPU*
14:    **end for**
15:    $pt = pt \cdot e^{-at}$, *parallelized by CUDA kernel on GPU*
16: **end for**
17: $GPU \rightarrow CPU$ copy of $pt$ vector

---

be compared. We also test HYB because in [14] this format was presented as the fastest format for the multiplication of a sparse matrix by a vector for huge matrices on GPU.

## 4   Numerical Experiments

In this section we evaluate parallel code computing the transient probabilities vector with the use of the uniformization method – in three following versions.

- A multithreaded implementation on the CPU (denoted TCPU) – it is an implementation of Algorithm 1. All computations are carried out on the CPU and the parallelism is achieved using a multithreaded implementation of BLAS (here: MKL) and OpenMP. For storing the matrices we use only the CSR format, because it is efficient format for general sparse matrices on CPU and is supported by MKL, in contrast to HYB.
- A CPU-GPU hybrid implementation with the use of the CSR format (denoted HCSR) – an implementation of Algorithm 2. Computations (as marked in Algorithm 2) are conducted both on the CPU (in parallel, if possible) and on the GPU (always in parallel). The MKL implementation of the BLAS were used for operations on the CPU.

  – A CPU-GPU hybrid implementation with the use of the HYB format (de-
    noted HHYB) – analogous to HCSR but with the use of the HYB format for
    storing the matrices.

## 4.1   Testing Environment

The codes were compiled by NVIDIA nvcc with the following packages: CUDA,
CUSP, MKL. The multithreaded fragments on the CPU (the routines of MKL
and the OpenMP directives) used all (that is twelve) available cores. Table 1
offers details on the hardware and on the software.

**Table 1.** Specification of the hardware and the software used in the experiment

| CPU | 2x Intel Xeon X5650 2.67 GHz (6 cores with HT) |
|---|---|
| Host memory | 48 GB DDR3 1333 MHz |
| GPU | 2x Tesla M2050 (515 Gflop/s DP, 3 GB memory) |
| OS | Debian GNU Linux 6.0 |
| Libraries | CUDA Toolkit 4.0, CUSP 0.2, MKL 10.3 |

We implemented the algorithms in C++ using double precision floating point.
We computed the transient probabilities vector for $t = 10$ and $t = 100$ over
45 matrices with $x(0) = (1, 0, \ldots, 0)^T$ [7] of appropriate length. We run the
experiments with the accuracy tolerance parameter values: $\varepsilon = 10^{-5}$, $\varepsilon = 10^{-7}$,
$\varepsilon = 10^{-10}$.

## 4.2   Results

Note that all the processing times are reported in seconds. Double precision
arithmetic is used in all numerical experiments. The results of accuracy are not
reported as all the solvers produce similar results up to relative machine precision
dependent on $\varepsilon$, not the implementation.

   Figure 1 reports the speedup of the algorithms HCSR and HHYB – both
relative to TCPU (that is the ratio of the speed of a given algorithm to the
speed of TCPU – or, equivalently, the ratio of the processing time in TCPU to
the processing time in a given algorithm). These speedups were computed for
all 45 investigated matrices for $t = 100$ and $\varepsilon = \{10^{-5}, 10^{-7}, 10^{-10}\}$. Such an
investigation allows us to find the threshold for which HHYB becomes better
than HCSR.

   For different values of $\varepsilon$ we achieve the same threshold. Moreover, Fig. 1 shows
that the speedup grows with the growth of the matrix size.

   In Table 2 we present the processing time of our implementations TCPU,
HCSR, HHYB for two selected matrices – the smallest one (9 581) and the largest
one (2 463 506).

   Some conclusions from Table 2 are following:

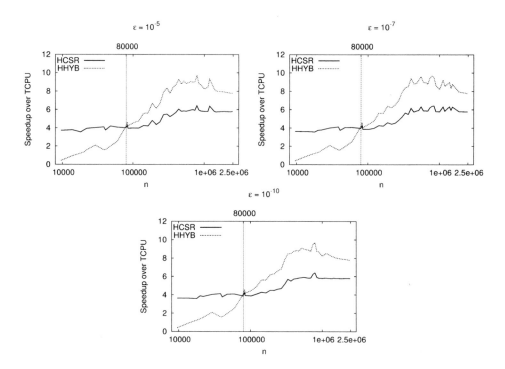

**Fig. 1.** The speedup of the algorithms HCSR and HHYB – both relative to TCPU for all 45 investigated matrices for $t = 100$ and $\varepsilon \in \{10^{-5}, 10^{-7}, 10^{-10}\}$ – note that the horizontal scale (sizes of the matrices) is logarithmic. The vertical line about 80 000 marks the threshold size for which HHYB becomes faster than HCSR for the larger matrixes.

**Table 2.** Processing times in seconds for investigated algorithms with various input parameters (best performance: bold)

| $n$ | $nz$ | $\varepsilon$ | $t$ | TCPU | HCSR | HHYB |
|---|---|---|---|---|---|---|
| 9 581 | 50 555 | $10^{-5}$ | 10 | 0.38 | **0.11** | 0.85 |
| | | | 100 | 3.76 | **1.01** | 8.69 |
| | | $10^{-7}$ | 10 | 0.39 | **0.11** | 0.85 |
| | | | 100 | 3.87 | **1.06** | 9.15 |
| | | $10^{-10}$ | 10 | 0.42 | **0.11** | 0.97 |
| | | | 100 | 4.13 | **1.14** | 9.79 |
| 2 467 506 | 13 851 622 | $10^{-5}$ | 10 | 130.901 | 21.36 | **16.26** |
| | | | 100 | 1261.31 | 217.81 | **162.41** |
| | | $10^{-7}$ | 10 | 131.42 | 22.79 | **17.28** |
| | | | 100 | 1329.77 | 230.22 | **171.63** |
| | | $10^{-10}$ | 10 | 140.99 | 24.47 | **18.57** |
| | | | 100 | 1426.77 | 246.73 | **183.98** |

- For all values of $\varepsilon$ the processing times are similar for the same algorithm (and the same matrix and $t$, of course).
- The processing time depends on the storage format – for small matrices CSR is better, for large ones HYB is better (see also Fig. 1).
- For largest $t$ the processing time is longer (proportionally to $t$). For the largest matrix, TCPU operates about 23 minutes, HHYB – about 3 minutes (for $t = 100$).

We also present a breakdown analysis of our hybrid algorithm for two test matrices (the smallest one and largest one) from Table 2. We show the fastest algorithms for these matrices (HCSR and HHYB, respectively), both for $\varepsilon = 10^{-10}$. Table 3 shows how much time the steps of the Algorithm 2 take – as % of the whole performance.

**Table 3.** Breakdown analysis of HCSR for the smallest matrix and HHYB for the largest matrix for $\varepsilon = 10^{-10}$

| Step # | 9 581 | | 2 467 506 | |
|---|---|---|---|---|
| | $t = 10$ | $t = 100$ | $t = 10$ | $t = 100$ |
| 1 | 8.06 | 0.86 | 0.05 | 0.00 |
| 2 | 0.05 | 0.01 | 0.09 | 0.01 |
| 3 | 0.04 | 0.00 | 0.03 | 0.00 |
| 4–5 | 0.00 | 0.00 | 0.00 | 0.00 |
| 6 | 0.35 | 0.04 | 1.84 | 0.18 |
| 7 | 0.07 | 0.01 | 0.03 | 0.00 |
| 8–16 | 78.39 | 97.68 | 97.90 | 99.79 |
| 17 | 13.00 | 1.39 | 0.03 | 0.04 |

Looking at table (Table 3) we can observe that with the increase in matrix size time consumed by all the steps except 8–16 are decreasing. We can say computations in steps 8–16 are hiding other operations, including CPU-GPU memory transfer. Considering smaller matrices, CPU-GPU transfer has a significant impact on overall performance.

## 5   Conclusion

We have presented a new approach substantially reducing the cost of the uniformization method. We have developed a parallel uniformization-based algorithm that computes transient probabilities in Markov chains and which proved suitable for wireless network models.

The main motivation for studying the CPU-GPU hybrid approach of the uniformization method is the fact that the operation of multiplying a sparse matrix by a vector is performed many (that is: $L \cdot l$) times and takes a huge part of the whole time (see Table 3). We have studied how a heterogeneous CPU-GPU environment can accelerate the computations.

Usually, when we use a GPU, quite a lot of time is consumed by sending data from the CPU to the GPU and back. However, in the uniformization method we need only one transfer from the CPU to the GPU (the matrix) before computations, then we make $L \cdot l$ matrix-vector multiplications on GPU, and eventually have to send our result (the vector) from the GPU to the CPU.

The method has been validated using Markovian models of wireless devices, for various sizes of matrices. For small ones the best speedup is achieved by HCSR and for larger matrices – by HHYB. The finer granularity of the CSR format is advantageous when applied to smaller matrices [8]. For large matrices – of more than 2.5 millions rows – we achieved speedup of about 8 times. The best speedup is for the matrix of the size 788 114 – about 10 times.

**Acknowledgement.** This work was partially supported within the project N N516 479640 of the Ministry of Science and Higher Education of the Polish Republic (MNiSW) "Modele dynamiki transmisji, sterowania zatłoczeniem i jakością usług w Internecie".

# References

1. IEEE Standard for Wireless LAN Medium Access Control (MAC) and Physical Layer (PHY) Specifications, P80211 (November 1997)
2. Bianchi, B.: Performance Analysis of the IEEE 802.11 Distributed Coordination Functioni. IEEE J. on Selected Areas in Communications 18(3), 535–547 (2000)
3. Bylina, J., Bylina, B.: A Markovian Queuing Model of a WLAN Node. In: Kwiecień, A., Gaj, P., Stera, P. (eds.) CN 2011. CCIS, vol. 160, pp. 80–86. Springer, Heidelberg (2011)
4. Sidje, R.B.: Expokit: A software package for computing matrix exponentials. ACM Trans. Math. Software 24, 130–156 (1998)
5. Stewart, W.J.: Introduction to the numerical solution of Markov chains. Princeton University Press, Princeton (1994)
6. Dingle, N.J., Harrison, P.G., Knottenbelt, W.J.: Uniformization and hypergraph partitioning for the distributed computation of response time densities in very large Markov models. J. of Parallel and Distributed Computing 64, 908–920 (2004)
7. Sidje, R.B., Burrage, K., MacNamara, S.: Inexact Uniformization Method for Computing Transient Distributions of Markov Chains. SIAM J. Scientific Computing 29(6), 2562–2580 (2007)
8. Bell, N., Garland, M.: Efficient Sparse Matrix-Vector Multiplication on CUDA. NVIDIA Tech. Report No. NVR-2008-004 (2008)
9. Bylina, J., Bylina, B., Karwacki, M.: A Markovian Model of a Network of Two Wireless Devices. In: Kwiecień, A., Gaj, P., Stera, P. (eds.) CN 2012. CCIS, vol. 291, pp. 411–420. Springer, Heidelberg (2012)
10. NVIDIA Corporation. CUDA Programming Guide. NVIDIA Corporation (2009), http://www.nvidia.com/
11. http://code.google.com/p/cusp-library/
12. http://software.intel.com/en-us/articles/intel-mkl/
13. http://www.cs.purdue.edu/ellpack/
14. Bylina, B., Bylina, J., Karwacki, M.: Computational Aspects of GPU-accelerated Sparse Matrix-Vector Multiplication for Solving Markov Models. Theoretical and Applied Informatics 23(2), 127–145 (2011)

# A Markovian Model of a Network of Two Wireless Devices

Jarosław Bylina, Beata Bylina, and Marek Karwacki

Institute of Mathematics, Marie Curie-Skłodowska University
Pl. M. Curie-Skłodowskiej 5, 20-031 Lublin, Poland
{beata.bylina,jaroslaw.bylina}@umcs.pl, marek.karwacki@gmail.com

**Abstract.** The authors present a Markovian queuing model of two wireless devices sharing a sending channel and its behavior during the work. The authors investigate some measures (that is: the average queue length, the system utilization and the probability of packet loss). To find these values we compute transient probabilities of all the states of a Markov chain with the use of uniformization method. The Markov chain represents a queuing model of a system of two devices mentioned above. The probabilities are used in some analytical formulas for the investigated measures.

**Keywords:** Markov chains, Markovian models, uniformization method, queuing models, wireless network models, DCF (distributed coordination function), 802.11 protocol.

## 1 Introduction

The contemporary Internet demands a lot of flexibility and more and more throughput. That is the reason why wireless networks are highly investigated nowadays. Moreover, they need much more research and improvement because of their unreliability.

A lot of contribution to the research was done (for example [1,2]). Here we present a Markovian queuing model, which represents the behavior of two wireless stations sharing a common channel in the situation of sending data. The model is somewhat similar to a (not queuing) Markovian model of one station from [3] and is based on a queuing model of one station from [4]. That is because the queuing model has such an advantage that some number of such models (here: two of them) could be composed quite easily and investigated together as a model of a system of many wireless nodes.

We present the model of behavior of 802.11 protocol medium access mechanism called distributed coordination function (DCF) and investigate the results of the model. We are interesting in finding how using a queuing model can help us to observe how sharing the channel by two devices can influence their efficiency.

The article outline is as follows. Section 2 recalls the main traits of the distributed coordination function (DCF). Section 3 presents some previous solutions of the problem. Section 4 describes a new Markovian queuing model. Section 5 shows some numerical results and Sect. 6 concludes the paper.

A. Kwiecień, P. Gaj, and P. Stera (Eds.): CN 2012, CCIS 291, pp. 411–420, 2012.

## 2  Real System

The mechanism modelled in the paper, namely DCF, is a part of 802.11 standard [5]. We present here its details important for the paper.

DCF is a mechanism solving a problem of the concurrent access to a shared sending bus. When a new packet is going to be sent, the medium is checked if it is busy for a fixed period of time (distributed interframe space, DIFS). If the channel is free, the packet is transmitted. However, if the channel happens to be busy, a collision control mechanism is employed. It consists in three simple steps.

- First, a backoff time is randomly chosen as an integer number of fixed time intervals $\sigma$, from $\langle 0; W - 1 \rangle$ (where $W$ is a minimal value of the contention window).
- Second, the device waits the chosen time (freezing when the channel is busy).
- Third, if the channel is free after this countdown, the packet is being sent. If it is not free, the whole procedure is repeated but the range for randomized backoff time is $\langle 0; 2 \cdot W - 1 \rangle$. If the transmission also fails this time, the drawing range becomes $\langle 0; 2 \cdot 2 \cdot W - 1 \rangle$ and so on, up to $\langle 0; 2^m \cdot W - 1 \rangle$ (that is: $2^m \cdot W$ is a maximal values of the contention window).

There are various values of the parameters for different media (for instance: $\sigma = 50\mu s$, $W = 16$, $m = 6$ for FHSS, or $\sigma = 8\mu s$, $W = 64$, $m = 4$ for IR).

## 3  Some Previous Models

The mechanism was already modelled by various means from which we recall two.

### 3.1  Bianchi's Model

Bianchi's model [3] is a two-dimensional Markov chain. The first dimension describes the number of failed transmission attempts, the second describes the number of the current back-off stage (which maximum number is determined by the first component of the chain). For a given state $(k, f)$ the set of possible transitions include the following ones:

- to $(k + 1, f')$ for $f = 0$ – with some probability $p$ if the transmission is impossible (the channel is occupied); the new number $f'$ can be any from $\{0, \dots, 2^k \cdot W - 1\}$ (all with the same probability);
- to $(0, 0)$ for $f = 0$ – with the opposite probability if the transmission is possible (the channel is free);
- to $(k, f - 1)$ for $f > 0$.

The model allows for some analytical considerations and also numerical solutions [6].

## 3.2   Queuing Model of a Single Device

A Markovian queuing model of such a device is presented in [4] – see also Fig. 1. The state of the model can be described with four integers $(c, k, f, s)$:

- $c \in \langle 0; C \rangle$ is the current number of packets in the system ($C$ being the maximum capacity of the system);
- $k \in \langle 0; m \rangle$ is the number of failed transmission attempts ($m$ being the maximum number of counted sending attempts);
- $f \in \langle 0; 2^m \cdot W - 1 \rangle$ is the current number of time slots left to the moment of the next transmission attempt ($W$ being the minimum contention window);
- $s \in \{0; 1\}$ is a flag equal to 1 if and only if the current packet is being sent.

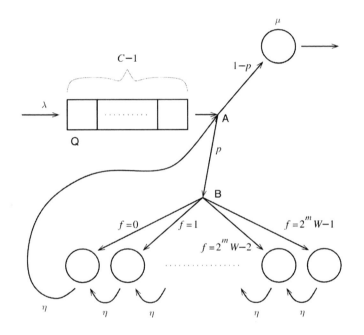

**Fig. 1.** The Markovian queuing model of one device with the DCF mechanism

Besides the parameters $C$ (maximum capacity of the system), $m$ and $W$ (these two were described in Sect. 2), the model has also some more:

- $\lambda$ is the rate of the new packet appearance in the device;
- $\mu$ is the rate of packets' transmission;
- $p$ is the the collision probability;
- $\eta$ is the rate of the transition between slots $f$ and $f-1$ ($\eta = 1/\sigma$, see Sect. 2).

Behavior of the model is as follows.

- Packets appear with an exponential distribution with the rate $\lambda$ and they get to the queue Q which has a capacity $C - 1$ (there can be $C$ packets in the system; and if there is only one, it is being processed, not waiting in the queue). It is only possible if the queue is not full, so the state of the system changes from $(c, k, f, s)$ to $(\min(c + 1, C), k, f, s)$.
- Every packet getting out of the queue (the fork A) can be sent (with the service rate $\mu$) if the channel is free (the probability $1 - p$) and then the state changes as follows: $(c, k, 0, 0) \rightarrow (c, 0, 0, 1) \rightarrow (c - 1, 0, 0, 0)$.
- Otherwise, the channel is busy (the probability $p$) and the packet is suspended (the fork B). It is done by randomly (and equally) drawing a backoff time (see Sect. 2) $f \in \langle 0; 2^k \cdot W - 1 \rangle$, and then the system's state changes from $(c, k, 0, 0) \rightarrow (c, \min(k + 1, m), f, 0)$.
- The last case where the packet can be in our model is the backoff loop, and here the state of the system goes from $(c, k, f, 0)$ to $(c, k, f - 1, 0)$ with the intensity $\eta$ – and eventually to $(c, k, 0, 0)$, and then back to the fork A, when the next attempt of transmission is being made.

## 4    Queuing Model for Two Devices

Here we present a new model of two devices. It is realized as a connection (see Fig. 2) of two queuing models of a single device with some minor modifications.

The state of the model can be described with a vector $(c_1, k_1, f_1, c_2, k_2, f_2, s)$. Here, the meaning of $c_i$, $k_i$, $f_i$ is quite the same as for the single device model (for the first and second station, respectively). Transitions are also similar. The differences are:

- the meaning of $s$ – in this model $s \in \{0, 1, 2\}$, where 0 means that the channel is free and other values indicate which station is using the channel at the moment;
- there is no need for the probability $p$ of the channel occupation – we have both stations directly in the model and now it is known whether the channel is free or occupied by one of them;
- the process of sending a packet is broken down into two stages (represented in Fig. 2 by serving stations with intensities $\mu_0$ and $\mu_1$, respectively) – the first stage (separate for each device) represents a short time when the station is checking the channel occupation (so $\mu_0 = 1/\text{DIFS}$, see Sect. 2); the second stage (common for both) represents actual sending.

## 5    Numerical Experiment

### 5.1    Assumptions for Tests

In our experiments we investigated our model for the following conditions:

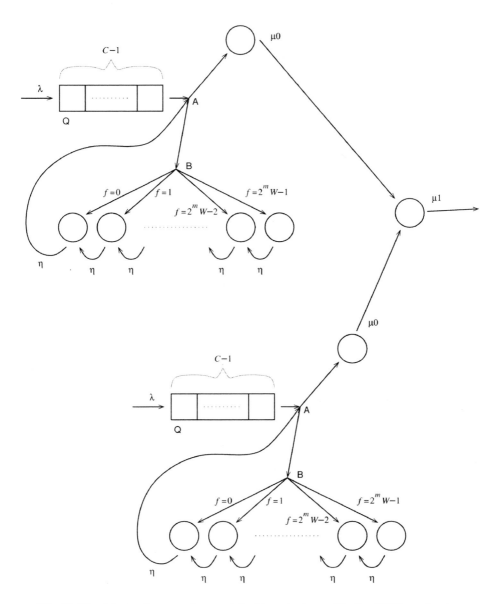

**Fig. 2.** The Markovian queuing model of two devices with the DCF mechanism

- the devices and the traffic were hypothetical, not based on the parameters of real ones, but rather chosen to check some general cases;
- both devices were the same, so was the arriving traffic; thus, we have: $c_i \in \langle 0; C \rangle$; $k_i \in \langle 0; m \rangle$; $f_i \in \langle 0; 2^m \cdot W - 1 \rangle$ (for $i \in \{1, 2\}$);
- for all the tests some parameters were fixed ($\lambda = 1$, $\mu_0 = 100$, $\mu_1 = 5$, $\eta = 90$, $m = 4$);
- other parameters were used with various combinations of their values ($C \in \{10, 20, 30, 40, 50\}$, $W \in \{4, 6, 8, 10\}$).

We tested our model by finding the transient probabilities (that is: dependent on time) of its states and computing with them some characteristics (also dependent on time) of a modeled device. The initial state of the system was: no packets in the system. All the characteristics were computed for the first device (they are identical to the characteristics of the second one because of the model symmetry). These are:

- the utilization of the device: $\displaystyle\sum_{c_1 > 0} Prob(c_1, k_1, f_1, c_2, k_2, f_2, s)$ – that is: what is the fraction of the hole time when the device works actually,
- the packet loss probability: $\displaystyle\sum_{c_1 = C} Prob(c_1, k_1, f_1, c_2, k_2, f_2, s)$,
- the average queue length: $\displaystyle\sum c_1 \cdot Prob(c_1, k_1, f_1, c_2, k_2, f_2, s)$.

The transient states were computed with the use of the uniformization method implemented as a hybrid CPU-GPU algorithm [7] and executed on Intel Xeon X5650 2.67 GHz / NVIDIA Tesla M2050 machine.

## 5.2   Results

The results of our computations are presented in Figs. 3–6.

Our model clearly shows that all the investigated measures grows with time to some asymptotic value – what is expected. The fair stabilization is achieved about 5th–10th time unit (the accurate moment depends on the parameters and considered characteristics).

All the characteristics are quite low because of the low ratio of the incoming traffic to the service rate ($\lambda/\mu_1$).

We can see that with the growth of the minimum value of the contention window $W$, the utilization of the device, the average queue length and the probability of packet loss grows (Figs. 3–5). It also seems expectable, because for a longer potential backoff time (maximum backoff time $2^m \cdot W$ is proportional to $W$), there is a bigger chance that some new packets arrive and so the investigated characteristics grow.

Quite the contrary, with the growth of the capacity $C$ of the device, the probability of the packet loss wanes (what is completely obvious).

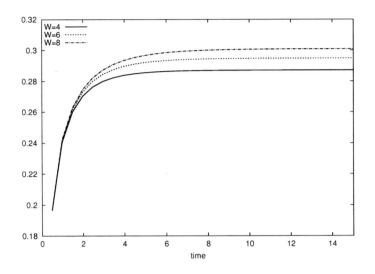

**Fig. 3.** Device utilization (in time) for $C = 10$ and $W \in \{4, 6, 8\}$

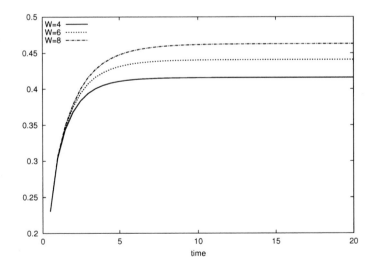

**Fig. 4.** Average queue length (in time) for $C = 10$ and $W \in \{4, 6, 8\}$

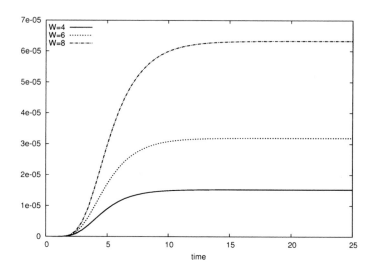

**Fig. 5.** Probability of packet loss (in time) for $C = 10$ and $W \in \{4, 6, 8\}$

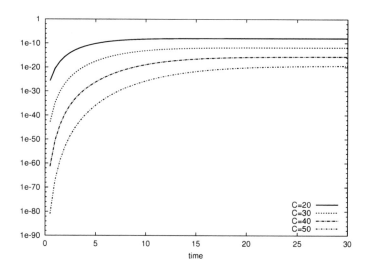

**Fig. 6.** Probability of packet loss (in time) for $W = 10$ and $C \in \{20, 30, 40, 50\}$ – note that the vertical scale (the probability) is logarithmic

# 6   Conclusion and Future Work

In this article a model of two wireless devices sharing a common channel was presented. The devices coordinate the access to the channel with the use of DCF protocol. The model is a Markovian queuing one and it can be numerically solved with the use of general methods used for Markov chains – as the uniformization method. With the use of our model we could compute transient probabilities of states of the model and with them we could analyze a lot of characteristics of the system – just like we did with exemplary characteristics: the utilization of the device, the packet loss probability and the average queue length.

Through a combination of two models of a single device into one model we could get rid of a troublesome analysis of the probability of the channel occupation by other devices – as in [3,2]. Other advantages include ease of investigating coexistence of different devices and traffic characteristics (we could change the parameters of one device independently of the other one).

On the other hand, there are some problems. One of them is a rapid growth of the number of states when we attach other devices to our model (our largest model had about $2 \cdot 10^6$ states, for three devices with the same parameters it would be about $3 \cdot 10^9$ states – the growth is exponential). We could try to overcome this problem with the use of stochastic bounds [8].

**Acknowledgement.** This work was partially supported within the project N N516 479640 of the Ministry of Science and Higher Education of the Polish Republic (MNiSW) "Modele dynamiki transmisji, sterowania zatłoczeniem i jakością usług w Internecie".

# References

1. Czachórski, T., Grochla, K., Nycz, T., Pekergin, F.: A diffusion approximation model for wireless networks based on IEEE 802.11 standard (to appear)
2. Wang, L.-C., Huang, S.-Y., Chen, A.: On the Throughput Performance of CSMA-based Wireless Local Area Network with Directional Antennas and Capture Effect: A Cross-layer Analytical Approach. In: WCNC 2004/IEEE Communications Society, pp. 1879–1884 (2004)
3. Bianchi, G.: Performance Analysis of the IEEE 802. 11 Distributed Coordination Function. IEEE Journal on Selected Areas in Communications 18(3), 535–547 (2000)
4. Bylina, J., Bylina, B.: A Markovian Queuing Model of a WLAN Node. In: Kwiecień, A., Gaj, P., Stera, P. (eds.) CN 2011. CCIS, vol. 160, pp. 80–86. Springer, Heidelberg (2011)
5. IEEE Standard for Wireless LAN Medium Access Control (MAC) and Physical Layer (PHY) Specifications, P80211 (November 1997)
6. Manshaei, M.H., Hubaux, J.P.: Performance Analysis of the IEEE 802.11 Distributed Coordination Function: Bianchi Model. Mobile Networks (March 2010), http://mobnet.epfl.ch

7. Bylina, B., Karwacki, M., Bylina, J.: A CPU-GPU Hybrid Approach to the Uniformization Method for Solving Markovian Models – a Case Study of a Wireless Network. In: Kwiecień, A., Gaj, P., Stera, P. (eds.) CN 2012. CCIS, vol. 291, pp. 401–410. Springer, Heidelberg (2012)
8. Fourneau, J.M., Mokdad, L., Pekergin, N., Youcef, S.: Stochastic Bounds for Loss Rates. In: 3rd Adv. Int. Conf. on Telecomunications (AIXT 2007). IEEE Computer Society, Mauritius (2007)

# Cost-Oriented Recommendation Model for E-Commerce

Grzegorz Chodak[1] and Grażyna Suchacka[2]

[1] Institute of Organisation and Management,
Wroclaw University of Technology, Wroclaw, Poland
`grzegorz.chodak@pwr.wroc.pl`
[2] Institute of Mathematics and Informatics,
Opole University, Opole, Poland
`gsuchacka@uni.opole.pl`

**Abstract.** Contemporary Web stores offer a wide range of products to e-customers. However, online sales are strongly dominated by a limited number of bestsellers whereas other, less popular or niche products are stored in inventory for a long time. Thus, they contribute to the problem of frozen capital and high inventory costs. To cope with this problem, we propose using information on product cost in a recommender system for a Web store. We discuss the proposed recommendation model, in which two criteria have been included: a predicted degree of meeting customer's needs by a product and the product cost.

**Keywords:** recommendation method, recommender system, e-commerce, B2C, Business-to-Consumer, web site, web store, cost.

## 1 Introduction

Over the last years many businesses have been moved to the Internet, taking advantage of offering products to a worldwide e-customer population. As a result of the ubiquitous information overload on the Web, as well as a multitude of products offered by online stores, recommendation methods came into play.

Online stores are implemented through Web sites, which allow Internet users (also called "customers" in this context) to browse and purchase products online. Recommender systems are essential components of such sites as they help customers to cope with a huge amount of items available in the store. Various recommendation methods have proved to be successful in enhancing e-commerce sales in the following ways [1]:

- converting page views into transactions through helping customers find products they wish to buy,
- increasing cross-sell by suggesting additional products and increasing the average order size,
- building customer loyalty by creating a value-added relationship between the site of the store and the customer.

A. Kwiecień, P. Gaj, and P. Stera (Eds.): CN 2012, CCIS 291, pp. 421–429, 2012.

The idea of product recommendation is simple: a user interacting currently with a Web store site (who is a "target user") is presented additional information on products in which they may be interested. To predict the most relevant recommendations different techniques have been proposed so far – they are based on user demographic data, user transactional behavior, product and user characteristics, and so on. We propose introducing a novel criterion into a recommender system, namely the information on product cost. To the best of our knowledge, such a criterion has not been considered in the context of recommendation systems so far.

Cost-oriented recommendation for a web store is based on the main purpose of each online retailer, which is profit maximization. The revenue level is one of profit determinants but the second one, also very important, is cost. Up-to-date recommender systems for e-commerce sites take into consideration only the aspect of increasing the revenue through additional sales whereas the cost aspect has been forgotten. As we analyze the information environment of a Web store, we can build very detailed behavioral profiles of customers based on their shopping carts, products' views, and data obtained from tracking systems. The Web store's databases combined with server logs and a data mining system can give us a lot of information about customer needs and tastes. To maximize revenues it is important to propose customers products in which they can be interested. For example, complementary products offered to a customer can increase the value of the shopping cart and the resultant revenue from e-business. However, from the cost-side point of view, which is crucial to online retailers, it is important to sell products which generate the highest cost in inventory.

Motivated by the aforementioned observations, the authors propose applying a multicriteria analysis in a Web store system in order to recommend products in which an e-customer may be interested, on the one hand, and which generate the highest cost, on the other hand. The general idea of the proposed recommendation model is the following: for the online retailer it is more profitable to recommend to a customer products which generate much higher cost than other products, even if the probability of buying them is a little bit smaller than for other products generating much lower cost.

The rest of the paper is organized as follows. Section 2 outlines the state of the art in recommendation methods for e-commerce Web sites, with a focus on the kind of information used and criteria applied. Section 3 discusses the proposed recommendation model, which aims both at meeting user's needs and limiting online retailer's cost while recommending products to a target user. Finally, Section 4 concludes the paper.

## 2   The State of the Art in Recommender Systems for E-Commerce

Recommender systems for online stores may use different kind of knowledge and apply various techniques to generate product recommendations. In this section, we overview these approaches with special respect to data sources used by them.

## 2.1   General and Impersonalized Recommendations

Basic recommendations in most of contemporary Web stores include novelties, bestsellers, and products selected at random. Such recommendations are easy to implement and do not involve complex computations.

## 2.2   Approaches Using Context-Free Information on Associations Between Products

Some products may be related to each other, e.g. they may be complements or substitutes. Such information may be very useful in predicting customer needs and offering to them products with high probability of being purchased together. In practice, this kind of knowledge is usually provided by experts and is hand-coded. However, knowledge on relations between products may be automatically discovered by using association rules or collaborative filtering. An association rule describes the relationship that one product is often purchased with other products. Collaborative filtering methods are capable of computing similarities between items, e.g. based on product text description or keywords [2,3,1].

## 2.3   Approaches Using Information on Products in the Context of a Single Customer

Customer behavioral data connected with products viewed, purchased, or rated by a customer may be recorded during multiple customer visits to the site. Then, at the next customer visit, the information may be used by some of content-based recommendation methods. For example, the utility of a new item for a target user may be estimated based on ratings assigned by this user to other items which are "similar" to the estimated item. In such a case the recommender system tries to discover similarities among the items which have been rated highly by the user in the past and then, new items matching user preferences the most are recommended.

Some methods use a customer profile based on user characteristics, such as age, gender, interests, etc. This information may be given explicitly by the user, e.g. through a questionnaire, may be assessed based on ratings assigned by the customer to specific products, or may be learned from user transactional behavior over time.

Techniques for content-based recommendation may use traditional heuristics based mostly on information retrieval methods and calculate predictions based on heuristic formulas, such as a cosine similarity measure. Other approaches for content-based recommendation are based on a user model. Such model is often built using statistical learning techniques, e.g. Bayesian classifiers or machine learning techniques: clustering, decision trees, and artificial neural networks [2].

## 2.4   Approaches Using Information on Products in the Context of Multiple Customers

Some data mining techniques, such as association rules between co-purchased products, are applied to discover associations between products purchased in the past not only by a target customer but also by other customers [4]. Such recommendations are offered by popular Web stores. An example may be a leading online bookstore, amazom.com [5], in which a user entering the page describing a selected book faces the following recommendation sections:

- *Frequently Bought Together* – a group of a few books that have been often purchased together, along with a common price for all of them;
- *Customers Who Bought This Item Also Bought* – a more numerous group of books often purchased together, with each item being described with a degree of association with the selected book, graphically presented by the appropriate number of colored stars;
- *Customers Also Bought Items By* – a group of other authors' books, often bought with the selected book;
- *What Other Items Do Customers Buy After Viewing This Item?* with links to the corresponding items.

More advanced, personalized techniques try to assess a degree of similarity between users based on their behavioral data over multiple visits and make recommendations connected with a new item for a target user based on the items viewed, purchased or rated in the past by users which are "similar" to the target user. Thus, these methods are particularly useful when the information on a particular user is limited.

To build models of similar user groups collaborative filtering methods are typically used. They may be either memory-based (heuristic-based) or model-based methods.

1. *Memory-based collaborative recommendation* approaches are based on the collection of previously viewed, purchased or rated items by similar users. Utility of a new product for a target user is usually computed as an aggregate of utilities of the same product according to the most similar users. The aggregation may be the simple average, the weighted sum, or the adjusted weighted sum [2,4].

   Example approaches include nearest neighbor algorithms based on computing the distance between customers based on their preference history. A prediction concerning a given product is based on the weighted average of opinions for the product given by nearest neighbors of the target customer. In practice, nearest neighbor algorithms use heuristics to search for neighbors, as well as use opportunistic sampling in the case of very large customer population [1].

   Various approaches are used to compute similarity measures between two users in collaborative recommender systems, including a correlation-based approach, a cosine-based approach, and other approaches.

2. *Model-based collaborative recommendation* approaches use different kind of data (e.g. product ratings provided by a user) to learn a user model, which is then used in a recommender system. A probabilistic approach to collaborative filtering may use Bayesian networks or cluster models to estimate

a probability that a target user will give a particular rating to an item given ratings assigned by that user to other items [6].

A Bayesian networks model is based on a graph, in which nodes and edges represent some customer information described in terms of probabilities. Cluster models are created by using different clustering techniques, which make it possible to identify groups of customers with similar preferences. A recommendation decision for a target customer is made by averaging opinions of other customers in the corresponding cluster. Some clustering techniques may assign a customer to several clusters with different degrees of participation and then, recommendation is made as a weighted average across the clusters [1].

There is also possible to combine machine learning techniques (e.g. artificial neural networks) and feature extraction techniques (e.g. singular value decomposition) in collaborative filtering [7]. Furthermore, there have been proposed hybrid approaches to product recommendation, e.g. by combining content-based methods and collaborative filtering [8,9,10,11].

## 3    Description of the Cost-Oriented Recommendation Model

We consider a set of $n$ products available in a Web store, $\{X_1, X_2, \ldots, X_n\}$. Each $i$-th product $(i = 1, 2, \ldots, n)$ can be described with a pair of attributes $(\mathbf{Q}_i, C_i)$, where:

- $\mathbf{Q}_i$ is a vector of recommendation quality ranks of the $i$-th product being recommended with other products, $\mathbf{Q}_i = [Q_{i1}, Q_{i2}, \ldots, Q_{i\,n-1}]^T$,
- $C_i$ is a cost of the $i$-th product.

We now discuss the proposed way of including these attributes in a recommendation model for an e-commerce site.

### 3.1    Recommendation Quality Ranks

Let us consider the first attribute of the $i$-th product, i.e. vector $\mathbf{Q}_i$. The $j$-th element of the vector, $Q_{ij}$ $(j = 1, 2, \ldots, n - 1)$ is a recommendation quality rank of the $i$-th product being recommended with the $j$-th product. In general, $Q_{ij}$ may be computed using various techniques, some of which have been discussed in Sect. 2. We assume that $Q_{ij}$ is a function of three variables:

$$Q_{ij} = F(NV_{ij}, NO_{ij}, CM_{ij}) \; , \tag{1}$$

where $NV_{ij}$ is the number of $i$-th product's views generated by other customers who bought the $j$-th product; $NO_{ij}$ is the number of $i$-th product's orders made by customers who bought the $j$-th product; $CM_{ij}$ is a complement measure from the range $[0, 1]$ representing the objective level of complements, which is an attribute of the pair ($i$-th product, $j$-th product).

Values of $NV$ and $NO$ for a given Web store can be computed by applying statistical techniques to analyze historical data on customer viewing and purchasing behavior.

Values of $CM$ for product pairs are given by an expert (e.g. a manger) and stored in database. For example, lens which suits the camera are complements with a high value of $CM$ near 1. In theory, the number of stored complement measures can be huge, even up to $n(n-1)/2$. However, in practice usually only a small percentage of all products are complements. Thus, a complements matrix has mainly 0 values because only a few pairs of inventory items are complementary products.

There is also a possibility to simplify the model by using boolean values of $CM$; in such a case two products $i$ and $j$ can be complements ($CM_{ij}$ equal to 1) or not ($CM_{ij}$ equal to 0).

$CM$ is a symmetric relationship whereas $NV$ and $NO$ are not symmetric.

There is also a question why not to show customers not only complementary products but also substitutes. It does not complicate the model, because $CM$ could be treated as a general value indicating whether two products need to be shown together to a customer, or not.

To build the multicriteria recommendation function we should transform non-zero values of $NV_{ij}$ and $NO_{ij}$ to a comparable base. We propose the standard normalization of each of these values according to the formula:

$$y_{ij}{}_{\_norm} = \frac{y_{ij} - min_{k=1,2,...,n-1}\{y_{ik}\}}{max_{k=1,2,...,n-1}\{y_{ik}\} - min_{k=1,2,...,n-1}\{y_{ik}\}} , \qquad (2)$$

where $y_{ij}{}_{\_norm}$ is a normalized value of the measure under consideration (i.e. $NV_{ij}$ or $NO_{ij}$) for the $i$-th item relative to the $j$-th item, $y_{ij}{}_{\_norm} \in [0,1]$; $y_{ij}$ is the measurement of $NV_{ij}$ or $NO_{ij}$ for the $i$-th item; $min_{k=1,2,...,n-1}\{y_{ik}\}$ is the minimum value of $NV$ or $NO$ measures for the $i$-th item; $max_{k=1,2,...,n-1}\{y_{ik}\}$ is the maximum value of $NV$ or $NO$ measures for the $i$-th item.

The shape of the recommendation quality rank function is another problem that requires deeper research. We propose computing the recommendation quality rank of the $i$-th product recommended with the $j$-th product as a simple weighted sum:

$$Q_{ij}{}_{\_norm} = w_1 \cdot NV_{ij}{}_{\_norm} + w_2 \cdot NO_{ij}{}_{\_norm} + w_3 \cdot CM_{ij} , \qquad (3)$$

where $w_1$, $w_2$, $w_3$ are weights assigned to the corresponding elements, $\sum_{k=1}^{3} w_k = 1$; $NV_{ij}{}_{\_norm}$ and $NO_{ij}{}_{\_norm}$ are normalized values of $NV_{ij}$ and $NO_{ij}$, respectively, according to (2); $\bar{CM}_{ij}$ is a complement measure representing the objective level of complements.

The weights in the recommendation quality rank function reflect the importance of each criterion. Establishing proper values of these weights is one of the most important problem in this model. We propose an exogenous approach, in which a decision maker has to rank the criteria according to their importance.

## 3.2   Product Cost

Let us consider the second attribute of the $i$-th product, i.e. the cost of the product, $C_i$. From the economic point of view one can distinguish many kinds of cost connected with products for sale. We propose choosing only two kinds of cost which suit the best the purpose of the model, namely the cost of purchasing the product (*purchase cost*) and the *inventory cost*.

On the one hand, the purchase cost of the product indicates the frozen capital which decreases the cash flow and that is why it is very important to take this kind of cost into consideration in our recommendation model.

On the other hand, the inventory carrying cost is significant. It can be calculated as the operational cost of carrying the product on stock for one month (or year). This cost is the function of the product size and additional product features connected with requirements of special inventory environment (temperature, humidity, etc.).

Therefore, the cost of the $i$-th product can be computed according to the following formula:

$$C_i = PC_i + IC_i \; , \tag{4}$$

where $PC_i$ is the purchase cost of the $i$-th product, calculated as the net purchase price; $IC_i$ is the inventory cost of the $i$-th product, calculated as the inventory carrying cost.

After calculating the cost of the $i$-th product we need to normalize its value taking the cost of other products into account:

$$C_{i\_\text{norm}} = \frac{C_i - min_{k=1,2,\ldots,n}\left\{C_k\right\}}{max_{k=1,2,\ldots,n}\left\{C_k\right\} - min_{k=1,2,\ldots,n}\left\{C_k\right\}} \; , \tag{5}$$

where $C_{i\_\text{norm}}$ is the normalized value of $C_i$, i.e. the cost of the $i$-th product calculated according to (4); $min_{k=1,2,\ldots,n}\left\{C_k\right\}$ is the minimum value of all $n$ products' costs; $max_{k=1,2,\ldots,n}\left\{C_k\right\}$ is the maximum value of all $n$ products' costs.

Furthermore, we propose taking into account a decline in the inventory cost factor for the $i$-th product, denoted by $LV_i$. This factor indicates changes in time of the $i$-th product's value: the higher $LV_i$, the faster the $i$-th product loses its value. This factor is included in the recommendation function (6) so that it affects the value of weight $w$; thus, the faster a product loses its value, the higher pressure we put on selling it. We propose $LV_i$ being a percentage annual loss of value of the $i$-th item.

## 3.3   Recommendation Function

The idea of our recommendation model is to offer products which are likely to be purchased by a target customer (i.e. products with high recommendation quality ranks) and which generate high cost in inventory at the same time. Depending on a specific Web store, these two criteria (the recommendation quality ranks and the product costs) can be combined together in many possible ways. For example, one could use any recommendation technique to obtain a set of $N$ top recommendations as a first step and then use product costs to obtain the final set of recommendations.

We decided to use both criteria in one step. That is why the normalized values of $Q_{ij}$ and $C_i$ are included in a multicrieria recommendation function.

Taking into consideration weights of the criteria two different kinds of models are possible. In some models the weights are endogenous whereas in others they are automatically generated when the model is optimized (e.g. in [12]). In our model the weights are partially exogenous: firstly, a decision maker (e.g. a manager) has to rank the criteria according to their importance and determine the value of weight $w$; then, $w$ is corrected by $LV_i$, the factor reflecting a decline in the inventory cost of the $i$-th product. The weights obtained are common to all items in the product population of a given Web store.

The value of recommendation function for the $i$-th product in relation to the $j$-th product is computed according to the following formula:

$$RF_{ij} = LV_i \cdot w \cdot Q_{ij\_norm} + (1 - LV_i \cdot w) \cdot C_{i\_norm} , \qquad (6)$$

where $RF_{ij}$ is the recommendation function value for the $i$-th product, estimated to take decision on showing the $j$-th product together with the $i$-th product or not; $LV_i$ is the inventory cost factor indicating a percentage annual decline in the value of the $i$-th product; $w$ is the weight established by a manager, $w \in [0, 1]$; $Q_{ij\_norm}$ is the normalized value of recommendation quality rank $Q_{ij}$, $Q_{ij\_norm} \in [0, 1]$; $C_{i\_norm}$ is the normalized value of the $i$-th product's cost, $C_{i\_norm} \in [0, 1]$.

Values of the recommendation function are computed for all $n$ products periodically and stored in database. The proposed recommender system should point out a given (parametrized) number of products with the highest values of $RF$. Products determined for the $i$-th product will be presented on the Web page describing the $i$-th product in a recommendation rectangle, which is usually placed under the product description.

## 4   Concluding Remarks

Motivated by problems of inventory management and high inventory costs encountered by online retailers, we propose a novel recommendation method for a Web store. We discuss the proposed model, which includes a predicted degree of meeting customer's needs and product cost while considering each product to be recommended with a product selected by a customer at a given moment. The innovativeness of the proposed recommendation model is including product cost, which has not yet been taken into consideration in recommender systems for e-commerce. We propose including the following product costs: purchase cost, inventory cost, and the inventory cost factor indicating an annual decline in the product value.

Our future work will concern a detailed design and implementation of the proposed model. We also plan to introduce some enhancements to the model. Firstly, the problem of establishing the most proper weight value should be explored; one of the possibility is optimizing the weight using the total profit value as the criterion. It is also important to compare efficacy of the proposed method with efficacy of other recommendation methods based on association rules, collaborative filtering, etc.

# References

1. Schafer, J.B., Konstan, J.A., Riedl, J.: E-Commerce Recommendation Applications. Data Mining and Knowledge Discovery 5(1-2), 115–153 (2001)
2. Adomavicius, G., Tuzhilin, A.: Toward the Next Generation of Recommender Systems: A Survey of the State-of-the-Art and Possible Extensions. IEEE Trans. Knowl. Data Eng. 17(6), 734–749 (2005)
3. Deshpande, M., Karypis, G.: Item-Based Top-N Recommendation Algorithms. ACM TOIS 22(1), 143–177 (2004)
4. Sarwar, B., Karypis, G., Konstan, J., Riedl, J.: Analysis of Recommendation Algorithms for E-Commerce. In: 2nd ACM Conference on Electronic Commerce (2000)
5. Amazon.com, http://www.amazon.com
6. Breese, J.S., Heckerman, D., Kadie, C.: Empirical Analysis of Predictive Algorithms for Collaborative Filtering. In: 14th Conference on Uncertainty in Artificial Intelligence. Morgan Kaufmann, San Francisco (1998)
7. Billsus, D., Pazzani, M.J.: Learning Collaborative Information Filters. In: International Conference on Machine Learning, pp. 46–54. Morgan Kaufmann, San Francisco (1998)
8. Burke, R.: Hybrid Web Recommender Systems. In: Brusilovsky, P., Kobsa, A., Nejdl, W. (eds.) Adaptive Web 2007. LNCS, vol. 4321, pp. 377–408. Springer, Heidelberg (2007)
9. Jung, K.-Y., Jung, J.J., Lee, J.-H.: Discovery of User Preference in Personalized Design Recommender System through Combining Collaborative Filtering and Content Based Filtering. In: Grieser, G., Tanaka, Y., Yamamoto, A. (eds.) DS 2003. LNCS (LNAI), vol. 2843, pp. 320–327. Springer, Heidelberg (2003)
10. Kazienko, P., Kolodziejski, P.: Personalized Integration of Recommendation Methods for E-Commerce. International Journal of Computer Science and Applications 3(3), 12–26 (2006)
11. Schein, A.I., Popescul, A., Ungar, L.H., Pennock, D.M.: Methods and Metrics for Cold-Start Recommendations. In: 25th Annual International Conference on Research and Development in Information Retrieval, pp. 253–260. ACM, New York (2002)
12. Ramanathan, R.: ABC Inventory Classification with Multiple-Criteria Using Weighted Linear Optimization. Computers and Operations Research 33(3), 695–700 (2006)

# Uncertainty-Dependent Data Collection in Vehicular Sensor Networks

Bartłomiej Płaczek

Faculty of Transport, Silesian University of Technology,
Krasińskiego 8, 40-019 Katowice, Poland
bartlomiej.placzek@polsl.pl

**Abstract.** Vehicular sensor networks (VSNs) are built on top of vehicular ad-hoc networks (VANETs) by equipping vehicles with sensing
devices. These new technologies create a huge opportunity to extend the
sensing capabilities of the existing road traffic control systems and improve their performance. Efficient utilisation of wireless communication
channel is one of the basic issues in the vehicular networks development.
This paper presents and evaluates data collection algorithms that use
uncertainty estimates to reduce data transmission in a VSN-based road
traffic control system.

**Keywords:** VSN, VANET, data collection, road traffic control.

## 1 Introduction

Vehicular sensor network (VSN) combines wireless communication provided by
vehicular ad-hoc network (VANET) with sensing devices installed in vehicles.
Sensors available in vehicles gather data sets including localisations, speeds, directions, accelerations, etc. Thus, the vehicles participating in VSN can be used
as the sources of information to determine accurately the traffic flow characteristics.

Monitoring of road and traffic conditions becomes an important application area of VSNs. This new technology creates a huge opportunity to extend
the road-side sensor infrastructure of the existing traffic control, management
and safety systems [1,2]. A major drawback of the current-generation traffic
monitoring systems is a narrow coverage due to high installation and maintenance costs. It is expected that the VSNs will help to overcome these limitations.

Unlike traditional wireless sensor networks, VSNs are not subject to major
memory, processing, storage, and energy limitations. However, in dense urban
road networks, where number of vehicles uses the same transmission medium
for many purposes, the periodic transmissions of all sensed data may consume
the entire channel bandwidth resulting in excessive congestion and delays in
the communication network. These effects are a major impediment for the time-
constrained control tasks and safety related services. Therefore the efficient use
of the wireless communication channel is one of the basic issues in VSNs applications development [1].

A. Kwiecień, P. Gaj, and P. Stera (Eds.): CN 2012, CCIS 291, pp. 430–439, 2012.

This paper presents and evaluates three data collection algorithms that use uncertainty estimates to reduce the data transmission in a VSN-based road traffic control system. The uncertainty-dependent data collection algorithms were inspired by an observation that for many cases the scope of real-time vehicular data potentially available in VSN exceeds the needs of particular traffic control tasks. The underlying idea is to detect the necessity of data transfers on the basis of uncertainty evaluation. The advantage of the introduced approach is that it uses selective on-time queries instead of periodical data sampling.

The rest of the paper is organised as follows. Related works are reviewed in Sect. 2. Section 3 describes the VSN-based road traffic control system. Algorithms for traffic data collection are presented in Sect. 4. Section 5 contains the results of an experimental study on data collection for the traffic control in a road network. Finally, in Sect. 6, conclusion is given.

## 2   Related Works

The emergence of VSNs technologies has made it possible to use novel, more effective techniques to deal with the problems of road traffic control. Several traffic control algorithms have been developed in this field for signalised intersections. These adaptive signal control schemes use real-time sensor data collected from vehicles (e.g. their positions and speeds) to minimise travel time and delay experienced by drivers at road intersections. Most methods are based on wireless communication between vehicles and road-side control nodes (e.g. [3,4]). In few proposed solutions, vehicle-to-vehicle communication is used for implementing the traffic control [5,6].

In the above cited studies, the real-time sensed data are assumed to be collected continuously from all vehicles in a certain area. Such periodical data sampling scheme may cause excessive congestion and latency in the communication network due to the bandwidth-limited wireless medium. Therefore, more research is needed to determine required input data sets as well as sampling rates that are necessary for the traffic control.

In the literature several methods have been introduced for wireless sensor networks that enable the optimisation of data collection procedures. Suppression based techniques have been demonstrated to be useful in reducing the amount of sensor data transmitted for monitoring physical phenomena [7]. Temporal suppression is the most basic method: sensor readings are transmitted only from those nodes where a change occurred since the last transmission [8]. Spatial suppression methods aim to reduce redundant transmissions by exploiting the spatial correlation of sensor readings [9]. If the sensor readings of neighbouring sensor nodes are the same or similar, the transmission of those sensed values can be suppressed. Model-based suppression methods use divergence between actual measurements and model predictions to detect the necessity of data transfers [10]. Implementing this approach requires a pair of dynamic models of the monitored phenomenon, with one copy distributed in the sensor network and the other at a base station.

Another effective approach to the optimisation problem of data collection in sensor networks is the model-based querying approach, in which the sensor data are complemented by a probabilistic model of the underlying system [11]. According to this methodology, sensors are used to acquire data only when the model is not sufficiently rich to answer the query with an acceptable confidence. Each query has to include user-defined error tolerances and target confidence bounds that specify how much uncertainty is acceptable in the answer.

In [12] an uncertainty-dependent data collection method was proposed for the VSN-based traffic control systems. In this method, the necessity of data transfers is detected by uncertainty evaluation of traffic control decisions. The sensor data are transmitted from vehicles to the control node only at selected time moments. For the remaining periods of time the sensor readings are replaced by results of an on-line traffic simulation. The effectiveness of this method was verified in an experimental study on traffic control at isolated intersection.

## 3    VSN-Based Road Traffic Control

The purpose of the VSN-based traffic control system is to manage the traffic flow by controlling traffic signals (Fig. 1). VSN senses positions and velocities of vehicles in a road network. The control loop includes data collection module, which sends selective on-time queries to retrieve necessary traffic data from the VSN. At each time step, the set of data that has to be acquired is determined taking into account the uncertainty estimated during decision-making procedure. Traffic model is an important component which uses the acquired data for approximation of the current traffic state as well as prediction of its future evolution. A task of the decision making module is to select an optimal control action on the basis of the information delivered by traffic model, according to the control strategy.

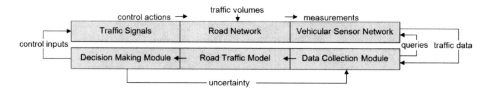

**Fig. 1.** VSN-based road traffic control system

### 3.1    Traffic Control Strategy

In the presented study, a decentralised self-control strategy [13] was applied to minimise travel times in a road network. The self-organised traffic control is based on an optimisation and a stabilisation rule. Both rules are executed in parallel for all intersections in the network in order to adapt the traffic control to local flow conditions.

According to the self-organised traffic control strategy the consecutive control decisions are made in time steps of one second. A particular control decision determines which traffic stream should get a green signal at an intersection. The decision is made using the following formula:

$$\sigma = \begin{cases} \text{head } \Omega & \text{if } \Omega \neq \emptyset \\ \arg\max_i \pi_i & \text{otherwise} \end{cases} , \tag{1}$$

where: $\sigma$ indicates the traffic stream which will get green signal, $\Omega$ is an ordered set containing indices of the traffic streams that have been selected using the stabilisation rule, $\pi_i$ denotes priority of stream $i$, which is calculated on the basis of the optimisation rule.

The aim of stabilisation rule is to assure that all traffic streams will be served at least once in $T_{\max}$ period. To this end, for each traffic stream a service interval $Z_i$ is predicted as the sum of preceding red time $r_i$ for stream $i$, intergreen time $\tau_i^0$ before switching the green signal for stream $i$, and green time $G_i$ required for vehicles in lane $i$ to pass the intersection:

$$Z_i = r_i + \tau_i^0 + G_i . \tag{2}$$

The index $i$ of traffic stream joins the set $\Omega$ as soon as $Z_i \geq T_{\max}$.

Optimisation rule aims for minimising waiting times by serving the incoming traffic as quickly as possible. According to this rule a traffic stream with the highest priority index $\pi_i$ gets green signal, provided that the set $\Omega$ is empty. The priority index for stream $i$ is defined as

$$\pi_i = \frac{N_i}{\tau_{i,\sigma}^{\text{pen}} + \tau_i + G_i} , \tag{3}$$

where: $N_i$ denotes number of vehicles in lane $i$ that are expected to pass the intersection in time $\tau_i + G_i$, $\tau_{i,\sigma}^{\text{pen}}$ is a penalty for switching from stream $\sigma$ to $i$, $\tau_i$ denotes intergreen time after green signal for stream $i$ and $\sigma$ is the index of currently served traffic stream. For more detailed information on the self-organised traffic control strategy see the paper by Lämmer and Helbing [13].

### 3.2 Traffic Model

Traffic model in the traffic control system is used to estimate the numbers of vehicles approaching an intersection ($N_i$) and to predict the required green times ($G_i$). In contrast to the original self-organised control method, which uses a macroscopic (fluid dynamic) traffic model [13], the proposed approach is based on the microscopic fuzzy cellular model [14]. This modification enables a better utilisation of the data acquired from VSN, concerning the parameters of particular vehicles.

The fuzzy cellular traffic model was formulated as a hybrid system combining cellular automata and fuzzy calculus. It was based on a cellular automata approach to traffic modelling that ensures the accurate simulation of real traffic phenomena [15]. A characteristic feature of this model is that it uses fuzzy numbers to represent vehicles positions, velocities and other parameters. Moreover, the model transition from one time step to the next is based on arithmetic of the ordered fuzzy numbers. This approach benefits from advantages of the cellular automata models and eliminates their main drawbacks i.e. necessity of multiple Monte Carlo simulations and calibration issues [16].

A traffic lane in the fuzzy cellular model is divided into cells that correspond to the road segments of equal length. Road traffic streams at an intersection are represented as sets of vehicles. A vehicle $j$ in traffic lane $i$ is described by its position $X_{i,j}$ (occupied cell) and velocity $V_{i,j}$ (in cells per time step). The maximum velocity is defined by the parameter $V_{max}$. The velocities and positions of all vehicles are updated simultaneously in discrete time steps of one second. All the above mentioned variables are expressed by fuzzy numbers.

In this study it was assumed that the fuzzy numbers have trapezoidal or triangular membership functions, thus they are represented by four scalars and the notation $A = \left(a^{(1)}, a^{(2)}, a^{(3)}, a^{(4)}\right)$ is used. Arithmetic operations are computed for the fuzzy numbers using the following definition:

$$o(A, B) = \left(o(a^{(1)}, b^{(1)}), o(a^{(2)}, b^{(2)}), o(a^{(3)}, b^{(3)}), o(a^{(4)}, b^{(4)})\right) , \qquad (4)$$

where $A$, $B$ are the fuzzy numbers and $o$ stands for an arbitrary binary operation.

The application of fuzzy calculus helps to deal with incomplete traffic data and enables straightforward determination of the uncertainty in simulation results [17]. The main advantage of the fuzzy cellular model relies on the fact that the prediction of the parameters $N_i$ and $G_i$ is computationally efficient and the results are also represented by means of fuzzy numbers, thus their uncertainties can be easily evaluated.

## 4     Data Collection Algorithms

This section introduces three data collection algorithms for VSNs as well as defines the uncertainty estimates that enable reduction of data transmission. The data collection algorithms are presented as components of the traffic control procedure, which was discussed in Sect. 3. It should be also noted here that the control procedure is executed independently for each intersection in the road network.

At first, some basic operations will be explained, which are common to the three proposed algorithms (see the pseudocodes below). The aim of the model *update* operation is to approximate the current state of the traffic flow i.e. current positions of all vehicles approaching an intersection ($X_{i,j}$). This approximation is based on both the real traffic data acquired from VSN and the results of real-time simulation. During the real-time simulation the traffic model is used to estimate the missing positions of vehicles that were excluded from direct data acquisition. Besides the data on vehicle positions, the model update operation has to take into account also the real-time status data of traffic control operations (i.e. current traffic signals).

As it was mentioned in the previous section, the traffic model is used to *predict* the numbers of vehicles approaching an intersection ($N_i$) and the required green

```
for each time step
    update traffic model
    for each lane i=1..m
        for each vehicle j=1..n(i)
            if unc(X_{i,j})>ut_pos then acquire X_{i,j}
    if new data collected then update traffic model
    for each lane i=1..m predict N_i, G_i
    make control decision
    execute control decision
```

**Fig. 2.** Pseudocode of data collection algorithm 1

times ($G_i$). Values of these parameters for all lanes are predicted by faster than real-time simulation using the approximation of current traffic state to determine initial conditions. The prediction results are used to *make control decision* that determines which traffic stream should get a green signal at an intersection. Finally, traffic control system *executes the control decision* by switching the appropriate signals.

Figure 2 shows pseudocode of the first data collection algorithm. According to this algorithm the vehicle position is acquired from VSN only if uncertainty of the position $unc(X_{i,j})$, approximated by traffic model, is higher than a predetermined threshold $ut_{pos}$ (in cells). This approach is similar to the concept of model-based querying, which was mentioned in Sect. 2. Note that the vehicle position is represented by a fuzzy number $X_{i,j} = (x_{i,j}^{(1)}, x_{i,j}^{(2)}, x_{i,j}^{(3)}, x_{i,j}^{(4)})$. In order to estimate its uncertainty, the definition was adapted, which is based on determination of the area under membership function. Using this definition, the following formula was derived:

$$unc(X_{i,j}) = 0.5 \left| x_{i,j}^{(1)} - x_{i,j}^{(2)} \right| + \left| x_{i,j}^{(2)} - x_{i,j}^{(3)} \right| + 0.5 \left| x_{i,j}^{(3)} - x_{i,j}^{(4)} \right| . \qquad (5)$$

The second data collection algorithm (Fig. 3) estimates uncertainty of the predicted green times $unc(G_i)$ using a similar measure to that defined in (5). If for a given lane ($i$) the prediction uncertainty of $G_i$ is higher than a threshold value $ut_{pred}$ (in seconds) then the positions of vehicles in that lane are acquired. During the data acquisition only those vehicles are taken into account whose positions cannot be precisely determined by the traffic model.

Uncertainty of control decisions is used for detecting the necessity of data transfers in the third data collection algorithm (Fig. 4). The decision uncertainty is estimated as the maximum of uncertainties associated with the two rules of the control strategy i.e. the stabilisation and the optimisation rule: $unc(decision) = \max(unc_{stab}, unc_{opt})$.

The decision rules include comparison operations that have to be executed for fuzzy numbers and thus the probabilistic approach to fuzzy numbers comparison [18] is employed. This approach enables estimation of the probability $P$ with which one fuzzy number is less, greater or equal to another fuzzy number.

```
for each time step
    update traffic model
    for each lane i=1..m predict Nᵢ, Gᵢ
        if unc(Gᵢ)>ut_pred then
            for each vehicle j=1..n(i)
                if unc(Xᵢ,ⱼ)>0 then acquire Xᵢ,ⱼ
    if new data collected then
        update traffic model
        for each lane i=1..m predict Nᵢ, Gᵢ
    make control decision
    execute control decision
```

**Fig. 3.** Pseudocode of data collection algorithm 2

```
for each time step
    update traffic model
    for each lane i=1..m predict Nᵢ, Gᵢ
    make control decision
    if unc(decision)>ut_dec then
        for each lane i=1..m
            for each vehicle j=1..n(i)
                if unc(Xᵢ,ⱼ)>0 then acquire Xᵢ,ⱼ
    if new data collected then
        update traffic model
        for each lane i=1..m predict Nᵢ, Gᵢ
        make control decision
    if unc(decision)<=ut_dec then execute control decision
```

**Fig. 4.** Pseudocode of data collection algorithm 3

The probabilities are used for uncertainty estimation of traffic control decisions. Detailed discussion of the decision uncertainty estimation method can be found in [12], here only the resulting formulas are given with short comments.

Let $\sigma$ denote the result of the control decision i.e. an index of traffic stream which will get a green signal. The stabilisation rule determines $\sigma$ when $Z_\sigma \geq T_{\max}$. It was assumed for this study that the condition $Z_\sigma \geq T_{\max}$ is satisfied if the probability $P(Z_\sigma \geq T_{\max})$ is above 0.5. Thus, in opposite situation the optimisation rule is activated. Such assumptions lead to the following definition of the stabilisation uncertainty:

$$unc_{\text{stab}} = \begin{cases} 2P(Z_\sigma < T_{\max}) & \text{if} \quad P(Z_\sigma \geq T_{\max}) > 0.5 \\ 2P(Z_\sigma \geq T_{\max}) & \text{if} \quad P(Z_\sigma \geq T_{\max}) \leq 0.5 \end{cases} . \tag{6}$$

The optimisation uncertainty corresponds with the comparisons that are necessary for finding the highest priority value $\pi_\sigma$. This uncertainty equals zero if the stabilisation rule determines the control decision:

$$unc_{\text{opt}} = \begin{cases} 0 & \text{if} \quad P(Z_\sigma \geq T_{\max}) > 0.5 \\ \max_i 2P(\pi_\sigma < \pi_i) & \text{if} \quad P(Z_\sigma \geq T_{\max}) \leq 0.5 \end{cases} . \tag{7}$$

The symbols used in (6) and (7) were defined in Sect. 3.1. Resulting values of the uncertainties $unc_{stab}$, $unc_{opt}$ and $unc(dec)$ are in range between 0 and 1.

## 5  Experimental Results

The proposed data collection algorithms were applied to the VSN-based traffic control in a road network. The experiments were performed in a traffic simulator which was developed for this purpose on the basis of Nagel-Schreckenberg stochastic cellular automata [15]. Structure of the simulated network is presented in Fig. 5. Roads are unidirectional, thus each intersection has two incoming traffic streams. Links between intersections consists of 40 cells that correspond to the distance of 300 m. Maximal velocity of vehicles is 2 cells per time step i.e. 54 km/h (the simulation time step is one second). Deceleration probability $p$ is 0.15. For the above settings of the Nagel-Schreckenberg traffic model, the obtained saturation flow at intersections is about 1700 vehicles per hour of green time.

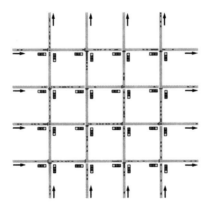

**Fig. 5.** Simulated road network

The self-organised traffic control was simulated assuming the intergreen times $\tau$ of 5 s and the maximum period $T_{max}$ of 120 s. Architecture of the considered traffic control system consists of two types of VSN nodes: control and vehicle nodes. The fixed control nodes installed at intersections collect sensor data from the vehicle nodes and execute the traffic control procedure. Each vehicle in the system is equipped with a wireless communication unit and uses a GPS device to determine its position. Every time a vehicle enters the road network, it has to register itself by sending a hello message. The data collection operation is initialised by the control node which generates queries to acquire positions of vehicles approaching an intersection.

Simulation results for the three data collection algorithms are compared in Fig. 6. The comparison takes into account the performance of traffic control and

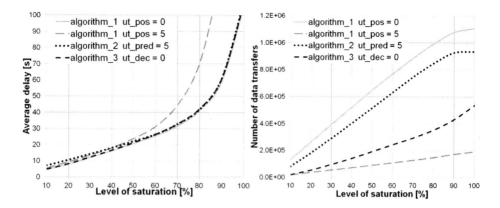

**Fig. 6.** Simulation results: average delay (left) and number of data transfers (right)

the number of data transfers from particular vehicles to the control nodes. The average delays of vehicles and data transfer counts were determined from 3 hour traffic simulations. In this experiment, the traffic flow volumes were changed gradually in order to reproduce saturation levels (demand-capacity ratios) from 0 to 100%.

The highest accuracy of the traffic information in the control system was obtained using the first data collection algorithm with threshold value $ut_{pos} = 0$. This scenario results in lowest delays and highest number of data transfers. The delays grow drastically for algorithm 1 if a threshold value above 0 is used (see the plot for $ut_{pos} = 5$). In comparison, the second algorithm for $ut_{pred} = 5$ provides low delays and reduces the data transfer counts. However, for algorithm 2 with higher threshold values an increase of the delays is observed especially at low saturation levels. The best results were obtained for algorithm 3, which enables significant reduction in the data transfers and does not decrease the performance of the traffic control.

## 6    Conclusion

In this paper three data collection algorithms were proposed for VSN-based traffic control systems. Effectiveness of the introduced algorithms was evaluated in an experimental study on the traffic control in a road network. Experiments were carried out using a simulation environment. The tests confirmed that the proposed algorithms enable reduction in the data transmission for a wide range of traffic conditions. The most promising results were obtained for the algorithm using decision uncertainty to detect the necessity of data transfers.

## References

1. Lee, U., Gerla, M.: A survey of urban vehicular sensing platforms. Computer Networks 54(4), 527–544 (2010)

2. Bernaś, M.: VANETs as a Part of Weather Warning Systems. In: Kwiecień, A., Gaj, P., Stera, P. (eds.) CN 2012. CCIS, vol. 291, pp. 456–466. Springer, Heidelberg (2012)

3. Abishek, C., Kumar, M., Kumar, P.: City traffic congestion control in Indian scenario using wireless sensors network. In: Proc. of WCSN 2009, pp. 1–6. IEEE (2009)

4. Wenjie, C., Lifeng, C., Zhanglong, C., Shiliang, T.: A realtime dynamic traffic control system based on wireless sensor network. In: Proc. of ICPP 2005, pp. 258–264 (2005)

5. Gradinescu, V., Gorgorin, C., Diaconescu, R., Cristea, V., Iftode, L.: Adaptive Traffic Lights Using Car-to-Car Communication. In: Proc. of VTC2007-Spring, pp. 21–25. IEEE (2007)

6. Maslekar, N., Boussedjra, M., Mouzna, J., Labiod, H.: VANET based Adaptive Traffic Signal Control. In: Vehicular Technology Conf., pp. 1–5. IEEE (2011)

7. Kulik, L., Tanin, E., Umer, M.: Efficient Data Collection and Selective Queries in Sensor Networks. In: Nittel, S., Labrinidis, A., Stefanidis, A. (eds.) GSN 2006. LNCS, vol. 4540, pp. 25–44. Springer, Heidelberg (2008)

8. Reis, I., Câmara, G., Assunção, R., Monteiro, M.: Suppressing temporal data in sensor networks using a scheme robust to aberrant readings. Int. J. of Distributed Sensor Networks 5(6), 771–805 (2009)

9. Min, J.K., Chung, C.W.: EDGES: Efficient data gathering in sensor networks using temporal and spatial correlations. J. Syst. Software 83(2), 271–282 (2010)

10. Chu, D., Deshpande, A., Hellerstein, J., Hong, W.: Approximate data collection in sensor networks using probabilistic models. In: Proc. of ICDE 2006, pp. 48–60 (2006)

11. Deshpande, A., Guestrin, C., Madden, S., Hellerstein, J., Hong, W.: Model-driven data acquisition in sensor networks. In: Proc. of VLDB 2004, pp. 588–599 (2004)

12. Płaczek, B.: Selective data collection in vehicular networks for traffic control applications. Transport. Res. Part C (2012), doi:10.1016/j.trc.2011.12.007

13. Lämmer, S., Helbing, D.: Self-control of traffic lights and vehicle flows in urban road networks. J. Stat. Mech., 04019 (2008)

14. Płaczek, B.: Fuzzy Cellular Model for On-Line Traffic Simulation. In: Wyrzykowski, R., Dongarra, J., Karczewski, K., Wasniewski, J. (eds.) PPAM 2009. LNCS, vol. 6068, pp. 553–560. Springer, Heidelberg (2010)

15. Maerivoet, S., De Moor, B.: Cellular automata models of road traffic. Physics Reports 419(1), 1–64 (2005)

16. Płaczek, B.: Fuzzy cellular model of signal controlled traffic stream. arXiv:1112.4631v1 [cs.DM] (2011)

17. Płaczek, B.: Performance Evaluation of Road Traffic Control Using a Fuzzy Cellular Model. In: Corchado, E., Kurzyński, M., Woźniak, M. (eds.) HAIS 2011, Part II. LNCS, vol. 6679, pp. 59–66. Springer, Heidelberg (2011)

18. Sevastianov, P.: Numerical methods for interval and fuzzy number comparison based on the probabilistic approach and Dempster-Shafer theory. Inform. Sciences 177(21), 4645–4661 (2007)

# Numerical Calculations for Geophysics Inversion Problem Using Apache Hadoop Technology

Łukasz Krauzowicz, Kamil Szostek, Maciej Dwornik, Paweł Oleksik,
and Adam Piórkowski

AGH University of Science and Technology,
Faculty of Geology, Geophysics and Environment Protection,
Department of Geoinformatics and Applied Computer Science,
al. A. Mickiewicza 30, 30-059 Kraków, Poland
{szostek,oleksik,pioro}@agh.edu.pl,dwornik@geol.agh.edu.pl
http://www.geoinf.agh.edu.pl

**Abstract.** There are considerations on the problem of time-consuming calculations in this article. This type of computational problems concerns to multiple aspects of earth sciences. Distributed computing allows to perform calculations in reasonable time. This way of processing requires a cluster architecture.

The authors propose using Apache Hadoop technology to solve geophysics inversion problem.This solution is designed rather for analyzing data, but it also enables to perform computations. There is an architecture of solution proposed and real test carried out to determine the performance of method.

**Keywords:** parallel computing, distributed computing, cluster, numerical computing.

## 1   Introduction

Earth sciences, and especially geophysics, are an area of intense research. The considerations are accompanied by a number of modeling and data analysis, which are tasks that require high computational power. A single computer is often not able to perform these calculations in a reasonable time. In such cases it is necessary to use clusters [1].

In the paper [2] the problem of ground vibration modeling is presented. The authors used a very large size of the model. The numerical calculations were time-consuming, therefore calculations were made in parallel on an effective computer cluster.

Seismic wave field modeling is a topic of [3,4]. This is another time-consuming problem in geophysics. It can reveal the nature of an analyzed wave phenomenon. This modeling is often a part of complex and extremely time consuming methods with almost unlimited needs of computational resources, therefore computations are dedicated for academic centers, especially with support from oil and gas companies. The GPU-PC cluster and cluster of component environments were tested.

A. Kwiecień, P. Gaj, and P. Stera (Eds.): CN 2012, CCIS 291, pp. 440–447, 2012.

Another geophysical phenomenon is geothermal field [5,6]. Heat transfer modeling is very important in solving physical problems in Earth science as volcanoes, intrusions, earthquakes, mountain building or metamorphism. This kind of calculations requires high computational power that exceeds the capabilities of a single PC. There is the ability to use a high performance cluster. A solution based on the component technologies was set, but it was not fault-tolerant and did not support a load balancing.

# 2   Inverse Problem for Vertical Transverse Isotropy Geological Medium

Knowledge of velocity distribution in geological medium is one of the most important things in mining exploration. Process of reconstruction velocity distribution is named an inverse problem. Solving inverse problem, especially in anisotropic medium, is a difficult process, because of the non-linear relationship between distribution of value of the elastic parameters and received travel times of wave propagation. This relationship implicates that deterministic method of inversion is useless. One of the methods to obtain velocity distribution is a stochastic inversion. In this paper Monte Carlo method was used to obtain parameters values.

Stochastic inversion is based on generating a huge set of models, calculating theoretical travel times for each seismic ray and evaluating solution. To compare the estimated and received travel times, a following equation was used:

$$L = \frac{1}{N} \sum_{i=1}^{N} |T_i^{\text{est}} - T_i^{\text{rec}}| \; . \tag{1}$$

## 2.1   Seismic Anisotropy

Travel times were estimated using The Shortest Time Method (e.g. [7,8,9]). In this method geological medium was divided into several velocity cells described by Thomsen parameters [10]:

$$v_{\text{P0}} = \sqrt{\frac{c_{33}}{\varrho}} \tag{2a}$$

$$v_{\text{S0}} = \sqrt{\frac{c_{44}}{\varrho}} \tag{2b}$$

$$\gamma \equiv \frac{c_{66} - c_{44}}{2c_{44}} \tag{2c}$$

$$\varepsilon \equiv \frac{(c_{11} - c_{33})}{2c_{33}} \tag{2d}$$

$$\delta \approx \frac{(c_{13} + c_{44})^2 + (c_{33} - c_{44})^2}{2c_{33}(c_{33} - c_{44})} \tag{2e}$$

where $c_{11}$, $c_{13}$, $c_{33}$, $c_{44}$, $c_{66}$ are coefficients of the stiffness tensor. Using this parameters, velocity of seismic wave in $\theta$ direction in vertical transverse isotropy medium is described by following equation [10]:

$$v_P(\theta) \approx v_{P0} \left[1 + \delta \cdot \sin^2 \theta \cdot \cos^2 \theta + \varepsilon \cdot \sin^4 \theta\right] \tag{3a}$$

$$v_{SV}(\theta) \approx v_{S0} \left[1 + \left(\frac{v_{P0}}{v_{S0}}\right)^2 \cdot (\varepsilon - \delta) \cdot \sin^2 \theta \cdot \cos^2 \theta\right] \tag{3b}$$

$$v_{SH}(\theta) = v_{S0} \left[1 + \gamma \cdot \sin^2 \theta\right] . \tag{3c}$$

## 2.2   Seismic Inversion

Crosswell tomography is a method for reconstruction Thomsen parameters' distribution. In one well seismic wave is generated and received by geophones from next well. To reconstruct velocity field 31 shot points and 76 received points were used. In this case only P-wave was used for Thomsen parameters reconstruction, so it was impossible to obtain $\gamma$ and $v_{S0}$ value. Geological medium was divided into 24 velocity cells. That gives 72 independent values to obtain.

This hyperspace excludes regular sampling. Typical method is to generate huge numbers of models and remember only a few of the best solutions. This method does not give information of under-sampled areas and perspective areas (region with small value of error, which can be sampled more). First problem will not be discussed in this work. On the other hand, the second problem can be partially solved by sampling space near the best solutions.

## 3   The Apache Hadoop

The Apache[TM]Hadoop[TM]technology provides a framework for data sets distributed processing. It is meant to be run on clusters of computers, from single servers up to thousands of nodes and to process millions of megabytes. This library is designed to be independent from any hardware failures – it implements malfunction detection on application layer [11,12]. Apache Hadoop technology is widely used in cloud computing systems directed to process large amount of heterogeneous data in a distributed environment [13].

The Apache Hadoop works on three main subprojects:

- Hadoop Common, which supports other subprojects,
- Hadoop Distributed Files System (HDFS) – a high-throughput data access distributed file system,
- Hadoop MapReduce – large data sets processing framework, dedicated for clusters.

Hadoop technology was invented to process large amount of data. In 2009 this framework won the one minute sort: 500GB was sorted in 59 seconds on 1406 nodes. Then 100 terabytes sort was performed in 173 minutes on 3400 nodes [14]. Many companies and organizations use Apache Hadoop for research and production, i.e. Amazon, Yahoo!, Google, Facebook and more [15].

## 3.1  Hadoop Distributed File System and MapReduce

Apache Hadoop technology takes advantage of MapReduce programming model. It allows user to write his own Map and Reduce tasks in Java or, by use of wrappers, in other programming languages, like C++, Ruby or Python. As Apache Hadoop itself is written in Java, using this language for MapReduce tasks is the fastest option, because otherwise Hadoop Streaming or Hadoop Pipes have to be used, i.e. for C++ these wrappers executes C++ Map or Reduce class and communicate through sockets (Fig. 1).

The second strongest side of the Hadoop technology is highly failure-tolerant HDFS, designed to cope with large amount of data. As the data is distributed along all cluster nodes in HDFS, MapReduce can easily process it in parallel.

## 3.2  Apache Hadoop Job Workflow

The Figure 1 shows default Apache Hadoop job workflow, which is:

1. The program is executed by client users. In this paper Job Tracker is used, which is the main node.
2. Client receives Job's ID from Job Tracker.
3. Client copies data, that is program and input data, to the HDFS.
4. Client confirms the job.
5. Job Tracker initiates the job.

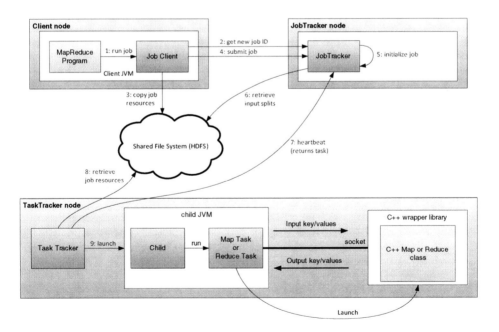

**Fig. 1.** Conventional job submission diagram in Apache Hadoop framework with C++ wrapper

6. Job Tracker receive input data from HDFS, then split it into parts and sends back to HDFS.
7. Task Tracker sends a heartbeat signal to the Job Tracker periodically to inform Job Tracker about its presents and readiness to work.
8. All working Task Trackers receive split data from HDFS.
9. Task Trackers start the task.
10. In child JVM the job is executed. If C++ program was submitted wrapper is used to run C++ MapReduce libraries. It communicates through the socket. Otherwise Java classes are used directly.
11. Task Trackers return results to the HDFS.

## 4 Tests

The main idea of the Apache Hadoop utilization is to search perspective regions by analysis of the generated data with them evaluate value. In order to take advantage of Apache Hadoop results, they should be sent back to inversion algorithm and tested to produce more accurate solution. Process should be repeated until satisfactory results achieved.

The algorithm used to process this data was written in Java and C++, thus Apache Hadoop executes the second using wrappers, as mentioned before. The algorithm itself is not complicated, because most of the complex work is moved to Apache Hadoop framework. This significantly reduces time for code writing.

The algorithm is performed in two stages. First, in the Mapper stage, it gathers parameters with similar compare estimator and produces key-value tuplets, where key is the rounded estimator and value is the set of parameters. Next, in the Reducer stage, all sets of tuplets with the same key are processed. At this moment, for each parameter of one set, average values are calculated and emitted as output. The results of this approach are sufficient, but may be extended in future for more accurate results.

Tests were run on cluster of varying number of nodes: 1, 2, 4 and 8 nodes. Each of them was a low-cost PC running openSUSE 11.4 (Linux kernel 2.6.37), with Intel(R) Pentium(R) 4 CPU 2.8 GHz, 1 GB RAM and 100 Mbit/s Ethernet connection. The impact of network speed in the cluster computations is discussed in the article [16]. Nodes were connected into star network using 100 Mbit/s switch. The Job Tracker was also Task Tracker and was exactly the same machine as all other Task Trackers (Fig. 2).

Input data was collected in 8 files that consist of the compare estimator and 72 values, 1.5 GB in total. Every single test was repeated 30 times, measuring times of three stages: copying the input data to HDFS, performing calculations and copying results back from HDFS.

## 5 Results

Tests show that increasing number of nodes affects computation time, but the results are not impressive (Fig. 3 and 4). However, time required for data distribution to all nodes increases with number of nodes as well as computation

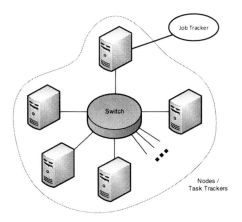

**Fig. 2.** The network configuration used in the tests. The number of nodes varying from 1 to 8

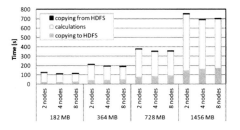

**Fig. 3.** Times of data analysis for different number of nodes

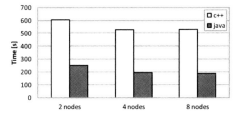

**Fig. 4.** Times of data analysis for different number of nodes and programming languages

time decreases slightly. What is more, time for copying data back from HDFS is inconsiderable, as it takes less than 1% of whole processing time, because of the size of results. For more accurate results the Reducer task should generate more data, which might significantly increase processing and copying time, but on the other hand will take more advantage of all Apache Hadoop's features.

## 6 Conclusion and Future Works

In this paper the Apache Hadoop technology was used to perform geophysical numerical analysis. As this technology is directed to process large data sets, algorithm to reconstruct Thomsen parameters' distribution was implemented and tested. The main advantage of using Apache Hadoop technology appears to be high scalability and ease of MapReduce code writing. Unfortunately, to benefit fully from this framework it is necessary to use more powerful computers in increased number as well as faster network configuration.

The future works will be focused on increasing number of nodes and amount of data, hence the inverse problem will be extended to more accurate models, which in consequence produce more data – more suitable for Apache Hadoop framework. What is more, to increase speed and accuracy of the inverse problem, forward solution will be implemented as a part of MapReduce class. This will enable the usage of the MapReduce tasks' results in next Thomsen parameters estimation in more convenient way. As the small-file problem might by significant for analysis speed in presented configuration, certain optimization should be considered in future work [17]. Furthermore, next tests should be performed using a slightly different configuration. As it is mentioned in Apache Hadoop documentation, when network consist of more than four nodes it is better to set up Job Tracker and Name Node on separate machines.

**Acknowledgments.** The study was financed in part by the statutory research project No 11.11.140.561 of the Department of Geoinformatics and Applied Computer Science, AGH UST and by grant No. N N525 256040 from Ministry of Science and Higher Education.

This work was co-financed by the AGH – University of Science and Technology, Faculty of Geology, Geophysics and Environmental Protection, Department of Geoinformatics and Applied Computer Science as a part of statutory project.

## References

1. Onderka, Z.: Stochastic Control of the Scalable High Performance Distributed Computations. In: Wyrzykowski, R., Dongarra, J., Karczewski, K., Waśniewski, J., et al. (eds.) PPAM 2011, Part II. LNCS, vol. 7204, pp. 181–190. Springer, Heidelberg (2012)
2. Pięta, A., Danek, T., Leśniak, A.: Numerical modeling of ground vibration caused by underground tremors in the LGOM mining area. Gospodarka Surowcami Mineralnymi – Mineral Resources Management 25(3), 261–271 (2009)
3. Danek, T.: Parallel and distributed seismic wave field modeling with combined Linux clusters and graphics processing units. In: IEEE International Symposium on Geoscience and Remote Sensing, IGARSS, pp. 2588–2591 (2009)
4. Kowal, A., Piórkowski, A., Danek, T., Pięta, A.: Analysis of selected component technologies efficiency for parallel and distributed seismic wave field modeling. In: Proceedings of the 2008 International Conference on Systems, Computing Sciences and Software Engineering (SCSS), part of the International Joint Conferences on Computer, Information, and Systems Sciences, and Engineering, CISSE 2008, Bridgeport, Connecticut, USA (2008); In: Sobh, T. (ed.) Innovations and Advances in Computer Sciences and Engineering, pp. 359–362. Springer (2010)

5. Piórkowski, A., Pięta, A., Kowal, A., Danek, T.: The Performance of Geothermal Field Modeling in Distributed Component Environment. In: Proceedings of the 2009 International Conference on Systems, Computing Sciences and Software Engineering (SCSS), part of the International Joint Conferences on Computer, Information, and Systems Sciences, and Engineering (CISSE 2009), Bridgeport, Connecticut, December 4-12 (2009); In: Sobh, T. (ed.) Innovations in Computing Sciences and Software Engineering, pp. 279–283. Springer (2010)
6. Kowal, A., Piórkowski, A., Pięta, A., Danek, T.: Efficiency of selected component technologies for parallel and distributed heat transfer modeling. Mineralia Slovaca 41(3) supl. Geovestnik, 361 (2009)
7. Moser, T.J.: Shortest path calculation of seismic rays. Geophysics 56, 59–67 (1991)
8. Fischer, R., Lees, J.L.: Shortest path ray tracing with sparse graph. Geophysics 58, 987–996 (1993)
9. Dwornik, M., Pięta, A.: Efficient algorithm for 3D ray tracing in 3D anisotropic medium. In: 71st EAGE Conference & Exhibition Incorporating SPE EUROPEC 2009, Extended Abstracts, Amsterdam, Holland, p. 138 (2009)
10. Thomsen, L.: Weak elastic anisotropy. Geophysics 51, 1954–1966 (1986)
11. http://hadoop.apache.org/
12. White, T.: Hadoop: The Definitive Guide, 2nd edn. O'Reilly Media (2010)
13. Kim, H., Kim, W., Lee, K., Kim, Y.: A Data Processing Framework for Cloud Environment Based on Hadoop and Grid Middleware. In: Kim, T.-H., Adeli, H., Cho, H.-S., Gervasi, O., Yau, S.S., Kang, B.-H., Villalba, J.G. (eds.) GDC 2011. CCIS, vol. 261, pp. 515–524. Springer, Heidelberg (2011)
14. Sort Benchmark Home Page, http://sortbenchmark.org/
15. http://wiki.apache.org/hadoop/PoweredBy#G
16. Wrzuszczak-Noga, J., Borzemski, L.: Comparison of MPI Benchmarks for Different Ethernet Connection Bandwidths in a Computer Cluster. In: Kwiecień, A., Gaj, P., Stera, P. (eds.) CN 2010. CCIS, vol. 79, pp. 342–348. Springer, Heidelberg (2010)
17. Mohandas, N., Thampi, S.M.: Improving Hadoop Performance in Handling Small Files. In: Abraham, A., Mauri, J.L., Buford, J.F., Suzuki, J., Thampi, S.M. (eds.) ACC 2011, Part IV. CCIS, vol. 193, pp. 187–194. Springer, Heidelberg (2011)

# DCOM and CORBA Efficiency
# in the Wireless Network

Zdzislaw Onderka

Dept. of Geoinformatics and Applied Computer Science,
AGH, Univ. of Science and Technology, Krakow, Poland
`zonderka@agh.edu.pl, onderka@ii.uj.edu.pl`

**Abstract.** The communication efficiency in the wireless network was analysed for the distributed objects which were realized the client and the server functionalities. The communication based on CORBA standard and DCOM standard was compared in order to show the possibilities of its application to the distributed computation or client-server systems. It was analyzed transmission times for the character and double tables and the structures. The tests shows easier and more efficient use of standard CORBA in the wireless network[1].

**Keywords:** distributed application, client-server, wireless network.

## 1 CORBA and DCOM in Wireless Network

Given the increasing availability of low-cost PCs with high computing power in particular portable computers (for example popular Intel Core i5 processor – with 2 or 4 cores or Intel Core i7 with 4 cores and 4–8 GB RAM, NVIDIA graphic card with CUDA technology) with wireless network cards, it becomes possible to implement distributed applications in a wireless local network (for example at home). This includes the implementation of distributed applications which require large computational power for example: simulations, computations of the CAE problems, numerical computations of data received from database, or others [1] and client-server applictions [2,3].

Modern distributed applications are often designed based on distributed object model. Therefore, presented study extends results presented in [4] for popular standards CORBA and DCOM [5]. Analysis presented in [4] concerned the case in which client and the server were executed on the same computer. Presented study shows not only the efficiency these standars but also the stability of communication time in the heterogeneous environment i.e. for two compuetrs with different MS Windows operating systems (which in case of Windows systems has significant importance) connected via wireless network and different hardware platforms.

CORBA (Common Object Request Broker) is based on defined in the middleware layer of the software object exchange mechanism ORB (Object Request

---

[1] In the statutory framework of the Dept. of Geoinform. and Applied Comp. Science.

A. Kwiecień, P. Gaj, and P. Stera (Eds.): CN 2012, CCIS 291, pp. 448–458, 2012.

Broker) [6]. It can also be seen as an interface between the hardware and software components for different manufacturers. Another important feature of CORBA is usage the standard of communication between the ORB-s i.e. IOOP protocol (Inter ORB Protocol) based on TCP/IP and applied since version 2.0 [7].

One of the major advantages of the CORBA standard is the ability to implement distributed objects using different programming languages (e.g. C/C++, Java, Pyton) and the possibility of using the various system platforms in contrast to the standards dedicated for specific programming languages (like RMI, EJB) or system environments (like .NET Remoting, COM/DCOM, WCF).

DCOM standard (Disributed Component Object Model) [8] is dedicated for Windows systems and it defines a protocol that allows COM components to cooperate in the environment of computer networks. The client communicates with the object using only interfaces which are not a classes or objects, but a set of functions that can be used to access the COM object methods.

For some applications DCOM has been replaced by the mechanisms of the .NET Framework environment and through support for Web services by WCF (Windows Communication Foundation) [9,10]. However, DCOM objects can be used with any language of the .NET environment through technology .NET COM Interop.

For the DCOM technology, Microsoft has introduced a new set of call interfaces at the low level called Object Remote Procedure Call (ORPC). ORPC is located on top of the standard Distributed Computing Environment RPC (DCE RPC) and expands the procedural programming model to accommodate distributed objects [5].

In both DCOM and CORBA standards, the interactions between a client process and an object server are implemented as object-oriented RPC-style communications. So, it is possible to use not only wired network but the wireless too. In CORBA, to call the remote object method one should give the proper IP address of the remote machine in the wireless network. Using DCOM one can give proper IP address in wireless network or the name of the remote computer specific for the Windows environment.

Summarizing, both standards: CORBA and DCOM can be utilized to implement the distributed and client-server applications in the wireless local network.

## 2   Project of the Sample Application

The project assumes the implementation of two client-server applications sending among the client and the server through wireless network the test data. First one applies the CORBA communication standard and the second one the DCOM standard. In order to obtain the most reliable results, both applications should be similar to each other as much as possible. Therefore, both of them were built on the same project templates of the development environment, and for both applications, the interface of the server has been implemented in IDL language (Interface Definition Language). But he implementation of the server interface in an IDL language for the environment MS Visual C++ differs from the IDL language for CORBA. The full analysis of needs and the object oriented project are presented in [4], where the project provided more test functions (for simple data types too) and the tests are made on the single machine, so the system internal mechanisms were tested.

Hovever, in case of heterogeneous network it was necessary to implement new client and server applications based on DCOM standard. Applications properly communicating on the same computer not working in the network of computers.

The measurements of the time required to transfer a fixed amount of structural data (arrays and structures) were provided to estimate the communication efficiency. The project provides functions for one-dimensional tables transfer and structure data transfer from the client to the server and from the server to the client. The definition of data types in IDL language is as follows:

```
typedef   char     ct[1000];
typedef   double   dt[1000];
struct    Test_Structure {
char c;
double d;
ct char_tab;
dt double_tab; } ;
```

- *void getCharTab( inout ct charTab )* //gets character table from the server,
- *void getDoubleTab( inout dt doubleTab )* //gets double table from the server,
- *void putCharTab( in ct charTab )* //sends character table to the server,
- *void putDoubleTab( in dt doubleTab )* // sends double table to the server,
- *void putStructData( in char c, in double d, inout ct charTable, inout dt doubleTable )*
  //sends to the server components of the structure each one as a parameter,
- *void putStruct( in Test_Structure ts )* //sends the whole structure to the server,
- *Test_Structure getStruct()* //gets the whole structure from the server.

## 3    The Environment and the Tests

The local network consists of two laptops and one rooter. First laptop (TOSHIBA) worked under the Microsoft Windows 7 Home Premium operating system (Intel® Core™ i7 CPU Q740 @1.73 GHz with Tubo Boost mode to 2.93 GHz, 4 GB RAM, 64-bits operating system, WIFI card: Broadcom 802.11n, karta graficzna GeForce GT 330M – 48 rdzeni CUDA) and the second one (ASUS) under the Microsoft Windows XP Professional operating system with Service Pack 3 (AMD Turion™ 64 Mobile, 1.6 GHz, 1 GB RAM, 32-bits operating system, WIFI card: ASUS 802.11g). The management of the wireless network is realized by the broadband rooter Linksys WRT54GL – 2.4 GHz (CISCO system) which serves the wireless 802.11g standard with maximum data transfer rate 54 Mb/s (Fig. 1).

In order to obtain quantifiable results of single test the client implementation has used multiple calls of a method of a remote object in a loop (1000 iterations).

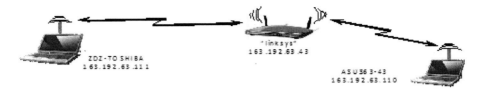

**Fig. 1.** The wireless network

The time is measured just before the loop and immediately after the loop. Each test is repeated 100 times. There were performed seven complete tests for each type of data and the type of communication In each test the sever was executed on the ASUS laptop and the client was executed at the TOSHIBA laptop.

### 3.1 Tests for CORBA Standard

The application used the CORBA standard for communication between client and server is implemented in the Microsoft Visual C++ 2008 as Win32 Console Application using Object Request Broker – omniORB (ver 4.1.4) free software.

The client has to use the only IP address of the computer with running server. The server and client were executed like normal console. There were no any problems with communication between client and the server except the need to define exceptions for Windows firewall. The presented below results are in milliseconds.

The results of sending the character table to the server presented at the Fig. 2, shows good stability, but an average of about 11% of calls varies significantly with a leading trend, wheres sending the character table from the server to the client presented at the Figure 3 shows weak stability, but only an average of about 5% of calls varies significantly with a leading trend.

The results of sending the double table to the server presented at the Fig. 4 shows good stability although an average of about 21% of calls varies significantly with a leading trend, whereas the results of sending the double table from the server to the client presented at the Fig. 5 shows weak stability and an average of about 13% of calls varies significantly with a leading trend.

The results of sending the structure from the server to the client (the complete structure is a parameter of remote function) presented at the Fig. 6 shows weak stability but an average of about 5% of calls varies significantly with a leading trend, whereas sending the structure to the client presented at the Fig. 7 shows quite good stability but an average of about 18% of calls varies significantly with a leading trend.

The results of sending the structure to the client (each individual field of the structure is a parameter of remote function) presented at the Fig. 8 shows weak stability but an average of about 7% of calls varies significantly with a leading trend.

For additional review, it was chacked if there is the impact between the obtained results and the number of lost packets. The test is done using system

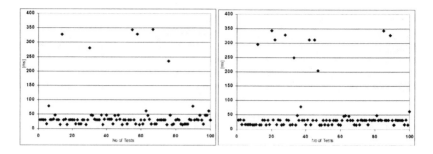

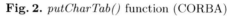

**Fig. 2.** *putCharTab()* function (CORBA)

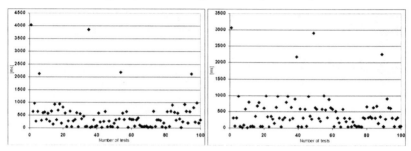

**Fig. 3.** *getCharTab()* function (CORBA)

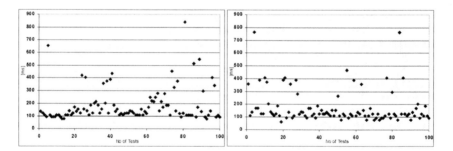

**Fig. 4.** *putDoubleTab()* function (CORBA)

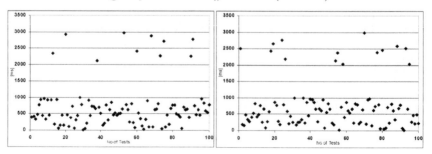

**Fig. 5.** *getDoubleTab()* function (CORBA)

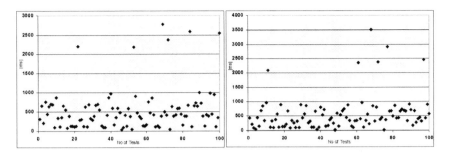

**Fig. 6.** *getStruct()* function (CORBA)

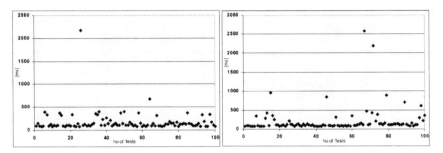

**Fig. 7.** *putStruct()* function (CORBA)

*ping* command which sending 32-byte packets between the client and the server used simultaneously with the descibed above tests. For 1000 sending 32-byte packets an average maximum transfer time was about 51 miliseconds (minimum less then 1 ms) and there were about 11 lost packets (1%). So, the lost packets could have influence on results which significantly vary from the main trend.

Summarizing, sending data to server gives better results than sending the same data from the server to the client except the structure data type. Moreover, it is significantly better efficiency for sending structure to server in way of only one parameter then putting each structure field separately as function parameter.

## 3.2 Tests for DCOM Standard

The application that used the DCOM communication standard was implemented using the Microsoft Visual C++ 2008 environment. Implementation of the server was based on a template project using ATL (Active Template Library) – recommended by Microsoft and the client application is the dialog-based project.

The generated IDL file is dependent on the MS Visual Studio environment. Each interface class and data type is assigned a globally unique identifier (GUID) Similarly, each object class is assigned a unique class ID (CLSID). Moreover, each function in the interface has a result of type HRESULT (predefined long) and all data passed to function and returned from the function, are passed

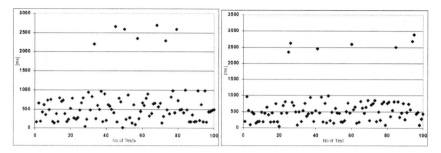

**Fig. 8.** *putStructData()* function (CORBA)

by the parameters. Input parameters have the attribute *[in]*, the input/output parameters have *[in, out]* attributes. The returned value from the function is passed by the parameter with the *[retval]* attribute. Therefore, in DCOM there is different definition (IDL file) and implementation of the *getStruct()* function then in CORBA. Other functions differ only in the type of function i.e *HRESEULT* (DCOM) and *void* (CORBA).

The server was registered as a service in Windows system and the DLL module (generated from the application project) of the server had to be registered using *regsvr32* system application. During the implementation of the communication there were a lot of practical problems (at the execution phase) with establishing of the connection between client and the server. The exceptions for the Windows firewall and a lot of additional settings had to be done for Windows services, system and test service and users using the *dcomcnfg* system application (practicaly it is requiered to remove the majority of system security settings).

Function *getStruct( TestStructure* pStruct )* invoked from the service running on different machine in comparison with the same function implemented and tested in [4] on only one computer has more complex implementation. Because the server in [4] is running in the same RAM memory as client, so there was sufficient to assign to the passed parameter *pStruct* only the address of the server structure. But when server and client are executed on different computers and consequently on different address spaces, so implementation of that function is required to copy each field of server structure to each field of the structure pointed by the parameter (passed remotely from the client). Therefore, the performance for that function will be worse. The presented below results are in milliseconds.

The results of sending the character table to the server are presented at the Fig. 9 Sending character data to the server shows good stability, but an average of about 20% of calls varies significantly with a leading trend.

The results of sending the character table from the server to the client presented at the Fig. 10 shows good stability, but an average of about 15% of calls varies significantly with a leading trend.

The results of sending the double table to the server presented at the Fig. 11 shows weak stability and an average of about 14% of calls varies significantly with a leading trend.

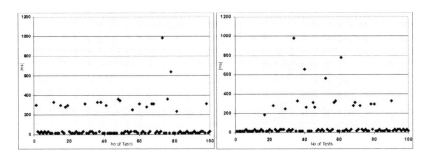

**Fig. 9.** *putCharTab()* function (DCOM)

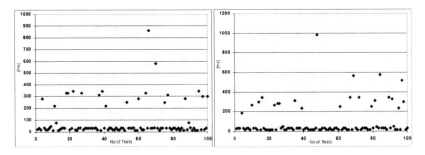

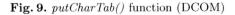

**Fig. 10.** *getCharTab()* function (DCOM)

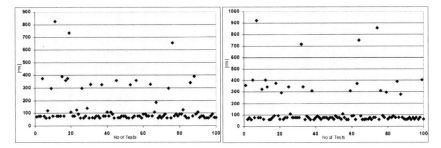

**Fig. 11.** *putDoubleTab()* function (DCOM)

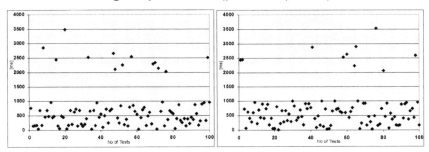

**Fig. 12.** *getDoubleTab()* function (DCOM)

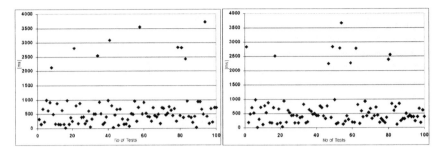

**Fig. 13.** *getStruct()* function (DCOM)

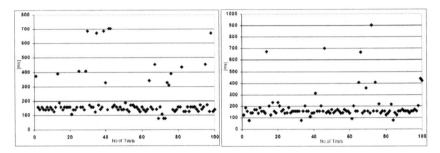

**Fig. 14.** *putStruct()* function (DCOM)

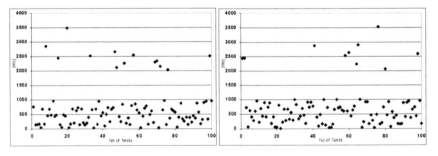

**Fig. 15.** *putStructData()* function (DCOM)

The results of sending the double table from the server to the client presented at the Fig. 12 shows quite good stability but an average of about 21% of calls varies significantly with a leading trend.

The results of sending the structure from the server (the whole structure is a parameter of the remote function) to the client presented at the Fig. 13 shows weak stability but an average of only about 9% of calls varies significantly with a leading trend, whereas the results of sending the structure to the client presented at the Fig. 14 shows good stability and an average of about 16% of calls varies significantly with a leading trend.

The results of sending the structure to the client (each field of the structure is a parameter of remote function) presented at the Fig. 15 shows good stability but an average of about 20% of calls varies significantly with a leading trend.

Similarly to CORBA tests there were examined the results of packet (32-byte) transfer using system command *ping*. For 1000 sending 32-byte packets an average maximum transfer time was about 52 milliseconds (minimum less then 1 ms) and there were about 25 lost packets (2%) (test is done simultaneously withe communication tests). So, the lost packets could have influence on results which significantly vary from the main trend.

## 4    Conclusions and Future Works

The presented above study, tested the performance of wireless communication for the two client-server applications under Windows system using CORBA standard and dedicated for Windows systems DCOM standard. Seven tests were performed. The tables (Table 1, Table 2) show the average times for 100 calls of each remote function. Results are given in milliseconds.

First observation is that in the most test cases the CORBA standard gives similar or better results then DCOM. In comparision with the tests performed in [4], CORBA gives worst results than DCOM. Moreover, in cases of *putCharTab()* function and *putStructData()* the DCOM application is only about 3 times better then the CORBA application. In comparision with tests performed in [4], in case of *putCharTab()* function the CORBA is about 9 times worst than DCOM and in case of *putStructData()* CORBA is about 10 times worst than DCOM.

Other interesting observation is the comparision between sending and getting the same data structures, for the same communication standard. Generally, getting data from the server is more efficient than sending the same data structure to the server with the exception of structure data type transmission. In this of the structure data type transfer via wireless network the situation is reversed.

Above observations show that the marshaling/unmarshaling functions at the client or server side have a large amount of time to carry out for data conversion and that the method invocation in DCOM is more complicated then in CORBA.

Received results for the communication in the wireless networks implyes that CORBA is more efficient then was received in [4]. Moreover, taking into account

**Table 1.** CORBA: 1 – *putCharTab()*, 2 – *getCharTab()*, 3 – *putDoubleTab()*, 4 – *getDoubleTab()*, 5 – *putStruct()*, 6 – *getStruct()*, 7 – *putStructData()*

| Test No | 1 | 2 | 3 | 4 | 5 | 6 | 7 |
|---------|-------|------|-------|-------|-------|-------|-------|
| Test 1 | 475.2 | 70.0 | 702.9 | 187.8 | 249.3 | 555.8 | 615.8 |
| Test 2 | 379.0 | 59.7 | 750.4 | 182.5 | 179.6 | 559.3 | 634.4 |
| Test 3 | 451.4 | 47.3 | 785.5 | 185.9 | 160.9 | 478.6 | 591.9 |
| Test 4 | 453.4 | 54.6 | 868.2 | 155.9 | 172.5 | 537.5 | 703.4 |
| Test 5 | 370.7 | 61.1 | 762.5 | 201.5 | 215.5 | 637.0 | 576.3 |
| Test 6 | 409.1 | 66.5 | 753.7 | 206.8 | 176.9 | 537.6 | 626.7 |
| Test 7 | 509.4 | 73.1 | 812.5 | 175.8 | 152.6 | 528.9 | 548.6 |
| Average: | 435.4 | 61.7 | 776.5 | 185.2 | 186.7 | 547.8 | 613.9 |

458    Z. Onderka

**Table 2.** DCOM: 1 – *putCharTab()*, 2 – *getCharTab()*, 3 – *putDoubleTab()*, 4 – *getDoubleTab()*, 5 – *puttStruct()*, 6 – *getStruct()*, 7 – *putStructData()*

| Test No | 1 | 2 | 3 | 4 | 5 | 6 | 7 |
|---|---|---|---|---|---|---|---|
| (1) | 123.1 | 51.4 | 869.9 | 150.3 | 225.6 | 686.6 | 215.4 |
| (2) | 164.1 | 91.0 | 840.4 | 151.1 | 218.2 | 586.0 | 201.7 |
| (3) | 88.5 | 72.6 | 710.0 | 139.1 | 206.1 | 624.6 | 241.0 |
| (4) | 96.8 | 74.0 | 805.8 | 125.2 | 226.2 | 665.4 | 229.2 |
| (5) | 113.2 | 35.0 | 821.8 | 196.7 | 193.0 | 655.3 | 209.8 |
| (6) | 97.8 | 43.5 | 719.4 | 139.5 | 198.6 | 571.4 | 198.8 |
| (7) | 89.0 | 89.9 | 944.6 | 188.6 | 227.5 | 695.3 | 271.0 |
| Average: | 110.4 | 65.3 | 816.0 | 155.8 | 213.6 | 640.6 | 223.8 |

the the duration of process of implementaion and launching client-server application in the heterogeneous wireless network the CORBA standard is much more usfeul than DCOM standard.

# References

1. Piórkowski, A., Pieta, A., Kowal, A., Danek, T.: The Performance of Geothermal Field Modeling in Distributed Component Environment. In: International Joint Innovations in Computing Sciences and Software Engineering, pp. 279–283. Springer (2010)
2. Onderka, Z., Piórkowski, A.: Project and Implementation of the Geological Database in the Internet. In: Pochopień, B., Kwiecień, A., Grzywak, A., Klamka, J. (eds.) Wysokowydajne Sieci Komputerowe, Zastosowanie i Bezpieczeństwo. WKiŁ, Warszawa (2006) (in Polish)
3. Onderka, Z.: The Efficiency Analysis of the Object Oriented Realization of the Client Server Systems Based on the CORBA Standard. In: Schedae Informaticae 2011, vol. 20, pp. 181–193. Sci. Booklets of the Inst. of Comp. Science of Jagiellonian Univ. (2011)
4. Cichy, M., Onderka, Z.: The comparision of the communication performance for the CORBA and DCOM standards in the client-server systems. Studia Informatica 32(3A(98)) (2011) (in Polish)
5. Chung, P.E., et al.: DCOM and CORBA: side by side, step by step, and layer by layer. C++ Report (1998), http://www.research.att.com/~ymwang/papers/HTML/DCOMnCORBA/S.html
6. Siegel, J.: CORBA Fundamentals and Programming. J. Wiley, New York (1996)
7. Catalog of OMG CORBA®/IIOP® Specifications, http://www.omg.org/technology/documents/corba_spec_catalog.html
8. Eddon, G., Eddon, H.: Inside Distributed COM. Microsoft Press (1998)
9. Kowal, A., Piórkowski, A., Danek, T., Pieta, A.: Analysis of selected component technologies efficiency for parallel and distributed seismic wave field modeling. In: Innovations and Advances in Computer Sciences and Engineering, pp. 359–362. Springer (2010)
10. Piórkowski, A.: The Methods of Implementation Distributed Database Applications in the .NET Technology. In: Sieci Komputerowe: Aplikacje i Zastosowania, vol. 2, pp. 195–202. WKiŁ, Warszawa (2007) (in Polish)

# VANETs as a Part of Weather Warning Systems

Marcin Bernaś

Faculty of Transport, Silesian University of Technology,
Krasińskiego 8, 40-019 Katowice, Poland
marcin.bernas@polsl.pl

**Abstract.** The Road Weather Information Systems used nowadays are
a vast network infrastructure which covers local weather stations, regional and national management centres. Despite that, its response time
still takes up from 5 to 10 minutes. During that time vehicles are vulnerable to rapidly changing weather conditions. Integrating VANETs into
this structure could significantly improve a system's effectiveness. The
paper describes the possible solutions and compares the network issues
connected with the system integration. The security issue and traffic
influence were also taken under consideration.

**Keywords:** VANET, GSM, data transmission, weather information
system.

## 1 Introduction

Road weather information systems are based on a long-term (12 h–24 h) weather
forecast. The forecast is made basing on numerical models as well as traditional
meteorologist assessments. The numerical weather model, which merges data
and generates weather forecast, is treated as a supporting tool. Final estimation
of the threat level still belongs to the specialist or to the trained expert system.

Numerical forecast model, based on Navier-Stokes equations, is non-linear
one and therefore, due to its complexity, is not applicable in single onboard
unit (OBU) or roadside units (RSU). However, performed research shows that it
is possible, basing on obtained forecast, to implement vehicle onboard weather
threat level estimation system [1].

The generated forecast together with warning threshold values can be enriched by the data from vehicle sensors. Additionally, more accurate data can be
obtained via road meteorological weather stations (RMS) extended with infra-
structure-to-car (I2C) communication protocols. RMS provides wide range of
data, which is unavailable for a car sensor to acquire. Nevertheless RMS quantity is thousand times smaller than the number of vehicles.

To estimate the road weather threat level, the system should take into account
the following features: road and air temperature, their changes in time, rain
levels, condensation of salt in water, wind direction/speed and sunlight intensity.

A. Kwiecień, P. Gaj, and P. Stera (Eds.): CN 2012, CCIS 291, pp. 459–466, 2012.

There are several models to predict weather conditions. The first group covers global weather models which are too inaccurate for road usage [1]. However, they serve as a base for regional models' development. The most common regional models used at present are: Fifth Generation Penn State/NCAR Mesoscale Model (MM5), High Resolution Limited Area Model (HIRLAM), ALADIN, COSMO model (also known as COSMO/LM or LAMI). Aforementioned models are able to predict the weather conditions in locations organized as a symmetrical mesh. Mutual distance between locations varies from 2.5 km to 4 km. Even if provided forecast is enhanced with road meteorological stations data, the threat level estimation and warning propagation time is still not satisfying [2]. The paper proposes using VANET and Global System for Mobile Communications (GSM) network as weather data carrier, while leaving its evaluation to an intelligent vehicle. Section 2 presents related works concerning weather warnings on roads. Section 3 provides the model variations to connect regional weather network with VANET. Finally Sections 4 and 5 provide the simulations results and their discussion.

## 2    Related Works

Vehicular ad-hoc networks are becoming a more and more reliable tool to exchange data between vehicles. The vehicles and road side stations can communicate to create unified network. There are many standards and propositions how to store, secure and process the information in vehicular networks [3,4,5,6]. According to Dedicated Short Range Standard (DSRC) to be able to create the network its nodes must be within each other's range. In case of Japan it is 30 m, in Europe it is 15–20 m and in the USA it is up to 1000 m [7]. In Europe and the United States of America 5.8 GHz bandwidth is dedicated to the vehicular communication, providing 7 or 4 channels with transmission rate 250 kbit/s for upload and 500 kbit for download (in USA: 1–4 Mbit). One of the channels is reserved for security and safety purposes. Second standard is based on 802.11p transmission protocol and WAVE/ IEEE1609.3 specification [8]. Its effective transmission range is usually estimated at 100 m. However, some research shows outdoor usage for range from 400 m up to 1200 m [9].

Dynamic changes of the structure (reconstruction approximately every 6 s [10]) are a characteristic feature of vehicular networks. Network is usually build with redundant connections in order to stay consistent.

VANET offers a lot of opportunities for application's development. Modern OBUs with connection interfaces are not only able to communicate with traffic participants, send or receive warnings, but also to analyse complex messages and to generate their own safety assessments. The application possibilities of OBU are vast: avoiding traffic jams and accidents, warning about the weather threats like ice, fog or strong wind. GPS systems, used in several routing algorithms, can utilise and validate the received data.

VANETs are built basing on the following communication standards: Wi-Fi IEEE 802.11p, WAVE IEEE 1609, WiMAX IEEE 802.16, Bluetooth, IRA or

ZigBee. The trials were undertaken to incorporate WiMAX standards and GSM standards into VANET [11].

VANET is considered as a vital part of modern Intelligent Transport System (ITS). In this case we can distinguish: Inter-vehicle communication (IVC) and road-to-vehicle communication (RVC) [10].

There are researches [12] and applications [10] concerning warning about slippery road for preceding vehicle. However, without the data from regional weather systems, they are unable to verify them or make independent assessment. The paper presents the possibility of adopting VANET together with road side weather networks and OBU to create the onboard weather warning system.

## 3    Proposed Network Architecture

Regional models, together with roadside metrological station, can become a base for road information systems provided via VANET network. Figure 1 presents the basic model structure.

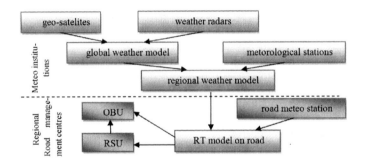

**Fig. 1.** Data transmission with weather forecast centres

The data, which is sent to the management centre, can be redirected to vehicles and RSUs. Vehicle OBU, based on provided data, can estimate current weather threat level and send warning to a driver. The generated warning will not be limited to Anti-Lock Braking System (ABS), Electronic Stability Program (ESP) or Anti-Slip Regulation (ASR) systems readings.

Proposed solution reduces the probability of reacting on a faked warning that could be forged in VANET. Figure 2 presents the model structure, with and without road units.

The proposed model is flexible and it can be adapted to any standards e.g. WAVE or DSRC. The model allows vehicles to gain extended data either from RSUs or from external network via GSM. In this case, the additional layer is added. Furthermore, sole inter-OBU communication was also taken under consideration.

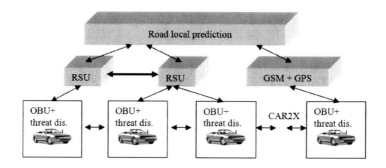

**Fig. 2.** VANET data limitation model

## 3.1  Data Frame

Basing on the author's weather threat detection surveys [2,1,12] the data set required to create the weather onboard warning system was defined. It cannot be obtained via car sensors only. The considered parameters obtained from weather stations are:

- road's slipperiness basing on air temperature in the range from $-6°C$ to $+1°C$, rain level, relative humidity greater than 85%, variable road surface temperature from positive to negative, concentration of the brine on the road,
- visibility based on humidity, light sensor or passive camera sensor,
- strong side wind blows based on wind speed and its direction.

The vehicle sensors' data are obtained from traction control systems such as ABS or ESP. Additionally, air temperature and rain level can be obtained via sensors available in most vehicles. The following data were divided into two frame types: infrastructure-to-vehicle frame (I2Vf) and vehicle-to-vehicle frame (V2Vf). The frames are presented in Fig. 3.

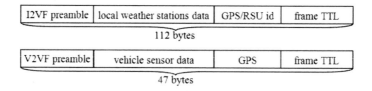

**Fig. 3.** Frames types for onboard weather system prediction

To simplify the comparison the GSM data received via general system are formatted as I2Vf frame. The onboard warning unit is able to make a full assessment, if the data from a previous car are received and up-to-date information about weather in the given area is provided. The size of the frames is used to calculate the overall data transmission requirements.

## 3.2   Simulation Model

For simulation purposes Nagel-Schreckenberg cellular automata model was used. The model was enriched to simulate road traffic conditions with average speed of 90 km/h and urban traffic with uncongested speed equal to 50 km/h. To fulfil the requirements, following rules of the cellular automata were extended as follows:

$$v_i(t) = \{v_i(t-1) + 1, g_i(t), v_{\max}\} \; , \tag{1}$$

$$g_i(t) = x_j(t) - x_i(t) - 1, \xi < p \quad \text{then} \quad v_i(t) = \{0, v_i(t) - 1\} \; , \tag{2}$$

$$x_i(t) = x_i(t-1) + v_i(t) \; , \tag{3}$$

where: $i$ – no of vehicle in $t$ time step, $j$ – no of vehicle directly in front of vehicle at $i$-th position, $x_i(t)$ – no of cell occupied by vehicle $i$ at time step $t$, $\xi$ – statistical variable with normal distribution (values within range $[0, 1]$).

Traffic lights simulation was included by adding to Equation (1) following rule:

$$\text{if} \; s(t) = 1 \; \text{and} \; x_i(t) < x_s \quad \text{then} \quad v_i(t) = \min\{v_i(t), x_s - x_i(t) - 1\} \; , \tag{4}$$

where: $s(t)$ – defines position of vehicle at $i$-th time step $t$ $s(t) = 1$, while red traffic lights, $x_s$ – is a position of cell directly after the full stop line.

The model was further extended by introducing RSU position vector $Y$, where: $y_i$, $i = 1, 2, \ldots, \max$, which defines the positions of RSU offering weather data. GSM data can be received on request anytime in this model. Signal propagation from RSU to other cars via car-to-car communication takes place every second. The transmission's range was defined basing on DSRC standard to 20 m (100 m for 802.11 standard), which is equal to 4 (20 for 802.11 standard) cells in the presented model. The lifetime of a package in the network was estimated at 5 seconds. The algorithm processes data provided from car-to-car communication, RSU or paid GSM communication. The communication algorithm is processed separately by every vehicle OBU:

1. Decrease TTL value of frames by 1.
2. If TTL<=0 remove I2Vf and V2Vf frames.
3. Check for C2C data from preceding vehicle. If the data contains also I2Vf data, set I2Vf TTL value as car TTL value decreased by 1.
4. If more than one I2Vf or V2Vf were acquired, calculate each parameter as a weighted mean, where weights are estimated basing on relative distance.
5. Broadcast V2Vf data and I2Vf data if present.
6. Check for RSU data. If I2Vf data are present, store data and set TTL to default value.
7. If GSM connection is present and I2Vf data are not present, request regional weather data. Store the requested data as I2Vf data.
8. If I2Vf and V2Vf data are present, estimate full weather threat level.
9. If only one of I2Vf and V2Vf data are present, respond to sent warnings.

## 4    Simulation Results

The defined model was adapted to GSM and VANET protocols specification. The model functionality is limited by the structure presented in [8]. Considered road lane is unidirectional. The road during simulations had length from 40 to 130 cells, which corresponds to the real distance range from 300 m up to 1000 m. Maximal velocity of vehicles was divided into two categories: 2 cells per time step for urban traffic (54 km/h) or 4 cells per time step for suburb traffic (90 km/h). Deceleration probability $p$ is 0.15. To measure effectiveness of car-to-car communication different traffic volumes were selected. Respectively the traffic volumes were analysed ranging from 100 vehicles per hour (veh/h) up to 2200 vehicles per hour (veh/h).

Architecture of the considered model was tested within RSUB security protocol [10]. While performing simulations the overhead of this standard were taken under consideration. Each vehicle in the system is equipped with OBU with GSM and VANET connection possibility. Additionally, vehicle is able to localize its position (GPS device) for position based routing algorithms. The GSM communication is used on request (effectiveness based on GSM network coverage on average: 92%). OBU is programmed to work in real-time and communicate every second. Firstly the sensitivity of a model was measured based on the lane length parameter. The received average speed, number of generated vehicles as well as successfully received message rate remain unaffected (deviation less than 2%), therefore 1000 m lane length was used in further simulations. Basing on the proposed model three connections types were compared: Car2Car, RSU and GSM communication. Figure 4 presents ability to inter-vehicle communication based on traffic volumes and selected communication standard. The simulation data show strong correlation between average cars speed and its ability to communicate especially in DSRC standard. Decrease in communication ability is caused by bigger gaps between vehicles while driving with higher velocity.

Next simulation was aimed to strike a balance between a number of RSU and usage of commercial GSM networks. Their relations were presented in Fig. 5. According to simulation results to minimise the need of using commercial GSM

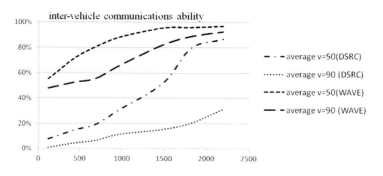

**Fig. 4.** Car to car message forwarding ability

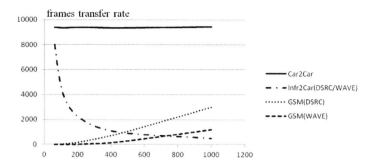

**Fig. 5.** Relation of GSM and RSU data transfers

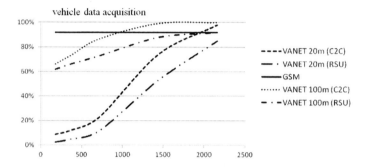

**Fig. 6.** Threat level estimation ability using various technologies

networks the optimal distance between separate RSU units is from 100 m to 200 m for DSRC (400 m for WAVE standard). Additionally, transfer rate for RSUs on 100 m distance equals 1.2 kB/s. Average transfer rate between OBU units are equal 5.4 kB/s. Reducing the distance between RSUs below 100 m increases channel's occupation rapidly while not increasing system effectiveness. Estimated 100 m (400 m for WAVE standard) distance is able to reduce the GSM support data to minimum.

Finally, based on defined RSU distance and defined communication algorithm, the simulation was performed (Fig. 6). Figure 6 present the data acquisition from RSU and perceiving vehicles. The data analysis shows that for default 20 meters broadcast radius, connected with traffic volume below 1400 veh/h, RSU transmission is insufficient to maintain high weather thread detection rate and must be supported by GSM network. However, using WAVE standard detection rate is above 60% even for traffic volume lower than 100 veh/h. Increasing the OBU transmission range can improve the data acquisition significantly.

## 5  Summary

The paper introduces an approach for using the VANET structure in development of advanced on-board car weather threat level estimation engine within

OBU. For verification purposes the network with standard 20 m and 100 m transmission range was used. Simulations firmly show that in low congested one-lane traffic the provided data from DSRC is insufficient to make threat assessment every time. For traffic volume of 1400 veh/h it can fully asses threat in 60% of cases and it can do it partially in 93% of cases. In case of external GSM support and WAVE standard the effectiveness of fully threat estimation is increased to 92%. Presented values are gained when RSUs are equally distributed. In practice, RSUs tends to be situated close to dangerous spots which can increase safety.

Future research will cover data fusion analysis and more complex simulation environments like SUMO with NS3 support.

# References

1. Bijleveld, F., Churchill, T.: The Influence of Weather Conditions on Road Safety. In: SWOV Institute for Road Safety Research, p. 11 (2009)
2. Bernas, M.: Implementation of fuzzy model as an embedded system for transport safety improvement. In: Transaction on Transport Systems Telematic & Safety, Poland, Gliwice, pp. 91–102 (2011)
3. Harsch, C., Festag, A., Papadimitratos, P.: Secure position-based routing for VANETs. In: Proceedings of IEEE 66th Vehicular Technology Conference, pp. 26–30 (2007)
4. Lochert, C., Scheuermann, B., Mauve, M.: A probabilistic method for cooperative hierarchical aggregation of data in VANETs. Ad Hoc Network 8, 518–530 (2010)
5. Kulik, L., Tanin, E., Umer, M.: Efficient Data Collection and Selective Queries in Sensor Networks. In: Nittel, S., Labrinidis, A., Stefanidis, A. (eds.) GSN 2006. LNCS, vol. 4540, pp. 25–44. Springer, Heidelberg (2008)
6. Płaczek, B.: Selective data collection in vehicular networks for traffic control applications. Transport. Res. Part C (2012), doi:10.1016/j.trc.2011.12.007
7. Bhakthavathsalam, R., Nayak, S.: Operational inferences on VANETs in 802.16e and 802.11p with improved performance by Congestion Alert. In: Consumer Communications and Networking Conference (CCNC), pp. 467–471 (2011)
8. Festag, A.: Global standardisation of network and transport protocols for ITS with 5 GHz radio technologies. In: Proceedings of the ETSI TC ITS Workshop, pp. 117–127 (2009)
9. Leung, K.K., Clark, M.V., McNair, B., Kostic, Z., Cimini, L.J., Winters, J.H.: Outdoor IEEE 802.11 Cellular Networks: Radio and MAC Design and Their Performance. IEEE Transactions on Vehicular Technology 56(5) (2007)
10. Xiong, H., Chen, Z., Li, F.: Efficient and multi-level privacy-preserving communication protocol for VANET. Computers & Electrical Engineering (2011), doi:10.1016/j.compeleceng.2011.11.009
11. El-Tawab, S., Abuelela, M., Gongjun, Y.: Real-time weather notification system using intelligent vehicles and smart sensors. In: Mobile Adhoc and Sensor Systems, pp. 627–632 (2009)
12. Jonsson, P.: Road condition discrimination using weather data and camera images. In: Intelligent Transportation Systems (ITSC), pp. 1616–1621 (2011)

# Author Index